DATE DUE			
Aug. 8			
APR 2 9 2004			
SEP 2 3 2008			
			DISCARD
GAYLORD			PRINTED IN U.S.A.

Indoor Air Quality

Indoor Air Quality

BEAT MEYER

Department of Chemistry
University of Washington
Seattle, Washington

1983

Addison-Wesley Publishing Company, Inc.
Advanced Book Program/World Science Division
Reading, Massachusetts

London • Amsterdam • Don Mills, Ontario • Sydney • Tokyo

363.73
M575i d.d.

Library of Congress Cataloging in Publication Data

Meyer, Beat.
    Indoor air quality.

    Bibliography: p.
    Includes indexes.
    1. Air—Pollution, Indoor. 2. Air quality management.
I. Title.
TD883.1.M49 1982   363.7'392   82-8911
ISBN 0-201-05094-3

Copyright © 1983 by Addison-Wesley Publishing Company, Inc.
Published simultaneously in Canada.

All rights reserved. No part of this publication may be reproduced, stored in a retrieval system, or transmitted, in any form, or by any means, electronic, mechanical, photocopying, recording, or otherwise, without the prior written permission of the publisher, Addison-Wesley Publishing Company, Inc., Advanced Book Program/World Science Division, Reading, Massachusetts 01867, U.S.A.

Manufactured in the United States of America

ABCDEFGHIJ-HA-898765432

Contents

Preface		xi
1	**Introduction**	1
	1.1 Basic Needs	1
	1.2 Building Factors	4
	1.3 Ill Buildings	7
	1.4 Pollutants	10
	1.5 Monitoring of Air Quality	11
	1.6 Air Quality Control	13
	1.7 Health and Building Standards and Regulations	14
2	**History**	15
	2.1 Respiratory Needs	15
	2.2 Stale Air	19
	2.3 Air Pollution Legislation	21
	2.4 Health and Air Quality at the Workplace	22
	2.5 Indoor Air Comfort	24
	2.6 Architecture and Building Materials	26
3	**Comfort and Climate**	29
	3.1 Comfort and Sensory Perception	29
	3.2 Life Support	33
	3.2.1 Chemical Balance	34
	3.2.2 Heat Balance	38
	3.2.3 Thermal Comfort	41
	3.2.4 Radiation and Electric Fields	47
	3.3 Thermal Properties of Air	48
	3.4 Climate	53
	3.4.1 The Global Climate	54

		3.4.2	Diurnal and Seasonal Cycles	55
		3.4.3	Vertical Mixing	56
		3.4.4	The Microclimate	56
		3.4.5	The Role of Humidity	59
		3.4.6	The Indoor Climate	60

4 Buildings 62
4.1 Building Design 63
- 4.1.1 Ventilation 65
- 4.1.2 Thermal Comfort 67
- 4.1.3 Thermal Efficiency of Buildings 68

4.2 Building Types 70
- 4.2.1 Residential Housing 71
  - 4.2.1.1 Single-Family Dwellings 71
  - 4.2.1.2 Apartments and Condominiums .. 72
  - 4.2.1.3 Manufactured Housing and Mobile Homes . 72
- 4.2.2 Commercial Buildings 80
  - 4.2.2.1 Office Buildings 80
- 4.2.3 Public Buildings, Schools, and Arenas .. 81

4.3 Building Materials and Components 83
- 4.3.1 Heat Transfer 84
- 4.3.2 Wood Products 84
- 4.3.3 Insulation 86
- 4.3.4 Wind Resistance 88
- 4.3.5 Moisture 88
- 4.3.6 Furnishings 91
- 4.3.7 Windows 91

5 Air Impurities 93
5.1 Metabolic Products 93
- 5.1.1 Water 93
- 5.1.2 Other Metabolic Contaminants 94

5.2 Ambient Air Pollutants 96
- 5.2.1 Particulates and Aerosols 96
- 5.2.2 Sulfur Dioxide 101
- 5.2.3 Ozone 103
- 5.2.4 Carbon Monoxide 103
- 5.2.6 Nitrogen Oxides 105
- 5.2.7 Other Ambient Air Pollutants 105

5.3 Microbes 105
5.4 Metals 109
5.5 The Halogens 112
5.6 Organics in Air 113
- 5.6.1 Methane 113

		5.6.2	Ethane, Propane, Butane, Octane and Other Hydrocarbons	114
		5.6.3	Halocarbons	114
		5.6.4	Aerosol Sprays	114
		5.6.5	Pesticides	116
			5.6.5.1 Insecticides	119
			5.6.5.2 Fumigants	123
			5.6.5.3 Rodenticides	123
			5.6.5.4 Herbicides	124
			5.6.5.5 Fungicides	125
	5.7	Ionizing Radiation		126
		5.7.1	Cosmic Radiation	127
		5.7.2	Terrestrial Radiation	127
		5.7.3	Nuclear Explosions	133
	5.8	Odors and Stale Air		134
6	**Air Monitoring and Analysis**			**137**
	6.1	Standard Methods		137
	6.2	Equipment for Monitoring Physical Properties of Air		138
		6.2.1	Thermal Comfort Measurements	138
		6.2.2	Temperature	138
		6.2.3	Humidity	139
		6.2.4	Air Velocity	139
	6.3	Measurement of Air Pollutants		140
		6.3.1	Ambient Air	140
		6.3.2	Indoor Air	141
		6.3.3	Monitoring Devices	142
			6.3.3.1 Field Monitoring Laboratories	142
			6.3.3.2 Personal Monitoring Devices	143
	6.4	Analytical Methods		143
		6.4.1	Ambient Air Gases	143
		6.4.2	Other Air Components	145
		6.4.3	Organic Vapors	146
		6.4.4	Particulates	149
		6.4.5	Microbial Sampling	153
		6.4.6	Radioactivity and Radon	154
		6.4.7	Body Burden	156
	6.5	Interpretation of Data		157
7	**Exposure Levels**			**159**
	7.1	Human Activity Patterns		159
	7.2	Pollutant Concentration		162
		7.2.1	Humidity	165
		7.2.2	Carbon Dioxide	166
		7.2.3	Sulfur Dioxide	167

		7.2.4	Carbon Monoxide	169

- 7.2.4 Carbon Monoxide ... 169
- 7.2.5 Nitrogen Oxides ... 173
- 7.2.6 Ozone ... 175
- 7.2.7 Particulates and Dust ... 177
  - 7.2.7.1 Particle Resuspension ... 179
  - 7.2.7.2 Indoor Particulate Levels ... 181
  - 7.2.7.3 Smoking ... 182
  - 7.2.7.4 In Transit ... 184
  - 7.2.7.5 Asbestos ... 186
  - 7.2.7.6 House Dust ... 188
  - 7.2.7.7 Microbial Dust ... 190
  - 7.2.7.8 Radioactive Dust and Radon ... 193
- 7.2.8 Organic Vapors ... 197
  - 7.2.8.1 Aerosol Propellants ... 198
  - 7.2.8.2 Formaldehyde ... 201
  - 7.2.8.3 Hydrocarbons, Halocarbons, and Pesticides ... 207
- 7.2.9 Odors ... 212

7.3 Exposure Level Models ... 212
- 7.3.1 Exposure Models ... 215
- 7.3.2 Formaldehyde Release Models ... 217

7.4 Twenty-four-Hour Total Exposure ... 218
- 7.4.1 Ambient Pollution Exposure ... 218
- 7.4.2 Carbon Monoxide ... 219
- 7.4.3 Formaldehyde ... 221
- 7.4.4 Radon ... 227
- 7.4.5 Asbestos ... 227

8 Health ... 228

8.1 Dose-Response Relationship ... 230
8.2 Toxicity ... 235
8.3 The Respiratory System ... 241
8.4 Climate and Health ... 245
- 8.4.1 The Indoor Climate ... 245
- 8.4.2 The Outdoor Climate ... 247

8.5 Particulates ... 249
- 8.5.1 Inhalation of Dust ... 250
- 8.5.2 House Dust ... 252
- 8.5.3 Microbes ... 253
- 8.5.4 Allergens ... 256
- 8.5.5 Metals ... 257
- 8.5.6 Sulfate ... 258

8.6 Potential and Recognized Carcinogens ... 258
- 8.6.1 Classification of Carcinogens ... 260

		8.6.2	Radon Daughters	260
		8.6.3	Cigarette Smoke	261
		8.6.4	Other Organics	263
		8.6.5	Inorganic Carcinogens	263
			8.6.5.1 Pneumoconiosis	264
			8.6.5.2 Asbestosis	265
			8.6.5.3 Silicosis	265
			8.6.5.4 Talcosis	265
			8.6.5.5 Coal Miner's Pneumoconiosis	266
			8.6.5.6 Siderosis	266
			8.6.5.7 Toxic Chemical Pneumonitis	266
		8.6.6	Factors Influencing Cancer	266
	8.7	Gases and Vapors		268
		8.7.1	Sulfur Dioxide	268
		8.7.2	Nitrogen Oxides	270
		8.7.3	Carbon Monoxide	271
		8.7.4	Ozone	272
		8.7.5	Ammonia	273
		8.7.6	Hydrocarbons and Halocarbons	273
		8.7.7	Formaldehyde	276
	8.8	Ocular Pollutants		279
	8.9	Risk Assessment		280
		8.9.1	Smoking	281
		8.9.2	Formaldehyde	284
		8.9.3	Asbestos	285
		8.9.4	Radon	287
9	**Control**			289
	9.1	Active versus Passive Control		289
	9.2	Temperature Control		293
	9.3	Humidity and Moisture		294
	9.4	Ventilation		295
	9.5	Air Pollutant Control		299
		9.5.1	Control of Outdoor Air Pollution	299
		9.5.2	Metabolic Pollutants	300
		9.5.3	Indoor Air Contaminants	301
	9.6	Building Materials		304
10	**Legislative Control, Regulation, Codes, and Guidelines**			306
	10.1	Indoor Air		306
	10.2	Ambient Air		308
	10.3	Occupational Air Quality Standards		310
	10.4	Laws Regulating Occupancy		312
	10.5	Smoking Regulations		312

10.6	Ventilation	315
10.7	Energy Conservation Standards	316
10.8	Building Codes	320
	10.8.1 Mobile Home and Manufactured Housing Regulations	320
	10.8.2 Federal Building Energy Performance Standards (BEPS)	322
	10.8.3 Codes Developed by Private Initiative	323
	10.8.4 Building Temperature Regulation	324
10.9	Material and Installation Standards	325
	10.9.1 Building Materials	325
	10.9.2 Pesticides, Toxic Substances, and Solid Wastes	328
	10.9.3 Textiles and Clothing	330

Appendix I   Units and Conversion Factors 331

Appendix II   Acronyms and Abbreviations 333

References and Bibliography 337

Author Index 399

Subject Index 415

Preface

Our life depends on incessant access to an adequate supply of viable air. However, the indoor air we breathe is not always clean, and amazingly little is known about the nature and concentration of indoor pollutants. In fact, when it comes to judging the viability of air in a poorly ventilated home, in a stuffy hotel, in a crowded meeting room, or in a smoky office, we are left to our intuition and may wonder how many of the recent advances in the sciences, medicine, engineering, and public education have been implemented.

The goal of this book is to describe some of the factors that determine indoor air quality, to provide a review of the status of our knowledge, and to supply an updated list of publications that describe the frontiers of research.

The book is intended for engineers, scientists, architects, builders, health specialists, lawyers, regulators, legislators, politicians, and the consumers of indoor air who jointly influence and control the many factors that determine whether indoor air is fresh and wholesome or stale and acutely or gradually detrimental to our health. Since it is hoped that the book will be useful to both specialists and nonspecialists, the text includes summaries of data in the form of tables and figures, as well as references to more detailed publications, so that readers can choose the depth of their involvement.

Because of the recent rapid development in this field, the literature is not well balanced yet and much of the data is only available in the form of government contract reports which are hard to identify and locate. To date there are only a very few articles that deal explicitly with indoor air pollution. However, there is a vast and well established literature on subjects that relate to indoor air quality without explicit reference to the topic. For example, the extensive literature on architectural design (which determines the air movement, air exchange rate, humidity, and temperature of the built environment) covers several thousand years and would by itself easily fill a library. Likewise, the literature on respiratory physiology and medicine is extensive. Obviously, full coverage of all related subjects would be impossible in one volume; thus, an arbitrary selection had to be made. One decision was geographical. It was decided to emphasize air quality problems in the most common building types in North America, since such housing presents a model that has general validity and can be applied to housing forms in other countries. Likewise, the regulatory system of the United States of America is used as a model for current regulatory trends in other countries. Furthermore, this book emphasizes long-term exposure and chronic health effects because only a thorough coverage of these subjects makes it possible to understand the concern of public health researchers about the threat of gradual health deterioration induced by trace pollutants and balance it with the concern of industry that apprehension about unproven adverse health effects may erect excessive social and economic barriers to the implementation of new technology.

The introductory chapter provides an overview of the indoor air problem; Chapter 2 sketches a history of the development and discontinuities in the discovery of air quality problems and their health effects. Chapter 3 discusses the parameters that define the health and welfare of building occupants, and Chapter 4 surveys a few selected building factors that form the framework for indoor air control. Chapter 5 lists the prevailing air pollutants and their sources, and Chapter 6 describes current air monitoring and analysis methods. Chapter 7 reviews the current knowledge of indoor pollutant concentrations, exposure levels, and 24-hour total exposures; it also delineates the trend of measurements and criteria for judging indoor air quality that are likely to dominate research and regulatory attitudes during the coming decade. Chapter 8 presents a very broad review of the complex and vast field of health effects, ranging from dose–response measurements to the correlation between climate and health. Chapter 9 guides the reader to some of the extensive control literature and recalls some old methods, and Chapter 10 describes the legislative and voluntary regulatory situation in the United States of America, as a model of current worldwide trends.

This book reflects personal views derived from my work in several related fields. In the area of air pollution abatement chemistry, I owe much to my students and coworkers at the University of Washington. In the field of high-temperature and combustion chemistry, my work has been shaped by more than twenty years of association with Leo Brewer at the University of California and the Lawrence Berkeley National Laboratory. The chemistry of urea-formaldehyde resins and wood adhesives was introduced to me by Bill Johns of Washington State University, and by Ben Bryant of the Forestry College at the University of Washington. The development of my perspectives of the complex relationship among industries started with my work as director of industrial research of The Sulphur Institute of Washington DC and London. My views on the regulatory relationship between industry and the federal government were greatly enhanced by a recent nine month assignment to the

U.S. Environmental Protection Agency and by work with the U.S. Consumer Product Safety Commission.

This book would not have been possible without extensive discussion and help from many people. Heinz Baumann and Fritz Kramer have explained to me the technical problems of the UFFI-Industry; Charles Morschauser, Jim Hackett, and John Emery explained the problems faced by the formaldehyde-using Forest Products industry and its trade organizations; Peter Breysse has shared his concern, active since 1961, for the potential public health hazards in mobile homes, and the late Craig Hollowell freely shared with all of us his excellent and imaginative research on the frontiers of monitoring and control of indoor air quality.

Many scientists, managers, and regulators in the federal government and in several states made it possible for me to learn about the specific charges of individual agencies and their regulatory processes. Among them are Bill Cain, Clyde Dial, Bob Hartley, Jim Repace, Kurt Riegel, David Stephan and Lance Wallace of the U.S. Environmental Protection Agency; Jim McCollom of the U.S. Department of Housing and Urban Development; Janet Haartz and Larry Doemeny of the National Institute of Occupational Safety and Health; Kay Dally of the Wisconsin Department of Public Health; and especially Joe Fandey, Harry Cohen, Jim Keenan, and Andrew Ulsamer of the U.S. Consumer Product Safety Commission. I was ably assisted through all stages of the manuscript preparation and handling by Ms. Marilee Kapsa.

This book is dedicated to all those who through the current scientific, technical, and social revolution struggle to maintain a rational approach to solving indoor air quality problems.

BEAT MEYER

Indoor Air Quality

1. Introduction

If the temperature drops below 67° or if it is drafty, people do not feel comfortable at home, office workers are not fully productive, and the frequency of influenza increases significantly. Likewise, if it is stuffy or smoky, or if the temperature and humidity are high, we feel sluggish. In fact, to function properly we require an environment whose physical, chemical, and biological properties are narrowly delimited. Since nature does not provide us with such a climate, we must create an artificial indoor habitat which better fits our needs.

The indoor environment is determined by the outdoor climate, building design, building management, and the actions of a building's occupants. Each of these factors is governed by an overwhelming number of variables. Therefore, each room of the many millions of buildings in the United States has its own distinct indoor climate. It would be presumptuous to attempt a comprehensive definition of these indoor climates. Instead, this book focuses on indoor air, because (a) indoor air is the medium through which people, buildings, and climates interact; (b) our health and well-being are determined by the physical, chemical, and biological properties of indoor air; and (c) indoor air quality can be readily defined and rationally controlled.

The purpose of the following chapters is to define the chemical, physical, and biological parameters that characterize indoor air quality, to review the current knowledge of the interactions of the components of air quality in buildings, to explore their relationship to all factors influencing the indoor environment and our well-being, and to consider how our current understanding can be used to improve the quality of life.

1.1 Basic Needs

In order to stay alive, a sedentary person has to inhale about 10,000 breaths of air each day. This adds up to about 10–20 m^3/day of air and about 1.5 lb of oxygen, which is used to fuel metabolism. The air should be fresh and clean because the human respiratory system is a very sensitive and efficient transmitter of gases, benign or poisonous, and of fine dust. For example, one bacillus in 50 m^3 of air is sufficient to cause pulmonary tuberculosis. Likewise, allergic people will react violently to just one microspore of ragweed pollen. Fortunately, however, the body is tolerant of many other pollutants, and our lungs can easily handle the 10–40 $\mu g/m^3$ of respirable dust that are present even in pure ambient air. This air is exhaled at 37°C and 100% humidity. We further need to dissipate waste heat at a rate of 80 W in order that the heat content of the body remain constant and its core temperature be maintained at 37 ± 0.3°C. This stability is achieved partly by perspiration, which contributes about 2 liters of water to indoor air. The exchange depends on activity (Figure 1.1).

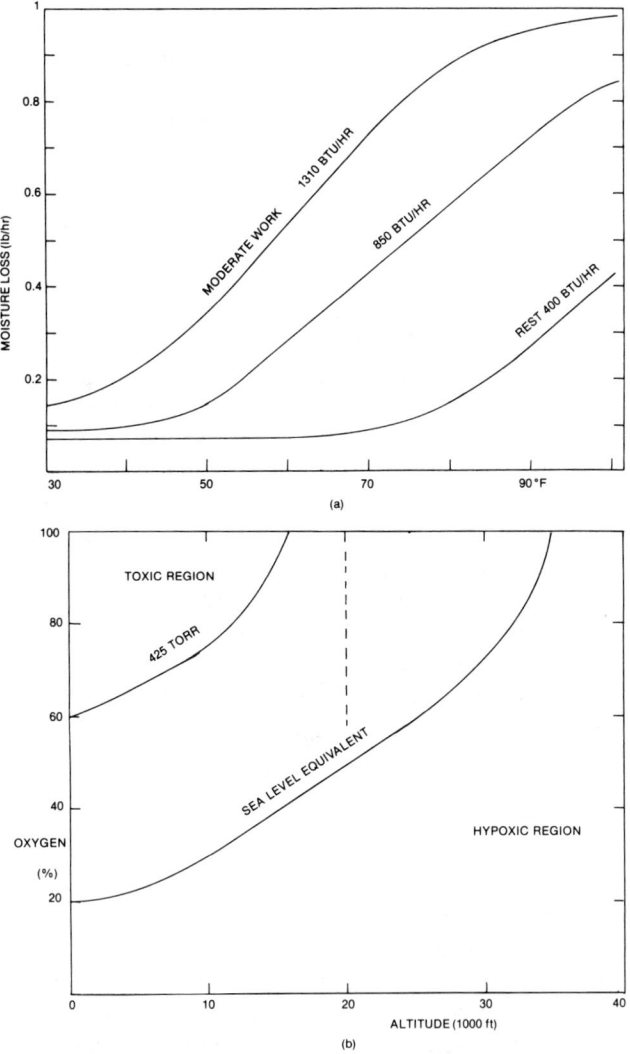

Figure 1.1. (a) Water vapor loss from the human body as a function of temperature and exercise; (b) human response to oxygen.

1.1 Basic Needs

Furthermore, each day we need to dissipate about 1 lb of carbon dioxide and traces of some 200 chemicals; the latter are responsible for normal body odors. A lightly clothed sedentary person feels most comfortable at 24.5°C, 40% humidity, and an air flow of about 0.25 m/sec, but individual comfort differs considerably, depending on activity level. If the temperature is not suitable, the body can adjust its level of heat dissipation by such defense mechanisms as increased metabolic activity and sweating or shivering. Comfort also depends on cultural expectations. Table 1.1 shows the temperature proposed by the war restoration board in England in 1945. In 1931 60% of U.S. residences had unheated outdoor privies to keep odors away from the house. Today, a pleasant year-round temperature of about 75°F is expected in all residences. In 1947 Markham wrote: "Air conditioning has been introduced as a feature of the house of the well-to-do in the U.S. and Canada, but even this is not sufficient to justify us in considering the control of high temperature as within the power of the masses of any country." In July 1981 the California Public Utility Commission unanimously voted to expand its program of low energy rates for minimal cooking, heating, and cooling services for the needy in the hot and cold parts of the state, the so-called lifeline program, to include six months annually of air conditioning in the warm belt of northern California at a rate of up to 200 kWh/month.

Obviously, life-styles in industrialized nations are changing rapidly. The average life expectancy in the United States has increased from 38 years in 1850 to 47 years in 1900 and 71 years today. In 1908 only 8% of U.S. residences had electricity; today, 98% of the homes have a television set, radio, electric coffee maker, and vacuum cleaner; 82% have color TV; 60% have electric blankets and a clothes dryer; 43% have a freezer; and 10% have a microwave oven. Over the same period urbanization grew. According to the

Table 1.1

British Standard of Warmth in Homes[a]

Room	Temperature (°C)	(°F)	Humidity (%)	ET[b] (°F)
Living room	16–20	60–68	30–65	62–66
Bedroom	12–14	55–57	30–65	50–55
Kitchen	15.5	60	70	60
Hallways	7.2–10	30–65	30–65	50
Bathroom	13–14	55–57		

[a]Data from the Heating and Ventilating Reconstruction Committee, Department of Scientific and Industrial Research, British Standards Code of Practices, London, 1945.
[b]Equivalent temperature; see Section 3.1.

1980 census, 37% of all the people in America live in cities with a population of 1 million or more; 42% live in suburbs and 25% in central cities. The number of nonfamily households has doubled since 1960 and now makes up 25%; families with children make up 30% and couples without children 30%. One-parent households account for one third of households with children. The percentage of married women working outside the home is 50% in the United States and is expected to be 70% in 1984. Figure 1.2, showing typical time budgets for family members, discloses that on the average everybody spends 20 or more hours each day indoors.

1.2 Building Factors

The purpose of the built environment is to provide for our basic chemical, physical, and biological needs. This is achieved by ventilation and, as mentioned above, heating or cooling are necessary. Figure 1.3 shows the ventilation requirements for maintaining oxygen and for removing carbon dioxide, body odors, and wastes from other activities such as smoking.

Heating requirements vary according to climate, building size, and the energy efficiency of the building. Efficiency of energy use varies tremendously. In 1980 the following fuels were used in American residences: 55% gas, 22% oil, 15% electricity, 1.8% wood, and only 0.7% coal; 68% of the dwellings

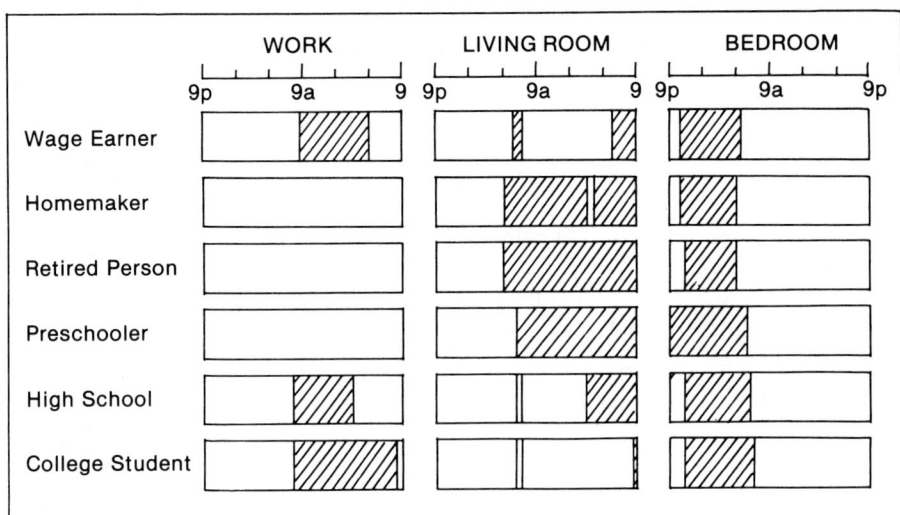

Figure 1.2. Time budget of seven family groups.

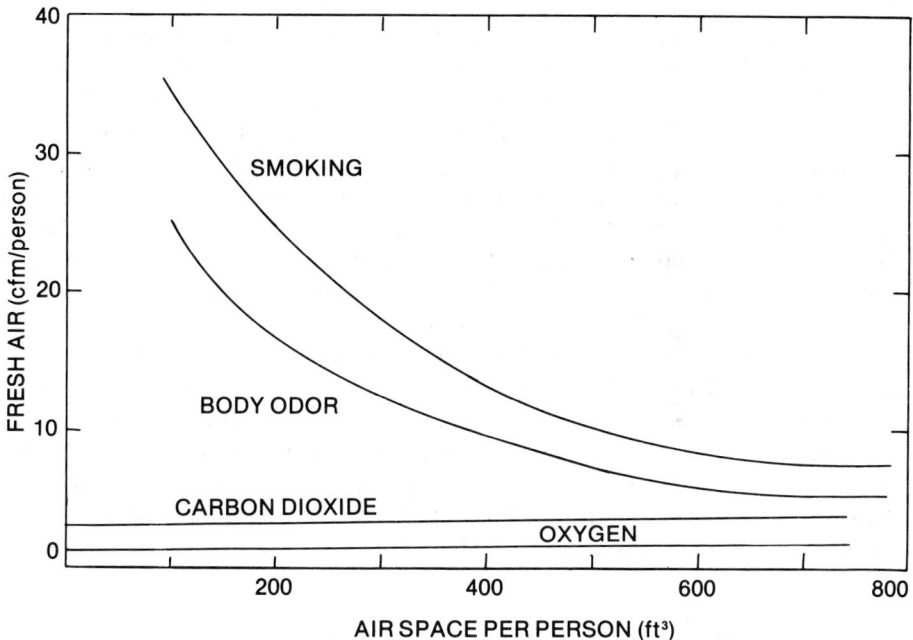

Figure 1.3. Ventilation requirements for maintenance of oxygen, control of carbon dioxide, control of odor, and control of tobacco smoke (after Yaglou, 1937).

had forced warm air heating, 8% electric baseboard heating, 8% vented room heaters, and 6% unvented room heaters. In 1960 coal was used to heat 35% of the homes, gas only 25%, and oil 23%. The trend in commercial buildings was similar.

The overwhelming majority of the current housing stock was built before 1970 when energy was freely available. At that time, it made sense to save on insulation and capital investment because heating and cooling buildings cost relatively little. These facts encouraged the construction of tall office buildings gleaming with steel and glass, full-sized indoor swimming pools and ice skating rinks, domed sports arenas accommodating a football field and more than 100,000 spectators, and giant structures that comprise hotels, shopping centers, entertainment facilities, and gardens, i.e. an entire city center under one roof.

It also made possible the construction of some 20 million large and very conveniently equipped single-family dwellings. Over this period the North American middle class became the largest and most affluent of any society ever. Most of the homes were built by developers in batches, on site, according to increasingly standardized plans and with uniform materials that minimized costs and ensured high acceptability by the purchasers, who changed jobs and residences on an average of about once every three years. These homes are poorly insulated boxes that leak air and heat at such a rate that a comfortable indoor environment can be created only by recycling 100% of the indoor air through a heater, or in summer through a cooler. All the fresh makeup air comes from leaks and cracks in the structure. The same heating and cooling concept was used in the 4 million mobile homes (Section 4.2.1.3) built during the last 25 years. Typical residential infiltration rates are shown in Figure 1.4. Commercial buildings recirculate 80–100% of the indoor air; 0–20% is fresh air and an additional 20–30% is supplied by uncontrolled infiltration.

The average heating energy use per ranch-style suburban dwelling is 90 GJ/year in the United States and 120 in Canada, as compared to 45 GJ in Scandinavia and 52 in England (Chaddock, 1976). The energy loss occurs in every area of the building. About 25% is lost through floors and walls, and about the same amount is lost through the windows, but the specific quantity depends on the climate. About 24% of the lost heat could be saved by simply lowering the indoor temperature 12°F at night. Window caulking could save

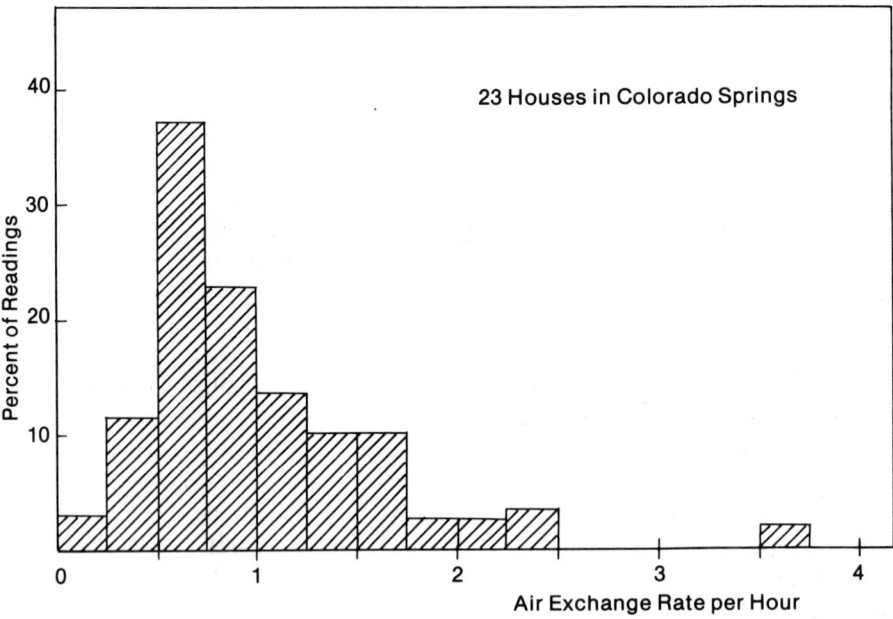

Figure 1.4. Natural air infiltration rates for 23 houses in Colorado Springs. The average of 114 readings in 23 homes is 0.82 air changes per hour (after Grot, 1979).

up to 20%, because most windows have single panes. Air leaks are also evenly distributed throughout the house: 25% of the total infiltration is through the floor, 20% through electric wall outlets; 12% through the windows, 14% through heat ducts, and 5% through doors and fireplaces. Dryer vents, kitchen vents, and bathroom vents account for 3, 5, and 1%, respectively. (Potential energy savings are shown in Figure 4.1.)

1.3 Ill Buildings

During the last 10 years the cost of energy has increased by almost a factor of 10. The U.S. government, private industry, and the commercial sector have all developed plans for better designed buildings and for retrofitting the current building stock with better insulation and other energy conservation devices. The sudden pressure for change has brought about piecemeal changes that can adversely alter the sensitive indoor habitat unless those responsible for the changes and those living in the buildings fully understand, and control all the ramifications of their actions. Since the indoor environment is extremely variable and complex, it is not always technically possible to do this, and often it is not economical. For example, it is easier to reduce heating costs by cutting the influx of fresh air than by installing insulation. As a result, many people currently live in tightly sealed structures in which 100% of the air is recirculated. This has led to complaints of eye irritation, headaches, dizziness, fever, nausea, sleepiness, and poor concentration from every segment of the population and every type of building, as well as from the passengers and crew aboard airplanes because airlines can achieve a 1% fuel savings by reducing ventilation.

Ventilation regulations have undergone substantial changes during the last 160 years (Figure 1.5). The initial increase was due to the discovery that air was the carrier of the organisms that cause tuberculosis, measles, chicken pox, and other diseases, and that ventilation could greatly inhibit their spread. When general hygienic measures reduced the incidence of these diseases, the ventilation rate was cut back to the level necessary to control tobacco smoke and body odors (Figure 1.3). This made possible air-conditioning and other indoor extravagances that soon came to be taken for granted. During the last five years, ventilation standards have been lowered to 1830 levels in the belief that current technology, building materials, personal hygiene, life-style, and smoking laws permit a narrower safety margin. The rash of complaints mentioned earlier indicates that this belief is not warranted.

Episodes occurred in 1976 in the newly inaugurated U.S. Department of transportation building in Augusta, ME; in 1978 in the Social Services building in San Francisco, and the new Oakland high school library. At the same time the University of Massachusetts at Amherst had to temporarily close a 17 story 3 tower complex because of student, staff, and faculty complaints. During the

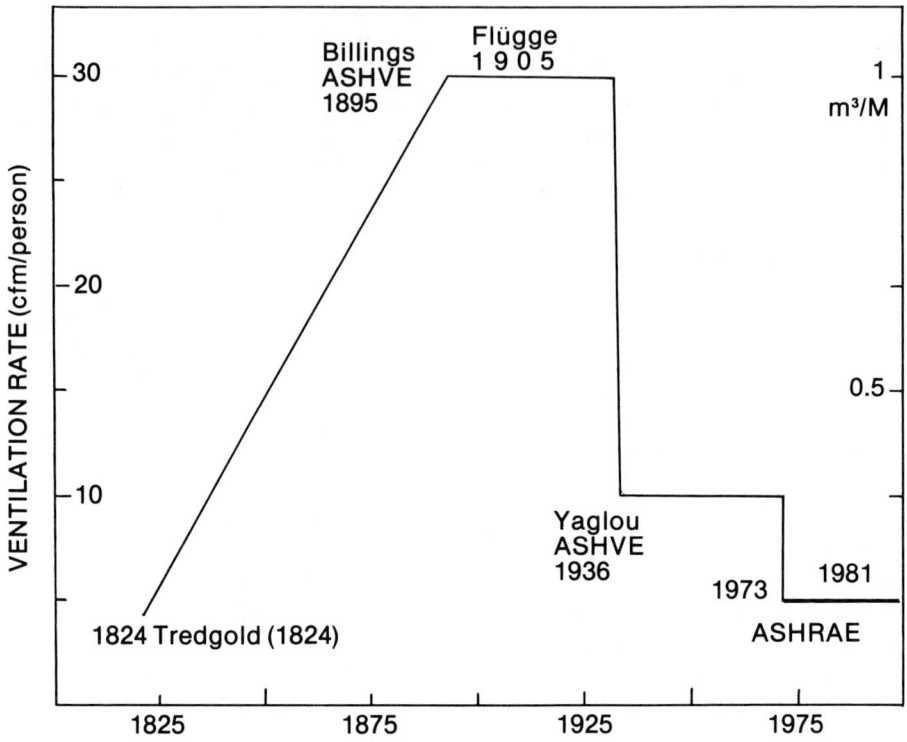

Figure 1.5. Ventilation standards from 1820 to 1981. The current standard for the United States is summarized in Table 9.2, and 9.3.

last five years numerous new school houses in the United States and Canada have had to be closed because of complaints by teachers, staff, students, and parents of "burning lungs and eyes" and similar symptoms. In May 1981, two months after moving into its new building, the Tax Assessor's office in Fort Worth, Texas, was forced to close down because 15 of its 18 employees were ill. A pilot study of 960 chemical workers in Boston and Cleveland showed that 70% felt that fresh air was inadequate (Gregory, 1981). The New York City environmental protection department receives daily complaints about inadequate office air quality. These complaints do not include episodes of acute air poisoning by photocopy liquids, cleaning fluid spills, and spreading of respiratory bacteria, described in the next section.

1.3 Ill Buildings

Private testing laboratories, state public health offices, the National Institute for Occupational Safety and Health (NIOSH), the Center for Disease Control, and the Environmental Protection Agency (EPA) have investigated several hundred complaints in every part of the United States and have discovered that indoor air problems are often complex and can be due to distinctly different causes. Although most complaints about stale air in residential and commercial structures did not involve life-threatening situations, the conditions in several buildings were distinctly not healthy. In some, this was due to poor maintenance of air conditioning devices.

According to the annual housing survey of the U.S. Department of Housing and Urban Development and the Bureau of the Census, about 14% of all residences have water leaks through the roof; 10% have inadequate heating; and a similar percentage suffer from open cracks, broken plaster, and peeling paint. In newly purchased homes, defective doors and windows, leaky pipes, and roof leaks (in decreasing order) have led to complaints.

However, problems in new buildings are also due to shortcomings in design. Figure 1.6 shows an absentee scattergram for the staff of an organization that moved into a newly remodeled building. Complaints are summarized in Figure 1.7. The problem was due to an unbalanced ventilation system, to an insufficient supply of fresh air, and to exhaust air from a neighboring restaurant that was entering the building via the ventilation system.

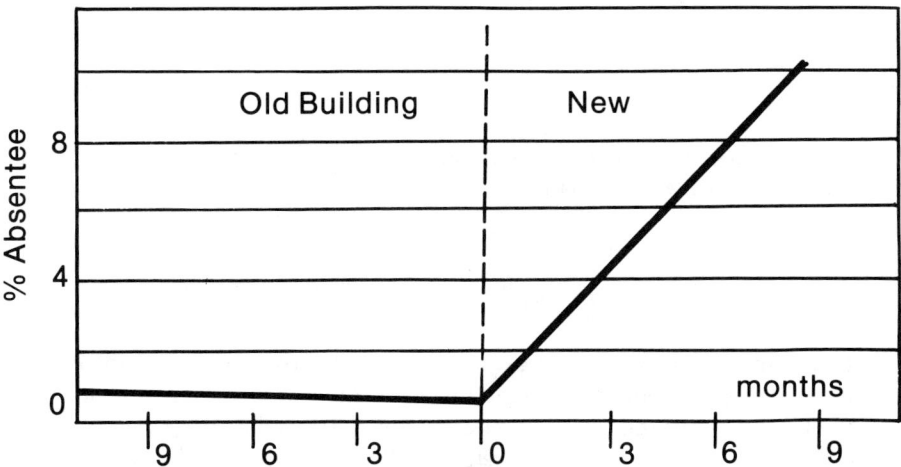

Figure 1.6. Absentee rate scattergram for office workers before and after move to an ill building (after Sterling, 1980).

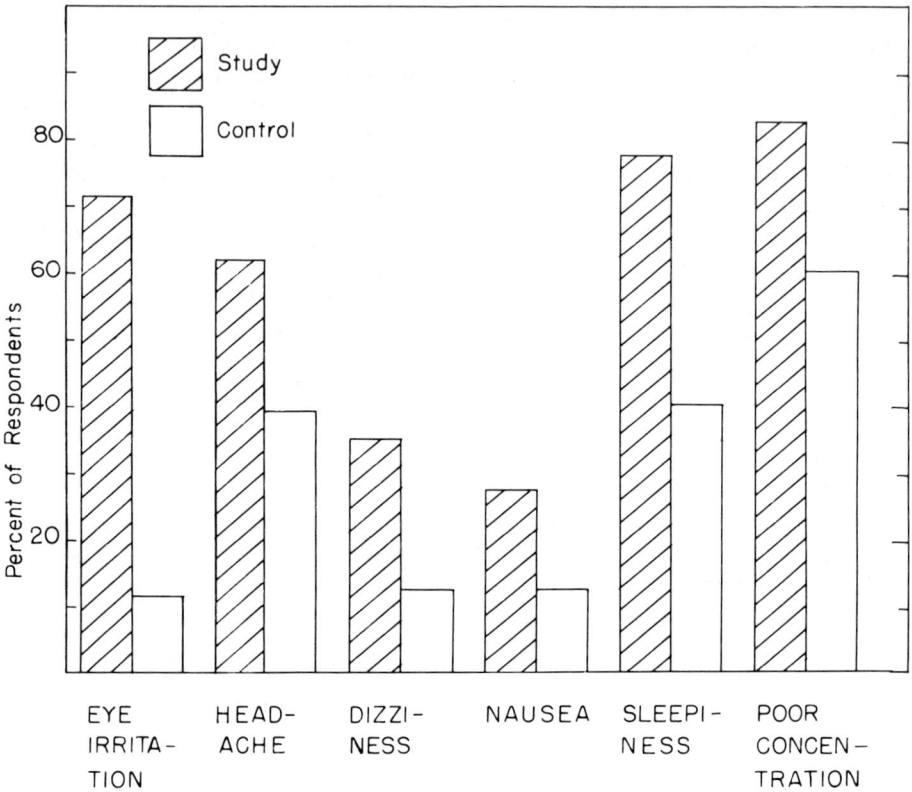

Figure 1.7. Symptom distribution in study group and among controls. All are nonsmokers (after Sterling, 1980).

1.4 Pollutants

Field monitoring and chemical analysis have shown that discomfort or ill health is caused not only by mismatched heat, humidity, and air velocity, or by acute chemical contamination, or odors, but also by hardly perceptible concentrations of various other persistent pollutants that, singly or combined, over a period of hours or days can add up to a significant accumulated exposure. Sometimes the resulting dose is comparable to or above industrial safety standards, but in many cases the levels are so low that they are only noticed by those who suffer from allergies or chemical sensitivity, whereas healthier occupants of the building experience reduced functions and diffuse symptoms.

Recent studies have identified over 250 indoor air components at concentrations between 1 ppb and 1 ppm.

Such pollutants can be due to outdoor sources, building materials, building maintenance, or indoor activities. Examples of indoor sources are ozone from photocopiers, methanol from wet-copy machines, pesticides from fumigation, and of course tobacco smoke. Examples of pollutants originating from building management practices include ammonia from cleaning liquids and organic vapors from floor waxes.

Poor maintenance practices can cause dust and a buildup from microbial growth in air conditioner filters. One of the first such incidents occurred in 1965 at the St. Elizabeth Hospital, a chronic care home in Washington, DC, where 81 people fell ill and 14 died from a bacterium that is now called *B. legionella,* after an incident in Philadelphia which caused 29 deaths at an American Legion convention. In a similar incident in 1968, 95 of 100 employees of the Public Health Department in Pontiac, MI, came down with high fevers.

Building materials can release certain agents, such as radioactive radon from cement or formaldehyde from particleboard, or they can act as a sump for moisture and serve as breeding ground for fungi. Outdoor sources of pollution are chemical vapors from waste dumps, pesticides from agricultural operations or home fumigation, and radioactive gas from the soil or water supply, and combustion gases from automobiles or near-by fossil fuel burning electric power plants.

Problems in homes can also involve formaldehyde, microbes, radioactive radon, and pesticides, but the major indoor problem is combustion gases, especially carbon monoxide, nitric oxide, and smoke from tobacco smoke, kitchen stoves, heating ranges, and wood stoves and fires.

1.5 Monitoring of Air Quality

Normally, building occupants have to rely on their five senses and such signs as odor, dry bronchus, and burning eyes to detect indoor air pollution. Unfortunately, our senses are better at judging comfort than at judging the threshold of toxicity. They are reliable for only a few molecules, such as ozone, nitric oxide, or formaldehyde, whereas the more dangerous carbon monoxide, particulates, asbestos, and sulfate are undetectable far beyond safe limits. Furthermore, almost all buildings host odors (Figure 1.8). In commercial buildings, these odors are occasionally so persistent that building ventilation systems are perfumed to mask them. Odors cannot yet be measured objectively and so must be assessed by a cumbersome procedure involving human panels, which makes it impossible to regulate odors.

During the last 10 years the analytical sciences have improved to the extent

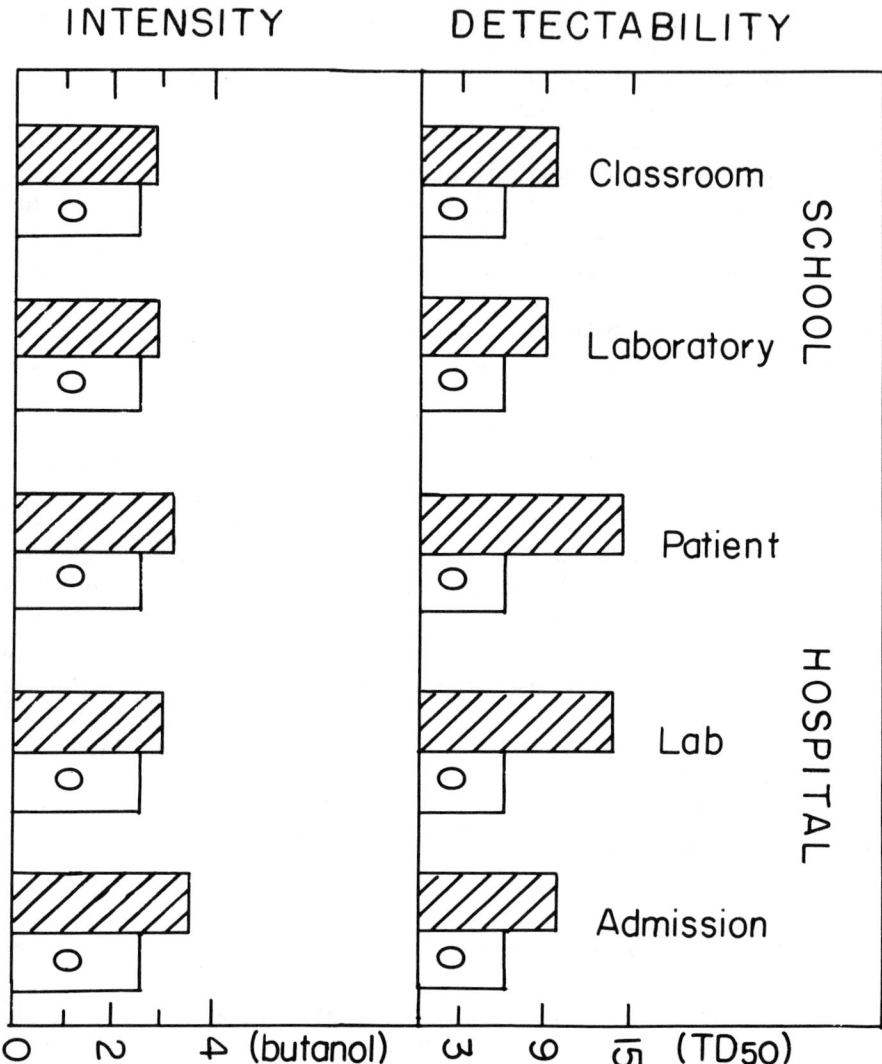

Figure 1.8. Odor detectability and odor intensity for West Coast schools and hospitals (O means outdoors; from Cain et al., 1981).

that we can now detect 10^{-12} g of pesticide in the air and in human adipose tissue. Gas chromatograph–mass spectrometer combinations can analyze one part per billion (1 ppb) of aromatic vapors in air or in human breath in a mixture of several hundred components. Finally, pocket-sized meters are now available on which the concentration of carbon monoxide, formaldehyde, and other substances is immediately displayed in digital form. On the other hand, it is still incongruously difficult to measure air humdity or many other important pollutants. Thus, our knowledge of air pollutants is distorted by the availability of analytical techniques: we tend to ignore problems until we can measure them accurately, but tend to overreact to and worry about potential effects of anything that we can measure and express in numbers.

1.6 Air Quality Control

Most indoor air pollution problems can be lessened or solved by increased air mixing, by ventilation, by eliminating indoor sources and adjusting our activities, or by cleaning recirculated air. There is no sure or easy way to find the best mix. The foremost consideration in all buildings is to provide adequate transfer of oxygen and metabolic products. In residential buildings, the main problem is humidity control. A family of four vaporizes about 4 gal of water from perspiration, respiration, and household activities per day. This volume requires 36,000 ft^3 (or 1300 m^3) of air at 75°F and 50% relative humidity (RH). At a rate of 30 ft^3/min it will take about a day to remove this heat, and the overall mass transfer of water would appear to be balanced. In reality, however, the water is produced during the two short mealtime peaks; instead of vanishing slowly, it migrates to cold spots where it condenses and remains stubbornly hidden in the form of moisture in the building materials, while most of the indoor air rapidly becomes normal or dry.

Condensation is a very serious problem if buildings designed for freely flowing air are suddenly retrofitted with insulation or vapor barriers to reduce heating costs. This condensation threatens the health of both the basic building structure and the occupants. The only adequate solution is to seal the building as tightly as possible and to provide intentional forced or natural ventilation at a rate adequate to mix air fully and remove excess moisture. Furthermore, buildings that rely on mechanical ventilation should not rely on uncontrolled infiltration, but should provide either natural cross-draft ventilation, or forced air circulation or both. Further, it is vital that such buildings have an appropriately placed air intake through which air may be admitted, either continually or in batches, as desired. In any case, natural ventilation should always be provided. In this regard, contractors, buyers, and owners should take advantage of the know-how available from the professional fields of engineers, meteorologists, geologists, chemists, physicists, biologists, physiologists, medical doctors, and specialists in the allied health and social sciences, and the architectural experience deriving from as far back as the public bath in Rome, and the famous temple of Ephesus, which had both central heating and fresh air ventilation more than 2000 years ago.

In mobile homes, a high priority should be to provide an option for natural ventilation and an air inlet for forced purging with outdoor air. In commercial buildings, air quality should be monitored instrumentally at critical spots to identify local levels of smoke, humidity, ozone, and other work-related wastes, so that the ventilation rate and the ratio of fresh air can be adjusted. The current status of microcomputer technology makes such control possible, and economical, in large buildings.

1.7 Health and Building Standards and Regulations

The need for fresh indoor air depends on a person's health and activities. The air quality at the workplace is regulated by public health agencies which are charged with providing adequate protection "for the majority of workers" (Section 10.3), but the private and government occupation safety and health agencies (OSHA) do not recognize non-industrial pollutants, and thus the comfort of office workers which during the last fifty years had been carefully improved to enhance productivity is not covered. Today, the American Society of Heating, Refrigerating, and Air-Conditioning Engineers' (ASHRAE) thermal comfort standards are meant to satisfy only 80% of the average population. Only the U.S. Environmental Protection Agency (EPA) ambient air quality standards are intended to protect the entire population, including the young and infirm. The U.S. General Accounting Office recently proposed to Congress that a federal agency be formally charged with establishing non-work-related indoor air quality standards for offices, public buildings, and residences. Such chemical standards are implicit in the recent ventilation standard of the American Society of Heating, Refrigerating and Air-Conditioning Engineers (ASHRAE), ASHRAE 62–1981 which has been accepted by almost all local governments in North America, but it is not certain how rapidly explicit air quality standards will be formulated.

2. History

2.1 Respiratory Needs

The human metabolism has fascinated philosophers and scientists through the ages. Plato and his contemporaries thought that heat, *timaeus,* supplied the body with the power to digest food, that the heart was the source of fire, and that the lung provided the necessary cooling. Aristotle, in *De generatio animalium,* recognized that animal heat was not the same as fire. Hippocrates correlated exercise with internal heat and with food consumption. During the second century Galen of Pergamon stated: "Food is used up by our bodies to produce heat as oil is by a flame," but even in the 11th century Avicenna, in his *Canon of Medicine,* still thought that food entered directly into the bloodstream. Only in the 17th century did Harvey abandon the concept of the heart as "a hot kettle, a warehouse and permanent fire place." René Descartes considered the body a large mechanical machine. He thought that blood expanded in the hot heart and was recirculated from the air-cooled lung. In contrast, Jean Baptista Van Helmont believed that acid fermentation provided the body with heat. In the late 17th century Robert Boyle, John Mayow, and Robert Hooke all speculated about metabolism and recognized that air contained some active "aerial nitre," but the next hundred years were dominated by Georg Stahl's phlogiston theory, which was ingenious but incorrect, and paralyzed the chemical sciences. During this time Hermann Boerhaave proposed and developed an internal friction model that was adopted by John Arbuthrot in his "Essay Concerning the Effects of Air on Human Bodies," published in 1730; by George Martin in his *De similibus animalibus et animalium colore,* published in London in 1740, the year before his death during an American expedition; and by Albrecht von Haller. All three still took for granted that the interior body temperature was hot as a coal fire. The only contribution made by chemists during this period was the discovery of phosphorus in animal tissues, which led to the belief that phosphorus was the animal source of heat.

However, a Scotsman, John Stevenson, was suspicious of the mechanistic views:

> Not content with the ingenious and useful application of levers, ropes and pulleys to the bones, muscles and tendons and other valuable mechanical and hydrostatis pursuits...millstones were brought into the stomach, flint and steel into the blood vessels, hammer and vice into the lungs, but all to no good purpose; there being certain bounds beyond which mechanical principles and demonstrations do not reach.

Benjamin Franklin wondered whether the body might draw heat from the surrounding cold air. The understanding of metabolism had to await Priestley and Lavoisier.

Joseph Priestley, a country pastor who gained fame as a chemist and eventually became a member of scientific academies of London, St. Petersburg, Paris, Turin, Italy, Haarlem, and Orléans, and of the French Medical Society, stayed stubbornly loyal to the phlogiston theory until his death in Philadelphia, to which he emigrated after he was evicted from his parish. However, Priestley recognized that air was not an element. He built on Boyle's discovery that gases could be produced from some solids, and differentiated between Boyle's *fixed air,* which he identified as "choke damp, long known before to all miners. It lies at the bottom of pits, extinguishes candles, and kills animals that breathe it," and "fire damp, lighter than common air, taking its place near the roofs of subterraneous places, and...liable to take fire and explode." The first is carbon dioxide, the latter methane. Priestley also correctly related the fixed air in the Grotta del Cane in Italy, which Cavendish had shown to be 50% heavier than air, to fermenting vegetables and to the acidic agent in hot springs. Priestley's extensive diary, published in 1790, reveals his brilliant experimental work and interpretations. His favorite source of oxygen was red mercuric oxide. It is still used for this purpose.

The major breakthrough in the understanding of human metabolic needs was made by Lavoisier, who carefully followed Priestley's work through publications and intermediaries. Lavoisier's research on respiration and perspiration gradually developed from his key paper on inorganic combustion, published in 1775 when he was 32; "Mémoire sur la nature du principe qui se combine avec les métaux pendant la calcination, et qui en augmente les poids." In this paper, Lavoisier reported that air contains 20% oxygen; that oxygen is an element; that combustion is a form of chemical oxidation; that carbon, sulfur, and several metals gain weight, rather than lose phlogiston, during combustion; and that the phlogiston theory was intrinsically wrong. As an aside, he showed that the new element was the respirable component of air. He immediately conducted his "Expériments sur la respiration des animaux et sur les changements qui arrivent à l'eau en passant par leur poumon" ("Experiments regarding the respiration of animals, and the chemical changes that air undergoes while it passes through the lungs"), published in 1777, from which

2.1 Respiratory Needs

he concluded that oxygen was transferred and used during respiration of people as well as animals. He then designed a calorimeter, and built respiratory as well as respirator–perspirator devices (Figure 2.1) that made it possible for him to measure the oxygen consumption of his butler as well as of a variety of animals at rest and while exercising. His paper on general combustion followed in 1778, and in 1780 the "Mémoire sur la chaleur," in which he and Laplace laid the foundation for chemical thermodynamics. They established quantitative correlation between oxygen consumption, carbon dioxide (CO_2) release, and heat release; correlated the reaction to the heat of formation of carbon dioxide; and developed, as an aside, the principle of conservation of matter. In 1784 Lavoisier published experimental proof that water, one of the four elements recognized since antiquity, was not a chemical element. In 1787, he used guinea pigs to establish that the absolute oxygen pressure, rather than the gas mixture, was decisive for respiratory survival. His experiments were as accurate and brilliant as his theoretical work. He developed an entirely new chemical nomenclature, *Méthode de nomenclature chimique,* and in 1789 published the textbook necessary to explain the new chemistry, *Traité élémentaire de chimie,* and in 1785 "Altération qu'éprouve l'air respiré," an experiment that prepared the way for the famous paper, "Mémoir sur la respiration et la transpiration des animaux," which appeared in 1789 and 1790, with the help of Seguin.

According to their experiments, a man breathes 5 ft^3 of air per hour, from which only 2160 ounces of oxygen are absorbed, while 1645 ounces of CO_2 are exhaled, with 514 ounces of water as a by-product. Skin perspiration amounts to about 0.15 ounces per hour, which is equivalent to the heat of 18 ounces of water obtained by combustion of alcohol. The paper ends with the remark:

> This type of experiment was designed to establish a relationship between quantities which had previously appeared to be not related in any way. With its help, we can now determine, for example, how much work, in pounds of weight lifted, corresponds to the effort necessary to deliver a speech, or of a musician who plays an instrument. One can even determine how much physical effort is hidden in the creative work of a philosopher, a writer, or a composer. This type of effort, formerly considered to be purely moral, clearly contains a physical and tangible component.

In 1801, seven years after Lavoisier's death, Seguin published the second manuscript on animal respiration, in which he further confirmed the earlier data.

For over 25 years, Lavoisier worked daily in his laboratory from 6 to 9 A.M. and 7 to 10 P.M. During regular work hours, he designed chemical factories, developed city projects, and served the intrigue-riddled French Académie des Sciences, as well as the court of Louis XV, while promoting projects such as the metric system, until he became a victim of the revolution that he had inadvertently helped bring about through his work. Abandoned by his

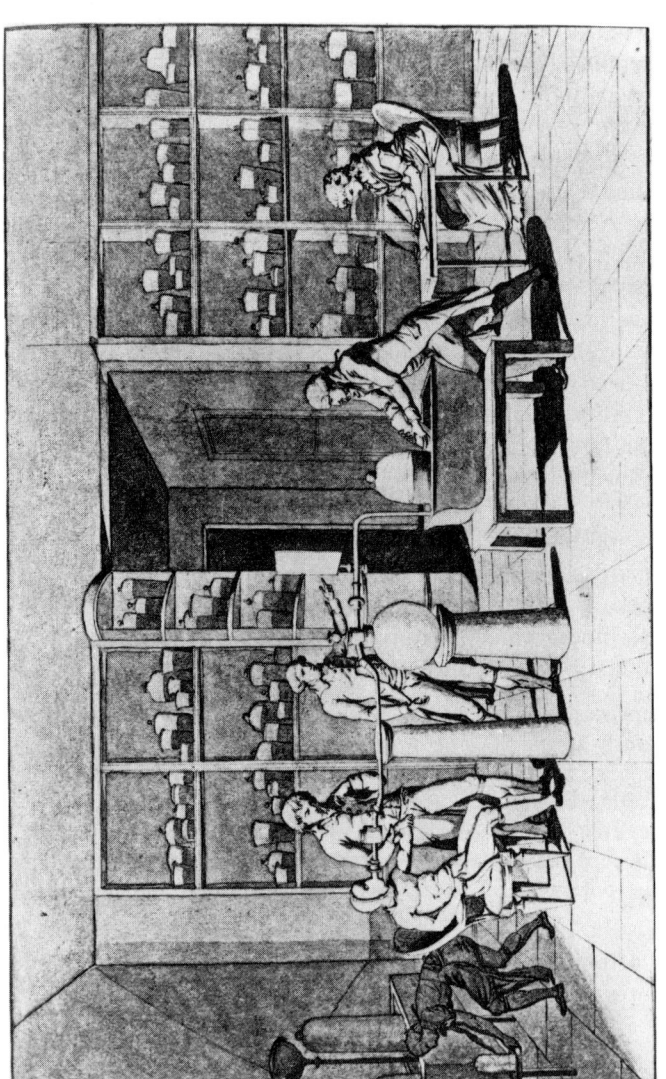

Figure 2.1. Lavoisier and his respiratory meter. Drawing by Madame Lavoisier, who also appears in the picture as record keeper.

colleagues, he was guillotined, after five months in prison, in 1794 at the age of 51. His name was misspelled on his death certificate, and despite the continued brave efforts of his brilliant widow, his name was purged from French scientific literature for the next 50 years. Ironically, French scientists promptly adopted his theories and rabidly defended them against the English, who, though admiring his aristocratic life, belittled his combustion theory.

2.2 Stale Air

For the next 70 years, it was accepted that the oxygen and carbon dioxide content of air determined whether it was fresh or stale. Then Max von Pettenkofer, a Munich researcher, used equipment of his own careful design to ascertain that the CO_2 levels in stale indoor air were insufficient to account for the discomfort people experienced. He recognized that humidity and temperature determined comfort and acted synergistically. Von Pettenkofer retained CO_2 as a tracer for comfort. He recognized that metabolic trace products were responsible for body odor and the odor of stale air, and he expounded on the important interaction between comfort and climate in many lectures and books, such as the one entitled *The relations of the Air to the clothes we wear, the house we live in, and the soil we dwell on* (1872).

Parallel with him, the Frenchmen Reignault and Reiset developed the concept of the "respiratory quotient," the ratio of CO_2 to O_2 volume. After von Pettenkofer, respirator design continued to improve until Atwater and Benedict (1905) at Wesleyan University, with support from the U.S. Department of Agriculture and advice from the National Bureau of Standards, perfected a respiration calorimeter that could host a human subject for several weeks, and yielded thermal, chemical, and weight data accurate to within 0.2%. Using a medical student as his guinea pig, Atwater measured a heat flux of 2113 cal/day in the calorimeter, as compared to a food energy uptake of 2088 cal/day computed by calorimetry of the food and feces. The corresponding oxygen consumption was 435.7 liters as opposed to a CO_2 expiration of 332.36 liters. Water from respiration and perspiration amounted to 838 g. A detailed description of the equipment, with detailed tables and photographs, the experiments, and the evaluation method, is contained in a monograph published by the Carnegie Institute (Atwater and Benedict, 1905); see Figure 2.2).

Effect of Air on Health. The Greeks and Romans were well aware of the adverse effects of polluted air, and those who could afford it spent summer at the Adriatic shore.

Aristotle described in detail why rural people looked healthier and stronger than their urban contemporaries, but during the Middle Ages many of these insights were lost, and rich as well as poor shivered through the winter in cold, smoky rooms, huddled around an indoor fire. When asked why his monas-

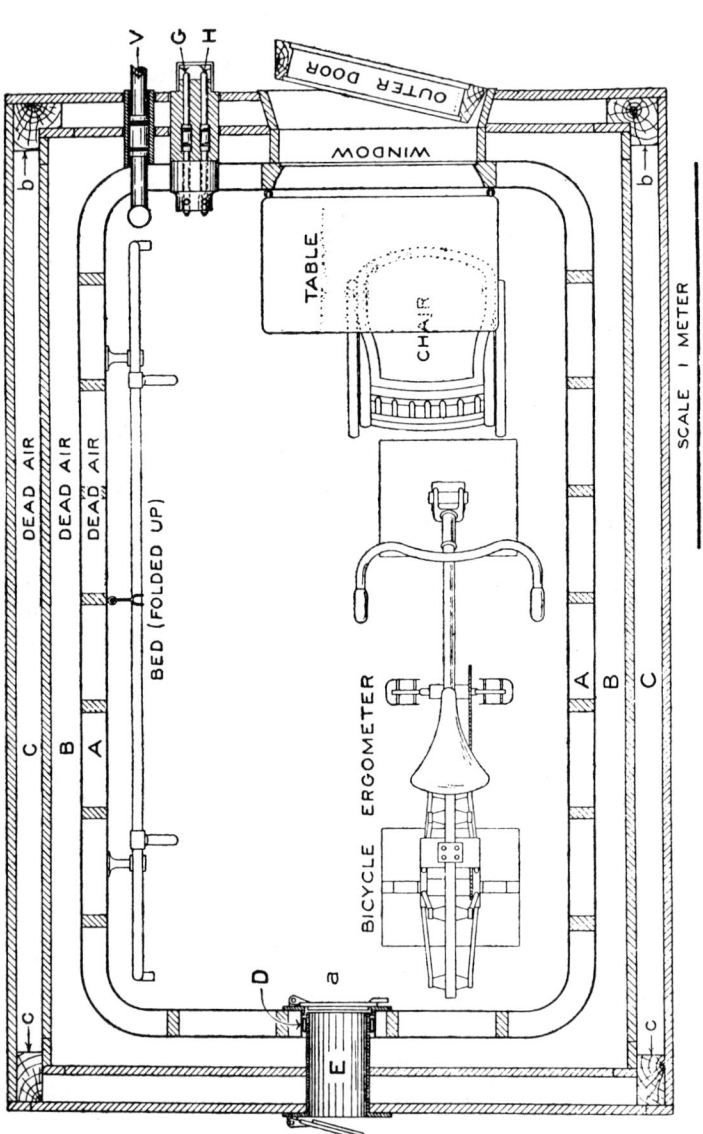

Horizontal Section of Respiration Calorimeter Chamber. Portions shaded are of wood.

A, dead-air space between Cu and Zn walls.
B, dead-air space between Zn wall and inside wooden wall.
C, dead-air space between inside and outside wooden walls.
D, pneumatic-packing air and heat insulated.
E, food-aperture tube.
a a, airtight ports (glass).
H, ingoing water for absorbing heat
G, outcoming water.
V, ventilating air current.

Figure 2.2. Respiratory apparatus of Atwater and Benedict (1905). The chamber was at Wesleyan University.

teries were built in damp cold areas, a French archbishop of the 16th century allegedly stated that the constant fear of infection helped remind his monks of the frailty of life and made it easier for him to control the spirit of his flock.

In the 17th century, the benign properties of fresh air were once again recognized, and it was rediscovered that rural people had a greater life expectancy than those dwelling in crowded cities. However, Rousseau's call for a return to nature did not change the aristocracy's reliance on strong perfumes and powders (rather than fresh air and ventilation) to cover up the stench of city life. Perfumes, which have no therapeutic value, consist of a mixture of 10-15 osmogenic species in a volatile base. Upon vaporization of this base, the osmogenic species mix with body sebum and produce a bouquet that alters the basic body odor (Section 5.1).

The situation in London was abominable: the particulate concentration in the air averaged about 150 $\mu g/m^3$ over several centuries. Aerial photographs, even in the 1940s, show entire neighborhoods veiled in dense residential coal smoke, a situation described in every book on air pollution. Humphrey Davies was among those who provided relief to the British Parliament by designing a ventilation system capable of delivering 10 ft^3 of outside air per member.

As early as 1661 John Evelyn had devoted an entire volume, *Fumifugium,* to the smoke problem. His proposal was to introduce parks with trees to absorb the soot and provide sunshine. In 1662 John Graunt wrote the *Natural and Political Observations upon the Bills of Mortality,* and spoke out against pollution, but society did not really respond until 200 years later. Robert Barr, returning from Canada in 1882, reedited Evelyn's book and wrote *The Doom of London,* in which he predicted that all life in London would cease in 1950 under the smog. This prophecy was almost fulfilled, and the killer fog of November 1950 finally led to legislative action.

The 18th and 19th centuries were dominated by the fear of tuberculosis and other diseases. Books of this period abound wth reports of such tragic episodes as the "Black Hole of Calcutta (the death, around 1700, of some 127 of 140 Englishmen confined overnight in conditions of extreme heat, humidity, and overcrowding) or the passage of the steamer *Londonderry* from Sligo to Liverpool (on December 22, 1748, Captain Johnston herded 150 emigrants into the $18 \times 9 \times 7'$ steerage cabin and then sealed it with tarpaulin to protect the vessel in the storm that was raging; 12 hours later 72 of the people had died of suffocation).

2.3 Air Pollution Legislation

In 1820 Germany and Austria introduced legislation that made the owner of the source formally liable for noxious air discharges. In 1863 England followed suit with the Alkali Works Regulation Act, which greatly influenced action in the United States: In 1876 St. Louis required chimneys to reach 20 ft above roofs, thus beginning municipal action in this country; in 1881 Chicago established a fine of $5–$50 for "dense smoke," leaving the definition deliberately vague, an approach still favored by politically oriented lawmakers. The international hostilities of the following 40 years overshadowed concern for public hygiene, and it took the international incident in 1920 at Trail, British Columbia (Robinson, 1974); the incident at Kucktown, TN; the 63 deaths in the Belgian Meuse Valley in 1930; the 1948 episode in Pittsburgh, PA; and the 4000 excess deaths in December 1950 in London to stimulate a world-wide effort to clean up the air.

All these events were intertwined with the unprecedented economic advances that followed World War II. Clean-air movements have become possible only because affluent societies could afford to replace coal with cleaner burning oil. Postwar development also brought with it new pollutants that replaced the 1000-year-old curse of sulfur dioxide with subtler, but far less predictable, synthetic chemicals. The discovery of slow-acting carcinogens is sketched in Section 2.4.

During the 1950s industrial effluents in the air reached such levels that synthesis occurred in mixed pollutant streams. For example, in Louisville, KY, certain weather conditions caused the effluents from a bromine plant and those from a styrene plant to combine, forming styrene dibromide, a very potent lachrimator, which inflicted considerable anguish on Louisville's residents until the source was finally identified.

In the United States air pollution control remained the prerogative of states and municipalities until the enactment in 1955 of the Federal Air Pollution Control Act. Twelve years later this act was reinforced by the pollutant criteria documents described in Chapter 10. Only Japan and Sweden have experience with government regulation of nonoccupational indoor air.

2.4 Health and Air Quality at the Workplace

Until two decades ago, interest in air pollution control centered entirely on the workplace. The ancient Egyptians had already perceived that the silicate dust produced by the cutting of construction stone caused respiratory disease. Hippocrates in 500 B.C. stressed that the air in mines and that encountered in certain trades was highly detrimental to health. Pliny the Elder noted in 23 A.D. that stonemasons and asbestos miners could not work without masks. Agricola (1494–1555) expressed compassion for mine workers, recording that the prisoners among them, who remained locked up in the tunnels during the night, suffered more and died much sooner than those who could recover outside overnight. The woodcuts in his book associate all chemical and mining work with dense fumes.

2.4 Health and Air Quality at the Workplace

Bernardino Ramazzini (1633–1714), the father of occupational medicine, not only observed his patients at his clinic in Milan but also visited their workplaces regularly.

Almost everything we know about the health effects of indoor air pollutants was learned "on the job." Moreover, until very recently illnesses associated with work were considered an inevitable, if not normal, concomitant of gainful employment. John Griscom (1850) vividly described the lot of a 19th-century apprentice:

> In very few trades or professions is [the Apprentice] subjected during the hours of labor, to any other atmosphere than such as is polluted, by either the respiration of numbers of persons in a small room, or by effluvia of materials of manufacture, increased and concentrated by the local situation and narrow enclosure of the house itself, by which all visitation of general atmospheric currents is prevented.

Except for house builders, shipbuilders, farmers, gardeners, and sailors, he finds that "he who decides upon himself, or his child, becoming an apprentice to any one of the great majority of trades must calculate upon a prostration of strength, paleness of countenance, languor of spirits and body, frequent sickness, and almost inevitably, as statistics show, a premature death." Griscom's detailed analysis established beyond doubt that the major causative factor was indoor air.

Especially deplorable conditions were encountered by chimney sweeps, often slim children 12 years old or younger (the better to fit the narrow chimneys). These young laborers slept in the workshops, usually in their work clothes, and few survived for long the chronic bronchial infections caused by soot irritation. Those who did survive often contracted cancer of the scrotum, a fact established and published by Dr. Percival Pott in 1778 in his *Chirurgical Works*. The exposure level and dose-response effects were even then well recognized.

While Griscom concentrated on case studies of various occupations and analyzed prevailing conditions, some of his contemporaries were already involved in very sophisticated epidemiological research. For example, by 1840 a Dr. Guy had reported to Griscom the finding that the incidence of spitting of blood among 500 letterpress printers was directly related to the size of the workroom. Of the 100 workers who had 500 ft^3 or less of air to breathe, 13% spat blood; of those with 500–600 ft^3 the value was 4.4%; and of those with more than 600 ft^3 the incidence was 3.96%. Dr. Guy also compiled data on other diseases, and found that their incidence remained a constant 17.31–17.82% among all groups that he studied. It was almost 50 years later when Rehn (1895) started more quantitative occupational epidemiology on cancer with his work on bladder cancer caused by aniline dyes. In 1906 Fischer identified azo dyes, such as scarlet red, as carcinogens. In 1918 Yamagiwa and Ichikawa applied suspected carcinogens, especially coal tars, to the ears of rabbits, a method that led to modern animal experiments. In 1935 L. Smith was the first to document that asbestos was retained in the lung and caused cancer; and in 1942 Schirz and Mehlinger established that metals could cause cancer, confirming suspicions that, in the case of chromium 3 had already been voiced by Pye in 1885.

For many substances long known to act as poisons, however, much has yet to be learned. Among these are nickel (Mastromatteo, 1967) and arsenic, first indicted by Hutchison in 1888. Many modern carcinogenic substances were first synthesized and manufactured well after World War I. A classic example of a suspect modern industrial bulk chemical is vinyl chloride (Creech and Mack, 1975); it is one of the reasons why the federal government now requires early testing of new chemicals that are earmarked for large-scale production. The first major accident in which impurities in a commercial product wrought havoc involved dioxin (2,3,7,8-tetrachloro-dibenzo-P-dioxin, or TCDD), some 2 kg of which were accidentally spilled, along with trichlorophenol, during the malfunction of a herbicide plant in Seveso, Italy, in 1976. Dioxin has also been found in significant quantities in the lipoid tissue of Vietnam War veterans, especially of those who loaded the chemical defoliant called agent orange (Section 5.6.5).

2.5 Indoor Air Comfort

It is virtually impossible for us to appreciate fully the incredible effort that was needed to construct and maintain buildings 2000 years ago. It is harder yet to accept that earlier cultures, without benefit of modern science and engineering, designed and produced more lasting edifices than anything we build today. Saeltzer (1872) analyzed the ventilating and heating systems of the Romans and explained in detail why the residences of the rich were so comfortable: These buildings had a basement, the hypocaust, where a fire was built so as to provide the greatest heat at the entrance door. The floor materials were conductive stone. Fresh air was admitted into the living space as necessary for the various activities, and the vertical temperature gradient was far smaller than in today's homes, thus providing a very pleasant feeling. Wall cavities served as chimneys and provided balanced radiative heat. This made it possible for citizens to gather in the public baths to socialize even in winter, when the temperature outside averaged 40°F. The subtle differences among the heating systems of Roman houses in Italy, England, and north of the Alps clearly show that the architects were in full control of their design.

The culture gap separating Roman civilization from that of the barbarians who swept through the heart of the Empire was too wide to allow the Romans' architectural and engineering technology to be preserved. In the west, the early English kings were hunters and horsemen, satisfied with shelters that were kept warm by an open fire. Throughout the Middle Ages European nobility lived in castles that in winter were drafty, stinking, humid caverns, until efficient chimneys were rediscovered and window glass was invented.

In subsequent centuries the value of fresh air was recognized anew, and the words "cross draft" became associated with good health. Accordingly, ventilation rates were increased among those sensitive to health issues (Figure 1.4).

John Griscom (1850) was keenly aware that all human activity depended on large doses of fresh indoor air:

> First, as a tender *Infant,* he scarcely has made his entrance into the great ocean of air, and uttered his plaintive petition for a portion of the new element to expand his little chest, ere, by the careful nurse, he is tucked away under the coverlid, with his head closely wrapped in a blanket-shawl, lest the already impure air of the chamber should be too strong for his weak organs.
>
> His next position is as the *Schoolboy,* and here we find him, with from five to five hundred others, immured between closed walls.... At night he occupies a room, rather larger than a prison cell, which contains sufficient air to allow him to breathe it in its purity for...thirty to sixty minutes...after which he inhales and re-inhales in larger and larger portion, the carbonic acid thrown off from his lungs as excrementitious poison, until, in the morning, he creeps from his bed in a dripping sweat and unrefreshed.
>
> The next period of his life is as the *Apprentice.* [They] pass their working hours in close, unventilated shops, redolent with the fumes of steam, smoke, white-lead, sompaste, alkalies, the gases from decomposing animal and vegetable matter. A ventilated workshop is almost an unknown thing.... Even housebuilders, farmers and gardeners [and sailors]...at night are generally subject to vitiated air, in small crowded chambers.... We find that evil is not confined to the laboring classes or the less intelligent portions of society. Wherever we find a civilized being domiciled, whether pursuing interest, pleasure, working or justice, above or in society, at home or travelling, we find him breathing an atmosphere more or less impure. Very often the effects are demonstrated at the moment in flushing of face, perspiration, drowsiness, headaches, vertigo, and fainting, and very frequently at more remote periods, by the development of more profound and serious diseases of the lung, heart, head and stomach.

Griscom then identifies churches, courtrooms, theaters, and lecture rooms, as well as private chambers, kitchens, offices, stores, and private soirees, as equally likely to offer nothing but foul air. However, he singles out bedrooms and dormitories as the worst offenders. As a New York surgeon, he was especially interested in consumption (i.e., tuberculosis), statistics for which indicated a high incidence in England, the highest incidence being noted among those exposed to dust inhalation and those places "where pit coal is burned to a great extent." The prevalent opinion at the time was that "deficient ventilation ...[is] more fatal than all other causes put together." (That tuberculosis is spread via the air had already been ascertained.)

The hundred years from 1830 to 1930 were dominated by a justified dread of tuberculosis and other diseases known to be contracted in crowded places. It was increasingly realized that infections were transmitted among people sharing the same room. Lister (1868) started to use phenol as an antiseptic treatment during surgery, and Miquel conducted a series of trend-setting studies at his Hopital de la Pitié in Paris (1883–1890) in which he followed the transmission of living organisms via the air. Parallel studies in Germany by Flügge (1897) and Cornet (1889) concerned themselves with the path of microbes.

At about the same time cities slowly gained better control over animal and human wastes, and Haldane engaged in a thorough study of air pollution that included the need to abate "the air of sewers." During the 19th century the quality of life among middle-class people reached unprecedented levels. Lippincott in Philadelphia (1887) wrote a book on house plants as sanitary agents, reflecting the gains made as a result of the industrial revolution. With daily environment cleansed of much putrefaction, the sense of smell was newly discovered, and people became far more discriminating. But the understanding of smell has proceeded slowly (Zwaardemaker, 1895; Henning, 1915) and the field is still full of riddles.

In the meantime, the habit of smoking tobacco went through several cycles of acceptance and rejection. Around 1880, smoking was widely identified as "a disgusting and pernicious habit," which was forbidden in 15 American states for moral as well as health reasons. During this time, the American Society of Ventilation and Heating Engineers (ASVHE) was founded and promptly contributed heavily to the basic understanding of the field of ventilation. This society quickly established itself as a leader in all technical matters; it has sponsored almost all of the leading academic research in the field, and it spearheaded the application of new findings in daily life in a systematic manner and over a time period unequaled by any other professional society in the field.

2.6 Architecture and Building Materials

Fresh indoor air is an intrinsic part of building design. The symbiosis of intuitive art and deductive science that characterizes architecture is reflected in every history of architecture, such as Sigfried Giedion's *Space, Time, and Architecture* (1954), René Alleau's *Histoire des Grandes Constructions* (1966), and Carl Condit's *American Building* (1968).

Mud, clay, trees, stones, reeds, and rushes have been used, according to availability, as construction materials by people up to the present. If available, lime was added as a plaster, and thatched straw has been used as roofing material in countries as remote as Russia and England, as well as elsewhere throughout the world.

Wood logs were rapidly recognized as a material with desirable physical, and therefore psychological, properties. The Japanese and other oriental temples show that, properly used, wood has a durability second to that of no other material, including marble. Moist wood, however, will rot within 10

2.6 Architecture and Building Materials

years, and is as vulnerable as limestone and marble, which do not last much longer if exposed to heavy air pollution.

The Egyptian pyramids are made of limestone. Concrete was developed and widely used by the Greeks; bricks were invented and fabricated independently by several cultures. Brick was extremely popular in the 17th century in what is now the State of Virginia. Because of its friability and weight, stone was of limited utility for buildings, even though early architects devised ingenious arches that made it possible to roof temples and churches and to construct domes. The pantheon built by Hadrian in 120 has a diameter of 38 m; the Byzantine Hagia Sophia Church in Constantinople, built in 532, spans 30 by 50 m.

Iron construction was initiated with St. Anne's church in Liverpool in 1772 and became very popular; but many buildings for which iron was used collapsed, and this material temporarily lost popularity in the United States after the Civil War until the Home Insurance Building and the Auditorium (with a span of 36 m) built in 1889 in Chicago started a local boom in skyscraper construction. Parallel with the Chicago steel buildings, monolithic concrete was developed and used in the Ponce de Leon Hotel in St. Augustine, FL, built in 1886. Concrete had been extensively used during antiquity, but knowledge of how to make and use it was lost for more than 14 centuries until John Smeaton reinvented it in England around 1600. Modern concrete construction was made possible by the patent of David O. Saylor in 1871. The first tall concrete structure is the Ingalls Building in Cincinnati, constructed in 1903. The combination of steel and concrete with modern material handling technology made possible the development of modern tall office buildings with glass curtain walls suspended by steel mullions from a reinforced concrete service core.

The development of factory-preassembled buildings could only be envisioned when it became economical to transport very heavy loads. The advantages of factory-prepared modular construction are obvious, but this technology has encountered so many nontechnical problems, such as tradition-bound local building codes and labor agreements, that the modular procedure has remained confined to large buildings—sports arenas, large hotels, or multipurpose indoor city centers—and to the mobile home industry.

Residential construction still relies heavily on wood, especially for flooring and roofing. The latter is usually set on trussed rafters, spaced 2 ft apart, that span 20–40 ft. Wood wall construction developed slowly from round logs to the modern hollow wall (described in Section 4.2). In modern sports arenas and public buildings wood is now used for roofs with spans up to 60 m; these spans achieved by laminating small wood sections with a formaldehyde-based adhesive invented during the 1920s (Meyer, 1979).

People have become so accustomed to thermal comfort that optimal indoor comfort is demanded not only throughout the home and workplace, but even in transit. Within 80 years, modes of transportation have metamorphosed from drafty horse-drawn carriages to fully enclosed automobiles, which were eventually heated and even cooled. Today the percentage of air-conditioned cars exceeds the percentage of homes that were air-conditioned a mere 5 years ago. Air conditioning is very wasteful in these vehicles, since they are poorly insulated, are used for short periods (which necessitate long pre-use and after-use heating or cooling periods), and have extremely high infiltration rates.

3. Comfort and Climate

3.1 Comfort and Sensory Perception

Comfort involves physical as well as physiological and psychological factors. For the chemist comfort means freedom from unwanted odors, and restriction of air contaminant concentrations below the levels at which irritation or toxic effects manifest themselves. For the heating and air-conditioning engineer comfort is a measure of the heat balance between the occupant and the indoor habitat. Thermal comfort depends on the operative temperature of the environment, the insulating value of the clothing, and the level of metabolism. An individual feels comfortable when metabolic heat is dissipated at the rate at which it is produced. Extensive experimentation has shown that for an average, sedentary, lightly clothed person this occurs most readily when the air in a standard room has a temperature of 24.5°C, a relative humidity of 40%, and an air velocity of 0.25 m/sec. According to thermal comfort standards currently valid in the United States (ASHRAE standard 55-1981), 80% of all adults dressed for winter indoor conditions find temperatures acceptable between 68°F and 74.5°F (20-23.5°C), a relative humidity of 30-60%, and the air velocity at 0.15-0.25 m/sec. Acceptable summer indoor temperature is between 73 and 79°F (20-26.5°C). Very lightly clothed people prefer a temperature between 79 and 84°F (26-29°C). The summer temperature can be increased to 83°F (28.5°C) if the air velocity is increased to 0.8 m/sec (160 ft/min). However, regardless of season and temperature, a vertical temperature gradient of more than 5°F is found by all to be unpleasant, as is radiative asymmetry exceeding 9°F in the vertical direction and 18°F in the horizontal direction. Thus, a person sitting 3 to 4 ft from an all glass outside wall in a tall office building may well feel uncomfortable on a very hot or very cold day, regardless of the indoor air temperature (Fanger, 1972).

An individual's assessment of comfort changes with metabolic rate, clothing, body temperature, and the degree of perspiration. Since all these factors, including metabolic level, change with activities and with circadian and diurnal cycles, a room that is kept at a constant temperature will at times be uncomfortable.

Moreover, two people will often choose a different approach for deriving comfort from a given situation. For example, for a tourist from the mainland, comfort on a tropical island might mean a day of basking, minimally clad, in the sun on a beach, whereas for a native it would more likely mean resting, fully dressed, in the shade of a tree. The tourist enjoys the very high rate of water transport through the skin without the wet feeling that a similar perspiration rate would produce in a clothed state in still air at home. The native prefers a more moderate heat flux between surrounding air and body. Analogous situations exist everywhere. In almost every large office at least one employee will need a sweater or vest in order to be comfortable in the same area where co-workers are content in lightweight garments regardless of the weather or season.

Since the human body does not register temperature, wind velocity, and humidity readings, per se, the question arises how the body registers and regulates comfort. The answer to this question is not known. In fact, we do not even know where human internal temperature sensors are located nor how they function. Obviously, these so-called temperature sensors respond not simply to the body temperature, but to a combination of temperature gradients and energy flux.

All five senses are active in the indoor habitat and have to be satisfied within a narrow range between deprivation (Solomon, 1971), and overstimulation or pain. Judgment of indoor air quality is, as we have indicated, highly subjective and related to many factors. At the dinner table, for example, food odors are as important as taste. In airplanes, however, food odors are generally perceived as inappropriate, at least in the crowded coach section; and in an office most people find osmogenic and thermal messages intrusive, so vigorous ventilation is needed to reduce the osmogenic levels.

Sensory response also varies greatly among individuals. For instance, the sensitivity of the eye depends on congenital factors. We tend to forget, however, that sensory recognition is an acquired skill—that it takes children several years to learn to recognize and correlate letters, words, and sounds; and that painters, sculptors, and musicians spend their entire lives improving and maintaining these sensory skills.

Temperature and Touch. Temperature is usually associated with touch (Weber, 1848; Green, 1978; Stevens and Green, 1978), although tactile contact is not necessary for the perception of radiative heat. Touch receptors are distributed over the entire body, but their density varies. Foster (1891) measured how far the points of a compass had to be separated in order to evoke two distinct responses on the skin. Values varied between 1.1 mm on the tip of the tongue and 6.6 cm on the back of the body (Table 3.1). The sense of touch is poorly developed in many people and operates on a very muted level in schoolchildren, who are notoriously insensitive to heat and cold.

The body's thermal system also includes an unknown number of the elusive internal receptors mentioned earlier. These receptors monitor the deep body temperature and serve as a reference standard that indicates whether trends in heat flow and skin temperature are comfortable or desirable under the given

Table 3.1

Distances at Which Two Compass Points Can Be Distinguished
When Applied to Various Parts of the Body[a]

Body Part	Distance (mm)
Tip of the tongue	1.1
Palm of last phalanx of finger	2.2
Palm of second phalanx of finger	4.4
Tip of the nose	6.6
Back of second phalanx of finger	11.1
Back of the hand	29.8
Forearm	39.6
Sternum	44
Back	66

[a]From Foster (1891).

conditions. As a result, this sense is capable of reacting differently to a given sequence of stimuli. On a cold day, for example, the radiation from a fireplace is perceived as comfortable, but on a hot day the same stimulus causes discomfort. Likewise, on a cold day the consumption of ice water or ice cream causes the blood to flow in such a way as to counteract the chill to the internal organs, whereas on a hot day it sends a stop signal to the skin glands responsible for regulatory sweating. Conversely, when hot tea or coffee is consumed, the internal sensors initiate sweating even on a cold day. This response is not paradoxical, since the function of the temperature sense is to maintain an adequate gradient, or to adjust the gradient, or heat flux, rather than to maintain a constant skin temperature.

Very little is yet understood about this amazing sense, which must keep the temperature of the core of the body stable within $\pm 0.5°C$. Cases of partial deficiency, common with the other senses, are not observed, because any error of this sense would induce fatal changes in metabolism.

Sight. The eye is the most thoroughly schooled sense. Thus, visual perception often overrides that of other senses. For example, we tend to equate sunshine with warmth, even if we are exposed to a chilly wind or air conditioner. Although the physical nature of light is well known and light can be accurately analyzed, we still know relatively little about the way the eye and brain translate frequencies into colors.

However, we know that color judgment is not based on the frequency of light alone. For example, the brain recognizes the color of clothing regardless of the spectral color of illumination and it perceives skin colors differently than a photographic film. Thus, photographic film must be adjusted to enhance red tones, and actors need makeup because we reject a physically accurate color rendering of normal skin. Similar differences between perception and physical fact exist for all senses and have a deeper influence than we, as a scientifically well schooled society, tend to realize. Until we better understand the meaning of this seemingly irrational difference, we should not reject the significance which the senses have for our comfort.

Hearing. A healthy young ear can recognize sound vibrations over a frequency range from about 16 to 40,000 Hz, a range of 12 octaves. In addition, the ear can voluntarily discriminate and deemphasize background noise, such as the gushing of the blood in the skull, the hum of an air conditioner or ventilator in a concert hall, or traffic noises. Furthermore, it can analyze a sound into individual components and correlate various overtones and noises so as to distinguish the individual musical instruments or voices in a group. Yet on an oscilloscope a sound signal appears as a simple bleep. It is not known how the brain deciphers it, but again psychologists claim that people function best if they experience a wide range of sound stimuli, fitting their circadian and diurnal cycles. Many features in buildings are designed to reduce noise and echo effects. If properly chosen and placed some, such as carpets and drapes, can enhance indoor air quality by serving as buffers for humidity; others merely collect dust and create suspended particulates.

Odor and Taste. Odor and taste are chemical senses, activated by contact between osmogenic molecules and geminal nerves (Henning, 1915) (Figure 8.5). The human nose is extremely sensitive and can discriminate among literally hundreds of different molecules; it can sharply distinguish between certain chemical functional groups such as acids, aldehydes, esters, and ketones, as well as among the various substituents of benzene and other aromatic compounds (Stahl, 1973; Turk, Johnston, and Moulton, 1974; Hershaft, Morton, and Shea, 1976). Few people have an opportunity to formally train their sense of smell, but for those who do the odors in the urban environment are as painful as urban sounds are to a sensitive musician.

Among the general population sensitivity to odors differs drastically. It is not uncommon for one person to detect 10^{-7} vol % of an osmogenic substance, whereas a 100,000–fold higher dose of the same scent is needed to evoke a response from someone else's odor nerves. The odor response is not linear (R.H. Wright, 1978).

There is no question that odor greatly affects comfort (NAS, 1980), but we do not yet understand how the brain classifies odors, nor can we measure odor by any instrumental method. As a result, it is still impossible to regulate odor in a meaningful way, and any legislative or legal arguments about the effects of odors on health necessarily remain vague. Furthermore, unlike the ear or eye, the nose cannot readily recognize more than one odor at a time, and even trained chemists find it difficult to analyze fragrances into their components.

Smell also differs from the other senses in that it is strongly time dependent: Steady odors become less obvious. Thus, people in hospital rooms rapidly become indifferent to odors that are so strong that they repel visitors. This characteristic, which enables people to live with their own internal and external body odors, deprives them of a defense against toxic gases that accumulate slowly. It is also this gullibility of the sense of smell that makes it possible to hide undesirable odors with perfume or air freshener. Unlike body deodorants, which chemically bind osmogenic substances, perfumes and air fresheners do not reduce the concentration of osmogenic molecules in the air, but merely overwhelm the response of our nerve buds.

The dislike for body odors seems to be acquired (Amoore, 1974). Osmogenic and thermal–tactile interaction used to be a normal part of all social activities from business meetings to formal dances. With the advent of air conditioners, this aspect of social conversation has been greatly deemphasized, but mass transit systems and energy conservation measures in crowded buildings are compelling urban people to learn anew how to cope with osmogenic encounters.

Summary. Although still very poorly understood, the human senses constitute a vital link between our environment, our comfort, and our behavior. One reason for this is that our senses serve life-support needs that cannot be expressed in simple physical parameters. They respond to differences, concentrations, and gradients of quantities which change periodically. Furthermore, an individual's perception of comfort changes as a function of body functions. Thus, stimuli that are perceived as pleasant at one time can be almost intolerable at others. Consider, for example, the gourmet who relishes raw onions at dinner, but finds their odor repulsive at the breakfast table; the executive who needs bright lights at work, but favors a dimly lit room in which to read a novel at home; or teenagers who thrive on loud music but are prevented from sleeping by the hum of an air conditioner.

Obviously, human beings require variety, and an evenly lit office with a steady noise level and constant temperature and humidity would be stultifying. Yet the environmental engineer, architect, and air-conditioning engineer have to find a range of stimuli appropriate for all employees, whether they perform physical or mental work and whether they smoke or not.

3.2 Life Support

In order to maintain life, the human body must be kept at a steady temperature of 37°C (98.6°F). Careful measurements on astronauts have shown that the normal internal base temperature varies by less than 0.3°C among different people. However, changes are observed during the normal daily activity cycle (Figure 3.1). Temperature regulation requires continual adjustment of breathing and metabolism to maintain the flux of oxygen, carbon dioxide, and water. This task must be performed day and night over the entire lifetime, and

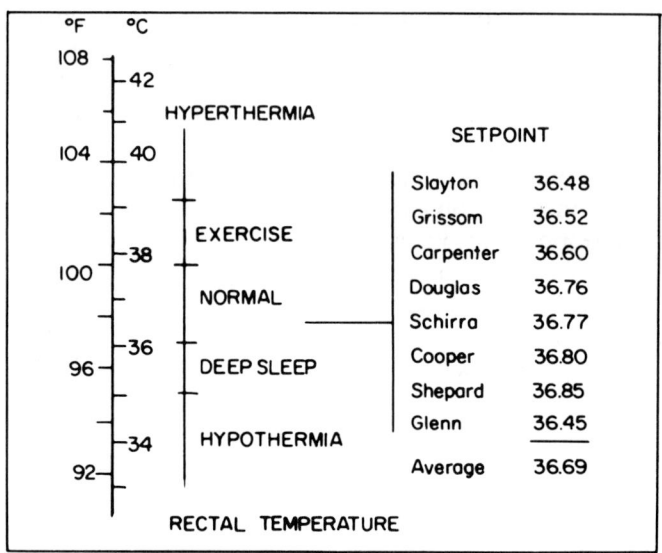

Figure 3.1. Rectal temperatures for eight activities, and variation of core temperature for 8 astronauts. (After Benzinger, 1979 and Fanger, 1979).

the flow of chemicals must adjust to every variation in metabolic rate and atmospheric conditions.

The human body copes with this constant challenge by a variety of straightforward and instinctive actions and reactions. As stated earlier, these involve a large number of complex physical, biological, and chemical mechanisms that defy simple scientific characterization. In fact, these interactions are so complex that apparently simple daily observations — the sensation of sudden cold that is experienced on stepping out of a cold shower into warm air, for example — provoke lengthy (and inconclusive) discussions among experts.

3.2.1 CHEMICAL BALANCE

The energy supply for the various life processes is provided by the chemical reaction of carbohydrates with oxygen. The body derives carbohydrates from food, and stores them until they are needed as fuel. In contrast, oxygen cannot be stored and must be extracted from air on a continuing basis at whatever rate is necessary to satisfy the energy demand. Air also serves to remove carbon dioxide from the body. Without air, and without continuous and vigorous respiration, the "Prometheus fire of life," as Lavoisier called it, would subside within a few minutes because the human body has a very limited ability to store oxygen.

3.2 Life Support

The total mass flow through man is now well known from experience acquired during extended space and nuclear submarine cruises (U.S. Navy, 1962; Roth, 1968; Thomas, 1968; Langton, 1969; Nefedov, 1972, 1976; Nakanishi, Pereira, Fan, and Hwang, 1973; Parker, 1973; Shepelev, Meleshko, Fofanov, and Tsitovich, 1974). The quantities of chemicals inhaled and exhaled are quite substantial. The sedentary six-man crew of the research submarine *Ben Franklin* used an average of 680 g (1.6 lb) of oxygen per man per day during a 30-day cruise in the Gulf Stream at the bottom of the Atlantic. This corresponds to about 5 ml oxygen per kg body weight per minute. In a jogger the oxygen consumption increases to about 45 ml/kg min, and in a marathon winner to about 80 ml/kg min. The latter corresponds to the consumption of about 240 g body fat at a rate of about 15 met. Depending on weather conditions, about half of the waste heat of a runner is dissipated by evaporation of perspiration.

Under normal conditions a sedentary person takes about 15–40 breaths a minute. During each air exchange, about 1 liter of air is replaced. This adds up to about 10–45 m³ (100–450 ft³) of air per day, which corresponds to the volume of a standard bedroom. Figure 3.2 shows the flow of the most important chemical components of air. The average person's daily indoor energy

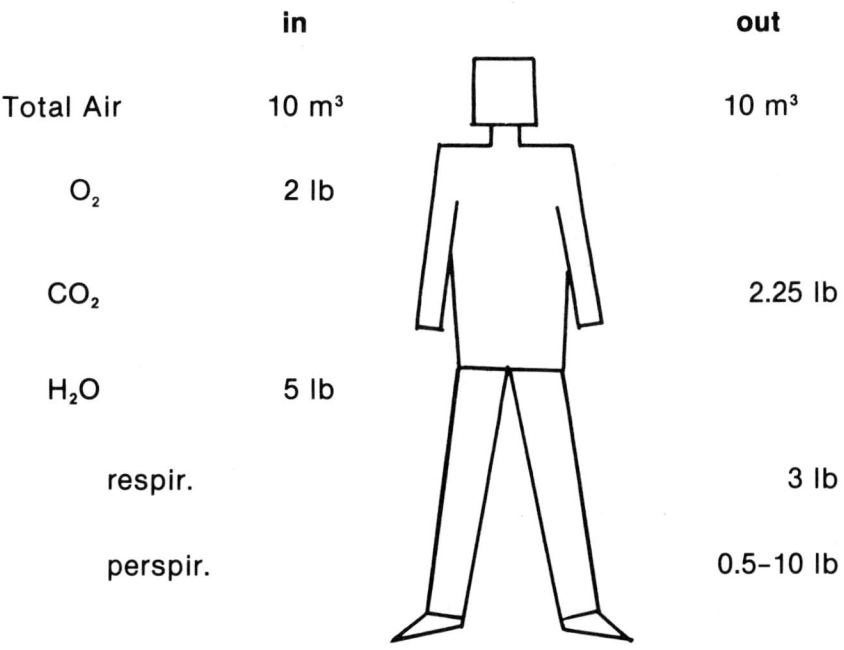

Figure 3.2. Mass balance for a man with 1.8 m² of skin surface, 50 m² of active lung surface, and 20 m² of bronchus surface.

need is about 1500–4000 kilocalories (kcal). Since 1 liter of oxygen is needed to produce 4.8 kcal, the daily oxygen demand is about 300–800 liters, or 20–40 ft^3. This corresponds to about 0.750–1.5 kg or 1–3 lb of oxygen, or 1500–4000 liters or 100–200 ft^3 of air.

Not all the oxygen is extracted from the air inhaled with each breath. Only about 4% is absorbed; exhaled air contains 16% residual oxygen. This surplus is necessary to maintain the oxygen gradient to which the human lung has adapted. Human beings are capable of living in air containing 12–60% oxygen. An atmosphere that is more than 60% oxygen is toxic; air with less than 12% oxygen will not maintain human life. However, our organism responds not to relative concentration but to the absolute oxygen pressure. The body can adjust to an oxygen pressure between 100 and 480 Torr, regardless of how much residual nitrogen or other inert gas is present. The U.S. *Apollo* astronauts did not breathe air at normal pressure, but pure oxygen at about 300 Torr, or 5 lb/in.^{2} (i.e., at about two fifths atmospheric pressure).

A comparison of the oxygen needs of various animals, on the basis of oxygen use per weight unit for a resting man, shows that mussels need 10%, crayfish 20%, carp 50%, and pike 170%; mice require 1000% while resting and 10,000% while running. During flight, butterflies use 500,000%; at rest they need only 300%.

In the lung oxygen is exchanged against carbon dioxide (CO_2). The CO_2 partial pressure in the lung is about 40 Torr. This corresponds to about 5.6 vol % of the air. Exhaled air contains 4% CO_2. This amounts to 0.7 g/min or 2.2 lb/day. If inhaled air already contains CO_2, the breathing ratio has to be increased to purge CO_2 at the rate at which it is produced. Figure 3.3 shows the CO_2 concentration that can be tolerated by human beings. A safe level for indoor living is considered to be 0.1%. Excessive carbon dioxide triggers increased breathing and eventually narcosis. High CO_2 levels can accumulate in deep wells and in wine cellars because CO_2 is heavier than air.

The efficiency of gas exchange in the breathing process is expressed as the respiratory quotient $RQ = V_{O_2(in)}/V_{CO_2(out)}$. This factor is normally 0.82 for humans. On a weight basis the coefficient is 1.375.

Exhaled air is almost fully saturated with moisture at the body temperature of 37°C regardless of the breathing rate. This corresponds to a water vapor pressure of 47 Torr H_2O or 6 vol % of air. The water exchange between air and the human body differs from that of oxygen and CO_2 in two fundamental ways. First, water is not only exchanged in the lung, but is also released via the skin; second, the water release does not serve a chemical purpose, but supplements conductive, convective, and radiative heat transfer as necessary to ensure disposal of metabolic waste heat. If inhaled air is already moist, water transfer in the lung is simply reduced. For room air at 21°C and 50% humidity, the water transfer amounts to about 1 lb/day per person. In a room with 25% humidity, 1.5 lb of water will be exhaled. This is only a fraction of the

3.2 Life Support

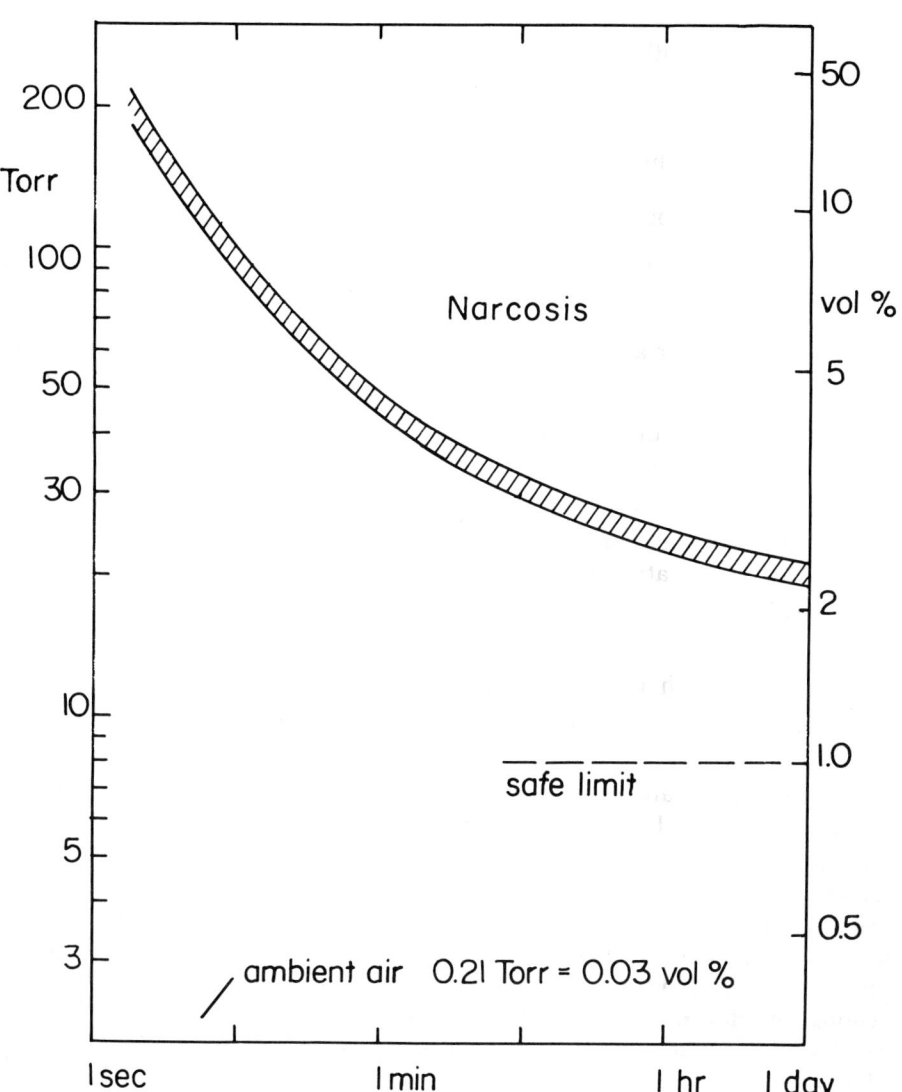

Figure 3.3. Carbon dioxide tolerance.

water that is released by perspiration through the skin. Typical vaporization losses are 2–10 lb/day (Figure 3.1). This water adds to the moisture already present in air. If several people share a small room, the water flux constitutes a significant burden on the capacity of indoor air. For example, three sedentary people watching TV in a living room will generate about 0.3 lb (130 g) of water per hour. During an exciting show, however, the release can easily triple. Since a $15 \times 15 \times 8$-ft living room at 70°F and 50% humidity can absorb only 34 g of water, air in a tightly sealed room will reach 100% humidity in about 15 minutes and water will condense on the coolest surface, usually the floor, windows, or walls. In order to maintain a stable room humidity of, say, 70%, a ventilation rate of 24–50 ft^3/min must be maintained. This corresponds to more than one air change per hour.

The production of metabolic water involves many different biochemical functions, but the purpose of sweating is physical temperature control. The large variation of water transfer between air and the human body does not reflect the full mass transfer of water, because the chemical mass balance of water involves the food chain as an important link, and thus involves water in the liquid phase. Almost all water enters the body as liquid. Whether water is excreted as liquid or as vapor is determined by physical needs. As far as the water flux through the skin and lung is concerned, the best comfort condition is achieved if the air humidity is about 40–60%, regardless of temperature. This level provides for continuous removal of metabolic products via respiration and perspiration.

From the viewpoint of the chemist, the human body is a reaction vessel in which the oxygen flux is adjusted to maintain a periodic concentration difference between reagents and products to fuel the driving force. Technically speaking, the respiratory system and much of the body is a batch reactor. Each breath of air is treated separately. In the chemical industry batch reactors are considered reliable but inefficient, because during the reaction the concentration of all the reagents decreases and the speed of the reaction also decreases. Furthermore, batch reactors cannot be as finely tuned as continuous-flow reactors, in which constant steady-state concentration gradients can be set at the optimum values. This raises the philosophical question whether there is some hidden reason why human life is based on a periodic process such as breathing rather than continuous oxygen osmosis, and generally whether people are best served by steady air conditioning or by periodic ventilation and heating. As far as climate is concerned, constantly sunny climates create deserts, which are as hostile to life as the permanently cold poles of the earth. Clearly, the physiology of human life requires periodic change between moderate contrasts.

3.2.2 HEAT BALANCE

A sedentary or slowly walking person has a metabolic efficiency of about 5%. A trained weight lifter can deliver about 20%. Thus, the metabolic efficiency of a human being in converting chemical energy into mechanical energy is about the same as that of an automobile engine. Both can only function if a steady negative temperature gradient is maintained toward their environment so that waste energy can be discarded. The human energy balance has been thoroughly studied during the last 200 years and it is now known that a sedentary adult dissipates heat at a fairly constant rate of about 50–80 W. This corresponds to the heat emitted by a normal incandescent light bulb. The heat produced by a sedentary person is defined as 1 met.

$$1 \text{ met} = 58.2 \text{ W/m}^2 = 50 \text{ kcal m}^{-2} \text{ hr}^{-1} = 18.4 \text{ Btu hr}^{-1} \text{ ft}^{-2} \quad (3.1)$$

The heat exchange between the human body and its surroundings can be represented as a sum of five major processes:

$$S = M + W + E + R + C \quad (3.2)$$

where S is the rate of heat storage, M the rate of metabolism, E the heat loss by evaporation, R the heat loss by radiation, C the heat loss by convection, and W the work performed by one person. The heats are expressed in units of W/m², kcal m^{-2} hr^{-1}, or Btu hr^{-1} ft^{-2}.

The three heat loss rates are proportionate to the body surface area. The latter can be determined by an empirical equation:

$$F = 0.202 \times w^{0.425} \times h^{0.725} \quad (3.3)$$

where F is the surface in square meters, w the weight in kilograms, and h the height in meters. An average-sized man—5 ft 8 in. and 155 lb (70 kg and 175 cm)—has a surface area of about 1.8 m² (19.5 ft²).

Clothing. The heat transfer is strongly influenced by the insulation value of clothing. Experimental work has shown that equation (3.2) is zero between 25–29°C (77–84°F) for lightly clothed people. Several equations are available to determine each factor in Eq. (3.2) (ASHRAE, 1978). A nude person with a fully exposed skin surface is most comfortable at an air temperature of 31°C. Heavily clothed people prefer an air temperature between 16 and 24°C.

Insulating values for typical clothing ensembles are listed in Table 3.2. These values are additive. Heat transfer equations have been proposed by Nishi (1979). He estimated, for the heat transfer by air between clothing and skin, a value of 0.035 W, using a body surface area of 1.6 m² and a trapped-air volume of 8 liters. For a room temperature of 20°C, the corresponding lung ventilation yields 10 W. Humphreys (1979) studied clothing habits and found that his subjects adapted their clothing to climate changes only slowly. That temperature preferences range between 17 and 30°C has been explained on the basis of different clothing preferences, which are often influenced by residential habits.

Medical doctors have recognized for several hundred years that vanity can make people insensitive to their thermal needs (Griscom, 1850). Thus, most

Table 3.2

Insulating Value of Typical Clothing[a]

Clothing	Insulation[b] (clo)
Men	
Cool socks, briefs, shoes, sleeveless short woven shirt, cool trousers	0.57
Cool socks, briefs, undershirt, shoes, short-sleeved woven shirt, warm jacket, cool trousers	1.12
Women	
Cool dress, pantyhose, bra, panties, shoes	0.27
Cool short-sleeved sweater, cool slacks, pantyhose, bra, panties, shoes	0.53
Warm long-sleeved sweater, warm skirt, warm long-sleeved blouse, pantyhose, bra, panties, shoes	0.98
Warm long-sleeved sweater, warm slacks, warm long-sleeved blouse, pantyhose, bra, panties, shoes	1.20

[a]From ASHRAE (1978).
[b]1 clo = 0.155 m^2 °CW^{-1}

business people invariably wear a three-piece suit, closed collar, and tie. Moreover, the suit is often made of synthetic fiber cloth, which is impermeable to moisture but an excellent conductor of heat. Thus, the wearer freezes in winter in the same clothing that acts as a sweat suit in summer. Ideally, fabrics should provide for steam transmission, steam absorption, and heat insulation. Thus, even modern campers find woolen pants one of the best all-weather garments (Table 3.2). As in the case of bedding, different clothing insulation factors will cause people to experience a different degree of comfort at a given temperature, humidity, and air movement rate. Since different activity patterns further alter human thermal gradients, it is little short of a wonder that workers in an office can peacefully coexist at any one chosen temperature.

Heat Dissipation: If a person performs physical work, the body produces waste heat in an amount proportionate to the physical load. A wrestler or a marathon runner produces up to 20 met. If the body temperature of a working person is to remain the same as that of a sedentary person, heat transfer to the environment must be increased. This may be achieved by regulating sweating, or by modifying the properties of the environment. Sweating is highly efficient because the heat of vaporization of water is 540 cal/g. Thus, 1 met can be dissipated by the evaporation of 5 g of water.

The radiative heat loss from the human body follows Stefan's law for blackbodies:

$$R = k(T_B^4 - T_2^4)$$

where k is the Stefan–Boltzmann constant, 5.7×10^{-8} W/m²·K⁴, T_B is the temperature of the skin surface in Kelvins, and T_2 is the bulk temperature of the environment. The radiative heat loss for a skin temperature of 27°C and a clothing temperature of 25°C is 21 W (i.e., about 20% of the metabolic heat). At an outside temperature of 20°C it is 74 W (i.e., 70% of the metabolic heat); at 10°C it would be 172 W (i.e., 1.5 times more than the normal metabolism); and at 0°C, 260 W (i.e., 2.5 times the metabolic rate). In a hot room the flow is reversed. At a skin temperature of 36°C the radiative transfer from a 40°C wall would be −29 W, and at 60°C it would reach −320 W.

The rate of convective heat transfer from the body surface to the air, factor C in Eq. (3.2), is determined by the temperature gradient, the humidity gradient, the air volume in contact with the skin, and the area of the skin that is wet. The most important environmental factor determining human comfort is a combination of humidity, temperature, and air movement: Hot dry air is comfortable because it enhances evaporation of perspiration; hot moist air is uncomfortable because the moisture gradient is insufficient for the air to absorb perspiration as it forms and accumulates on the skin. Air movement can offset the unpleasant effect of partly saturated moist air because of the increased volume of air that touches the skin.

The human body is capable of surviving very large air temperature gradients for considerable time. Some saunas are kept at 93°C (200°F). Direct outdoor sun exposure can produce body surface temperatures of 150°F or more, and the earth provides temperature extremes reaching from −80°F (−62°C) in the Yukon and −90°F (−65°C) in Siberia to 134°F (56°C) in Death Valley, California. The perception of high temperature gradients and the damage such gradients can inflict depend entirely on the rate of heat transfer. Often the heat flux through the air is largely determined by the heat capacity of air and by the wind velocity, which determines the volume of air involved in the transfer.

In contrast, the human body is extremely sensitive toward internal heat accumulation, and therefore requires continuous heat dissipation. Figure 3.4 shows how the requirement changes with activities.

3.2.3 THERMAL COMFORT

As mentioned above, conditions are very rarely ideal for heat transfer from the body, and the perception of comfort changes constantly as a person's needs change. Obviously, a hard-working athlete's perception of the comfort of a heated arena will differ from that of his sedentary audience.

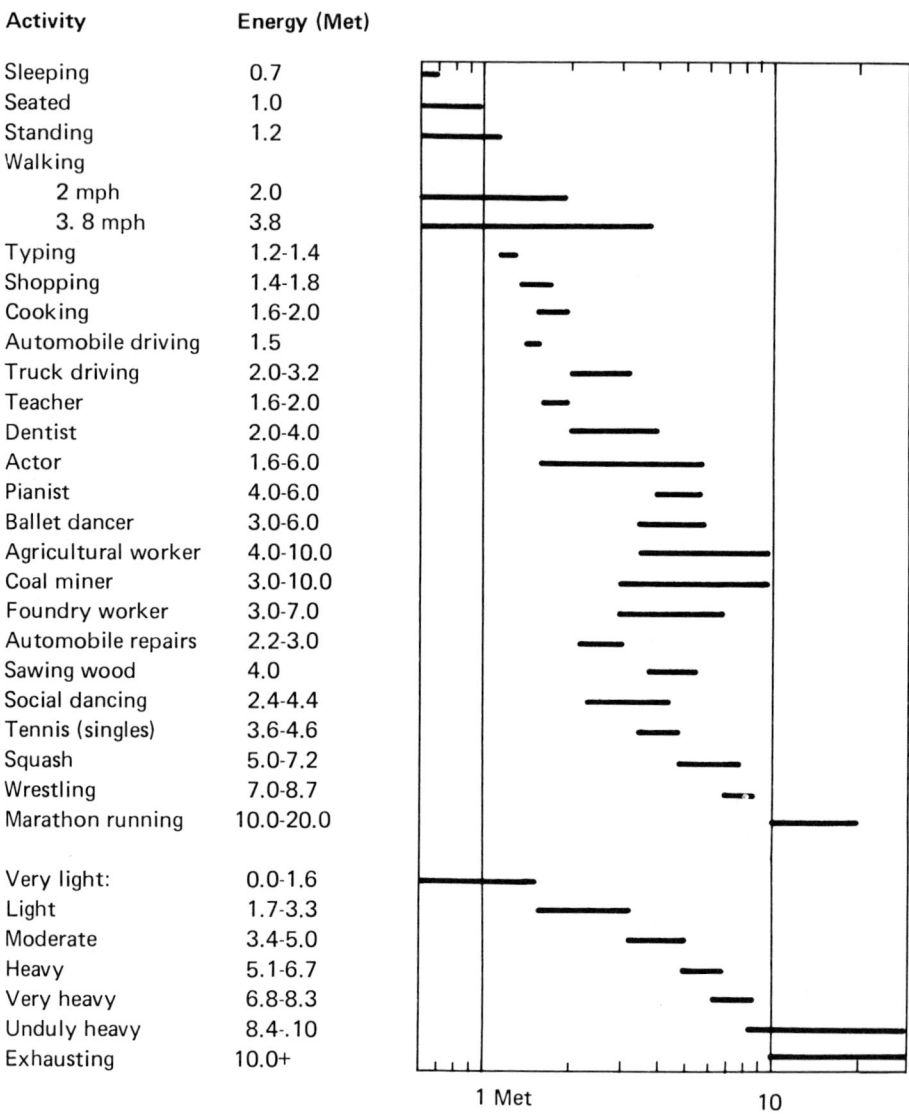

Figure 3.4. Metabolic rates for activities and professions. (Lehmann, 1950, and ASHRAE, 1977).

A substantial effort has been made to define comfort conditions so that buildings can be designed to provide optimal living and working conditions. Detailed coverage of this subject exceeds the scope of this book. To a rough approximation, outdoor comfort can be predicted from measurements of the wet-bulb globe temperature (WBGT):

$$\text{WBGT} = 0.7 t_w + 0.1 t_d + 0.2 t_g \tag{3.5}$$

where t_w is the wet-bulb temperature, t_d the dry-bulb temperature, and t_g the temperature of a black globe with a 6-in. diameter. Indoors, comfort follows the equation

$$\text{WBGT} = 0.85 t_w + 0.15 t_d \tag{3.6}$$

This equation neglects the windchill factor. The windchill index, a measure developed by the U.S. Army, is

$$\text{WCI} = (10.5 + 10V - V)(33 - t_a) \text{ kcal m}^{-2} \text{ hr}^{-1} \tag{3.7}$$

where V is velocity in meters per second; t is in degrees Celsius. The equivalent windchill temperature is

$$T_{wc} = -0.045(\text{WCI}) + 33 \text{ for } °C \tag{3.8}$$

Air movement enhances heat transfer between air and the human body and accelerates cooling of the human body. Exposure to wind has the same effect as lowering the temperature of quiet air. Thus, both processes cause the same sensation. This fact is used in hot climates and in airplane ventilation (Chapter 9). As early as 1923, ASHRAE sponsored the classic research of Houghton and Yaglou on thermal comfort. These authors introduced the term *effective temperature,* which includes temperature, humidity, and air movement. This scale was later found to overemphasize somewhat the effect of humidity in neutral and cool conditions. Gagge (1979), Stolwijk (1979), and Nishi (1979), developed the concept of the effective temperature scale (ET*). It is based on equal regulatory sweating as an indication of physiological response. Rohles (1978) has since further refined the ET* scale through studies with 1600 students at Kansas State University (Figure 3.5). Details on this complex subject, and a comparison of different comfort scales, can be found in the specialized literature (ASHRAE, 1963, 1974–1976, 1981).

Figure 3.6 lists the living or working conditions in several hot places. The heat exposure for 8 hr of moderate work (800 Btu/hr = 200 kcal/hr) or 4 hr of heavier work (14 Btu/hr = 350 kcal/hr) is indicated by the wet-bulb globe temperature line of 85°F. Clearly, almost all deep mine work exceeds the comfort limit. The danger line for heat stroke is at an effective temperature of about 34°C (95°F). If during a summer heat wave the dry-bulb temperature averages around 29°C and the wet-bulb around 23°C, then according to experiences with past heat waves, death rates in the general population will increase by up to 30% and then level off.

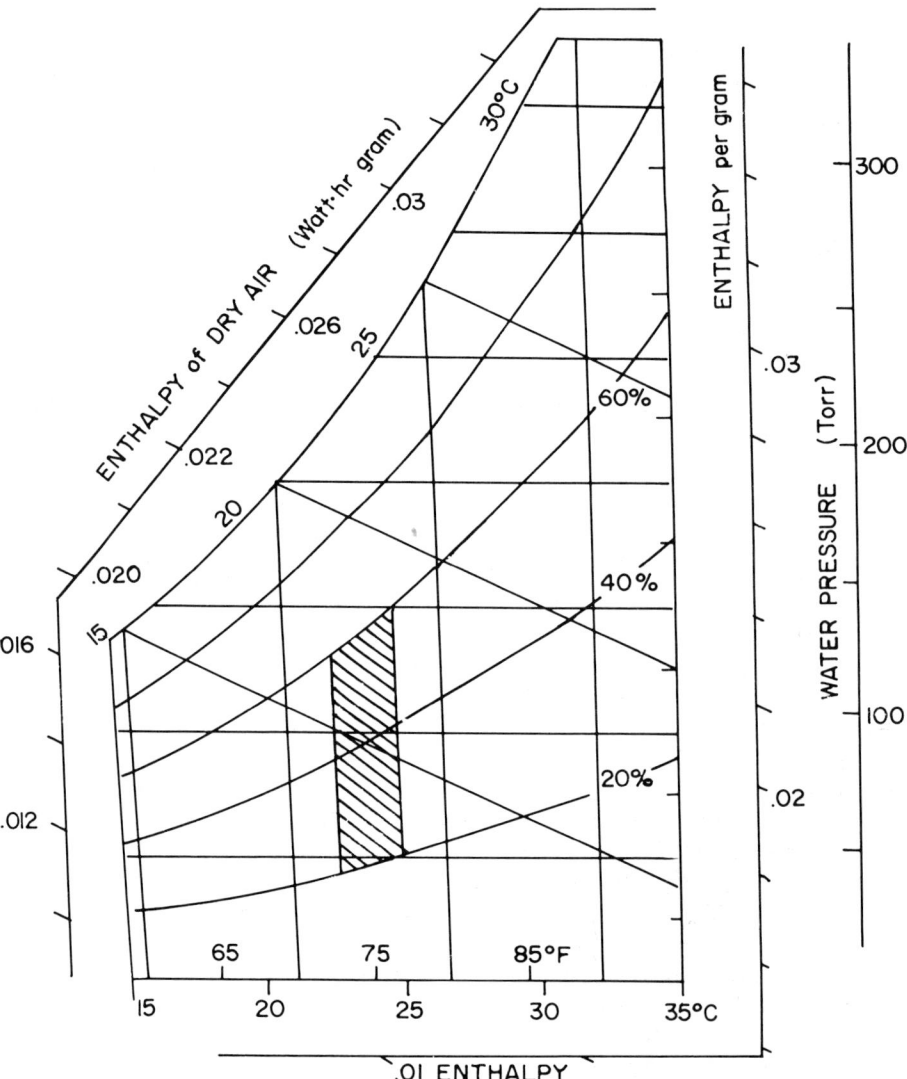

Figure 3.5. Effective temperature scale (Rohles, 1978, after ASHRAE 1977). The trapezoid corresponds to the ASHRAE Comfort Standard 55-1981. The curve applies for a lightly clothed, sedentary person in an area with low air movement.

3.2 Life Support

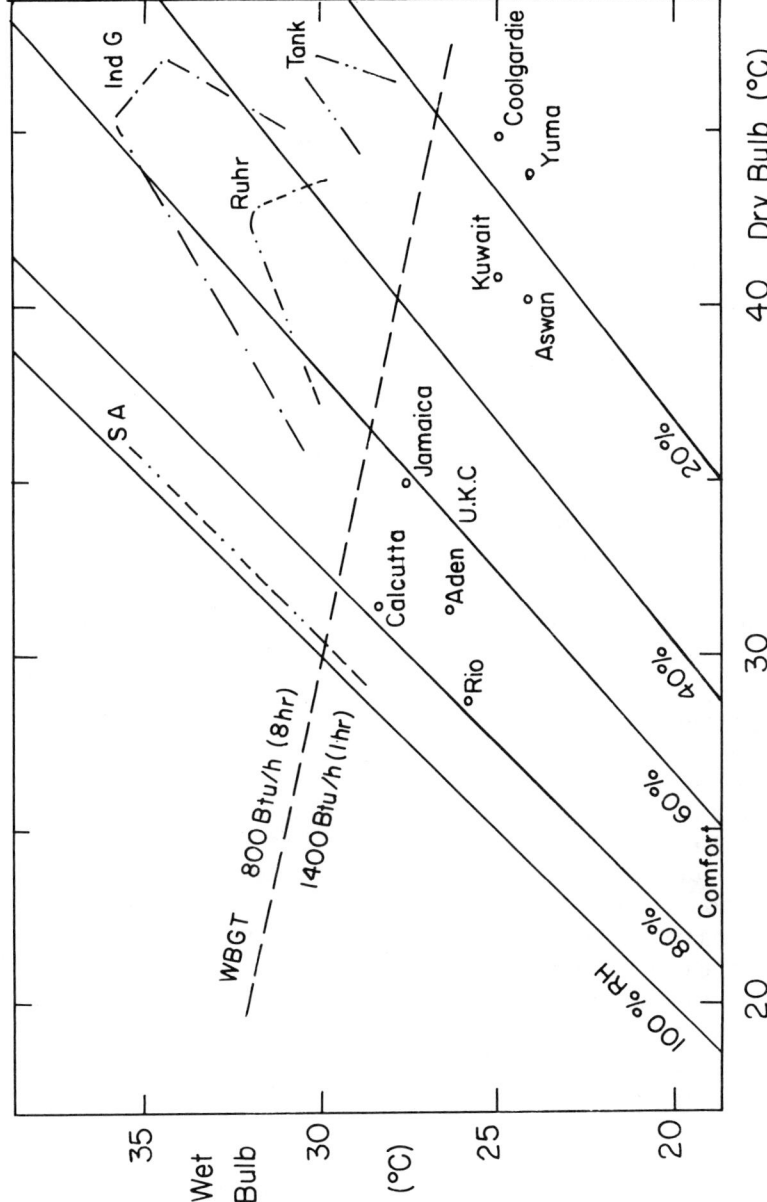

Figure 3.6. High-temperature working and living environments. The sloped line WBGT corresponds to the wet-bulb temperature curve for 85 °F at 800 Btu/hr during an 8-hr workday, or 1400 Btu/hr for a 4-hr workday. Key: SA, South African gold mines; Ind G, Indian gold mines; Tank, armored tanks in the desert during World War II; Ruhr, coal mines in the Ruhr Valley (after Angus, 1968).

Rohles observed a small but significant difference in initial response between men and women. The comfort range for most lightly clothed people working in an office falls within an area of

$$\begin{aligned}\text{relative humidity} &= 40\% \ (20\text{--}60\%) \\ \text{dry-bulb temperature} &= 24.5°C \ (76°F) \\ \text{air velocity} &= 0.23 \ m/sec \ (45 \ ft/min) \\ ET^* &= 24.5°C \ (76°F)\end{aligned}$$

Under these conditions, sweat vaporizes at the same rate at which it is excreted, which prevents the unpleasant sensation of wet skin. Indoor comfort depends on the over-all heat transfer between the occupant, the surrounding air and the building envelope. It is expressed in terms of the *operative temperature* which takes into account radiative and convective heat. Comfortable operative temperature ranges are listed in section 3.1.

Fanger and his co-workers developed many predictive equations that are invaluable for engineers and architects. Many extremely useful graphic presentations are contained in the ASHRAE *Handbook* (1978). Benzinger (1979) studied astronauts and found that central heat stimuli supersede hot skin stimuli in working and resting people, whereas cold skin stimuli may inhibit sweating. These observations differ from older assumptions that skin stimuli supersede the internal ones. Benzinger observed that warm discomfort was initiated by internal temperature increase, and cold discomfort by cold skin, and he concluded that ideal comfort was experienced when both were silent. This is only possible at rest, when the metabolic output is low. This physiological definition does not address the question whether such comfort is desirable or healthy.

The onset of sweating is regulated by a built-in threshold response that is activated at 36.9°C within less than ±0.5°C. Cold subjects entering a hot constant-temperature bath calorimeter will change their assessment from comfortably warm to unbearably hot as the internal temperature increases by less than 1°C. It is not yet clear how the body responds to hot and cold stimuli in the lung. Nielsen (1979) found that an increase in work load from 152 to 286 W/m^2 changed the temperature at which workers felt comfortable from 19.0 to 17.9°C, not because they preferred a larger skin-to-air temperature gradient, but because the lower air temperature caused skin temperatures to fall from 30.8 to 28.4°C. Demanding mental effort such as calculus can elicit similar responses. British subjects prefer a temperature of 22.0°C for office work. Evaluation of studies on 200,000 subjects indicates that the perception of comfort depends on climatic conditions. Collins (1979) and other recent researchers have found no difference in heat response between old and young people. Hypothermia is prevalent among old people because economic conditions frequently force the latter to live in excessively cool rooms.

On the average, men find heat less comfortable than women, but all can learn to respond with a more efficient sweat control.

Air movement has an important effect on comfort. Quiet air is not pleasant, nor is rapidly moving air. The preferred air speed differs among people. For some, a speed of 0.5 m/sec, such as is caused by walking, is already unpleasant. Vertical air movement, caused by the chimney effect along the body, has a similar effect, but temperature differences of 3°C between the feet and head are well tolerated by most people (Fanger, 1979).

Bedding. Temperatures above 26°C and below 13°C drastically reduce the deep sleep periods (Henane, 1977; Candas et al., 1979; Muzel, 1979). The sleeping habits of North Americans vary widely. Many people, especially those living in fully air-conditioned suburban ranch houses, have become accustomed to sleeping nude all year round under a light cover; others sleep under electrically heated blankets; some wear light pajamas; some float on heated waterbeds. Still others have the European habit of sleeping with open windows while wearing warm pajamas under down comforters. Furthermore, some sleep alone, whereas others sleep in pairs in double beds. It is hard to believe that all these people have the same objective, and that their body's thermal budget is the same (Thomson, 1979; Muzet, 1979; Candas et al., 1979). Since the body's heat budget does not greatly change between day and night, comfort in such differing bed microclimates can be achieved only if the room temperature is adjusted to offset all other factors.

The ideal air temperature for sleeping is determined by the thermal effect of bedding. There seems to be a heat redistribution between the core and the skin of a covered sleeping person that causes a gradual 7°C rectal temperature drop during the first 3 hours. Recent experiments (Candas et al., 1979) have shown that the most comfortable temperature inside the bed is 19–22°C for lightly clothed and covered people. If the outside temperature is 13°C, the bed temperature under two blankets rises to 26°C. Some people will tolerate temperatures as low as 16°C within the bed. If the outside temperature is 24°C, the bed temperature under two blankets will climb to 30°C (Henane, 1979). Hence, in lieu of a mandated lower residential thermostat setting, the same night energy savings could be achieved by promoting thermally appropriate bedding, which would impel those using it to lower the indoor air temperature. This would make possible a 10–15% energy savings in an average home.

3.2.4 RADIATION AND ELECTRIC FIELDS

Human activities and the human sleep cycle are influenced by light. Most buildings have windows that transmit visible light, but normal window glass absorbs wavelengths below 230 nm, and thus cuts out ultraviolet (uv) light, which the atmosphere transmits down to 185 nm. It is known that uv light can help sterilize air, but there is much controversy about the effect of or need for sunlight in buildings. Currently, the controversy centers mainly on the economics of excessive heat loss or gain through windows in relation to the cost of illumination. There have also been periodic controversies about the influence of electric fields on human well-being. Buildings usually contain electrical conductors in the building shell, and thus form field-free Faraday cages. König, et al. (1979), Albrechtsen (1979), Jonassen (1979), and others have studied human behavior in artificial fields simulating the earth's field, which is normally about 180 V/m. This corresponds to a potential of about 300,000 V between the ionosphere and the earth with a corresponding electric current of approximately 10^{-12} A/m^2 in the direction of the earth. The charge carriers are called air ions. It has been reported that such fields stimulate subjects to work harder and be more alert, but conclusive results are not yet available. Positive ions seem to be unpleasant. Negative ions seem to elicit positive feelings. Asthmatic patients seem to feel a slight relief if exposed to either positive or negative ions at concentrations below their subjective perception (Albrechtsen, 1979; Albrechtsen et al., 1979).

The problem with many earlier results is that voltage generators also produce ozone, which has its own effect on humans (Chapter 8.3).

Electric fields caused by friction between clothing, carpets, vacuuming, polishing, mopping, and so on can reach 30,000 V/cm. At this value atmospheric air breaks down, causing sparks. Static shorts below 1000 V are normally not sensed, but shocks from local potentials above 3000 V are noticeable and objectionable to nearly everybody. The accumulation of electrostatic charge is drastically reduced if the room humidity exceeds 40%. In dry climates, conductive floor polishes such as acrylic polymers can be used to bleed charges.

3.3 Thermal Properties of Air

The thermodynamics of air is important because human comfort depends on heat transfer via air. The specific heat and enthalpy of air can change by a factor of two, even at constant temperature, depending on the water content of air, because water vapor has a higher heat capacity—0.49 cal g^{-1} deg^{-1}—than any other air components. The heat of vaporization is also very high: 570 cal/g. The moisture content of air saturated with water is determined by the equilibrium water vapor pressure. The latter increases with temperature (Figure 3.7). If insufficient water is present to saturate air, the water content can assume any value between zero and saturation. This complicates the computation of thermodynamic data for ambient air.

The thermodynamic properties of moist air are well known, and are available in the form of psychrometric equations, tables, and charts (ASHRAE *Handbook of Fundamentals,* 1977; *Handbook of Chemistry and*

3.3 Thermal Properties of Air

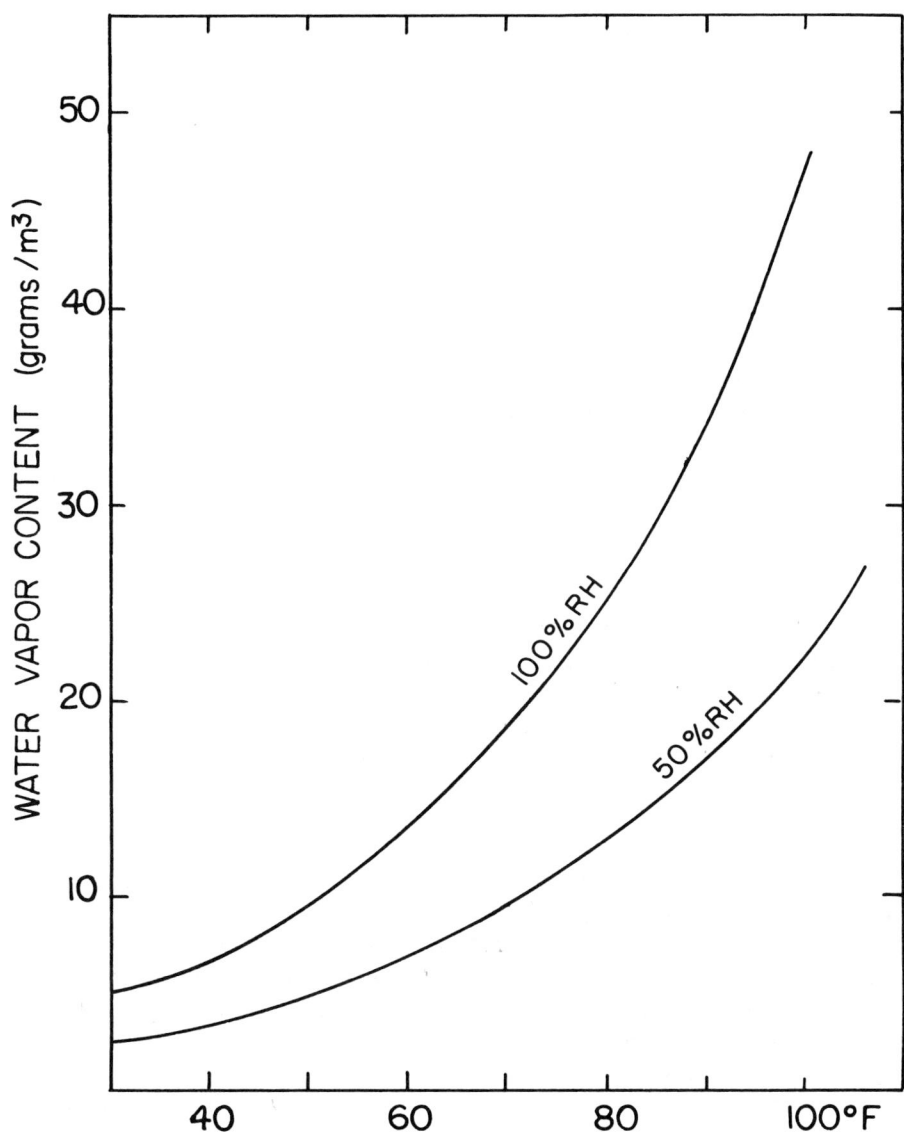

Figure 3.7. Water content of air at 50 and 100% relative humidity (RH).

Physics, 1981). For the purpose of this book, the air–water vapor system can be regarded as an ideal gas mixture.

The equilibrium vapor pressure of water above ice, in the range from $-100°C$ to $0°C$, is

$$\log p = -9.097(T-1) - 3.566 \log T + 0.877(1 - 1/T) - 2.220 \quad (3.9)$$

The equilibrium vapor pressure of water above liquid water is

$$\log p = 10.80(1-T) + 5.03 \log T + 1.50 \times 10^{-4}(1 - 10^a) \\ + 0.429 \times 10^{-3}(10^b - 1) - 2.220 \quad (3.10)$$

In Eq. (3.10) the coefficients are $a = -8.297(1/T - 1)$ and $b = +4.77(1 - T)$.

In both equations, all figures have been rounded and p is the water vapor pressure in atmospheres; $T = 273.16/T(K)$.

The relationship between the water vapor pressure p_w, air pressure p_a, quantity of water n_w and air n_a (both in molar units), volume, and temperature is

$$(p_a + p_w)V = (n_a + n_w)RT \quad (3.11)$$

The enthalpy of moist air is the sum of the enthalpy of air, h_a, and that of water, h_w: $h_a = 0.240T$ Btu/lb; $h_w = 1061 + 0.444t$ Btu/lb; if the temperature t is expressed in degrees Fahrenheit.

The enthalpy of moist air depends on the humidity ratio W, which is expressed as pounds of water per pound of dry air:

$$h = 0.240t + W(1061 + 0.444t) \quad (3.12)$$

In thermodynamic calculations it is important that all definitions and units be carefully matched. For humidity, for example, several different parameters may be used, such as the humidity ratio W (the ratio of mass of water to mass of air), the mole fraction (the molar ratio of water to air), the specific humidity (ratio of the mass of water to the total mass of moist air), and the absolute humidity or water vapor density (the ratio of the mass of water vapor to the total volume of the sample).

Another important quantity is the *relative humidity* (the ratio of the observed mole fraction of water to the mole fraction of an air sample saturated with water at the same pressure and temperature). The *dew point temperature* of moist air is the temperature at which the moisture content would correspond to 100% relative humidity.

A very useful practical parameter is the so-called *wet-bulb globe temperature* (WBGT). It is experimentally determined with a thermometer whose bulb is covered with a saturated wick held in an airstream. The observed temperature closely relates to the thermodynamic wet-bulb temperature, defined as the

3.3 Thermal Properties of Air

temperature at which moisture vaporizes at constant enthalpy. In Figure 3.5, the 70°F wet-bulb temperature curve corresponds to an enthalpy of about 0.022 W hr/g (34 Btu/lb) of dry air and connects points with 70°F–100% humidity, 80°F–60% humidity, 90°F–35% humidity, and 100°F–22% humidity. At 100% humidity, the dry-bulb, wet-bulb, and dew point temperatures coincide. At lower humidity, the wet-bulb temperature lies, very approximately, just below the middle of the dry-bulb and dew point temperatures.

The easiest way to calculate most of the practical thermodynamic properties of moist air is to use a psychrometric chart, in which enthalpy and the humidity ratio W are correlated to the temperature and relative humidity (Figure 3.8). This ingenious graph was first used by Mollier (1923). Different

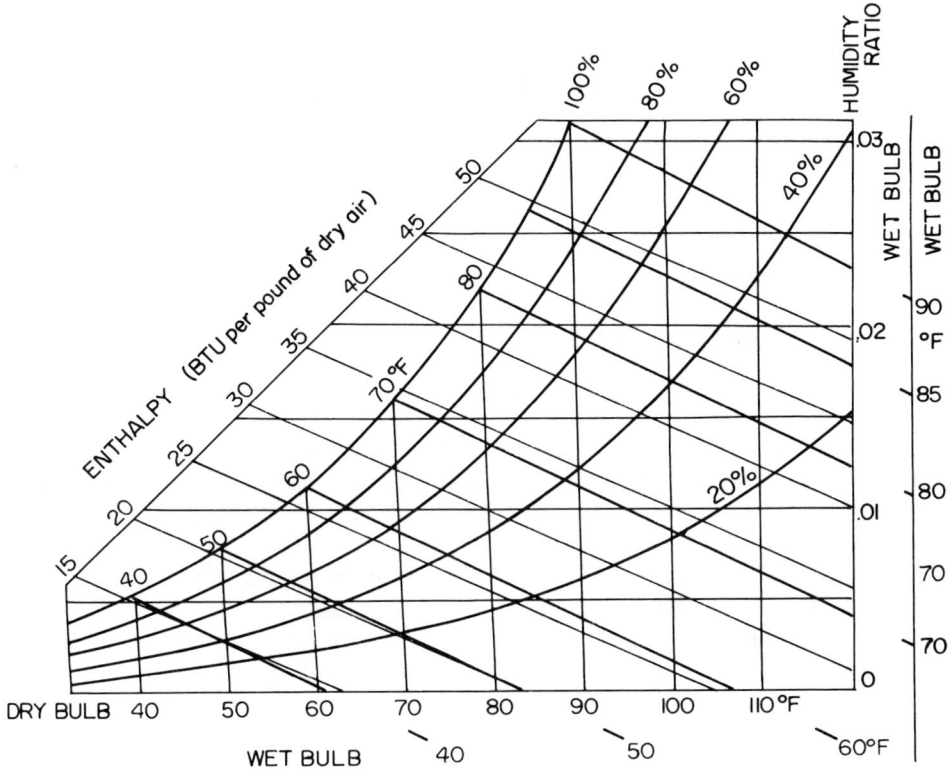

Figure 3.8. Abridged psychrometric chart for sea-level pressure (after ASHRAE, 1977).

charts have to be used for areas at high altitudes. Detailed charts are contained in the ASHRAE *Handbook of Fundamentals* (1977) and similar reference works. This chart can be used for many different computations. If the wet-bulb and dry-bulb temperatures are both known, this chart can be used to determine the dew point temperature, relative humidity, enthalpy, volume change, and other properties. The chart is routinely used by engineers to compute the energy needed for heating or cooling moist air. For such computations the conditions of steady-state mass and energy flow must be simultaneously fulfilled. The requirements for cooling moist air are usually expressed in tons of refrigeration, rather than Btu/min:

$$1 \text{ ton of refrigeration} = -200 \text{ Btu/min} = 3.5 \text{ kW} \qquad (3.13)$$

Heating of Cool Moist Air. By way of example, we compute the heat necessary to bring 30 ft³/min of moist saturated air at 35°F to 100°F. The process follows a horizontal line in Figure 3.8 from 35°F, with $h_1 = 13$ Btu/lb of dry air to 100°F with $h_2 = 29$ Btu/lb of dry air. The humidity ratio W remains unchanged at about 0.0043 lb of water per pound of dry air. The volume of entering air is 12.5 ft³/lb. The mass flow of dry air is

$$m_a = (30 \times 60)/12.5 = 143 \text{ lb/hr of dry air} \qquad (3.14)$$

The heat addition rate is

$$q = m_a(h_2 - h_1) = 143(29 - 13) = 2295 \text{ Btu/hr}$$
$$= 9.7 \text{ kWh} (38.3 \text{ Btu/min}) \qquad (3.15)$$

This is about the heat necessary to ventilate a living room with three people watching TV. Upon entering the room, the air will have a humidity of only 10% at 100°F, but upon mixing with room air to 70°F, it will have a relative humidity of 30%.

Cooling of Moist Hot Air. As a further example, we compute the heat that needs to be withdrawn from 30 ft³/min of air at 90°F and 80% relative humidity to bring it to 50°F. In this case $h_1 = 48.5$ Btu/lb and $h_2 = 20$ Btu/lb; $W_1 = 0.025$, and $W_2 = 0.0075$ lb of water per pound of air. The enthalpy of liquid water at 50°F is obtained from the literature as 18 Btu/lb of water. The specific volume of the hot air is $W_1 = 14.3$ ft³/lb. The mass flow of dry air is

$$m_a = 30/14.3 = 2.09 \text{ lb/min of dry air} \qquad (3.16a)$$

$$q_2 = 2.09[(48.5 - 20) - (0.025 - 0.0075)18.11]$$
$$= 2.09(28.5 - 0.317) = 58.9 \text{ Btu/min} = 15 \text{ kW} \qquad (3.16b)$$

Such conditions might be encountered on a hot humid day in Cincinnati. In contrast, on a day with the same dry-bulb temperature in Tucson, the air humidity might be only 20%. The cooling there would require less heat removal: $h_1 = 28$ and $h_2 = 19$ Btu/lb; $W_1 = W_2$ ½ 0.006 lb of water per pound of air and the specific volume would be 14 ft^3/lb:

$$m_a = 30/14 = 2.1 \text{ lb/min of dry air} \qquad (3.17a)$$

$$q_2 = 2.1(28 - 19) = 19.3 \text{ Btu/min} = 4.87 \text{ kW} \qquad (3.17b)$$

In fact, the cooling cost would be only one third of that in Cincinnati. The drastic difference in heat capacity also explains why diurnal temperature differences in a humid climate are only a fraction of those in a dry desert climate.

3.4 Climate

The chemical composition of air (Table 3.3) is well known. As indicated earlier, the humidity of air changes locally and depends on the temperature, and the degree of contact with bodies of water. At the breathing level, air always contains about 20–40 µg/m^3 of natural dust.

Table 3.3

Composition of Ambient Air at Sea Level and 70°F.

Constituent	Content vol %	ppm
N_2	78.084 ± 0.004	
O_2	20.946 ± 0.002	
CO_2	0.033 ± 0.001	
A	0.934 ± 0.001	
Ne		18.18 ± 0.04
He		5.24 ± 0.004
Kr		1.14 ± 0.01
Xe		0.087 ± 0.001
H_2		0.5
CH_4		2
N_2O		0.5 ± 0.1
H_2O	0 – 2.2	

3.4.1 THE GLOBAL CLIMATE

Atmospheric conditions are determined by the energy flux from the sun. On the ground, the earth's rotation converts this steady flow of radiation into daily and seasonal variations that constitute the thermal conditions and the driving force necessary for maintaining life.

One cannot overestimate the importance of these cycles for air quality and for life. Without periodic thermal variation, air pollutants would build up to the levels that we now experience during short-term temperature inversions, and would not be periodically washed out by cold fronts and other sinks, which provide the far lower equilibrium concentrations that normally prevail. Furthermore, the local temperature would reach an equilibrium value, temperature gradients would rapidly disappear, winds would die down, local humidity would be constant, and life would not be possible.

In reality, climatic contrasts are large: The temperature extremes on this planet range from $-83°C$ ($-127°F$) in Vostok, Russia, to $+57.7°C$ ($136°F$) in Al-Aziziya, Libya. On the North American continent they vary from $-63°C$ ($-81°F$) in Snag in the Yukon Territory to $56.7°C$ ($135°F$) in Death Valley, CA, and humidities cover the range from 20% in deserts to 100% in rainy areas. In addition, the earth's surface is belted with winds (Oliver and Mayhead, 1974), which can reach up to 170 mph (280 km/hr) in hurricanes at ground level, and pelted by rainstorms, which can dump up to 30 cm (12 in.) of precipitation in an hour, as was observed in Hollis, MO, in 1907. The development of cultures around the $21°C$ ($72°F$) isotherm shows that human beings flourish in a temperate habitat. The Vikings discovered this restriction 1000 years ago in Greenland, where in the long run the hostile climate frustrated their most valiant efforts at colonization.

On the average, the earth receives about 2 cal cm^{-2} min^{-1} of solar radiation. Depending on time of day, season, latitude, and weather conditions, about 1–22% of this radiation penetrates to the surface, and average surface air temperatures range between -40 and $30°C$. The average local diurnal, seasonal, and extreme temperatures vary between $3°C$ ($5°F$) on windward coasts to $23°C$ ($41°F$) in dry inland climates such as that of Afghanistan. Altitude reduces the temperature by an average of $6–7°C$/km, or $3–4°F$ per 1000 ft. Annual rainfall ranges from 250 cm/year in western Washington to 25 cm/year in deserts. The water evaporation rate ranges from 20 to 250 cm yearly. Climatic types have been classified by many methods. The most widely accepted, the Köpper system, distinguishes 10 basic types of climate that involve combined characteristics of temperature, rainfall, and humidity.

The most comfortable conditions for human life cannot be expressed by average values. Although an average annual temperature of $10°C$ braced most old cultures, it is more important that the climate remain within a range between $0°C$ ($32°F$) in winter and $24°C$ ($75°F$) in summer. Obviously, very few parts of North America meet this specification, and most of the 240 million residents of this continent truly depend on a climate-controlled home.

3.4 Climate

An in-depth consideration of climate is outside the scope of this book. Detailed maps of heating degree days and of nocturnal and diurnal solar radiation, as well as seasonal humidity and temperature curves, and all other pertinent data are maintained by, and are available in the form of climatic atlases from, the U.S. Department of Commerce and other sources. Local conditions are influenced by mountains, forests, lakes, and rivers; the orientation of slopes, altitude, and the difference between eastern and western exposure substantially influence local climates. Other local urban or rural activities, as well as vegetation, can influence the temperature, and especially the air quality.

3.4.2 DIURNAL AND SEASONAL CYCLES

Regardless of local climate in the United States, certain general trends hold. The coldest winter temperatures are reached between 6 and 8 A.M. and the hottest summer temperatures occur between 2 and 4 P.M. In Canada the times are shifted by about an hour. In order to determine suitable temperature selections for designing a house and its heating or cooling load, the local hourly weather reports must be reviewed. The standard procedure is to select the 15 hottest and coldest days of a 5-year period and compute the average dry-bulb and wet-bulb temperatures from these readings. These values will define the so-called typical hot and cold design day. Values for some 800 cities are tabulated in the ASHRAE *Handbook* (1977). However, detailed solar data are available for only a few of these places.

It is common to express local climate conditions in terms of heating and cooling degree hours or days (Figure 3.9). These values are computed by add-

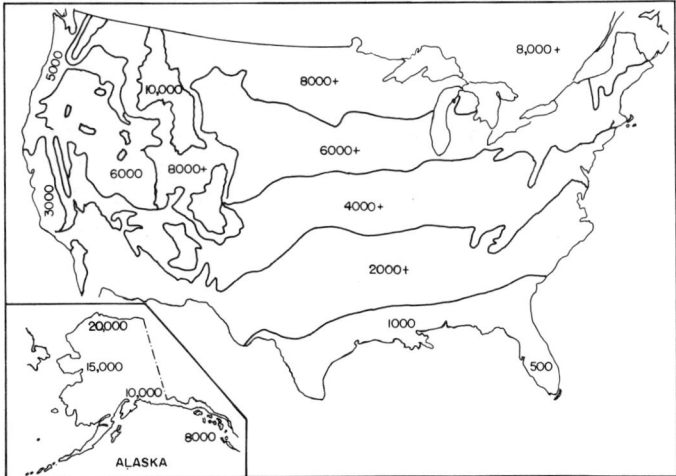

Figure 3.9. Average annual heating degree days (after *Climatic Atlas* of the U.S. Department of Commerce, June 1968.) The reference temperature is 65°F.

ing the difference between daily temperature and the chosen indoor temperature on an hourly or daily basis throughout the year. This indicates how long and how hot (or cold) a region is, but the values are not really suitable for comparison among different climates. For example, Cincinnati has 5070 heating degree days, and Seattle has 5185, but the winter design dry-bulb temperatures are 6°F and 27°F, respectively, while the summer design dry-bulb temperatures are 90°F–72% RH and 86°F–66% RH and the design wet-bulb temperatures are 74 and 65°F. The differences between the two climates are evident. Furthermore, Seattle is humid all year round, whereas Cincinnati is dry in winter and very humid in summer. In any case, the author has observed that heating bills for similar homes differ by a factor of almost two because during the short extreme-temperature season the radiative heat transfer component is more significant in Cincinnati.

3.4.3 VERTICAL MIXING

One of the most important factors in the mixing of airborne pollutants is the vertical temperature gradient. In the temperate zones the temperature decreases in the troposphere at a rate of about $-3.5°F$ per 1000 ft ($-6.5°C/km$) up to an altitude of about 33,000 ft (10 km). This temperature gradient is called the lapse rate. If the lapse rate is 1°C per 100 m, the atmosphere is adiabatic under normal pressure conditions and the air mass is at equilibrium. During a sunny day the earth's surface heats quicker than the air and increases the gradient (Figure 3.10). The lapse rate becomes "superadiabatic," and hot ground-level air rises and cool tropospheric air sinks at an accelerated rate. Thus, all vertical motions are accelerated and the atmosphere is thermodynamically unstable and mixes readily. The reverse situation occurs during a clear night, when radiation from the earth's surface exceeds radiative losses from the air; the ground temperature drops more rapidly than that of the air and the temperature gradient drops below that of the adiabatic rate. During the night and on a very hazy day the temperature profile may change direction. During such inversions, low-level air is trapped near the ground and higher air remains equally stably suspended. If horizontal air flow is limited by the local geography, temperature inversions can remain undisturbed for many days. The smog in the Los Angeles basin is an example of a local inversion that is regularly very stable. Paradoxically, the situation is stabilized by the coastal wind. In other areas similar effects are possible if all the climatic factors cooperate. Most serious air pollution episodes (such as the 3-day inversion in Donora, PA, in 1948) occur in a local climate that is predestined to inversions whenever the continental pressure conditions reduce laminar air flow.

3.4.4 THE MICROCLIMATE

The global and regional climates are continually monitored and recorded by local and federal agencies (USAF, 1978; EPA, 1970; WHO, 1976; U.S.

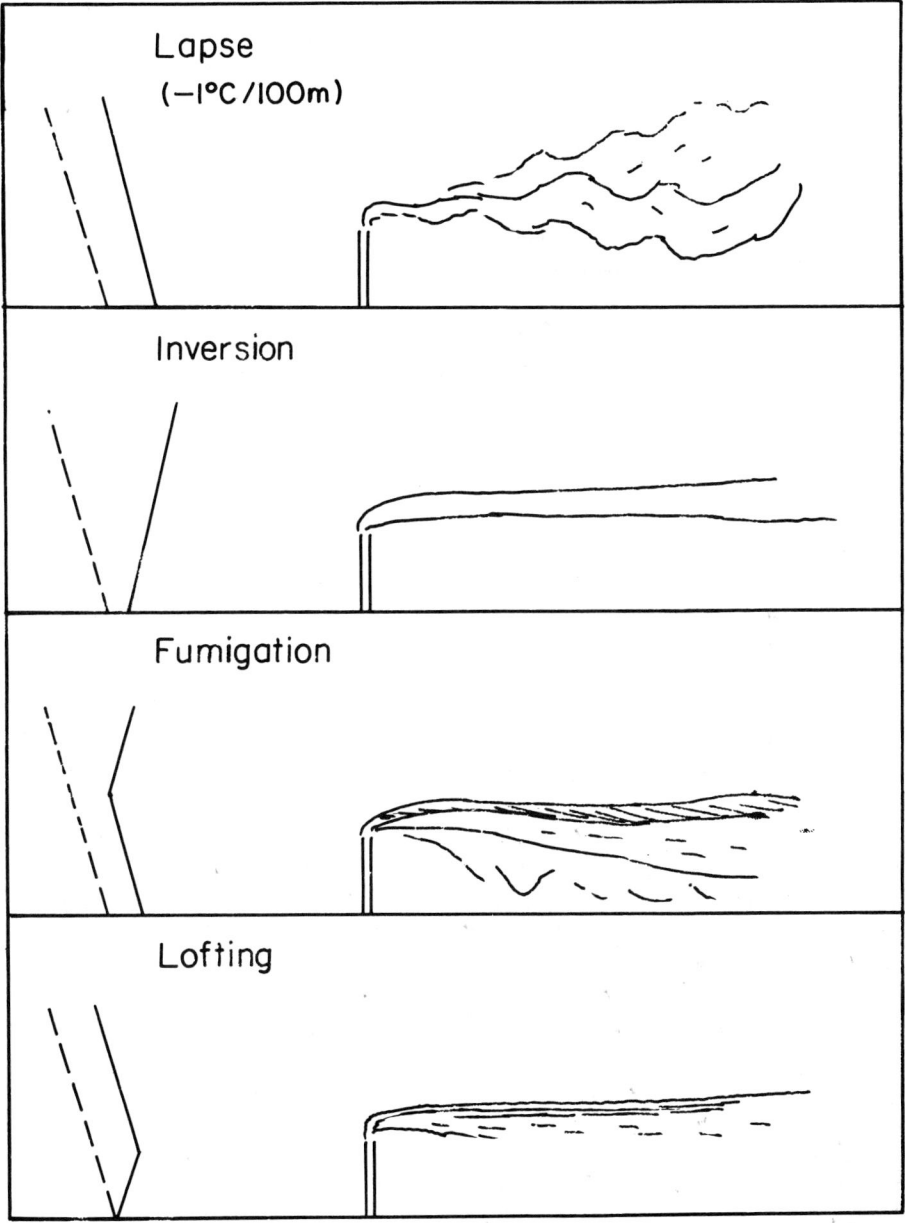

Figure 3.10. Four common vertical temperature gradients.

Department of Commerce, 1973; Bach and Daniels, 1975; National Council on Radiation, 1975; Morris and Barras, 1978), but the environment of individual buildings—the microclimates—can vary significantly within a small geographic zone or even within the same section of a town. Such differences can have a variety of causes, some obvious, others more subtle; these causes can range from the angle of the sun on an incline to the location of a brook or the presence of a large tree.

Since air diffusion is very slow, air mixing is very slow unless air motion is enhanced by convection or turbulence. Thus, the composition of ambient air is not as homogeneous as its visual appearance often suggests. This is demonstrated by the fact that episodes of radioactive rain in Troy, NY (Eisenbud, 1963), the acid rain in Scandinavia (Meyer, 1977) and Hawaii (Yoshinaga and Miller, 1981), and the Arctic winter and spring haze at Igloolik, Mould Bay, and Barrow (Barrie, 1981) contain pollutants at concentrations that are high above those expected in these locations. This is only possible because airstreams are capable of carrying pollutants across different climate zones with little mixing even though they travel for a week or longer over distances of 5000 km or more. Air mixing on a regional scale is also slow. In some cases local conditions can cause regular repetition of what might appear to be an unstable situation. An example on a microclimatic scale is provided by certain neighborhoods in San Francisco and across the Bay, for instance in Kensington and El Cerrito, which each summer periodically experience dense fog at the same time that adjoining areas—across the street or beyond a narrow ridge—enjoy bright sunshine. In fact, this phenomenon repeats itself in such a well-defined manner that it is reflected in the local property values.

Since buildings are within the surface friction layer of the atmosphere, which extends to about 100 m from the ground, horizontal laminar air flow is interrupted by landmarks and other surface features and is not as smooth as it can be in higher layers, say in clouds. Air mixing in this friction layer depends on horizontal wind speed and the buildings' surface structure. Figure 3.11 shows an example of ground-level air movement. When the wind direction changes, the situation is not simply reversed; the effect is so complex that it is best studied experimentally with scale models. Mixing velocity can span an enormous range, from molecular diffusion in stale confined air, which is 0.2 cm^2/sec, to a turbulent rate of 10^{11} cm^2/sec in the eye of a tornado. Even at the latter rate, however, the mixing of bulk air masses can be so slow that local dust has not yet settled when the storm funnel has crossed an entire nation. Superimposed on the problem of incompletely mixed air pockets are the periodic daily and seasonal variations caused by the changing intensity and angle of the sun.

Thus, a description of the overall regional climate and pollution levels does not reliably indicate the exposure a given person might encounter.

3.4 Climate

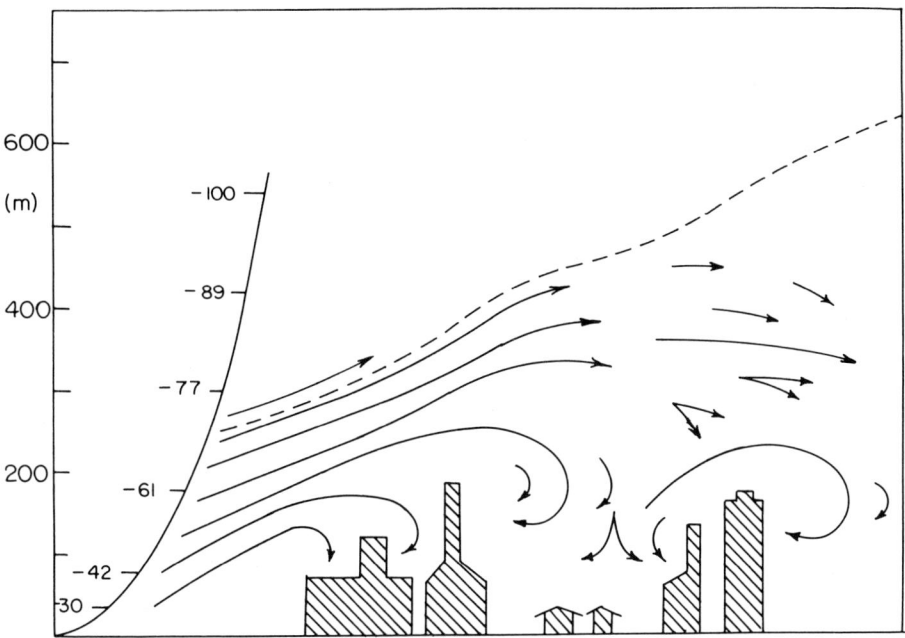

Figure 3.11. Effect of buildings on ground friction layer air movement.

3.4.5 THE ROLE OF HUMIDITY

Because of the temperature gradient (Figure 3.10), humidity decreases rapidly with altitude (Figure 3.12). The extremely low humidity at 40,000 ft is responsible for the uncomfortably low humidity in commercial aircraft. The ground-level outdoor humidity is constantly changing. During the day, water vaporizes from forests, fields, and lawns at about 1 mm/day, about the same rate as from lakes. As moist air rises to cooler air levels, clouds are formed. Also, there is always a humidity gradient between sunny and shadowy spots on the ground. Even a slight wind is effective in transferring moisture to a cooler spot, where condensation can occur. Traces of atmospheric dust or other matter, such as the leaves of some plant species, are capable of inducing from saturated air condensation that then drips to the ground and waters the roots.

During the diurnal cycle, the water content of air increases while the sun shines. During the night the temperature drops and dew forms and recycles moisture to the soil. In coastal areas and some areas of the midwest, the humidity approaches 100% at night on a regular basis. The result of these and

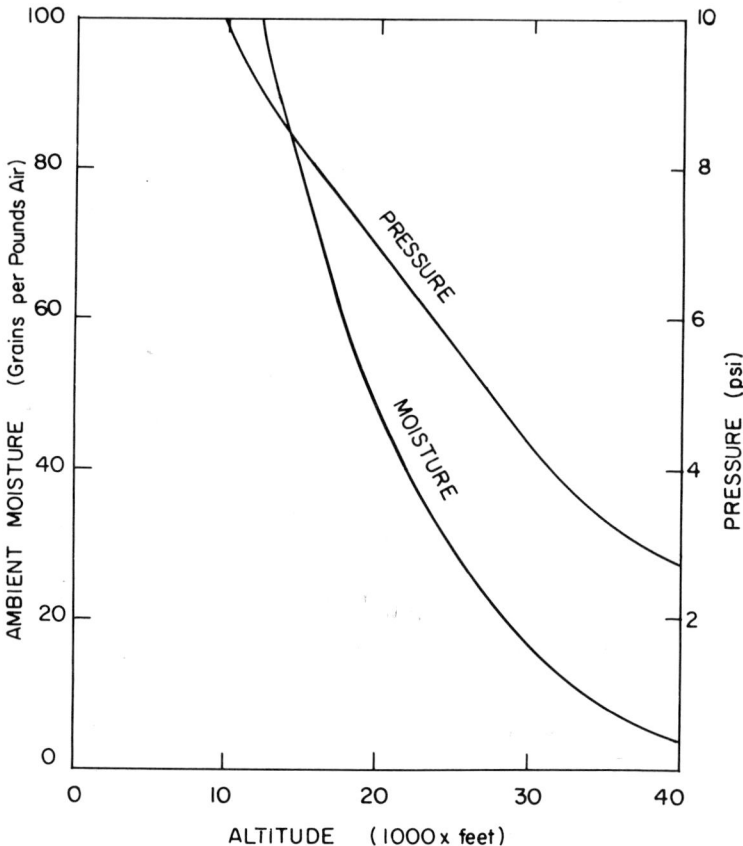

Figure 3.12. Ambient moisture and pressure from sea level to 40,000 ft (after ASHRAE, 1978).

other moisture transfer processes is a continuous flux of water. Plants and animals take advantage of these cycles and select the location best suited for their comfort. Although the atmospheric moisture flux is not readily visible to the eye, that the water vapor transport is substantial can be appreciated if we reflect that all the water in rivers and lakes was originally carried inland by the air.

3.4.6 THE INDOOR CLIMATE

The indoor climate differs in several fundamental ways from the outdoor climate, because air inside of buildings is confined in a comparatively small volume. Since the air exchange rates for modern buildings range from 0.4 to 10 air changes per hour — by contrast a mild breeze with an air flow of 5 mph is sufficient to replace the air on an open porch once every second — the active air volume available to a person indoors is several thousand times smaller. In fact, it is often insufficient to maintain human moisture and pollutant effusion below a noticeable level. Furthermore, indoor air is not part of the biological and climatic air cycles which purify ambient atmospheric air and disperse pollutants. As a result, the quality of indoor air undergoes tremendous variations in a short time. This is reflected in indoor humidity trends. The moisture content of a closed room increases rapidly because each occupant continuously adds moisture to the air in the form of perspiration and with every breath day and night at a rate of several liters of water per day. This water readily condenses on cold walls and windows, but since there is no soil or surface to absorb it, every effort must be made to replace the water-laden air by ventilation.

Another deleterious factor is that indoor spaces are kept at a constant temperature, which reduces air convection and turbulence drastically and prevents adequate air mixing. Lingering cigarette smoke in a poorly ventilated meeting room amply demonstrates this fact. Likewise, cooking and other human activities add odor and moisture to the indoor air. Thus, if buildings are not carefully ventilated, the indoor habitat becomes saturated with moisture, causing not only air quality problems, but eventually structural damage. Adequate air quality in the indoor environment can be maintained only if ventilation provides for an adequate flux of pollutants and water to the outdoors.

4. Buildings

Throughout the world, a comfortable indoor climate is possible only if the building tempers the prevailing outdoor temperature, humidity, and air speed. There are two ways of achieving these objectives. One is by constructing a tight box that prevents the flux of heat, air, and moisture, and within which the air quality is established with the help of an air-conditioning system that is independent of outdoor conditions; the other is by designing a building envelope that utilizes heat, air, and sunshine in such a manner that the flux of heat, air, and moisture can be modified as necessary to attain a comfortable level. In reality, every building includes some elements of both approaches, because all buildings are subject to involuntary infiltration.

Tight boxes with independent internal life-support systems are expensive to build and operate but they have the advantage that they can be designed and engineered for use anywhere without regard to the local climate. The only requirement is a lot of adequate size. Examples of such structures are skyscrapers, sports arenas, hotels, shopping malls, and mobile homes. The air quality of these structures depends on the quality of their air-conditioning equipment. If air is recirculated, it must be carefully filtered to remove dust and prevent the buildup of pollutants. At present this is done only in special environmental clean rooms in hospitals or industry.

The porous building is much harder to design because diurnal and seasonal variations, as well as other temporal weather vagaries, require separate, flexible adjustment of the air, humidity, and heat flux. This capability requires either expensive supplemental heating or cooling, or skilled architectural planning to correlate the heat and air flux with local weather conditions and other factors inherent in the site, design, and building materials (Olgyay, 1963; Givoni, 1969; Real Estate Research Corp., 1974).

During the last 50 years the practical trend has been toward structures that are impermeable to air and humidity, but porous to heat. Such buildings are popular because it is easy to build fairly attractive and convenient structures if costly and bulky thermal insulation can be omitted. Furthermore, the indoor climate in such structures can be easily controlled with central air conditioning as long as the energy supply is abundant and cheap. During the last 10 years a major reassessment of building design has aimed at reducing energy needs, but this effort has had mixed results because thermal insulation does not fit into the design of such structures and retrofitting is prohibitively expensive.

Efforts to reduce the cost of heating usually involve a dual approach. The building insulation is partly improved, and the intake of fresh air is reduced. As a result, despite high capital infusion, many retrofitted buildings offer poor air quality and high operating costs (Burch and Hunt, 1978). Obviously, a thorough analysis of land development and building practices using computer technology is necessary to determine the best procedures for insulating buildings so as to optimize passive thermal design.

The science of building construction and the art of architecture are based on thousands of years of experience. A good architect can combine an average lot and a simple house design to create a building that combines great comfort with high economy. The synergism among the environmental factors intrinsic in such a building is quite complicated and is outside the scope of this book. This art is a prerequisite for the successful design of solar homes (Sears, 1977; HUD, 1980).

Unfortunately, many people live in hastily erected buildings that were not designed and placed to suit local conditions, but rather to appeal to the anticipated expectations of an average buyer. As a result, many homes, regardless of price, contain environmentally adverse elements. The purpose of the following section is to point out some of the many factors that make even a simple row house a highly complex and sensitive system that requires thought and attention if its occupants are to enjoy maximal air quality and thermal comfort.

4.1 Building Design

All governments recognize the need for adequate housing. In primitive cultures, shelter consisted of temporary structures or of simple walls and roofs for protection at night or during bad weather, but in recent years most cultures have come to value permanent housing with sufficient space for a private area, with a private climate and comfort, to accommodate a family during 15 hours or more of its daily activities.

In affluent societies housing inevitably becomes extravagant (Table 4.1). The Romans knew 2500 years ago how to incorporate central heating into their villas, and they built public baths that enabled them to socialize in winter in a comfortable and stimulating environment while outside temperatures averaged only 40°F. Wealthy Romans owned second residences on the Mediterranean coast where they spent the summer. Although even the rich lived in unwholesome shelters during the Middle Ages, indoor comfort became an objective of Renaissance architects and builders. Postindustrial societies have added full

Table 4.1

Statistics on Residential Comfort in the United States[a]

Comfort	Number of Homes (millions)	
	1960	1980
Building Stock	52	82
Air conditioning	7.8	43
Electric bed covers	12	78
Coffeemaker	30	78
Clothes dryers	10	47
Freezer	12	35
Microwave oven		5.5
Radio	50	78
Television (color)	46[b]	66
Vacuum cleaner	38	78
Heat		
Warm air		53
Electric		5.6
Gas	26	55
Oil	23	22
Wood	11	1.8
Coal	35	0.6

[a] From HUD Annual Survey, 1980.
[b] Black and white.

air conditioning to almost everything, including automobiles; and the public markets and town squares of earlier times have become self-contained indoor city centers with hotels, supermarkets, shopping arcades, amusement centers, and sport arenas. These vast complexes occasionally resemble a town square from the Middle Ages placed inside a gigantic enclosure.

As described in Section 3.3, we have come to expect all such places to offer fresh air at a constant 24.5°C and 40% humidity with an air velocity of 0.25 m/sec. Furthermore, building ventilation and fire codes (Chapter 10) prescribe that adequate space be allotted for each person at the time of the highest possible occupancy. Often these spaces are conditioned 24 hours a day year round, regardless of the local climate and even if they are occupied only a few hours each week. Accordingly, per capita consumption of energy in North America increased exponentially from 1900 until the energy crisis of 1972.

The following subsections discuss building design elements that determine air quality, particularly with regard to minimizing energy costs.

4.1.1 VENTILATION

Indoor air quality, as we have said, depends on the quality of the outdoor air, the air volume available to each occupant, the ventilation rate, the air-cleaning or air-polluting properties of the building envelope, and the contamination of air by the occupants' activities and their metabolic state. (Air pollutant exposure levels and models are discussed more fully in Chapter 7.)

Engineers differentiate among three forms of air change: natural ventilation, infiltration, and forced ventilation. Natural ventilation includes air movement through open windows, the fireplace, and open doors. In most European countries natural ventilation is still widely provided for each room individually by vertical swing windows that are manually controlled to secure the microclimate desired by the occupants.

In addition, all real-life buildings are subject to involuntary infiltration through cracks, doors, windows, and chimneys. Thus North American residences, especially wood-stock homes, obtain most of their ventilation from leaks through cracks, which ensures that the wall components are well aired and thereby prevents mold and rot. (Figure 4.1 shows the average air leakage rate measured in 50 houses in Texas. Data for Colorado is shown in Figure 1.4.

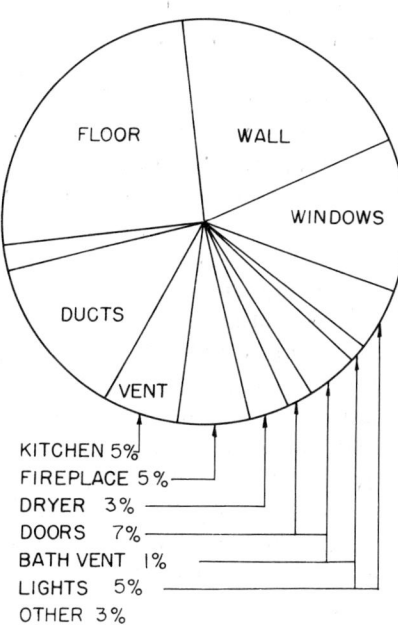

Figure 4.1. Infiltration in 50 Texas homes, each with an average floor space of 1780 ft² (after International Energy Agency, 1978).

Air infiltration depends on wind speed and temperature, the location of the building, and its design. (In tall buildings the internal chimney effect can cause large enough pressure differences to reverse the direction of the air leaks.) With infiltration accounting for up to 30% of the energy waste in older buildings, control of infiltration is one of the most powerful design tools with which to reduce energy loss.

Air leaks can be found by tracer methods (i.e., by mixing smoke or some chemical, say sulfur hexafluoride, with air and subsequently observing the change in concentration of the chemical in the air in the building). Another method used by energy auditors is to pressurize the entire building with a fan and to record the pressure drop. Both methods give good results.

In air-conditioned homes and in tall office buildings forced ventilation is provided by a central mechanical ventilation system. This method creates an integrated building system in which the microclimates of individual rooms become interdependent, which can lead to problems. For example, a change of air flow in any room will alter ventilation in all others. Thus, windows must be permanently sealed. During the development of tall-building technology in the 1960s, the cross-contamination problem was emphasized by the spread of smoke during some spectacular fires.

In the recent drive for energy-efficient design, ventilation rates are being reduced by 30% or more and the air exchange rate has returned to the levels established in the 18th century to control tuberculosis, and recognized by Yaglou in 1924 to be necessary to eliminate body odors (Figure 1.3). Accordingly, the new ASHRAE (1981) ventilation standard 62-1981 has been renamed "Standard for Ventilation Required for Minimum Acceptable Indoor Air Quality." It not only specifies minimum air change rates (Tables 9.2 and 9.3) for various places and activities, but also for the first time considers indoor air quality as an explicit goal and provides numerical limits for the maximum indoor concentrations of sulfur dioxide, particulates, ozone, nitrogen dioxide, lead, hydrocarbons, and carbon monoxide (the seven pollutants that are covered by the Clean Air Act (Table 10.1) for 27 other pollutants covered by the U.S. Department of Labor's Occupational Safety and Health Administration (OSHA), NIOSH, Canadian, and some sixty other indoor air pollutants covered by standards in Germany, Israel and other countries.

Various methods were used to set indoor pollutant limits. Ambient clean air values were accepted as published by the U.S. EPA. Occupational exposure levels were divided by 10, and foreign standards for pollutants not covered in North America were individually considered. In the case of unclassified odors, an agreement by a consensus of 30% of 20 untrained observers is proposed to determine corrective action. Although the goal of this standard has been widely lauded, problems remain with regard to tolerable levels for many substances. The U.S. Environmental Protection Agency, for one, fears that synergism among pollutants has not been sufficiently considered, and that the values favor energy savings over health. On the other hand, a trade association lost a formal appeal, claiming, in essence, that it is unnecessary and uneconomical to build homes free of formaldehyde odor.

The air in old-fashioned buildings with tall rooms and large temperature gradients along the walls is constantly in motion (Figure 4.2). This convection keeps air flowing along an occupant's body and carries odors and impurities away. This type of air circulation mixes air and transfers air pollutants to cold surfaces in the room where they may condense. Even mild air circulation will reduce humidity to the vapor pressure corresponding to the dew temperature of the coldest point in the room, i.e. 10-% below that in a better insulated room with the same air temperature.

4.1.2 THERMAL COMFORT

In any normal building indoor heat, air, and moisture are exchanged with the outdoor atmosphere and the building envelope acts as a buffer. Since the mass and heat capacity of air are small compared to the heat content of the

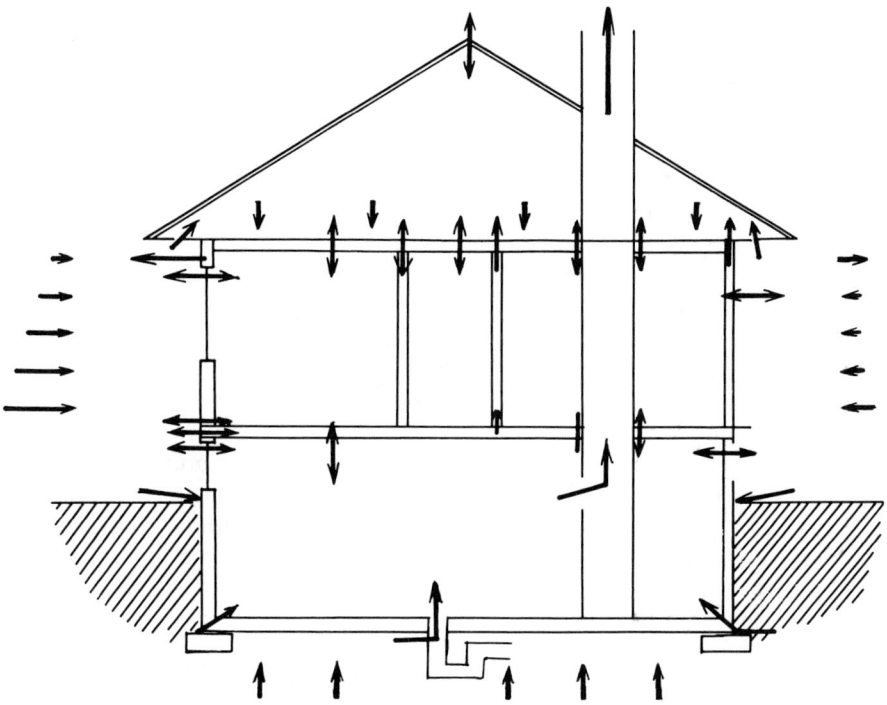

Figure 4.2. Air circulation and infiltration paths in closed rooms.

building envelope, the building plays a decisive and active role in indoor comfort and frequently dominates indoor conditions. The human thermal factors are described in Chapter 3.3. Since antiquity, the rich and powerful have always favored big buildings with large air spaces, and contemporary churches and temples are still often quite large. Bank lobbies reflect this predilection, which not only provides comfort but also inspires a sense of awe and safety. Heating such vast interior spaces is expensive because their occupancy is low and the building walls are comparatively flimsy. The dominant mass is the indoor air, and therefore all the heat escapes with the air. In contrast, in old churches and castles the dominant mass is the building, and the building itself sets and regulates the temperature. In summer such buildings are very comfortable because the high heat content smooths diurnal variations, but in winter the building temperature tends to drop close to, or below, the dew temperature, and the interior feels damp and cold all winter. That almost all old church organs suffer from the tin pest, an inorganic disease caused by an allotropic recrystallization of metal that occurs only below 18°C (64.5°F), confirms this fact.

In other settings, the occupants contribute the decisive heat content and supply. Such is the case in airplanes, crowded movie theaters, bars, nightclubs, and ballrooms. In all these places the main problem is to dissipate the heat load. Constant comfort can be provided only by vigorous ventilation.

Sometimes, comfort requires the maintenance of grotesque energy gradients. For example, urban audiences, accustomed to wearing light clothing in front of their television sets at home while observing events in all types of climates, demand that public auditoriums be kept at 75°F whether the event involved is ice hockey or boxing. This requires that different parts of the same hall be simultaneously heated and cooled.

The management of comfort at indoor public swimming pools presents a special challenge and has become an elaborate art as well as a science (Lammers and Hoen, 1979) because dry spectators, swimmers in the pool, and wet swimmers at poolside have different heat transfer efficiencies and thus different sensations and needs.

In the large indoor city centers and domed sports stadiums the energy balance is intrinsic in the truly colossal air volumes. These have a sufficient heat capacity and air mass to induce vertical air mixing, similar to the outdoor environment, so that ventilation is not needed.

4.1.3 THERMAL EFFICIENCY OF BUILDINGS

Energy conservation in the indoor environment has been mandated by several public laws (Chapter 10). Many thorough studies are under way or have been completed (American Institute of Architects, 1974; Limaye et al., 1975; Williams, 1977; Hollowell, Berk, and Traynor, 1978; O'Callaghan, 1978; Courtney, 1976; DOE, 1978; Roose, 1978; Sizemore, Clark, and Ostrander,

1979; Wolfe and Clegg, 1979; Booz et al., 1979; Kreith and West, 1980), but relatively little has been done to evaluate the impact on indoor air quality (Brun, 1974; Satish and Wanner, 1976; DOE, 1978; Morschauser, 1978; NAS, 1981). Thermal efficiency is a prerequirement for superinsulated and solar homes. However, the current housing stock consists largely of leaky homes and buildings. Thus, it is estimated that 40% of the heat in the 450,000 federal buildings in the United States is wasted; this amounts to 350,000 barrels of oil per day.

With the advent and dissemination of computer technology, it is now possible to ascertain rapidly and economically how optimal energy use can be achieved (Peatman, 1977; Dubin and Long, 1978; Kaluzny, 1980, Hollowell, et al., 1978–80). Many new projects, such as those of the World Health Organization, the International Energy Agency, and government and industry building institutes in the United States, England, Scandinavia, Holland, Japan, and other countries prepare computer programs for all sizes of buildings. In the meantime, many projects such as hotels, hospitals, and shopping malls are so large as to make such energy calculations profitable.

A major effort is under way to retrofit existing buildings with better insulation. This requires as much skill as designing new structures because any modification of the thermal flow in any part of a building disturbs the equilibrium of the entire building and causes feedback into the system, resulting in unexpected secondary changes. Among the problems are indoor moisture condensation and damage of the insulated building shells.

A combined effort of industry and government has resulted in a landmark set of standards called "Energy Conservation in New Building Design," ANSI-ASHRAE-TIES Standard 90-1980 (Chapter 10). The goal of this standard is to help save about 40% of the energy used to heat, cool, and illuminate buildings without reducing building utility or occupant comfort. This is achieved by detailed analysis of the design of the entire building, starting with the envelope. The standard, which is intended to encourage the exploration of new technologies, covers the design of heating, ventilation, and air conditioning; water heating; and energy distribution, including illumination. This ambitious standard covers all new buildings, including residential, institutional, mercantile, business, educational, and public assembly structures. The code explicitly exempts operation, maintenance, and use of buildings. These subjects are covered by other, largely local, codes and standards.

The standard limits the installation of room heaters in air-conditioned buildings, i.e., the so-called reheat systems. Cooling with outdoor air is recommended where suitable. Detailed provisions are made for such items as hot water insulation, pool covers for heated swimming pools, utilization of waste heat or solar energy, and lighting. The standard also proposes the use of a resource utilization factor, a multiplier that is weighted to reflect the availability and efficiency of locally delivered energy forms. This factor has created substantial controversy, since it disfavors coal as a source of electricity.

In the United States mobile home standards are the responsibility of the Department of Housing and Urban Development, but HUD works jointly with all segments of the housing industry and with ASHRAE. Ventilation requirements are prescribed in a separate code, 62-1981. The heating system must provide a range from 13°C, the temperature necessary to protect the unoccupied building, to 24°C (55–75°F) and 30% relative humidity. The cooling range is to be 21–29°C (70–85°F) with 30–70% RH.

Diurnal Temperature Cycling. A very simple method for saving energy would be to reduce the building temperature when the structure is not occupied. As shown in Section 7.2, most office buildings are used for only about 40 hours of the 148 hours in a week. Furthermore, many residences are empty during the day, and if the occupants sleep under covers, need not be at full temperature during the night. The U.S. Department of Energy has estimated that an automatic 8-hour nighttime reduction by 5°F could reduce the monthly heating bill by 10% in San Francisco, 9% in Washington, DC, and 12% in Miami, FL. A 10°F reduction might save 50% or more. However, the ambitious effort in 1979 by the U.S. Department of Energy and President Carter to convince the U.S. population that a general reduction of the indoor temperature by 5–8°F was necessary, was neither popular nor successful and was dropped in February 1981 (Chapter 10.8.4).

4.2 Building Types

The current U.S. building stock, consisting of approximately 72 million residences and 1.5 million commercial buildings, comprises structures of every type. This section describes only some very limited properties of a small and arbitrary selection of buildings. We distinguish three classes of shelters: residential buildings, office buildings, and public buildings. These classes differ not only in sophistication of indoor air quality design, but in mode of use and operation. Most residences are owner controlled and can be operated according to the wishes of the occupant, whereas large buildings are designed and operated by professional engineers.

Older buildings were usually built with local materials according to local tradition. Since World War II design and construction have been standardized and builders rely increasingly on 100% recirculated forced air heating, regardless of location.

In some cases, modern urban architecture has become internationalized. For example, tall buildings with hung glass walls, which are very attractive and energy efficient in the moderate climates for which they were originally designed, are now found throughout the world. Yet they are very unsuitable for warm climates, since the glass windows collect heat, and the entire air supply in such buildings must be cooled at the entry of the building even though part of it is reheated at the entry into individual rooms to provide local temperature control. On the other hand, in cold areas the glass on the outside walls is such a poor insulator that the heat loss accounts for 50% of the entire energy budget. In some areas the radiative heat transport through the glass ball is so large that the temperature asymmetry makes occupants feel uncomfortable, regardless of the indoor air temperature (Section 3.1).

4.2.1 RESIDENTIAL HOUSING

Table 4.1 shows that residential comfort has improved dramatically over the last 50 years. Very few homes still have unheated outside privies, and over 90% of today's homes have some form of heating in all rooms. In order to maintain such luxury, homes had to be tightened. Unfortunately, no historical statistics are available on the change in infiltration and ventilation rates. Until about 5 years ago, however, much of the improved comfort was not achieved by sophisticated thermal design, but by sheer forced air heating.

4.2.1.1 *Single-Family Dwellings*

Since residences change owners on an average of once every 5 years, the single-family dwelling has become increasingly standardized for the sake of resale value. In most parts of North America almost all homes built after 1960 are fully air conditioned and recirculate 100% of the indoor air. However, the ventilation depends strongly on the heating system. Older homes are heated with forced air or with hot water. Of homes built after World War II, about half are equipped with a gas heater and a quarter with an oil heater. About 7% are heated with electric baseboard heaters. Electrically heated homes are located mainly in the northwest and the Tennessee Valley, where local hydroelectric energy is comparatively cheap. The electrically heated and water-heated homes allow the occupants to heat individual rooms to their liking. Furthermore, in these homes air circulation is not tied to heating. Thus, ventilation can be regulated separately from heating. Each room constitutes an independent climate, which permits highly efficient air and heat management.

Occupants of homes heated by gas- or oil-fueled forced air (the majority of people) are not so lucky. In their homes, all rooms share the same air and heat. Often the kitchen, dining room, and living room are connected without doors, and kitchen odors and combustion products must be separately ventilated if they are not to spread throughout the house. On the average, family members spend about 4–6 hours daily in the living room and 8–10 hours in their bedrooms (Section 7.2). Both areas normally receive no outdoor air. The air quality and air temperature are the same in all rooms and are determined by the operation of a central thermostat. There is little personal freedom in adjusting the air temperature or flow to personal needs or preferences.

During the last two decades the restoration of older homes has become fashionable. The life-style in these buildings differs significantly from that in suburban ramblers. This becomes evident on hot summer days when people in remodeled urban homes sit on their porches and air their houses, whereas the suburban residents sleep with closed windows to the hum of air conditioners. Older homes have a higher heat capacity and provide a buffer for heat and humidity, but it is difficult to fully air condition these homes. Air quality in older homes must be managed with the help of windows and infiltration. The advantage is that air quality can be adjusted in each room to suit individual needs and preferences.

4.2.1.2 Apartments and Condominiums

Apartments and condominiums fall into the same age groups as residential homes: older buildings in which individual window ventilation is relied upon, and poorly insulated buildings erected since the 1950s in which air conditioning is a necessity in most regions. Since the construction cost of a high-rise condominium is less than 40% of that of a comparable suburban ranch house, and the local government can save about 30% of costs for roads and utilities, this type of housing is rapidly becoming popular. Most units are equipped with independent air conditioning, which is cheaper than in a suburban home because heat losses through the floor and ceiling are almost eliminated, except for the penthouse floor.

4.2.1.3 Manufactured Housing and Mobile Homes

Since it first became popular in North America around 1960, manufactured housing has encountered all types of resistance. However, the federal mobile home legislation of 1974 has greatly helped the industry. According to the original law, "a mobile home is a structure, transportable in one or more sections, which is 8 ft or more in width and 32 body feet or more in length and which is built on a permanent chassis and designed to be used as a dwelling with or without a permanent foundation, when connected to the required utilities, and includes the plumbing, heating, air conditioning, and electric systems contained therein." (U.S. Department of Housing and Urban Development, 24 Code of Federal Regulations, 280, 2 (16).) The revised legislation of August 1981 reflects the tremendous technical advances made by this industry. The title of the Code has been changed from "mobile home" to "manufactured housing," and the current code covers any "structure...larger than 256 sq ft. ...built on a permanent chassis frame complete with plumbing, electrical, and environmental control systems, which is capable of serving as a complete dwelling when properly set up on either a removable or permanent foundation, and connected to necessary utilities."

The current mobile home stock is about 4 million units. About 8% are located in Florida, 7.6% in California, and 6% in Texas; 3–5% are in each of the 11 states of Alaska, Arizona, Georgia, Illinois, Indiana, Missouri, New York, North Carolina, Ohio, Pennsylvania, and South Carolina. Hawaii lists fewer than 300 mobile homes.

Since the National Mobile Home Construction and Safety Standard Act of 1974 became law, the mobile home industry has worked under a federal regulatory system that supersedes all local regulations. As explained in Chapter 10, the mobile home industry is the only segment of housing that is regulated on a federal level. The code is comprehensive and prescribes everything from design, fire safety, body and frame construction requirements, testing, thermal protection, plumbing systems, heating, cooling, and fuel-burning systems, and electrical systems to transportation from the factory to the site.

According to the U.S. Department of Housing and Urban Development, modern mobile homes are built with much the same materials and techniques used in the construction of conventional homes, and when placed on a permanent foundation these units are frequently indistinguishable from site-built homes. It is likely that mobile homes will become an increasingly important component of the U.S. residential housing stock because they offer housing at a lower cost than other comparable accommodations.

Mobile homes are manufactured and outfitted for use in one of five climate zones. Since most local ordinances make it difficult to place these homes on residential lots, almost all are still located on special sites and in trailer parks, where they are placed without regard to the local microclimate. Thus, this type of home depends totally on internal power to produce the desired indoor climate. However, the public attitude toward manufactured housing is rapidly changing, and zoning laws in California and other states are being modified. As soon as their tax status is changed from personal property to real property, and conventional long-range real estate loans become available, the value of mobile homes is likely to increase rather than to depreciate, as they currently do.

While the detailed finish, furnishings, and comfort of individual homes vary and current prices range from $12,000 to $75,000, the basic mobile home system has become thoroughly standardized. The floor plans of a double-wide and single-wide mobile home are shown in Figure 4.3. A typical single-wide unit is $14 \times 66 \times 8$ ft and has a volume of approximately 6700 ft^3. The double- or multiple-wide units are built and shipped as 12×48-ft units and assembled on site. A standard house has three bedrooms, a kitchen, and a living room. The design requires at least 150 ft^2 of living area, 70 ft^2 of bedroom floor area for the first two people, and 50 ft^2 for each additional room. The ventilation system must provide for at least two air changes per hour. The mobile home is built on a chassis that is usually placed on a concrete foundation and tied down to secure it against wind and hurricanes. The requirement that each home be equipped with its own set of axles and wheels has been abolished.

Thermal Insulation. In the past, mobile homes as well as conventional homes were built on the premise that energy was cheap. The air quality in such homes depends strongly on the climate. If the weather is very hot or very cold and the walls are poorly insulated, a steady indoor temperature can be maintained only by pumping copious quantities of conditioned air through the building or by sealing the home tightly and recirculating all the air. Originally,

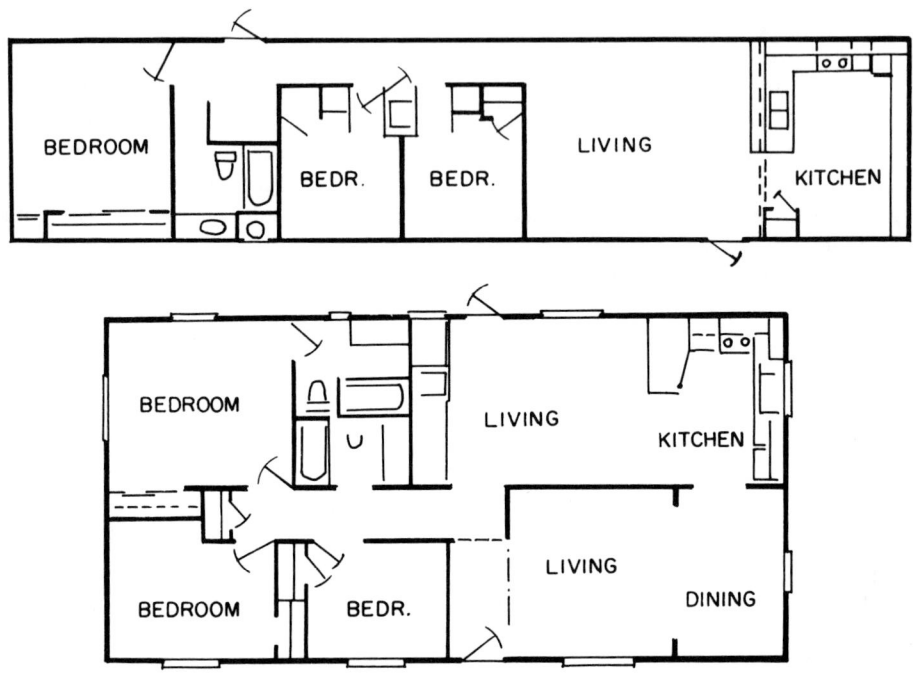

Figure 4.3. Floor plans of standard manufactured homes. Top, single-wide (14 × 66 ft); bottom, double-wide (24 × 48 ft).

all energy conservation programs used the latter approach because it is far easier to improve energy efficiency by reducing the intake of fresh air than by adding wall insulation. This has led to an undue buildup of humidity, odor problems, and in some cases to carbon monoxide poisoning. Extensive efforts are currently under way to determine a reasonable compromise between heat supply, fresh air, and insulation.

Since the mobile home standards are well defined and have been well tested in over 1 million homes in many different climates, they offer a unique opportunity to explore the interaction of the various components and parameters of indoor air.

Thermal Performance. Current regulations limit the heat transmission losses to 0.157 Btu hr^{-1} ft^{-2} °F^{-1} in the South and 0.104 in the cold zones where storm windows and insulating glass are required. The outdoor temperature rating must be marked on a readily visible certificate. The same holds for the cooling rating (National Fire Protection Association, 1977).

4.2 Building Types

The ceiling and exterior walls must have a vapor barrier rated at 1 perm or higher; wall cavities must be designed to allow dissipation of condensation within. The building envelope must limit air infiltration to reduce heat loss or gain. Therefore, envelope penetration for plumbing, heating ducts, and the likes must be sealed, and joints between connecting units must be caulked or sealed. The heat loss must be small enough to ensure that the factory-installed equipment can maintain 70°F in any of the three winter design temperature zones for which the unit is approved.

The thermal performance of two typical homes (Figure 4.3) was carefully analyzed by detailed field measurements under a variety of weather conditions with temperatures ranging from 1500 to 36,000 degree hours (HUD, 1978; 1980). In both houses, the energy consumption increased linearly with increasing temperature gradient. The energy consumption was 0.13 kWh deg^{-1} hr^{-1} for the double-wide and 0.09 kWh for the single-wide. The temperature profile is shown for the west wall in Figure 4.4a, the floor in Figure 4.4b, and the ceiling in Figure 4.4c. The thermal envelope response differed in each room with the heat shut off and an outside temperature of 5°F. The temperature in the master bedroom dropped from 80°F to 78°F in 10 min and to 63°F in an hour; in the living room it dropped from 80 to 68°F in 10 min and to 61°F in an hour. A thorough thermographic analysis was conducted on both houses. A comparison of the infrared and the normal photograph identifies cold spots in the envelope. Kitchen ventilation fans, electric outlets, door sills, studs, and of course, the windows, indicate heat leaks. Bottom and top plates also reveal thermal leaks. Details of the wall construction are shown in Figure 4.5a–d.

Sections of each wall type were laboratory tested to measure the insulation rating, indicated in Figure 4.5. The wall cavity section, Figure 4.5, is ventilated to control humidity; the second is sealed. The third, a so-called Arkansas-style wall, has 6-in. insulation with studs separated by 24 in. rather than the conventional 3-in. studs at 16 in. Details of the floor and ceiling design are shown in Figure 4.6a–d. The relative contribution of the various envelope segments to the heat loss is shown in Table 4.2. This study indicated that the tested homes were within the federal home standards, but the ventilated wall cavity in the single-wide performed poorly in the cold weather and the ventilated attic in the double-wide also performed poorly.

The infiltration rate was never less than 0.35 air change per hour. This is more than twice the 0.15 air change that corresponds to the 20-ft^3/min minimum standard established by the Department of Housing and Urban Development for indoor air control.

Analysis of the heating supply as a function of the solar load and daytime is shown in Figure 4.7 for a standard oil furnace in the single-wide mobile home. The performance of the furnace depends on its heating capacity. Oversized furnaces are not efficient. The temperature profile as a function of daytime is shown in Figure 4.7b for east and west walls. The time curves of the

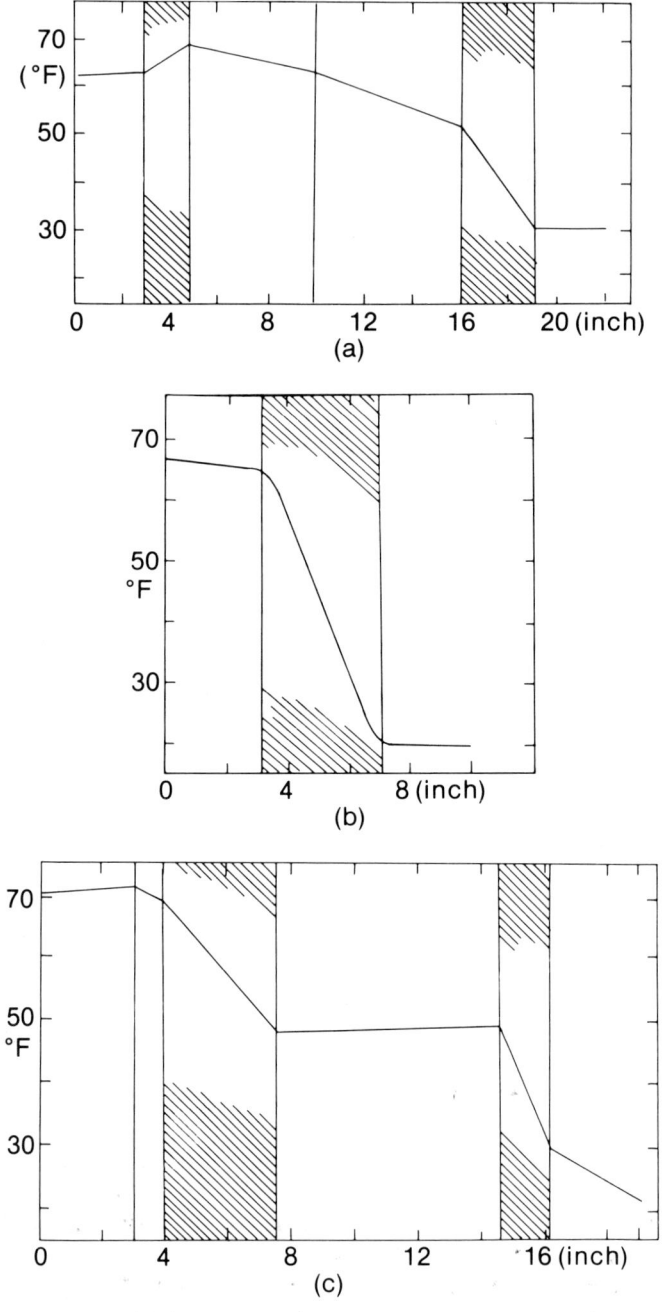

Figure 4.4. Temperature profile in a mobile home with the outdoor temperature at 20°F; (a) west wall, (b) floor, (c) ceiling.

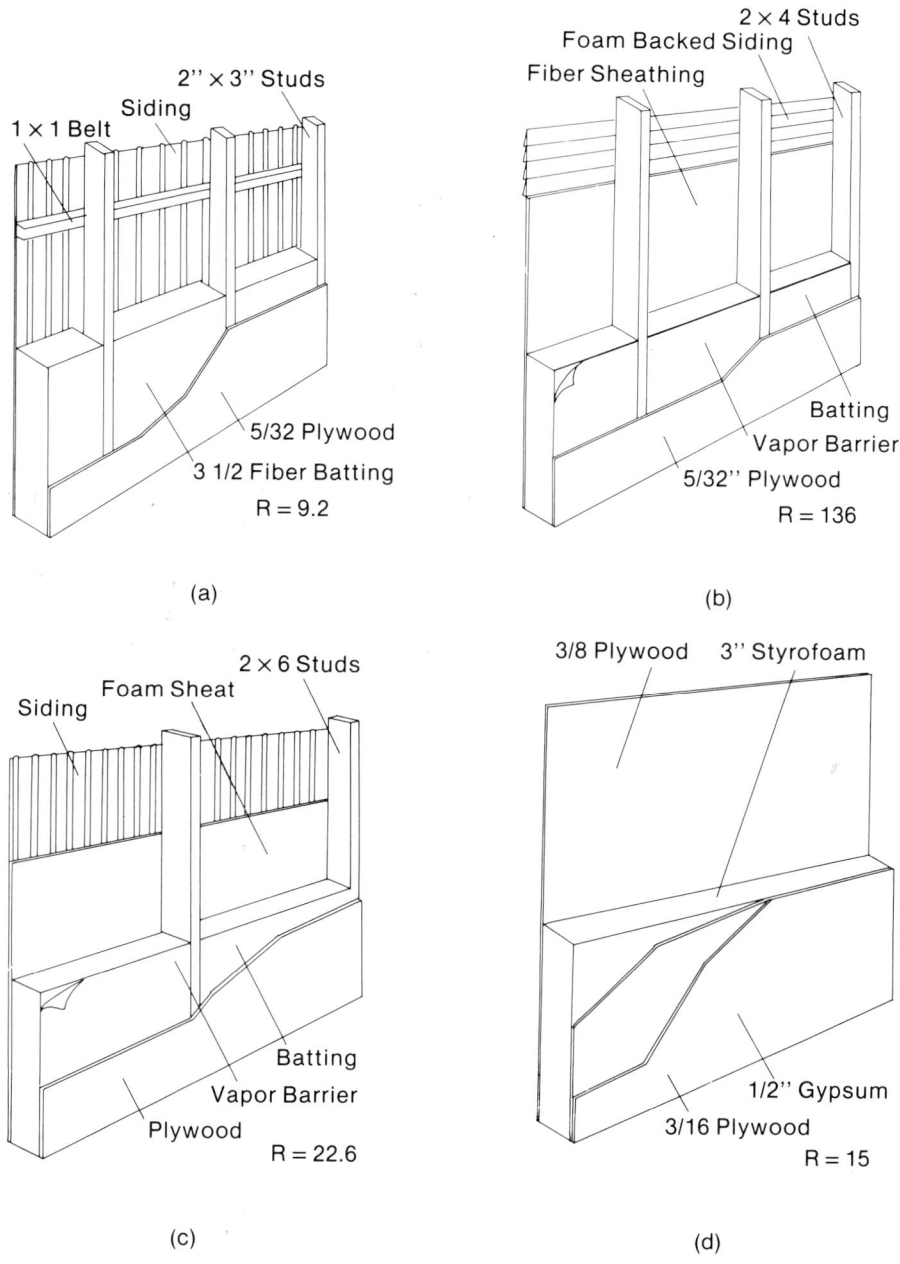

Figure 4.5. Wall detail of mobile home. (a) Standard single-wide; (U = 0.08 Btu hr^{-1} ft^{-2} °F^{-1}); (b) standard double-wide (U = 0.074 Btu hr^{-1} ft^{-2} °F^{-1}); (c) Arkansas type (U = 0.044 Btu hr^{-1} ft^{-2} °F^{-1}); (d) experimental styrofoam (U = 0.06 Btu hr^{-1} °F^{-1} ft^{-2}).

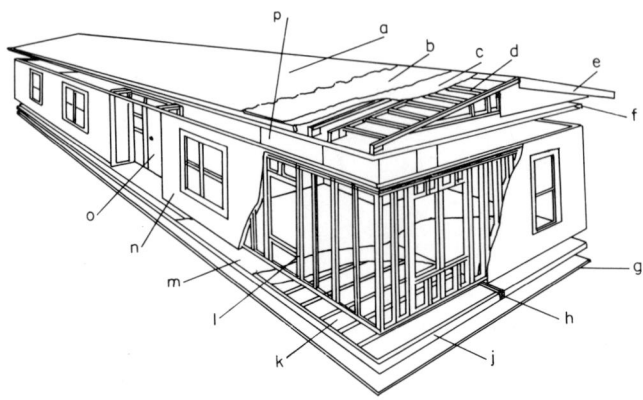

Figure 4.6. Construction details of a mobile home: a, shingle roof, 0.024-in. aluminum; b, roofing felt; c, 3/8-in. plywood sheeting; ridge beam made of structural plywood; e, overhanging eaves; f, plastic vapor barrier and 0.5-in. decorative particleboard ceiling; g, weatherproof bottom board and craft paper; h, joint between the two parts of the building; j, 2 × 6-in. transverse joists; k, glass insulation; l, 4-in. thick sidewalls with studs 16 in. on center; m, 5/8-in. thick particleboard decking with 3/8-in. rubber pad and shag carpet; n, 0.024-in. aluminum siding, 3.5-in. insulation, or 2-in. wood joist and 5/32-in. hardwood paneling; o, aluminum-faced cardboard core door; p, inside walls made from 5/32-in. hardwood plywood paneling.

Table 4.2

Heat Loss from a Standard Mobile Home in New York State with an Annual Heating Requirement of 7.8×10^7 Btu[a]

Component	Heat Loss (%)
Walls	
Cavity	24.0
Frame	8.9
Floor	
Cavity	18.0
Frame	3.1
Ceiling	
Cavity	14.7
Frame	3.7
Doors	3.3
Windows	24.0

[a] From HUD (1980).

4.2 Building Types

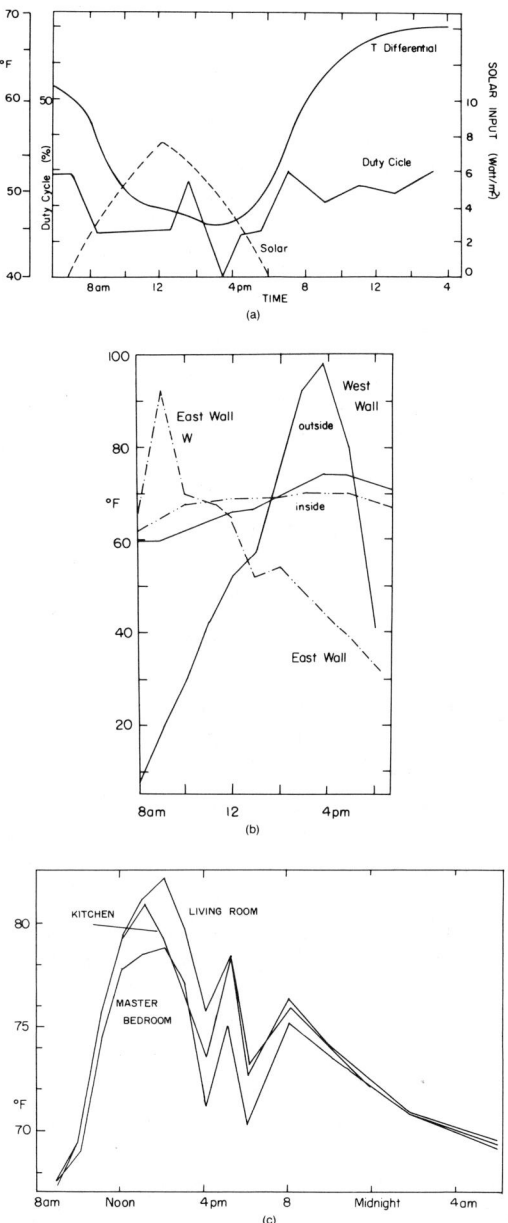

Figure 4.7. (a) Diurnal heating cycle as a function of the indoor/outdoor temperature difference and insolation; (b) diurnal temperature cycle on the West Wall and the East Wall for an outside temperature between 24°F and 39°F, and (c) summer indoor temperature with air conditioning set at 75°F (HUD, 1980).

two types of homes are consistent. The temperature profiles for summer air conditioning are shown in Figure 4.7c. The annual heat requirement of a mobile home in the severe climate of northern New York state was 7.7×10^7 Btu/year (1.9×10^7 kcal); this is eight times more than the experimental super-insulated home in Denmark.

These data show that temperature and temperature gradients in a typical North American home change rapidly and that different outside walls have very different gradients. Obviously, indoor air impurities will migrate in such a home with every heating cycle. The consequences of this migration are discussed in Sections 6.2 and 7.2.

4.2.2 COMMERCIAL BUILDINGS

The current inventory of commercial buildings comprises about 2500 km² of floor space. Public buildings, stores, and offices each account for 18% of this space. Commercial garages, hotels, and hospitals each account for approximately 7%. This inventory does not include manufacturing and production facilities that are covered by industrial regulations.

4.2.2.1 *Office Buildings*

Commercial office space accounts for 500 km², or 20% of commercial space. Currently about 20% of these buildings are classified as tall buildings, but the percentage of these buildings increases rapidly. Tall urban office buildings are integrated facilities with a variety of uses. They may include computer centers, stores, restaurants, recreational facilities, television studios, and observation decks. The basic building shell must be flexible in order to accommodate needs that change with tenants and with technological advances.

Because of their size, such buildings contain distinctly different climatic zones. The core area, or inner area, derives most of its heat supply from people, equipment, and lighting. The peripheral area may need heating on the weather side, or cooling on the sun side, or both at the same time.

The design of office buildings is strongly influenced by economic factors (Emerick, 1966; Courtney, 1976). The builder, owner, and occupant are often three different parties. The first two try to provide the most attractive looking building with the largest possible floor space and optimal convenience for the lessee at lowest possible capital. However, they are usually less concerned with operating costs, since these will be carried by the lessee. This separation between capital and operating costs is not conducive to energy-efficient design, and indoor air quality. In the early sixties, many large office buildings were built without light switches and without provisions for reduced nighttime and weekend heating or ventilation.

In office buildings the floor space allotted for each person varies from 75 ft² to 200 ft² and the energy balance depends on the local climate. The heat produced by light fixtures accounts for 25–66% of the energy use. Office machinery contributes about 30% or more. The major advantage of smaller office buildings is that adequate ventilation is possible through windows, whereas tall buildings often have sealed glass curtain walls.

In summer the cooling load depends greatly on the window area. If it constitutes 25% of the building surface, the window heat input increases up to 50%. People, windows, and air conditioning each contribute about a quarter of the total cooling load. In favorable cases heat conduction through the walls accounts for less than 5% (Courtney, 1976; Tishman, 1978).

Since cleaning and maintenance costs are high in office buildings, and since such costs are directly influenced by dust, air must be filtered carefully. During the last 15 years, the filter efficiency requirements have increased from 10% to 60%.

In large buildings the ventilation system is very carefully designed so that all parts balance, and the influence of the daily sun exposure and other seasonal variations can be corrected. During the spring and fall adequate air conditioning can be achieved by bringing outdoor air into the interior core without much recirculation. In summer and winter 100% of the air is recirculated throughout the building, and infiltration into the peripheral areas constitutes most of the fresh air supply. According to a recent study (Duffe and Jann, 1981), the ventilation rate could be reducd to 5 ft^3/min per person in most offices without noticeable body odor, but air contamination levels might dramatically increase for ozone, methanol, or other chemicals produced indoors. Cain (1981) proposed a minimum rate of 10 ft^3/min. However, a study by Sterling (1980) shows that ventilation systems can easily get out of balance and are very hard to optimize. Thus, an adequate safety factor should be included, which brings the value back to the one set 100 years ago, 30 ft^3/min.

Since the outside wall is the most important energy leak, square buildings use less energy than rectangular ones, and multistory buildings are more efficient than single-floor structures if both are made of the same materials. A typical 20-floor high-rise office building requires about 1.2 GJ per m^2 of floor space, whereas 20 single-story buildings with the equivalent floor space would consume 2.1 GJ/m^2 (Chaddock, 1978) because the ratio of building envelope area to floor area increases from 0.35 to 1.3, and the heat loss through the shell increases from 12 to 51% of total energy needs.

A study of 86 office buildings in New York City (Hittman, 1978) showed that energy consumption varied between 0.82 and 4.6 GJ/m^2 a year. There was a strong relation between the year of construction and energy use. In fact, in the period from 1950 to 1970, the electricity use per square meter increased by 60%, steam use increased 130%, and the total energy use more than doubled.

4.2.3 PUBLIC BUILDINGS, SCHOOLS, AND ARENAS

Public buildings, schools, and sports arenas are characterized by periodic changes in occupancy. Accordingly, the heat flux in these buildings undergoes discontinuities. Schoolrooms are specially critical, because schoolchildren are more active than adults and inhale almost twice as much air as adults, which explains why schoolhouses inevitably have a characteristic metabolic odor. Figure 4.8 shows the heat load for schoolrooms on the southern and northern sides of a building. Heating is only necessary to prepare the room for occupancy. Once the children arrive, the heat flux reverses. In a room with southern exposure, 35 students, a floor area of 1000 ft^2, a window area of 250 ft^2, an artificial light load of 3 W/ft^2, a heat transmission of $U = 0.15$ through the wall and $U = 0.25$ through the glass, no infiltration, and a solar intensity corresponding to 45° north latitude in December, the room must be cooled whenever the sun shines and the outside temperature is above 0°F. With artificial light and no sunshine, cooling is necessary if the outside temperature is above 40°F. Even without light and sun, the thermoneutral point is at 60°F. In a room with northern exposure the situation is the same, except that the sun contributes the equivalent of 30°F less to the heat flux. In interior classrooms the heat flux is shifted and the slope of the curve is reduced, yielding higher interior heat accumulations. Thus, fully occupied interior schoolrooms must be cooled all year round. The heat exchange in schools and auditoriums is made especially difficult because children want to be as close as possible to the black-

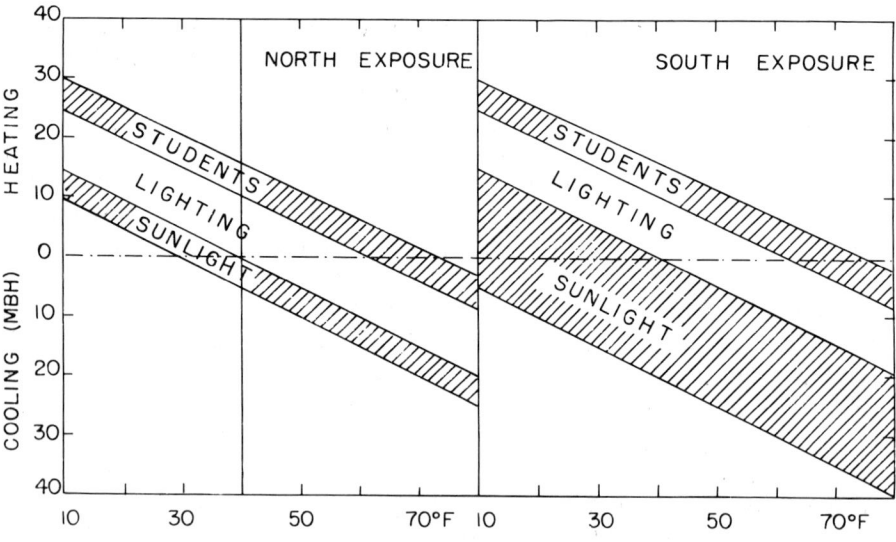

Figure 4.8. Heating and cooling requirements in 1000-ft^2 classroom with 35 students and a 250-ft^2 window area at 45° north latitude in December. (a) Northern exposure; (b) southern exposure (after ASHRAE, 1978).

board and teacher. Fire codes prescribe a minimum area of 20 ft^2; this is only a fifth of what is necessary in an office.

Obviously, ventilation is essential to establish an adequate heat balance in a schoolroom. To ensure adequate mixing, air motion is usually adjusted to somewhere between 20 and 50 ft/min. According to a recent study, an air exchange rate of 0.4 air change per hour is sufficient in schools to suppress all odors (Duffee and Jann, 1981). Similar observations were made by Young (1981) and Cain (1981).

In a ballroom, an assembly hall, and an arena each person is required to have at least 7 ft^2, and in bleachers 18 linear inches are required for each seat. In order to assure adequate breathing air and adequate heat flow, modern hotels and city centers have replaced one- or two-story ballrooms or conference halls with multistory open lobbies and spaces where vertical air movement provides natural ventilation, very much as in an open space. A comparison of fire codes with ventilation codes shows that floor space and ventilation needs are directly related (Table 4.3).

4.3 Building Materials and Components

The choice of building material has a substantial influence on indoor air quality, either by direct interaction with air or by defining the temperature and humidity of buildings. Traditionally, buildings have been built with locally

Table 4.3

Minimum Floor Space Requirements[a]

Facility	Floor Space/Person (ft^2)
One- or two-family unit (gross)	200
Penal facility (gross)	120
Business	100
Mercantile	30–60
Laboratory	50
Classroom	20
Conference room	15
Assembly	7
Standing and waiting area	3
Bleachers	18[b]
Mobile home code	97
City bus	20[b]

[a]From National Fire Protection Association Code 101-1976.
[b]Value in linear inches.

available materials. In North America wood is still far more abundant than in Europe or Japan. In western North America almost all homes are built of wood. In the east and middle west homes are either wood or brick, depending on the area. During the last three decades the construction and appearance of nonresidential buildings underwent great changes because large structures such as airport terminals, hotels, domed stadiums, downtown centers, tall office buildings, and mobile homes were increasingly financed, planned, and designed by large national or international corporations, and therefore incorporated similar materials and design everywhere, regardless of local climate or material availability. In fact, building materials and design features such as prestressed concrete, steel, or glass curtain walls are now more an indicator of when the building was built than of its location.

Each building material has different physical and thermal properties, and thus reacts differently to wind, temperature, humidity, and air impurities. Because of the wide variety of materials, predictions regarding the general thermal or air quality properties of buildings are difficult to make. Each component of the structure must be analyzed separately, as explained for mobile homes in Section 4.2.1.3.

4.3.1 HEAT TRANSFER

Heat transmission by the building envelope has a decisive effect on indoor air quality for several reasons. First, it determines the heat gradient between the wall and the indoor air. This determines thermal eddy currents, which are a key factor in mixing indoor air. Second, the heat transmission determines how much heat has to be supplemented or removed from the interior, and this in turn determines the rate of air exchange and the degree of heating, the amount of fuel used, and the amount of air pollutants released. All these factors affect air quality.

As indicated in Section 4.1, large, heavy stone or brick construction provides a large heat sink and smooth diurnal temperature cycles. As a drawback, large stone buildings, such as churches, cannot be quickly or economically heated for short peak-use periods. Wood, in contrast, is a good insulator, but wooden buildings have a greater tendency to develop cracks than brick buildings. Thus, wooden homes transmit heat faster and provide a shorter heat transfer delay. Aluminum-faced trailers and single-pane glass-paneled walls are essentially transparent to heat. Thermal properties are expressed in thermal resistance units, $R = \text{hr ft}^2 \, °F \, \text{Btu}^{-1}$ (ASHRAE 1977). Water and air transmission are listed in Table 4.4 and Figure 4.9.

4.3.2 WOOD PRODUCTS

Whole wood or timber has largely disappeared from the North American building market as floor or wall material except in structural applications. It has been replaced by plywood panels, which are made from veneer obtained by

4.3 Building Materials and Components

Table 4.4

Water Vapor Transmission of Building Materials

Material	Permeance[a] (perm)	Resistance[a] (rep)
Brick masonry (4 in. thick)	0.8	1.3
Concrete block (8 in. cored, limestone aggregate)	2.4	0.4
Tile masonry, glazed (4 in. thick)	0.12	8.3
Asbestos cement board (0.2 in. thick)	0.54	1.8
Plaster on metal lath (0.75 in.)	15	0.067
Gypsum wall board (0.375 in., plain)	50	0.020
Structural insulating board (interior, uncoated, 0.5 in.)	50–90	0.020–0.011
Hardboard (0.125 in., standard)	11	0.091
Plywood (douglas fir, exterior glue, 0.25 in. thick)	0.7	1.4
Polyester, glass fiber reinforced sheet, 48 mil	0.05	20
Aluminum foil (1 mil)	0.0	
Building paper, felt		
Duplex sheet, asphalt laminated, aluminum foil one side (43)	0.002	0.176
Single-kraft, double	31	42
Paint (three coats)		
Exterior paint	0.3–1.0	
Polyvinyl acetate latex coating, 4 ounces/ft^2	5.5	
Chloro-sulfonated polyethylene mastic, 3.5 ounces/ft^2	1.7	
Hot melt asphalt, 2 ounces/ft^2	0.5	

[a]One perm is equal to grain hr^{-1} ft^{-2} and per inch of water; a rep is a reciprocal perm, or 1/perm.

continuous peeling of a tree trunk along its axis with a long, sharp blade. The veneer is flattened and then sprayed with 2.5 wt. % glue, usually a phenolic aldehyde resin or ureaformaldehyde resin. Three, five, or more layers are cross-stacked and bonded at 350°F (Meyer, 1979). Exterior plywood bonded with phenolic resin resists water and moisture; indoor grades bonded with urea-formaldehyde resin resist water as long as it drips off, but they do not resist constant moisture. Plywood has a history of 75 years of commercial use and if used properly, in most applications it performs as well as or better than whole wood because it does not crack or shrink. During the last two decades plywood has found a keen competitor in particleboard, which after 30 years of commercial use has replaced plywood in several applications. Particleboard is made by spraying wood chips with 7–8 wt. % urea-formaldehyde resin and pressing panel-shaped cakes on huge cookie sheets, called cauls, at 350°F for

3–10 min. Particleboard makes an excellent flooring or underlayment on dry floors, and it has unexcelled properties for cabinet work and furniture, but currently available standard grades are not suitable for exposure to weather or moisture because untreated particleboard swells like sawdust if it absorbs moisture. Particleboard panels can be covered and sealed with paper or polyethylene sheets or metal foil during the manufacturing process. Almost all desks, tables, doors, and furniture are now made from particleboard covered with melamine-impregnated paper with an imprinted wood-grain design. The density and the thermal properties of forest products are almost indistinguishable from those of whole wood (Table 4.4). In Europe particleboard now accounts for almost 80% of all wood products; in the United States and Canada for about 40%.

Both plywood and particleboard have become mainstays of North American home construction. Each year about 100 million tons of both materials are produced, and their desirable properties have led to their being used in a variety of applications that were never envisioned by those who invented and formulated them (Meyer, 1979). For example, they are used to cover all the surfaces in mobile homes: A standard single-wide mobile home with a volume of 6700 ft^3 contains 2200 ft^2 of plywood paneling and 1300 ft^2 of particleboard. Such large exposed surfaces of resin-bonded materials in a small volume have caused unexpected problems with the odor of formaldehyde, which is trapped in new boards from the manufacturing process. The nature of the chemical mechanism that produces the odor is complex and an adequate description requires a book in itself (Meyer, 1979). This initial odor is amplified if ventilation rates are reduced. Furthermore, poor ventilation allows indoor moisture to build up to such a degree that the wood swells and eventually rots, and at high humidity levels the urea-formaldehyde adhesive can undergo surface hydrolysis. If the damaged product dries, it may release some formaldehyde vapor, which is noticeable at very low levels and can cause discomfort, as described in Chapter 8. It has been pointed out (Johns, 1980) that moisture condensation in homes in which well-meaning people have unduly reduced ventilation rates in the living room, kitchen, and bathroom is likely to cause many billions of dollars in damage to structural wood, whole wood, and wood products within the next decade, but recent progress in the application of wood adhesives has made the formaldehyde odor problem obsolete if wood products are properly used.

4.3.3 INSULATION

The heat transmission of structural and wall materials is so large that the building envelope must be equipped with separate layers of insulation material if the flow of heat between indoors and outdoors is to be significantly reduced. As long as fuels were cheap and abundant this was not always done, and many post-World War II buildings lack adequate space for retrofitting. The issue of insulation has substantial economic dimensions. The U.S. Department of Energy has estimated that if 40% of the 70 million residential housing units in the United States were upgraded to a wall insulation level of R-13, 300 million lb of cellulose and 200 million lb of glass fibers would be needed for retrofitting old buildings. This would result in a saving of 138×10^{12} Btu of heating fuel, which corresponds to a total energy use of about 20 quads/year for residential heating and cooking. The corresponding savings in commercial buildings would be about 4 quads. This would amount to a total annual fuel saving of over $1 billion. A 70% acceptance rate, and the addition of R-19 attic insulation, would bring the saving to over $2 billion per year (44 FR 64650)[a].

The problem with retrofitting is that the modification of thermal properties causes very complex secondary changes in the structure as well as the habitat. Insulation added to any part of a building changes the heat balance and profile of the entire building. Furthermore, poorly placed insulation can cause overheating in walls, or even fires; corrosion of electric wire and water pipes; fungal growth; and other air contamination. By far the worst consequence is that insulation constitutes a vapor barrier. The Veterans Administration estimated that in 1979 over $5 million worth of damage was caused by improper insulation vapor barriers in the State of Oregon alone (44 FR 66644).

The U.S. Department of Energy has approved and established standards for 11 thermal insulating materials: cellulose fibers, mineral fibers, mineral blankets and batts, vermiculite, perlite, cellular polystyrene, polyurethane and isocyanate, urea-formaldehyde foam, and aluminum foil. Many utility companies have participated in a program that gives customers a refund if they install these materials. Of all these, urea-formaldehyde foam insulation is potentially the best because it offers the highest insulating value for the lowest price. In fact, it is the only insulating material on the Department of Energy's list that is capable of properly insulating a 4-in. wall, and it has sufficient fire resistance to be used as fire protection around structural steel segments in some tall buildings. Finally, it can be foamed in place to fill any space and adapt to its shape. Thus, this material fills the bill for both new and retrofit installation. Unfortunately, it also has serious drawbacks. For one, it is moist when placed and must dry over a period of several days before it is fully cured. During this time the foam also releases traces of residual formaldehyde. Furthermore, it is susceptible to degradation if it is applied at a temperature below 50°F or above 85°F, and after it has dried in place it cannot withstand continued exposure to moisture. Finally, formulating and placing the material properly requires considerable knowledge and skill (Meyer, 1979). Thus, quality control and proper inspection of sites is crucial. In recent years chemically unsatisfactory formulations have been pumped into more than 1000 buildings

---

[a]Throughout this book reference to publications in the Federal Register of the United States is made according to the official abbreviation. The above quote is found in volume 44, page 64650.

in North America by installers who lacked a full understanding of the chemical properties of this material. Poor insulation resulted, and malodorous formaldehyde fumes were released that could, depending on the wall design, enter living areas. In December 1980, after several years of complaints by local authorities, the product was banned in Canada, and in March 1982, six years after the first petition by the Metropolitan Denver District Attorney, the U.S. Consumer Product Safety Commission also passed a ban (47 FR 14366), because formaldehyde levels in some homes had been reported at, or close to, occupational levels and foam samples supplied by eight U.S. manufacturers were releasing unexpectedly high formaldehyde concentrations, even after 18 months (Pressler, 1981). This very complex problem has been studied widely (HUD, 1978–1980; DOE, 1978–1980; U.S. Consumer Product Safety Commission 1979–1981; Long, et al., 1979; Meyer, 1979–1980; Pressler, 1981) and is discussed further in Section 7.2 and Section 10.9.1.

In the United States about 1.5 million homes have been insulated with urea-formaldehyde foam; in Canada about 125,000 home owners have obtained matching federal and provincial funds to encourage its use. Many of these homes are equipped with properly installed and manufactured material, which will provide the owners with safe and economic insulation, but it is likely that improperly prepared foam must be removed from several hundred homes.

4.3.4 WIND RESISTANCE

Metal- and glass-faced structures can be equipped with flexible plastic gaskets that render them impenetrable by the wind. Wood structures "work" and warp in temperature and humidity gradients, and thus tend to be susceptible to wind infiltration via cracks. Brick is so porous that one can blow out a burning candle through a funnel attached to a brick surface. Thus, brick surfaces must be finished with stucco or a similar surface coating. Figure 4.9 shows the air permeability of several construction materials as a function of wind pressure.

Air transmission accounts for a substantial fraction of winter heating energy. In most homes about 30% of the heat goes straight out through cracks around windows, doors, floors, and ceilings. However, this leakage also provides a comparable fraction of the fresh air, and prevents the buildup of moisture in the walls.

4.3.5 MOISTURE

Aluminum, steel, and glass faces are impervious to moisture, but whenever a building is exposed to an indoor–outdoor temperature gradient these materials tend to draw condensation because of their high heat transmission. In contrast, wood transmits moisture via cracks and wood absorbs moisture. Its moisture content is proportionate to the air humidity; see Table 4.5 (USDA

4.3 Building Materials and Components

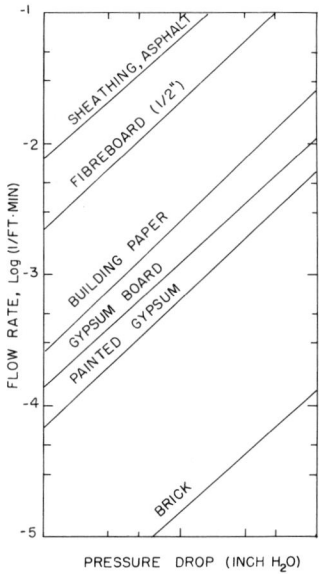

Figure 4.9. Infiltration of nine building materials as a function of air pressure differential.

Table 4.5

Moisture Content of Wood as a Function of Temperature and Humidity[a]

Dry-bulb temperature (°F)	Relative Humidity (%)									
	5	10	20	30	40	50	60	70	80	90	98
30	1.4	2.6	4.6	6.3	7.9	9.5	11.3	13.5	16.5	21.0	26.9
40	1.4	2.6	4.6	6.3	7.9	9.5	11.3	13.5	16.5	21.0	26.9
50	1.4	2.6	4.6	6.3	7.9	9.5	11.2	13.4	16.4	20.9	26.9
60	1.3	2.5	4.6	6.2	7.8	9.4	11.1	13.3	16.2	20.7	26.8
70	1.3	2.5	4.5	6.2	7.7	9.2	11.0	13.1	16.0	20.5	26.6
80	1.3	2.4	4.4	6.1	7.6	9.1	10.8	12.9	15.7	20.2	26.3
90	1.2	2.3	4.3	5.9	7.4	8.9	10.5	12.6	15.4	19.8	26.0
100	1.2	2.3	4.2	5.8	7.2	8.7	10.3	12.3	15.1	19.5	25.6
110	1.1	2.2	4.0	5.6	7.0	8.4	10.0	12.0	14.7	19.1	25.2
120	1.1	2.1	3.9	5.4	6.8	8.2	9.7	11.7	14.4	18.6	24.7
150	0.9	1.8	3.4	4.8	6.1	7.4	8.8	10.6	13.1	17.2	23.1
200	0.5	1.1	2.4	3.5	4.6	5.8	7.1	8.7	10.9	14.6	20.0

[a] From the USDA Forest Service *Wood Handbook* (1974). Values are wt. %.

Forest Service, 1974). This fact makes wood a pleasant buffer because it absorbs moisture from wet indoor air and releases moisture during dry weather and during heating periods. Brick and concrete have the capability of absorbing moisture as well as transmitting it. This can be advantageous under some conditions, but often it causes continuous moisture leaks through capillary action on the weather side of buildings or through a floor slab that comes in contact with moist soil.

Air Impurities. Gypsum, concrete, wood, and all building materials that absorb moisture have the capability of filtering or absorbing particulates, and they can serve as the host for the formation of airborne molds or fungi. Other materials act as carriers for such potentially airborne agents as paint thinner, lead pigment, and moisture.

Vapor Barriers. Insulation intrinsically provides vapor resistance, but in recent years it has become fashionable to provide some building components with separate vapor barriers to block both wind and moisture. Adding a vapor barrier to an older building can have the same effect on a building as a plastic sweat suit has on the human body because most older structures depend on the wall to maintain an adequate water vapor and air flux. In many homes built during the last 10 years, a 4-mil or slightly thicker polyethylene sheet is stapled to the wall studs and placed as an apron under the building. Such vapor barriers are useful only if ventilation is adequate to remove all moisture from indoor occupants and activities.

Furthermore, vapor barriers must be properly placed in order to be effective. In theory, this is simple because the barrier always belongs on the warm side of a wall and must be maintained above the dew point of the air from both sides. In practice, this is not always easily achieved because the wall temperature and the air temperature gradient change in vertical and horizontal directions as a function of sun exposure and as a function of time of day and season, indoor and outdoor activity, and occupancy. In many locations the thermal gradient in walls exposed to sunshine changes its sign not only seasonally but even daily, and it can happen that the dew temperature along a wall shifts horizontally from east to west. This effect pumps moisture across insulation or along the wall.

Vapor barriers must be either tightly sealed or properly vented; otherwise they invite disaster because moisture inevitably accumulates at the coldest place, and such trapped moisture will rapidly corrode, rot, or otherwise destroy wood, masonry, steel, insulation, or almost any material. The reason for the rapid corrosion is that condensation forms droplets each of which constitutes a closed microchemical system in which large chemical concentration gradients act as if the condensate were concentrated acid or a nutrient broth for microbes. Obviously, improperly placed vapor barriers damage both the structural integrity and the air quality of many buildings.

The proper installation of vapor barriers and the design of heated and unheated attic ventilation to prevent water penetration include provision for

4.3 Building Materials and Components

maintaining a proper ratio between the vent and roof surface. The vent surface should be about a thousandth or less of the insulated area.

4.3.6 FURNISHINGS

Indoor temperature, humidity, and air quality are influenced by home furnishings as much as by wall materials. Wall-to-wall carpeting placed on a rubber mat directly on concrete slabs or particleboard has almost totally replaced wood floors in commercial and residential construction. Such carpets now serve as integrated building elements and provide thermal and noise insulation. Although carpets have a large absorbing capacity for most gaseous, liquid, and particulate air impurities, this capacity is limited and carpets tend to saturate with air pollutants long before physical wear is evident. Thus, many carpets in commercial buildings are so malodorous that the building ventilation is perfumed to conceal their odor; but this device can be inadequate in heavily polluted places, such as hotel rooms and meeting rooms.

Carpet odors can be controlled by proper selection of the rag structure and the chemical blend of fibers, but carpets have to be carefully maintained by steam or vacuum cleaning. However, certain commercial rug shampoos can produce residual dust which may emanate formaldehyde and other organic vapors. In certain settings, such as schools with a great deal of student movement, carpets have caused so many problems that they have been replaced with other surface materials. These problems will be discussed in Chapter 9.

4.3.7 WINDOWS

Since 1850, when window glass became widely available, windows have been designed so that building occupants can open them to a greater or lesser degree in order to regulate the flow of light, heat, air, and humidity in a room. During the last 20 years, however, movable windows have been increasingly replaced with permanently sealed glass panes. This is necessary in air-conditioned buildings because open windows there throw the air flow in the entire building off balance.

Whether fenestration should include ventilation touches on philosophical as well as practical issues. Windows are still common everywhere in Europe. In fact, some European school laws still mandate that windows be opened for 3–5 min between classes for the vigorous airing of the classrooms. A similar practice is also observed in hospitals (Attia, 1979). In continental Europe vertical swing windows are common. They can be opened over the entire window area. In contrast, the traditional anglosaxon sash windows only slide half way open.

As far as heat transfer is concerned, all windows are an expensive investment because the choice is between single-pane glass, for which the radiative and conductive heat transfer is large, or double windows, which are costly. In tall office buildings windows account for 40% of the total heat loss. The construction of a tight and well-insulated window is shown in Figure 4.10b. To some extent, the investment in windows can be recovered by judicious use of the seasonal changes in the sun angle (Figure 4.10a), which makes possible the selective collection of sun radiation in winter. Household pets frequently become virtuosos at utilizing direct solar radiation to establish a large temperature gradient over all or part of their bodies so as to optimize comfort. Using the sun's intensity and angle as a factor in building design was once a routine part of architectural planning but it has been widely neglected in mobile homes, tall buildings, and all standardized construction, which ignore the angle of the sun in favor of orientation of houses toward streets and neighboring structures.

The impact of windows on indoor air is very significant: Windows increase air mixing and air circulation and establish a temperature gradient that can serve as a driving force to transfer and deposit pollutants within a closed room. Thus, a window can act as a solar distillation column for purifying indoor air. More information on fenestration can be found in architectural and building engineering literature, but very little is really known about the quantitative contribution of windows to air purification.

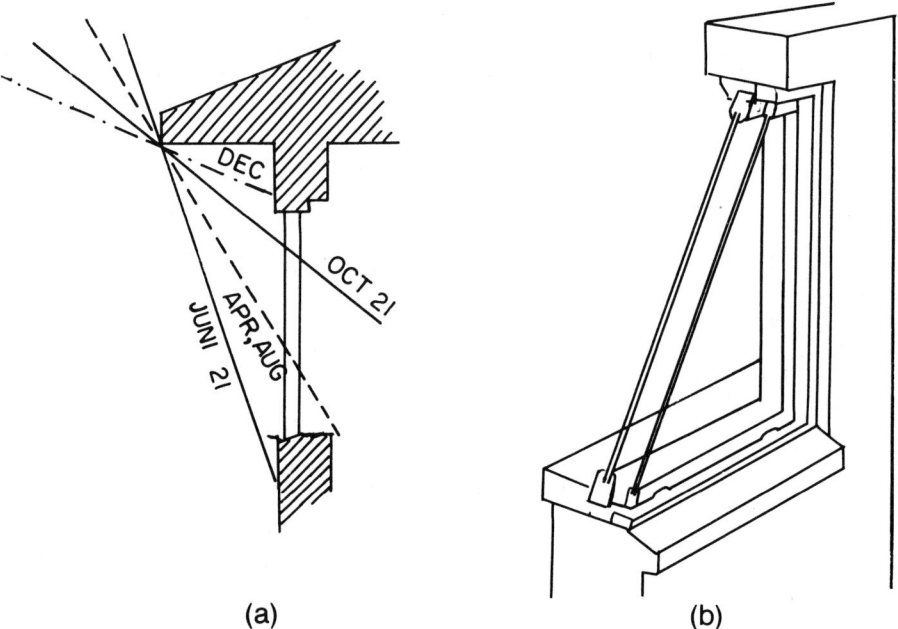

Figure 4.10. Window details: (a) shading of south window by roof overhang at 43° north latitude; (b) detail of thermal window design.

5. Air Impurities

Indoor air impurities can originate in the outdoor air or can stem from dust carried by indoor or outdoor air, or they can be generated by building materials and by a building's occupants and their activities. The most abundant indoor air contaminants are moisture and combustion gases. In this chapter we discuss some of these impurities and their properties. The composition of pure air is listed in Table 3.3. First we consider human metabolic impurities, and then some of the most important potential outdoor pollutants, starting with the classic primary ambient air pollutants. Agents that are known to cause discomfort are emphasized.

5.1 Metabolic Products

The human body continually releases water vapor, carbon dioxide, and traces of over 100 other pollutants that can reach noticeable levels in small confined areas.

5.1.1 WATER

The main metabolic product is water, which is dissipated in exhaled air and via skin perspiration. With each breath 1 liter or more of 37°C air saturated with moisture is exhaled. Depending on the humidity of the inhaled air, this can add up to 1000 g of water per day per person. Perspiration adds another 2–4 kg of water, depending on the indoor temperature and the work performed. If normal household activities such as cooking and washing are included, a family of four releases 4–10 gal or 25–100 lb of water into the indoor air. This is far more than the air in any home can absorb, and so ventilation is essential in order to avoid water condensation.

The capacity of air to absorb and carry moisture depends on the temperature. Figure 3.7 shows the water content of saturated air as a function of temperature. The vapor pressure changes from 5 g/m^3 at 32°F to 50 g/m^3 at 105°F and by a factor of more than 500 between the melting point of ice and the boiling point of water. Thus, hot air can hold 10 times more water vapor than cold air. If hot summer air at 90°F and 90% humidity is cooled to a comfortable level of 70°F and 50% humidity, the water content drops from 30 g/m³ to 8 g/m³. This means that 22 g of water have to be removed for each cubic meter of fresh air. At a ventilation rate of 30 ft³/min this amounts to more than 2 lb of water per hour. In winter, the reverse process occurs, and cold outdoor air picks up indoor moisture.

The indoor humidity is determined by the balance between outdoor and indoor moisture. In wood frame and brick buildings a substantial amount of moisture migrates with air through the walls. The permeability of gypsum and particleboard is 50 perm, concrete 2.4, masonry and plywood 0.8, and polyethylene sheet 0.08; 1 perm = 1 grain hr^{-1} ft^{-2} $in.^{-1}$ of water pressure. In summer in a hot, humid climate indoor and outdoor moisture build up and cause condensation. In a hot, dry climate indoor moisture can be readily controlled by ventilation. In winter the situation is more subtle. In poorly insulated buildings heated with forced fresh air, indoor air rapidly becomes too dry. This condition promotes nasal and sinus infections, such as are common in most parts of midwestern and eastern North America. If buildings are well insulated and fuel is saved by keeping buildings cool, indoor moisture builds up and condenses. This allows microbes to prosper on moist walls and causes bronchial problems, such as are common in England.

5.1.2 OTHER METABOLIC CONTAMINANTS

The nature and concentration of airborne human waste is now fairly well known because human wastes must be carefully managed in the confined atmospheres of nuclear submarines and spacecraft (Fomin, 1965; Johnson and Sargent, 1968; Nefedov, Savina, and Sokolov, 1972; Parker, 1973; Shepelev et al., 1974). The principal sources of air contaminants are expired air, skin scales, perspiration, flatus, urine, and feces (Roth, 1968). Exhaled air contains not only carbon dioxide, but also ammonia, carbon monoxide, and acetone, depending on health and diet. Further components reported are acetaldehyde, acetic acid, acetonitrile, acetylchlorine, alamine, ammonia, butyraldehyde, creatinine, diethyl ketone, dimethyl sulfide, ethane, ethylamine, ethyl chloride, ethyl ether, formamide, hydrogen, hydrogen chloride, isoprene, isopropyl alcohol, lysine, methane, methyl amine, methyl ether, methyl ethyl ketone, methyl thiocyanate, methyl urea, nitric oxide, nitrogen dioxide, nitrogen oxychloride, propionaldehyde, propyl alcohol, trimethylamine, urea, uric acid, and xylose, to mention only the most volatile components. The aerosol produced by sneezing may contain literally dozens of inorganic salts, organic acids, alcohols, amino acids, enzymes, esters, ethers, hydrocarbons, halogenated hydrocarbons, hormones, sugar, sulfur derivatives, and vitamins.

Regulatory sweat, for thermal control, contains 99–99.5% water and 1.2–1.6% solids, including measurable quantities of 11 cations, of which the most prominent is calcium, since it accounts for about 2 g per 100 ml. The chloride ion accounts for anywhere between 30 and 500 mg/ml. Sweat also contains some 14 amino acids, primarily arginine (14 mg per 100 ml), histidine

(8 mg per 100 ml) and threonine (6 mg per 100 ml). Ammonia may make up between 2 and 35 mg per 100 ml and urea between 5 and 40 mg per 100 ml. Lactic acid ranges from 5 to 500 mg. Some 10 vitamins have been identified in sweat at concentrations below 0.07 mg per 100 ml. Vitamin C is most abundant. Nonregulatory sweat and skin secretions contain up to 60% protein. Fatty acids may account for more than half of it. Butyric, valeric, and caproic acids, which exude the pungent odor identified with sweat, make up 1.2 wt. %. Glycerides account for the bulk of the rest. The most important other substances are 9 wt. % aliphatic alcohol, 8% hydrocarbons, and 2% paraffins. The exact composition depends on health and other factors. The composition also changes on different parts of the body. For example, it is known that the triglyceride content is only 16% on the scalp but 32% on the forearm and 44% on the forehead.

Hair, dried skin, and sebum contribute to indoor dust. About 3 g of dried surface skin is shed each day, and a T-shirt absorbs about 0.6 g of sebum during a workday. The forehead alone releases about 20 mg during the same time period. Sebum consists of many millions of fine scales, each of which carries on the average four or more bacteria (Speers et al., 1966; May and Pomeroy, 1973; Fanger, 1979). Dust from an electric shaver weighs about 0.3 g/day per male adult.

Combined, the human wastes contain about 160 mg of microorganisms, of which a fraction may be released into the air. Thus in nuclear submarines with large crews, a rash of respiratory infections that may last for a few weeks is often noted (Morris, 1972; Morris and Fallon, 1973). Very little is yet known about viral infections in spacecraft, but staphylococci and streptococci are known to prosper on the walls and furniture of spacecraft, and constitute a constant health nuisance, just as in humid homes. In their memoirs, Russian astronauts complain about the daily chore of washing down instrument panels with disinfectant to control microbes.

In confined areas the air must be renewed or purified to keep humidity and other components at the comfort thresholds for short or indefinite exposures. The worst problem in confined spaces is posed by particulates. It has been found that particulates will reach a level of 0.4 mg/m^3 after 100 hr in submarines. This is twice the level found on a smoggy day in Los Angeles. Furthermore, body odors may become noticeable if ventilation is limited and the available airspace is restricted because the odor threshold for several human waste products is at or below 10^{12} molecules/ml. Yaglou (1936) observed a threshold line that extends from 5 ft^3 of ventilation in an air volume of 1000 ft^3 per person to 40 ft^3/min of ventilation in a 400 ft^3 air volume. However, the odor sensation is lost upon extended exposure and remains only noticeable to visitors.

Urine contains at least 24 elements in measurable amounts. Over 100 organic nitrogen compounds have been identified; the leading component is urea, with 15–40 g/day; followed by creatinine, which accounts for about 1.6–2 g in the total volume of 60–600 ml of liquid discharged each day by a healthy adult. Hippuric acid averages 0.7 g/day, uric acid 0.6–1 g/day; all amino acids make up 1.1–2.8 g/day. Lactic acid, with 0.2 g, averages highest among the acids, but citric acid can reach peaks of up to 1.4 g/day. Some lighter hormones have been identified, with androsterone reaching up to 0.5 g/day. The vapor pressure of most of these compounds is small enough to make urine an unimportant source of indoor air pollution in all locations except poorly ventilated bathrooms.

The same holds for gases emanating from feces, except in rooms where baby diapers are stored or changed. Upon storage, feces can release up to 6 ml of carbon dioxide and about 1 ml of methane per gram of feces. Feces are rich in calcium and potassium. Arginine, lysine, and leucine account for half of the nitrogen, which may amount to 1.5 g/day.

Unventilated nurseries and barracks may accumulate measurable quantities of components of flatus, which may reach up to 3 liters per person per day, even with a cabbage-free diet. Nitrogen accounts for 60% of the gas, hydrogen averages 20%, carbon dioxide 9%, methane 7%, and oxygen 4%. Hydrogen sulfide averages 3×10^{-4}%, but may be up to five times higher.

Saliva contains 3–12 wt. % ammonia and 1.2% alarine, to list only the most volatile components. The ionic part consists of sodium and calcium bicarbonate, and chloride with phosphorus and sulfur accounts for the remaining inorganic content. The leading enzyme in saliva is lipozyme. Amazingly, tears, which are generated at a rate of about 1 g/day, also contain about 5% ammonia, but the bulk is protein.

5.2 Ambient Air Pollutants

This section deals with contaminants that enter the indoor air environment with ambient air through windows, ventilation shafts, doors, and inadvertent leaks. Many of these pollutants can also be generated by indoor sources but they are discussed here because their sources, occurrence, properties, and control have been studied extensively under the well-established regulatory programs of the Clean Air Act, which in the United States is currently administered by the Environmental Protection Agency, as described in Section 10.2. The list of primary ambient air pollutants is currently being reviewed. It comprises sulfur dioxide, nitric oxide, carbon dioxide, particulates, ozone, and lead. The concentration of these pollutants is periodically reviewed (Ferris, 1980).

5.2.1 PARTICULATES AND AEROSOLS

Suspended solids and liquid particles form dusts, smoke, smogs, and fogs that can obscure sunlight, corrode buildings, and attack the respiratory system. This section deals with physical and chemical particulates. Microbes are discussed in Section 6.3. Airborne particulates and aerosols are the result of turbulent mixing. Under chemical equilibrium conditions particulates and aerosols are not stable in air, and these species separate and settle in quiet air. As a result, they accumulate in buildings at aerodynamically favorable places until indoor air disturbances stir up the dust and resuspend the particles (J.E. Davies et al., 1975; Gregory, 1973).

Particulates or aerosols can contain almost any type of chemical. According to the U.S. government air quality criteria definition (HEW, 1969; EPA, 1980), particulates are "any dispersed matter, solid or liquid, in which the aggregates are larger than single gas molecules, but smaller than 0.5 mm." The "total suspended particulate" (TSP) matter contains species ranging from minute aggregates to large dust grains (Spirtas and Levin, 1970; Hesketh, 1977). The very fine particles, smaller than 0.1 μm, remain airborne for a very long time. They perform Brownian motions and cause Rayleigh scattering, which is responsible for the color of the sky. These particles can adsorb water or aggregate and form the nucleus for larger and heavier particles that may settle to the ground.

Particles with a size between 0.1 and 1.5 μm constitute the most important group for this book because they enter the respiratory tract and penetrate deep into the lung (Section 8.1). This particle size is commonly called fine suspended particulate (FSP) matter (Schrag and Rao, 1976) or respirable particles (RSP). The efficiency of the lung as a filter for RSP is shown in Figure 8.1. Depending on their biological, physical, and chemical properties, these particles may dissolve and enter the bloodstream, wash out with the mucus, or be carried to the lung lymph nodes where they may become embedded for life.

Aerosols consist of liquid droplets suspended in air (American Chemical Society, 1974). Their lifetime depends on their size. Depending on temperature and humidity gradients, droplets may evaporate and form dry particulates, or may aggregate until they form drops. Aerosols with a size from 4–40 μm form visible fogs. Particle sizes from 0.05 to 0.5 μm make fumes. Outdoor aerosols often condense upon entering homes. Thus, the aerosol level is normally lower indoors than outdoors. However, certain humidifiers and certain cooling techniques can introduce aerosols into shelters. A thorough discussion of aerosols can be found in the specialized literature (NAS, 1978; Perera, 1979).

The word particulate denotes both organic and inorganic materials. Small filaments and particles up to 0.01 μm in size are called smoke; particles from 0.1 μm to the size of sand grains are called dust. Household dust contains a variety of biological, organic, and inorganic agents; these may include such diverse items as mites, viruses, bacteria, asbestos fibers, or human sebum. The chemical composition changes with the seasons (Takeushi et al., 1974). Highway dust is largely inorganic and contains a variety of elements. The eight most important metals and their concentrations are lead, 3400 ppm; copper,

2300 ppm; zinc, 2850 ppm; iron, 12,860 ppm; magnesium, 190 ppm; chromium, 150 ppm; nickel, 150 ppm; and manganese, 15 ppm (Davison et al., 1974). Urban particulates contain inorganic material from many industrial sources as well as organic materials. The latter include pesticides and such biological materials as pollen, bacteria, and viruses. Particulate concentrations are discussed in Section 7.2.

The most important sources of indoor particulates are wind-borne dust and tobacco smoke (Spedding, 1974). Windborne dust may contain natural inorganic dust and biological particles, including pollen, road dust, and pesticides. Its composition depends on climate and local geography. There are substantial differences between urban and rural dust (National Research Council, 1979; Higgins, 1979). Depending on particle size, dust may settle locally or may spread over the entire hemisphere. Sulfate levels are normally between 1–17 $\mu m/m^3$; lipophilic organics between 0.8 and 8 ppm; lead between 0.01 and 1 ppm; and metals between 0.1 and 0.4 ppm. The size of some particulates is shown in Figure 5.1. The National Research Council estimates that 9 million tons of particulates are emitted from various sources each year (Higgins, 1979).

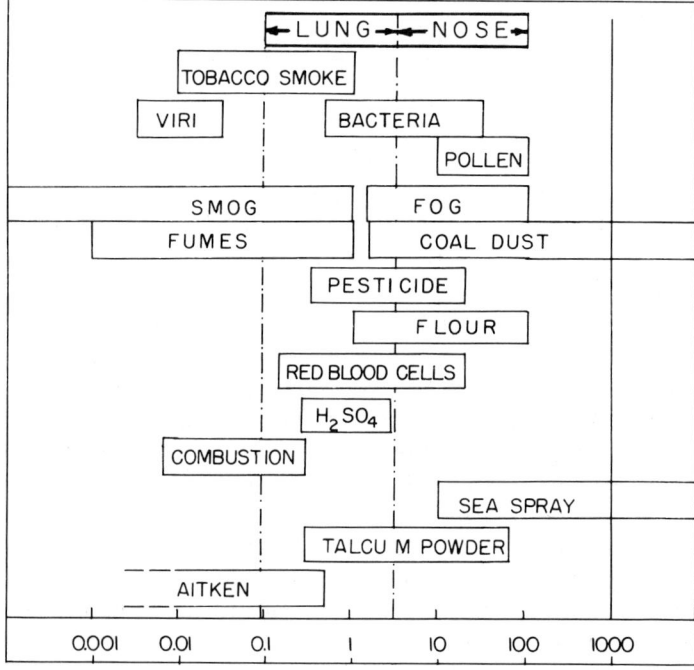

Figure 5.1. Particle sizes for particulates (in micrometers).

The most important industrial sources of respirable particles (in decreasing order of importance) are coal-fired electric power plants, paper mills, non-ferrous secondary smelters, cement plants, asphalt plants, lime plants, refineries, and oil-fired electric power plants. Automobiles, especially diesel engines, can cause substantial local accumulations of sulfate particulates and sulfuric acid aerosols, lead, nickel, cadmium, mercury, and other metals and organics. Among the organics are polycyclic aromatics similar to those present in soot and tobacco smoke.

Household dust (Joshi and Wanner, 1975; Moschandreas, 1978, 1979) acts as a carrier for microbes (Section 5.3) and outdoor pollutants such as pesticides (Davies et al., 1975). The combustion of dust in household appliances creates new toxic species (Satish and Wanner, 1975). House cleaning and vacuum cleaning drastically increase dust levels (Annis and Annis, 1973).

After aerial atomic explosions, particulates contain the radioactive isotope strontium 90.

The toxic effect of particulates depends on how deeply they penetrate the respiratory system, what fraction is retained, and how well the body can cope with the toxic agents. Particulates in the range of 0.5–5 μm are the most dangerous, because they will be retained deep in the lung. Smaller particles are normally exhaled. Larger particles are trapped in the larynx or the bronchial tubes before they reach the lung. The relationship between particle size and place of retention is shown in Figure 8.2. Particulates are often difficult to analyze, since they must be trapped on sticky surfaces, collected on quartz balances, or individually identified with an electron microscope; these tedious analytical techniques have delayed our understanding of them.

Particulates act synergistically with other pollutants because these small solids have enormous surface areas on which chemical reactions can be catalyzed by a variety of effects. Sulfur dioxide (Section 8.2.2) is but one example of a substance that has a more toxic effect in the presence of sulfate particles, soot, coal dust, tobacco smoke, and asbestos.

Asbestos. The term asbestos is used to designate actinolite, amosite, anthophyllite, crocidolite, tremolite, and chrysotile. These natural minerals form respirable fibers that can be readily identified under the electron microscope. Once an asbestos fiber is inhaled, it is trapped in the lung where it is stored in the lymph nodes. Asbestos exposure is cumulative, and high exposures cause cancer of the lung. Therefore, the use of asbestos has been severely restricted and friable asbestos products should be removed from buildings (Sawyer, 1979).

During the past 100 years, the use of asbestos has increased from 500 tons to 4 million tons worldwide. Asbestos is used to insulate everything from hair dryers to steel trusses against fire, and it is also used in the manufacture of building tiles and of shingles for roofing, and even as a road material. Further uses include brake linings in automobiles and for fire-protective garments. Since asbestos is very friable, it can easily enter the air. Asbestos concentrations in the air above natural deposits can reach 48 fibers/liter. Normal ambient air contains none. Water around natural deposits may contain up to 10^7 fibers/liter. Excellent reviews of asbestos are available (NRC, 1971; Chambre Syndicale de Paris, 1964; Sawyer, 1979). The primary exposure route is by inhalation. Typical integrated urban asbestos levels are on the order of 10^3 fibers/g of dry lung. Exposure to high levels causes acute asbestosis; pleural plaques, and eventually, mesotheliomas, an otherwise rare form of malignant tumor. The incubation time can be as long as 30 years. Synergism between asbestos and smoking is very strong.

The asbestos fibers cling to fabrics, which can cause cross contamination. Thus, family members of asbestos workers have a significantly increased body burden. The common method for identification of asbestos is direct observation of the characteristic fiber under the microscope.

Glass Fibers. Dust from glass insulation bats comprises particle sizes from 0.1 μm up. Thus, glass dust can be deposited in the lung. Glass dust contains a mixture of fibers and droplet-shaped particles. It is known that the droplet form is not as dangerous as the fibers. It is not yet known whether fibrous glass dust by itself is dangerous, or whether it poses a threat only in combination with other materials.

Tobacco Smoke. The most important source of indoor air particulates, other than house dust, is tobacco smoke. As with all forms of smoke, it consists of a mixture of solids, liquids, and dissolved gases. An average cigarette emits about 30 mg respirable particulates, a pipe also about 30 mg and a cigar about 68 mg (Repace, 1982). The composition of tobacco smoke is fairly well known (Stedman, 1968); it is quite similar to that of soot from wood fires, coal fires, kitchen grills, and diesel exhaust. High-pressure liquid chromatography is capable of separating out more than 100 different compounds in such mixtures. The best known component is benzo-α-pyrene (BaP), which is usually used as a measure of other aromatic components (Table 5.1). BaP is not carcinogenic, but the human metabolism transforms it into a highly carcinogenic form. Other components include ketones, quinone, aldehydes, and the dihydroxy derivative of polynuclear aromatic hydrocarbons (PAHs), as well as heterocyclic compounds. Indoor concentrations between 1.1 and 150 ppb have been measured by Bridbord (1975). The danger of such smoke to both the smoker and his companions is well documented (Chapter 8).

It is estimated that the chemical components of tobacco smoke include 27 different alkanes, 51 alkenes and alkynes, 96 aromatic hydrocarbons, 15 sterols, 24 alcohols, 272 esters, 45 aldehydes or ketones, 1 quinone, 30 nitriles, 57 acids, sulfur compounds, 55 phenols, 107 alkaloids, 14 amino acids, 17 inorganic salts, and other ashes. The particulate density is about 5×10^9 per milliliter, with an average size of 0.2 μm.

The composition of smoke depends on the tobacco leaf, the stuffing of the cigarette, the butt length, and the puff volume duration, pressure, and frequency. The tar content varies widely among different cigarettes; a typical range is between 4 and 30 mg. In filter cigarettes there is a large difference between the tar content of the mainstream smoke and of the unfiltered sidestream smoke that emanates from the burning leaves between puffs (Repace, 1982).

The gas phase smoke contains 73 mole % nitrogen, 10% CO_2, 4.2% CO, 1% H_2, and 0.5% methane, 0.3% acetone, and traces of other gases, among which methanol makes up 700 ppm; methyl ethyl ketone, 500 ppm; ammonia, 300 ppm; nitrogen dioxide, 250 ppm; methyl nitrite, 200 ppm, and acrolein, 150 ppm. Hydrogen sulfide and formaldehyde account for 30 ppm each. Nitrosamines are also present (Brunnemann, 1978). The metal content is in the order of micrograms.

The PAHs are an important indicator of the potential carcinogenicity of tobacco smoke. Some of these components are listed in Tables 5.1 and 8.1. Some 60 compounds with three or more fused rings are now known. Benzo-α-pyrene can be present in the leaf from atmospheric pollution during growth. Depending on the smoke temperature, the alkaloids thermally decompose and transform. The temperature ranges from 400–900°C.

Carbon monoxide (CO) concentrations from tobacco smoke can build up significantly. For example, in unventilated indoor arenas CO levels can exceed 10 ppm, the EPA ambient 8-hour level. Elliott (1975) and Selber (1977) measured levels up to 40 ppm; Fischer (Fischer et al., 1978) observed both CO and NO. The control of tobacco smoke is difficult because of resuspension and the high vapor pressure of the liquid and gaseous components on the particulate surfaces (Narasaki, 1976). The regulation of smoking is discussed in Chapter 10.

5.2.2 SULFUR DIOXIDE

Until 30 years ago sulfur dioxide (SO_2) from domestic coal combustion was the most obnoxious air pollutant in all urban and industrial areas. Sulfur dioxide from domestic and commercial coal burners was responsible for the dreadful London smogs that plagued that city between 1400 and 1960. It was also responsible for the infamous air pollution episodes in the Meuse Valley in Belgium, in Donora, WV, and in other places. Since domestic coal burning has virtually ceased, the main source of SO_2 is the combustion of coal in electric power plants. During the last 15 years urban SO_2 levels have dropped by a factor of 10 or more as a result of joint efforts by the electric utilities and the U.S. Environmental Protection Agency. Local sources of SO_2 are copper smelters, oil refineries, and domestic and industrial oil heating systems. Since SO_2 is heavier than air, most of it settles to the ground within a few miles of its source when it enters the natural sulfur cycle. On its path or in the soil the compound is oxidized to sulfate, a normal soil component. Sulfate is a vital plant nutrient and is artificially added to fertilizer compositions. However, some plants are

Table 5.1

Composition of Tobacco Smoke: Aromatic Fraction[a]

Compound	Concentration (ppm)
Carbon monoxide	42,000
Carbon dioxide	92,000
Methane, ethane, propane, butane, etc.	87,000
Acetylene, ethylene, propylene, etc.	31,000
Formaldehyde	30
Acetaldehyde	3200
Acrolein	150
Methanol	700
Acetone	1100
Methyl ethyl ketone	500
Ammonia	300
Nitrogen dioxide	250
Methyl nitrite	200
Hydrogen sulfide	40
Hydrogen cyanide	1600
Methyl chloride	1200

Major Classes	Number of Compounds
Alkanes	27
Alkenes and alkynes	51
Aromatic hydrocarbons	96
Sterols and oxygenated isoprenoid compounds	15
Alcohols	24
Esters	272
Aldehydes, ketones, and quinones	45
Nitriles, cyclic ethers, and sulfur compounds	30
Acids	57
Phenols and phenolic ethers	55
Alkaloids and other bases	107
Brown pigments (volatile bases and alkaloids)	11
Amino acids, proteins, related compounds	14
Miscellaneous components	
Inorganics	17
Agricultural chemicals (+ decomposition products)	12
Others	8
Total	841

[a] This fraction is usually identified by means of benzo-α-pyrene (BaP); from Stedman (1968).

sensitive to SO_2 (Meyer, 1977). During the last half century, tall stacks have been used to disperse the gas. In such cases, the SO_2 is carried beyond local and national boundaries, often for many hundreds of miles. During the trip airborne sulfate particulates form. The long-range transport of sulfur oxides is responsible for the acidic rain in Scandinavia and for similar effects observed in the United States and Canada (Meyer, 1977).

Indoor air levels of SO_2 are usually lower than outdoor levels (Figure 7.7) because SO_2 can be absorbed on many building materials. In fact, SO_2 can react with formaldehyde. Thus, it can serve as a scavenger on particleboard (Andersen, 1976).

Sulfur has an unstable oxidation state of $4+$ in SO_2. In the presence of air, it readily oxidizes to S^{+6} which converts to sulfate, which with calcium forms gypsum.

5.2.3 OZONE

Pure ground-level air contains no ozone. Ground-level ozone is the result of a combination of urban and industrial pollutants with sunlight. Ozone in indoor air is formed by fluorescent lights and by electrostatic copy machines. Normally, indoor air levels of ozone are lower than outdoor levels because ozone decomposes or reacts upon contact with many building materials (Section 7.2.4).

5.2.4 CARBON MONOXIDE

Incomplete combustion in fuel-rich flames produces carbon monoxide (CO). Concentrations above 50 ppm are poisonous because CO blocks oxygen respiration. Carbon monoxide is odorless and colorless, and is therefore an invisible silent killer. Carbon monoxide poisoning has been observed in parking garages, poorly ventilated school buses, police cars, automobiles with leaking exhaust pipes, poorly ventilated kitchens, near wood stoves, and almost everyplace where combustion occurs.

Ott (1980) has conducted a thorough study of CO and CO_2 and has shown that outdoor CO levels change rapidly and exhibit large local variations. For example, on a busy road CO can increase from 5 ppm to 100 ppm (Kahn et al., 1975; Godin et al., 1972; Ott, 1971, 1980) when automobiles stop at a traffic signal.

Carbon monoxide is one of the most ubiquitous indoor pollutants. It forms in poorly tuned heaters (White and Kragulski, 1972; Barrett et al., 1973; Hollowell, 1978), ranges and stoves, and garages (Yabroff, 1974). In ice rinks CO is produced by ice cleaning and finishing equipment and can reach significant levels (Breysse, 1965; Spengler et al., 1978; Koepsel, 1977).

5.2.5 LEAD

Lead has been smelted for over 4000 years. Lead pitchers poisoned millions of unsuspecting Romans who used lead 2000 years ago for cups and pitchers for water and wine. During the last century lead was used widely as white paint pigment. Since lead oxides and acetate have a sweet taste, unattended infants find it pleasant, nibble on it, and are poisoned. Therefore, the use of lead paint has been discontinued. During the last 50 years organic lead compounds have been used as antiknock additives for upgrading gasoline. The lead disperses with other combustion gases, and traces have been found in the blood of families living along freeways. In 1974 the Environmental Protection Agency completed a thorough evaluation (NRC, 1972; EPA, 1974, 1977; Bridbord, 1977; Boggess and Wixson, 1977) and added lead to the list of primary ambient air pollutants. Since 1974 all new automobiles have been built to operate with unleaded gasoline, and leaded gasoline is gradually being phased out. As a result, lead levels in air have diminished substantially during the last decade, except in indoor rifle ranges (Ganter et al., 1977), or in other special locations.

It has been speculated that the current per capita use of lead, 6 kg/year, is about the same as it was in Rome 2000 years ago. Currently, about 600,000 tons are mined in the United States; some 700,000 tons are recycled from lead batteries. Lead emission from smelters has been substantially reduced. Until 10 years ago printers were exposed to high lead levels. Today, newsprint contains only about 0.53 ppm lead. The Sunday color comic section contains about 0.72 ppm. This lead is released when the paper is burned in fireplaces or barbecue pits (Las and Blackwell, 1979). Chemical analysis of ice in Greenland showed that global air levels of lead increased from 0.001 ppm in 800 B.C. to 0.08 ppm in 1800. They have rapidly increased since; urban air now contains about 2 mg/m^3.

The human intake is approximately 0.45 mg/day; about a third is from the air and two thirds via food. A recent study (Rabinowitz et al., 1975) showed that about half of the absorbed lead is retained by the body. The average body burden varies from 80 to 200 mg depending on local pollution. Blood levels above 80 μg/ml correlate with symptoms of poisoning. Human bones contain about 200 mg with a residence time of 100 days; blood retains 2 mg for 35 days; soft tissue retains 0.6 mg of lead for 40 days; the rest is in the bile, pancreas, hair, and nails and is excreted with perspiration and urine. Lead poisoning can be recognized chemically by a decrease of the delta-aminolevulinic acid dehydratase, a key enzyme in porphyrin synthesis. Lead can be readily recognized by a spectrophotometric method described in Chapter 6.

The radioactive isotope lead 210 accounts for part of the natural background radiation. This isotope is discussed together with radon, Section 5.7.1.

5.2.6 NITROGEN OXIDES

Nitrogen is capable of forming several types of gaseous oxides. In nitric oxide, NO, nitrogen has an oxidation state of $+2$; in nitrogen dioxide, NO_2, it is $+4$, and in nitrous oxide, N_2O, it is $+1$. All these oxidation states are unstable relative to the fully ionized species N^{+5} or N^{-3}, which constitutes the rare gas electron configuration. Nitric oxide and nitrogen dioxide are found in very hot combustion flames and both are toxic (NAPCA, 1970). Nitrous oxide is not formed in normal combustion. This gas, called laughing gas, acts as a narcotic and is used in dental practice and surgical medicine to produce or maintain anesthesia.

Nitric oxide and nitrogen dioxide are formed as side products in hot oxygen-rich combustion flames of natural gas or of oil in kitchen stoves and heaters, and internal combustion engines, or by combustion of nitrogen-containing substances such as coal (EPA, 1973). Both are free radical compounds, very reactive, and can interconvert. This accounts for their catalytic action in gas mixtures. The concentration of NO_x in Los Angeles smog has been significantly reduced in recent years (Horie et al., 1977). The exposure levels of nitrogen oxide have been measured by Hollowell et al. (1976) and Macriss and Elkins (1977), Section 7.2.3.

5.2.7 OTHER AMBIENT AIR POLLUTANTS

Ambient air contains traces of many other gases. Carbon dioxide, a product of all combustion and of the human metabolism, is ubiquitous. Above 3 vol %, it acts as an asphyxiant, but normally it does not accumulate in air because it is absorbed by plants (Section 7.2.2).

Ammonia. Ammonia forms during putrefaction, and is found in human breath as well as in the air above forests. At high concentrations it is caustic and toxic. It is used as a component in cleaning agents and as a laundry additive. Ammonia dissolves readily in water and forms the ammonium ion and hydroxide. Household ammonia has a concentration of about 3 moles. Ammonia reacts with SO_2 and formaldehyde and is used as a fumigant to suppress the formaldehyde odor in mobile homes. The resulting hexamethylene tetramine is quite stable and inert, but this treatment is only a stopgap.

5.3 Microbes

In the context of this book the term microbe encompasses everything from bacteria and viruses to molds, fungi, spores, and pollen. Only species that can adversely affect health or welfare are discussed.

Wherever people go, they leave a trail of microbes, shed together with the sebum and spread by air. Microbes are present aboard aircraft and in subway stations, office buildings, homes, schools, hospitals, greenhouses, animal sheds — in fact, in every indoor space except some special, truly clean environmental chambers.

During the last century people were extremely conscious of the role of indoor air as a carrier of microbes because of the constant threat of pulmonary tuberculosis. Today, it is widely but incorrectly assumed that infections are transmitted only by personal contact, and the importance of airborne microbes is generally played down, even by hospital personnel, except by those suffering from allergic reactions. However, several recent epidemics of airborne infections in office buildings with low ventilation rates, and in buildings with poorly maintained air conditioner filters, have served as a reminder that airborne bacteria still are a constant threat (Pennington et al., 1966; Grieble et al., 1970; Rosenzweig, 1970; Scott and Jacobson, 1970; Hodges et al., 1974; Crowley, 1978; Rylander et al., 1978; Covelli, 1973). An episode of Legionnaire's disease in Philadelphia has been carefully investigated (Russenberger, 1974; Brunner, 1977; Thacker et al., 1978; Eickhoff, 1979; Broome and Fraser, 1979). Other airborne microbes discussed in Section 7.2.5.8 and 8.5.3, cause smallpox (Henderson, 1976), measles (Riley et al., 1978), pneumococcal pneumonia, pneumonic plague, foot and mouth disease, rhinovirus, adenovirus, respiratory virus infections (Banaszak et al., 1970; Fink et al., 1971; Flindt, 1969; Knight, 1973; Salvaggio and Karr, 1979; Granier et al., 1980), and many dozen allergic reactions. During the last 10 years some cases of Q fever (*Rickettsia burnetii*) transmitted by airborne microbes have been documented, one of them in Oakland, CA (Edmonds, 1979). Another half dozen airborne diseases are known (Gregory, 1973). Indoor microbes include not only viruses and bacteria, but also fungi (Schönborn and Winden, 1973; P.W. Smith, 1975; Tovey and Vandenberg, 1978).

Microbes may become airborne via several paths. In addition to the direct action of air turbulence, they may be activated by splash drops, drip-splash, and mist in air conditioners, air humidifiers, and air filters. Each human sneeze releases up to 40,000 droplets, with sizes of 5–100 μm, that travel at a speed of up to 40 m/sec (Jennison, 1942). Under these conditions the droplets travel about 1 m before they partly evaporate, forming particles of a size between 1 and 10 μm (Wells, 1955; Gregory, 1973) that remain airborne and viable for some time. Human sputum is loaded with microbes (Reznikov et al., 1971), as is vomitus (Eckmann et al., 1974).

The transport mechanisms for sneeze droplets and spores of this size have been well studied by aerobiologists (Seisaburo et al., 1959; Lewis et al., 1969). They may settle rapidly or travel for several dozen miles, depending on local air turbulence and prevailing winds. The deposition is enhanced by pressure gradients around trees and buildings (Spendlove, 1975), by the wind speed, and by the density and shape of the microbe.

The concentration of bacteria and molds has been studied for over 100 years. Miquel (1899) reported seasonal values in Paris. The lowest counts were around 200 microbes/m^3 in winter in the Montsouris park; the highest, near the city hall of Saint-Gervais, were 12,000 bacteria/m^3 and 3000 molds/m^3 in summer. Rain reduced the number to 10 or 20 microbes/m^3 within minutes. Two daily peaks were observed. Gregory confirmed the daily cycles in his studies of pollen and spores in England, and established the existence of a strong sedimentation effect. The counts decreased from 20,000 microbes/m^3 at 10 cm above ground and 19,000 microbes/m^3 at 30 cm to 7200 per cubic meter at 120 cm above ground. Airplane studies in the 1930s showed that *Puccinia graminis* uredospores can be caused by vertical air currents to 4000 m. Microbe levels in Japan have been reported by Kawarabayashi (1978).

The indoor fate of microbes is usually determined by the race between death and settling. However, air convection along walls and ceilings can be an efficient vehicle for resuspension of dust, and the microbes in it can cause irritation and allergies even in the inanimate state. Pollen such as the *Cladosporium* and ragweed pollen penetrate homes through cracks, windows, and doors as soon as the wind speed exceeds a mile or two per hour.

Microbes from damp walls and decaying moist timber can accumulate to levels of up to 360,000 spores/m^3 (Gregory, 1973). The prevailing indoor species, especially in winter, is the *Penicillium bacterium* (Levetin and Hurewitz, 1978). The air concentration depends on human activities. Walking across a floor, shaking beds, or brushing carpets will immediately increase the air concentration 20-fold or more. However, in a closed, evenly heated room microbes will settle, and indoor levels can drop to 10% or less of the outdoor count.

As mentioned earlier, air conditioners can serve as breeding grounds for bacteria. Other potential amplifiers are wall-to-wall carpeting (Rylander et al., 1974), dishwashers (Jopke and Hass, 1970), pets (Mantovani, 1978), water closets (Darlow and Bale, 1959), and house dust (Gip, 1966).

The main carriers for indoor microbes are the approximately 3 g of human skin scales that a person sheds each day in the form of particles with an average size of 8 μm (Speers et al., 1966; Sidorenko, 1967; May and Pomeroy, 1973; Dickgiesser, 1978). Such particles settle at a rate of approximately 2 mm/sec. How difficult it is to manage microbes is demonstrated by the fact that even hospitals have problems excluding infections from such clean rooms as burn centers and surgical theaters.

Allergens. Airborne allergens, the most famous of which is probably the *Ambrosia* pollen of ragweed, periodically plague about 15% of the population of the United States. It is not known yet how atopic people acquire their specific sensitivities, but it is firmly established that a single microspore of pollen suffices to cause violent reactions in sensitive individuals. Unfortunately, allergens remain active long after their microbial viability ends, until the proteins are thoroughly denatured. House dust, for example, remains active, even after the sugars are separated, until all amino acid has been hydrolyzed (Rimington et al., 1947). Hay fever (pollen rhinitis) occurs often in conjunction with eye irritations (conjunctivitis). Allergens have been extensively studied and reviewed (Newmark, 1968). The predominant species in North America are not the same as in Europe. The 20 most common pollens observed in New York City, together with their seasonal occurrence, are shown in Figure 5.2.

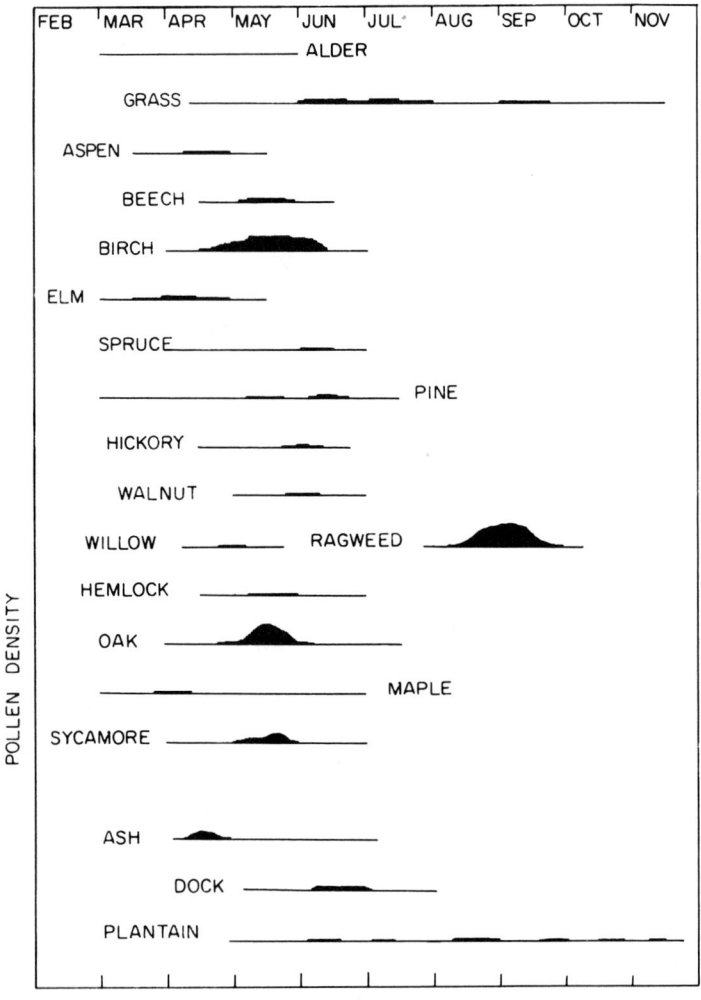

Figure 5.2. Seasonal frequency of 19 common pollens found in New York City (Edmonds, 1979).

Indoors, asthma can be caused by any one of a large number of microbes. The best known is *Aspergillus fumigatus*, a mold that grows over a temperature range greater than 20–50°C, and can infect people as well as birds and other pets. Other common fungal spores are *Cladosporium bulvum*, the dry rot *Serpula Merulius lacrymans*, the *Ustilago* smut, and the *Chaetomium*. The house mite produces *Dermatophagorides pteronyssinus*. When enzyme laundry detergents were marketed in the 1960s, some housewives suffered from the *Bacillus subtilis*. Smaller allergens, which penetrate more deeply into the lung, such as *Micropolyspora faerni*, can cause farmer's lung and other deadly infections. Allergies in old libraries were blamed on the fungus *Taxa*, but Burge (Burge et al., 1978) questions whether the assignment is correct.

Cattle feeding areas, chicken and turkey sheds, and flour mills generate allergen levels sufficient to cause infection among local office workers and housewives that is similar to that found normally only among cowboys and farmhands (Nikolov, Kambulin, and Kolupaeva, 1977).

5.4 Metals

Lead and mercury are the most plentiful metal contaminants, both in the body burden and in ambient air. Lead has been described in Section 5.2.5.

Mercury. Current indoor background concentrations of mercury are low (Foote, 1972) but it has long been known that mercury vapor and mercury compounds are poisonous; in fact, the widespread use of mercury to cure various human diseases was based on the assumption that diseased tissues were less resistant to poison than the healthy ones (i.e., that the infected tissues would be killed before the entire organism would die). It was also hoped that the poison would stimulate the formation of generally effective defenses. Thus, for several centuries mercury was one of the staples of doctors' bags, and as recently as 1940 mercury compounds were the only effective drugs available with which to combat syphilis. All but 5% of the mercury used in the United States is imported. The main sources of mercury in air are wastes and by-products. Until 10 years ago paper and chlorine producers discharged mercury waste into lakes. This practice has been virtually discontinued but the environmental burden will persist for many years. About 1 million lb of mercury in the air result from fuel combusion in the United States every year. Atmospheric mercury levels range from 0.6 mg/m^3 over the open ocean to 1200 mg/m^3 above mercury deposits. Urban areas contain about 5 mg/m^3 (NRC, 1978).

Nonoccupational exposure includes mercury vapors from mercury amalgam prepared in dental offices. The air concentration in such offices ranges from 0.01 to 0.18 mg/m^3 (Gronka, 1970). The upper limit was observed in places where careless handling caused mercury to accumulate in cracks in and below the floor. Similar concentrations can be observed in scientific laboratories when mercury spills are not promptly treated with elemental sulfur, which converts the metal into the more stable sulfide. High mercury levels can cause chronic poisoning.

Recently, the U.S. Department of Energy initiated research on fluorescent lamps for use in conventional Edison-type screw sockets in offices and homes. The most energy-efficient product emerging from this program so far is the Litek lamp, which would incorporate mercury as a starter. All experience in chemical laboratories indicates that if such a consumer device were introduced, mercury spills would be inevitable, and continued exposure to vapor from liquid trapped in cracks in floors would cause a significant health threat to people, especially in minimally ventilated energy-efficient homes.

Whereas organic mercury reaches people mainly through the food cycle, inorganic mercury affects them primarily via the respiratory tract. Once it has entered its host, mercury transfers via the circulatory system. It accumulates in the liver, kidney, and central nervous system. Organic mercury is retained longer than inorganic mercury. Mercury is excreted with the feces, urine, and bile. The effect of mercury poisoning is well known from the "mad hatters" of yesteryear, who worked in rooms saturated with mercury vapor.

Zinc. A correlation between zinc levels and cancer has been proposed by several independent researchers over the last 10 years, but it is quite possible that some other factors are responsible for the observed cancers and that high zinc levels are merely coincidental. According to Holleman, Ryon, and Hammons, (1980), zinc poisoning has never been conclusively established, whereas zinc deficiencies are well known to be harmful to both children and adults. Zinc is chemically determined by atomic absorption spectroscopy; it almost always occurs jointly with cadmium.

Cadmium. Occurring in association with zinc, copper, and lead, cadmium is used in electric batteries and fungicides, as well as in phosphorus and metal alloys. Cadmium air levels range from 0.001 to 0.350 $\mu g/m^3$. Cadmium is normally ingested with food, except by tobacco smokers, who inhale approximately 1 μg per pack of cigarettes. Welding and soldering activities can lead to inhalation of cadmium fumes. The respiratory retention efficiency is about 40%, much higher than the amount acquired from food.

Cadmium causes both acute and chronic illness. The most famous cadmium poisoning episode occurred among Japanese women, who developed symptoms that were called itai-itai disease. It has recently been suggested that the observed effects were not only due to cadmium but were enhanced by dietary problems.

Copper. Copper is ubiquitous in the urban environment and is essential to human health. Toxic effects are observed if the copper concentration in the body falls below or climbs above 80–120 mg. Copper deficiencies are rare but have been reported. Excess copper ingestion causes nausea, vomiting, and eventually death.

Magnesium. The human body contains an average of 20 g of magnesium; most of it is within cells. Neither deficiency nor excess are known to cause toxic effects.

Manganese. Ingestion of manganese seems to be harmless. However, chronic exposure to high manganese air levels can cause manganese pneumonia or bronchitis. Manganese is collected in the liver and excreted in the bile. A problem with manganese is that it could not be readily analyzed until 1975, when neutron activation was developed. Thus, our knowledge about manganese is limited.

Molybdenum. The environmental balance of molybdenum is fairly well known. Found in industrial smoke and oils, this element may be toxic to human beings. Sheep and cattle suffer weight loss and other ill effects from ingesting molybdenum. As in the case of manganese, our current knowledge is limited because until recently trace amounts of this element could not be confidently measured.

Selenium. The most toxic of the trace elements that are essential to human life, selenium is produced by industrial sources in amounts far less than those released by nature. The body contains an average of 15 mg. (Selenium is a component of the enzyme glutathione peroxidase.)

Administered as dust or vapor or in food, selenium has been shown to produce cancer in laboratory animals but no causal relationship with human cancer has been established. In fact, selenium may have some anticarcinogenic power (NRC, 1976).

It is possible that selenium will be found more frequently in indoor air because it is used as a component of electronic copying machines and in electronic calculators. It could become an important part of solar energy collectors. However, none of these applications cause alarming concentrations.

Tellurium. About 500 mg of this element are built into our bone, but tellurium is not found in normal indoor air.

Polonium. Polonium is viciously poisonous. A chemical relative of oxygen and sulfur, its effect on human beings is different from that of either, because all of its 27 different isotopes are radioactive.

Polonium is part of the uranium decay sequence, Figure 5.4, which also includes radon, a radioactive gas. Polonium accounts for much of the natural background radiation. Only minute traces are found in the normal environment but its radioactive decay is so intense that it can be readily observed.

Arsenic and Antimony. Both arsenic and antimony are moderately toxic. Arsenic occurs naturally and is found in the human body. It has been used in medicines and in poisons since antiquity, although it is no longer widely used. Solar energy collectors may reintroduce arsenic into our daily environment, but by and large, high arsenic exposure is restricted to the industrial environment, and to areas around the few remaining arsenic smelters. The greatest present-day use of arsenic is in pesticides and rodent poisons. Antimony poisoning is restricted to occasional episodes of inadvertent industrial fumes.

Thallium. Thallium forms viciously poisonous compounds and has been used as a rodenticide and ant killer. Early in this century thallium was still used to treat tuberculosis, but its use has been banned since 1972.

Chromium. This element and the other transition elements can occur in different chemical oxidation states. Trivalent chromium is less toxic then hexavalent chromium. The human body contains about 2 mg. The daily uptake is approximately 200 µg. Although widely used in industry, chromium does not seem to constitute a health threat to the nonoccupational population (NRC, 1974).

Cobalt. Cobalt dust in excess of 0.1 mg/m^3 is toxic, but traces of cobalt are essential to human life, since it forms part of vitamin B_{12}, cobalamin. If the daily intake drops below 0.0434 µg, death from pernicious anemia will result.

Nickel. The human body contains about 10 mg of nickel. Fumes containing more than 1 mg/m^3 are considered toxic (NRC, 1975). Occupational exposure to nickel can cause squamous carcinoma in the nose, similar to that observed in rats after prolonged and high exposure to formaldehyde (CIIT, 1980).

Vanadium, Titanium, Zirconium, and Niobium. The elements vanadium, titanium, zirconium, and niobium are all considered harmless in air, except as concentrated dusts.

Strontium. Except for the radioactive isotope strontium 90, which is one of the radioactive decay products resulting from atmospheric atomic explosions, strontium is harmless. The strontium-90 content of air is used as a measure of atmospheric radioactivity.

Beryllium. A relative of magnesium and calcium, the viciously toxic beryllium has a quite different impact on human life once it enters the bloodstream. The chief hazard of beryllium is by dust inhalation. Air levels above 0.01 µg/m^3 are considered dangerous to the general population. Beryllium inhalation is followed by a latent period, after which lung disease and chest pain and other cardiac insufficiencies develop. Rats and other laboratory animals develop cancer from beryllium inhalation. The health hazards of beryllium have greatly restricted its use in the metallurgical industry, where its light alloys would otherwise have value.

5.5 The Halogens

Fluorine. Fluorine is essential to human life. The fluorine level in drinking water should be close to 1 ppm; below that level bone deficiencies occur and above it bone malformation may occur. Water fluoridation programs have provided ample evidence that proper water fluoridation can more than double the resistance of teeth to decay.

Experience over the last 50 years has shown that uncontrolled fluoride emission from aluminum plants can burn vegetation, kill cattle, and cause illness to people but such uncontrolled exposures are no longer frequent. In the past it was extremely difficult to measure environmental fluorine levels (NRC, 1971) because a good analytical method was lacking.

Chlorine. Chlorine can occur in different forms, each of which has a different effect on man. The chloride ion in table salt is an essential dietary component. Elemental chlorine gas in concentrations above 1 ppm (3 mg/m^3) is poisonous. It is, however, used to sterilize drinking water; in the process it converts into harmless chloride but in contaminated water it may produce traces of volatile organic chlorocarbons (Chapter 7). Chlorine bleaches contain chlorine in an unstable oxidation state in which chlorine is poisonous and corrosive.

Bromine. A component of many drugs and agricultural chemicals, bromine has a biological half-life of about two weeks in man. Elemental bromine vapors are pungent and corrosive, but only chemical workers come in contact with them. It is noteworthy that not a single case of influenza has been reported among chemical workers in bromination plants, which have been in operation for more than 40 years.

Iodine. This element is vital for all animal life, since it is part of several hormones that control growth. Iodine is concentrated in the thyroid gland. Excess iodine is excreted. The vapor pressure of iodine is small. Except for isolated allergic reactions to alcoholic iodine solutions used as wound disinfectants, iodine has proven harmless to human beings.

5.6 Organics in Air

Indoor air contains several different organic compounds (Miksch et al., 1980; Jarke, 1979). Ambient rural air contains vapors, oils, and dusts of plants. In swampy areas putrefactive gases are common. Hydrogen sulfide and mercaptans are responsible for their odor, but they consist mainly of methane, carbon dioxide, and nitrogen. Urban air contains literally hundreds of different organics from refineries, solvents, gasoline, chemical plants, and many other sources. Only a small number can be identified in this section.

5.6.1 METHANE

Methane (CH_4) gas forms violently explosive mixtures with air. At high concentrations it acts as an asphyxiant. It is formed during the anaerobic decay of organic matter. It emanates from swamps and has been observed in the basement of homes built on abandoned cemeteries, but normally, methane is a component of natural gas used for heating or cooling buildings. It also occurs in coal mines, where federal mining laws require that the methane concentration be continuously monitored to avoid explosions. Since natural gas is odorless, traces of highly malodorous mercaptans are added to it when it is to be used for home heating and cooking so that leaks can be noticed long before the gas reaches dangerous concentrations.

5.6.2 ETHANE, PROPANE, BUTANE, OCTANE AND OTHER HYDROCARBONS

Volatile hydrocarbons are used in household spray cans in place of fluorocarbon propellants, which were recently banned. These chemicals must be used with caution, and only when room air is not recirculated, because they are highly flammable. Bottled propane and butane are used as fuel in vacation homes. Octane and heavier hydrocarbons are present in automobile fuels, and their fumes can reach substantial concentrations in residential areas around gasoline stations. Higher molecular weight hydrocarbons form waxes that are used to finish floors and polish furniture surfaces.

Hydrocarbon emission from oil refinery plants and the emission of incompletely burned hydrocarbons from automobile engines have become a major problem in urban areas, especially in cities where local geography hampers the movement of ground-level air. Even small concentrations of hydrocarbons are obnoxious in outdoor air, because sunlight induces photochemical reactions. The resulting products include aldehydes, ketones, organic acids, and many free radical intermediates that, even in small quantities, cause acute respiratory discomfort.

5.6.3 HALOCARBONS

Low molecular weight fluorocarbon compounds form very stable gases that have been used extensively as refrigerants, as propellants in the manufacture of cellular methane foam, and as propellants in spray cans whose contents range from deodorants and hairspray to pesticides and dust-removing agents.

5.6.4 AEROSOL SPRAYS

Since 1954 throwaway spray cans have made it easier to apply everything from deodorant, shaving cream, hair-setting lotion, surgical dressing, dog repellant, varnish, and window and oven cleaner to every type of pesticide. In 1974 Cote (Table 5.2) estimated that an average suburban household uses about 150 g of aerosol spray per week. This adds up to 15 lb of propellant per household per year, which enters the air and remains permanently airborne.

The inventor of the modern spray can was Erik Rotheim, who filed the first patent in his native Norway in 1929. He contemplated using low boiling point hydrocarbons and halocarbons as propellants for dispensing liquid soap, paint, insecticide, fire-extinguishing chemicals, and cosmetics. His preferred propellant was dimethyl ether. Some 15 years later L. D. Goodhue and W. N. Sullivan of the U.S. Department of Agriculture Bureau of Entomology developed the "bug bomb," of which 50 million were rushed into production

Table 5.2
Common Household Aerosols and Their Use in the United States in 1970[a]

Product Category	Households Using Product (%)	Frequency (uses/week)	When Used	Location	Average Usage Rate per Household (g/month)	Propellant Emission Estimate (g/use)
Deodorant spray	74	7	A.M.	Bathroom	112–140	1.0–7.2
Hairspray	71	3	A.M., P.M.	Bathroom	84–112	4.9–6.5
Shaving foam	45	7	A.M.	Bathroom	84–112	0.3–0.4
Air fresheners	26	1	A.M., P.M.	Throughout	28–56	5.6–11.2
Disinfectant sprays	63	3	A.M., P.M.	Kitchen, bathroom	112	7.5
Furniture polish sprays	84	1	A.M., P.M.	Living room, dining room	56	8.4
Dust sprays	18	1	A.M., P.M.	Living room, dining room	28–56	4.2–8.4
Oven cleaners	42	0.1	A.M., P.M.	Kitchen	84	20–25

[a] After Yocom (1973).

and distributed to U.S. troops in North Africa and in the Far East in 1942 to fight off insects. Their patent, issued in 1943, is based on Freon 12, dichlorodifluoromethane, which liquefies at a pressure of about 5 atm at 20°C. After the war, beer cans were modified and equipped with mass-produced plastic valves to gear up an industry that produced 6 billion cans a year at its peak in 1975. The lighter containers use Freon 11, trichlorofluoromethane, with a critical pressure of 1 atm at 30°C. The propellants are made from methane, chlorine gas, and hydrofluoric acid.

Since the chlorofluorocarbon propellants are chemically stable and not soluble in water or soil, they remain permanently in the air. Rowland, a chemistry professor at Irvine, California, realized in 1974 that these propellants would eventually have no place to go except into the stratosphere, where the sun would break their chlorine–carbon bond. The resultant chlorine radical would enter the natural upper-atmosphere chemistry, and catalyze decomposition of the earth's protective ozone layer. From laboratory experiments he predicted an 8% reduction in this layer by 1990.

Rarely has a single scientific paper had such immediate, significant impact (Dotto, 1979). The National Academy of Sciences, which had just formed several committees to challenge commercial supersonic air travel initiated by the French and English *Concorde* program, switched its emphasis from the study of stratospheric nitric oxide pollution from aircraft fuel (NRC, 1975) to that of pollution from aerosols; it issued three major reports, which found the risk double that estimated by Rowland and predicted a substantial increase in skin cancer. After a fierce but short battle, industry agreed to halt production of nonessential Freon, and prepared to switch to a butane–propane mixture with a pressure of about 3 atm (Sugden and West, 1980). Parallel with the curtailment of production of Freons, the use of chlorocarbon solvents (Section 7.2) was also greatly reduced, but the controversy still boils (NRC, 1982).

Table 5.3 shows the active ingredients of common sprays. Section 7.2 explains that indoor use of these sprays can result in acute toxic levels of many of the components.

5.6.5 PESTICIDES

Pesticides are chemicals that are meant to kill insects, rodents, and other pests.

If properly used, pesticides are a powerful tool for providing a healthy indoor habitat; if used indiscriminately they can cause acute and chronic suffering. The result depends on a careful choice of the proper chemical and dosage.

As late as 1939, only 30 chemical pesticides were registered in the United States. They were highly poisonous to man and mainly used by farmers and trappers. The situation changed abruptly during World War II with the introduction of DDT. Today several hundred pesticide formulations for indoor applications are available to homeowners and exterminators.

Table 5.3

Ingredients in Six Aerosol Spray Product Classes[a]

Housekeeping Product	Ingredients
Furniture polish	Dinitrobenzene, 1,1,1-trichloroethane, petroleum distillates, silicone, wax morpholine
Spot remover	Perchloroethylene
Oven cleaner	Sodium hydroxide, hydroxyethyl cellulose, polyoxyethylene fatty ethers
Drain cleaner	1,1,1-Trichloroethane, petroleum distillates
Lysol	o-phenylphenol, N-alkyl-N-ethyl morpholinium ethyl sulfates
Clorox	4-Chloro-2-cyclopentylphenol, diethanolamide-lauric acid amide
Tile cleaner	Tetrasodiumethylenediamine
Prewash treatment	Perchloroethylene, petroleum distillates
Window cleaners	Sodium nitrite, isopropyl alcohol, ethylene glycol, ammonium hydroxide
Disinfectant sprays	Triisopropanolamine morpholine
Air fresheners	Propylene glycol morpholine, ethanol
Personal use	
Deodorant spray	Hydrated aluminum chloride, isopropyl myristate talc, triglycerides
Hairspray	Vinyl acetate copolymer resins, polyvinyl-pyrolide resins, ethanol, lanolin
Shaving foams	Stearic acid, triethanol amine, menthol, glycerine
Paint sprays	
Protective coatings	Polyurethane
Stain	Hydrocarbons, ethers
Metallic plating	Hydrocarbons, resins
High-temperature coating	Hydrocarbons, resins
Craft spray	Petroleum distillates
Glass frosting	Toluol, xylol
Primer	Toluol
Adhesive	Petroleum distillates
Glaze	Toluol, xylol
Varnish	Petroleum distillates
Insecticides	
Wasp and hornet	o-isopropoxyphenylmethyl carbamate, 2,2-dichloro-vinyldimethyl phosphate, petroleum distillates
Ant and roach	Pyrethrins, piperonyl butoxide, n-octyl sulfoxide of isosafrole, petroleum distillates
Mosquito and fly	Dicarboxamide, petroleum distillates
Moth	Dicarboxamide, petroleum distillates
House and garden	Cyclopropane, petroleum distillates
General insect	Toluamide
Garden sprays	
Poison ivy control	Dichlorophenoxyacetic acid
Weather protector	Pinolene
Herbicide	Dichlorophenoxyacetic, propionic and benzoic acids
Pruning paint	Asphalt solids, petroleum distillates
Pet sprays	
Dog repellant	Paradichlorobenzene
Cat trainer	Paradichlorobenzene
Cat flea spray	Carboxymethylcarbamate

[a] From Yocom (1973).

The insecticidal power of DDT, first synthesized by Zeidler in 1874, was discovered in 1939 by Müller in Switzerland. It was extensively used during World War II and stopped a typhoid epidemic in Naples in 1943. In 1955 the World Health Organization started a program to eradicate several communicative diseases, among them malaria which during that year affected more than 200 million people. When DDT was used in Ceylon (to kill the mosquito *Anopheles* that transmits the protozoa Plasmodium) the rate of malaria infection dropped from 3 million in 1955 to a mere 17 cases in 1963. Similar success was also achieved everywhere against yellow fever (the vector of which is also a mosquito), trypanosomiasis, the so-called sleeping sickness (transmitted by the tsetse fly), the plague (carried by fleas that are hosted by rats), and, as mentioned above, against typhus (transmitted by the common body louse).

DDT proved equally effective in agriculture and forestry. As a result, its use increased rapidly, and in 1962 an estimated 650,000 tons of DDT were manufactured and used world-wide. The use of other pesticides grew equally fast. With such a large volume of applications, problems were bound to occur. The first was that some pests such as mites, scales, and aphids survived better than their natural predators, such as beetles. The second was that DDT residue persisted long beyond its intended use and rapidly became ubiquitous. DDT began to be found anywhere from the Antarctica to Greenland to the Olympic mountains. In some locations these residues caused massive fish kills and other problems vividly described in the popular book *Silent Spring* by Rachel Carson, published in 1962. It was soon discovered that DDT was stored in human body fat, and that large segments of the population carried up to 10 mg/kg body lipid weight. DDT was banned in the United States on December 31, 1972, only 24 years after Müller was honored with the Nobel Prize for discovering its effectiveness.

Today, we know that overexposure to some of the most effective pesticides can cause mutagenic (Kurzel, 1981), possibly carcinogenic, and other unexpected adverse health effects. The public sentiment against pesticides increased when it was learned that during the Vietnam war over 20 mil gal of agent orange and other herbicides were sprayed over 6000 sq miles to defoliate trees and crops, and that many millions of civilians and soldiers on both sides acquired a measurable body burden of these chemicals. Thus, in December 1980 San Jose, CA, banned all aerial spraying of pesticides locally and over adjoining cities. This caused the rapid spread of the Mediterranean fruit fly and much public debate, which ended with the decision that the risk involved in spraying malathion was preferable to economic loss. The levels used in this type of application are far lower than the levels of more poisonous substances that are used daily inside the home.

Although lead arsenate is no longer used — at a rate of 100 lb/acre per season — to spray apple orchards, lead and calcium arsenate, as well as arsenites, are still produced and used by the millions of pounds for pesticides.

Moreover, there are so many pesticides that only the most common agents can be discussed here. A detailed review of this subject can be found in the more specialized text of McEwen and Stephenson (1979). During the last fifteen years the goal has been to produce pesticides which after use rapidly convert into harmless end products.

5.6.5.1 Insecticides

Natural Products. One of the first agents found to act on the central nervous system was nicotine, introduced at the end of the 18th century. In 1940 1 million lb of it were used, but it has limited utility because it degrades rapidly in sunlight. The esters of chrysanthemum flowers have been used for the control of lice and fleas on household pets in Asia for more than a century, but the active agents, the pyrethrins, are also photosensitive. Furthermore, some insects can detoxify pyrethrins. However, the great difference between insect and mammal toxicity make these natural compounds attractive candidates for future development.

Elemental Sulfur. One of the most unusual pesticides, elemental sulfur has been used for several centuries as an insecticide against ticks, thrips, chiggers, and mites. It is nontoxic to mammals, and in fact has been used as a spring tonic for many centuries. This versatile agent can also be used as a fumigant, and it has recently been shown that ingested sulfur acts as an insect repellent in tropic regions (Meyer, 1977).

Chloroaromatic Compounds. A large number of insecticides contain chloroaromatic compounds. They are all quite specific, poorly water soluble, and highly fat soluble. Thus, they are all quite persistent in the environment and can accumulate in human adipose tissue (Meyers, 1981; Mes, 1982).

DDT. The best known member of the chloroaromatic class of compounds is clearly 1,1,1-trichloro-2,2-bis(chlorophenyl)ethane, or dichlorodiphenyl trichloroethane, which was first synthesized in 1873. Its low acute toxicity to man and plants precipitated a revolution in pest control. Since less than 1 ppm of DDT dissolves in water, this chemical remains effective for a long time. Thus, 1 g/m^2 on the floor of a hut protects its occupants against malaria for over a year. The toxicity ranges from 2 mg/kg for the housefly to 200 mg/kg for the rat. Its lipophilicity helps DDT to penetrate insect membranes but also accounts for its accumulation in human body fat.

In 1977 it was estimated that the average concentration in human adipose tissue was 7.88 mg/kg. The highest level reported was 1131 mg/kg for a worker in a pesticide formulation plant. Despite the high industrial exposure of the 1950s, no cancer cases have yet been demonstrated to have been caused by DDT, and DDT at the levels now current in the environment is not considered to be a significant hazard (Holleman et al., 1980). Because gas chromatography in conjunction with an electron capture detector can record as little as 10^{-12} g of DDT, it is one of the best studied pesticides.

The metabolites of DDT have been carefully studied. They depend on the host. DDT and related compounds such as DDD or Pethane and Dicotol reside in human tissues for many years. Thus, the use of this compound has been drastically cut back. The structure of this class of compounds is shown in Figure 5.3.

Methoxychlor. 1,1,1-Trichloro-2,2-*bis*(*p*-methoxyphenyl)ethane is less persistent and is widely used in dairy barns.

Chlordane. The cyclodiene chlordane forms a brown liquid of which only 5 ppm dissolve in water. Accordingly, it persists for up to 5 years in soil or in timber. It is highly effective against wood pests such as carpenter ants and termites. Almost 25 million lb were used for the control of such pests in 1975 in the United States, but its manufacture here was suspended in 1979. Technical-grade chlordane is a mixture of 70% cis and 25% trans isomer. Chlordane has been formulated as dust, granules, and wettable powders. It has a fairly high vapor pressure—10^{-5} atm at 20°C. Savage (1975) measured chlordane levels in crawl spaces and around houses equipped with forced hot air heating systems and detected 5 pg/ml in blood samples of volunteers. Chlordane toxicity is greatly enhanced by impurities that are present in the technical-grade pesticide, among which are heptachlor and oxychlordane.

Heptachlor. 1,4,5,6,7,8,8-Heptachloro-3,4,7,7-tetrahydro-4,7-methano-indene, a waxy solid, was first found as an impurity in chlordane. It has a lower vapor pressure than chlordane is far more effective than it. Heptachlor makes an excellent and persistent wood pesticide but it can enter the food cycle and appears in milk. It was banned in Canada in 1969. In the United States it is used for the control of fire ants.

Aldrin; dieldrin. Highly effective against ants and termites, aldrin and dieldrin both accumulate in the food cycle and human body fat. The production of both chemicals has been discontinued in the United States.

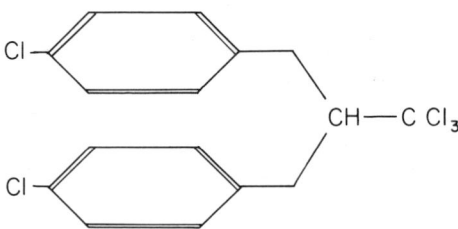

Figure 5.3. Structure of DDT.

Endrin. An isomer of dieldrin, endrin is the most toxic of all commercial cyclodienes. Thus, its use on potatoes, tobacco, and cotton has been discontinued. It has been used as an insecticide as well as a rodenticide, and is highly toxic to fish.

Mirex. Dodecachlorooctahydro-1,3,4-metheno-2H-cyclobuta[c,d]pentalene, or mirex, contains perchlorated contiguous carbon rings. Oxidation of mirex forms kepone.

Kepone. Chlorodecone, the pentalen-2-one keto analogue of mirex, is highly effective and selective against ants and cockroaches. In fact, mirex and kepone were believed to be the ideal answer to these pests until it was discovered that kepone can cause cancer in mice, adversely affects reproduction in laboratory animals, is highly persistent and accumulates in human adipose tissue. Furthermore, several workers in a kepone plant in Hopewell, VA, suffered severe and debilitating illness. When kepone was discovered in the James River and in the Chesapeake Bay, fishing in those areas was banned.

Endosulfan. Endosulfan ($C_{11}O_3SCl_6$), oxidizes readily. In soil it is metabolized to the sulfate. It is only moderately toxic to mammals but very toxic to fish. It has been used to eliminate trash fish before restocking.

Toxaphene. Probably the most widely used of the chlorinated terpenes, toxaphene is effective on cotton but is toxic to fish. Its toxicity is 100 mg/kg for rats.

Organophosphorus Compounds. Organophosphorus insecticides are esters of phosphoric acid. They do not accumulate in human tissue; their potential danger is that they inhibit the function of such enzymes as cholinesterase. Some of the most widely used compounds are the following.

Parathion. O,O-diethyl-O-nitrophenyl phosphorothionate, or parathion, is probably the most widely used compound in this class. Its action is based on its conversion to paraoxon, an anticholinesterase. In 1971 some 15 million lb were produced, but the pesticide has lost its effectiveness against the cotton pest, which has become resistant to it.

Methyl Parathion; Diazinon. Methyl parathion is used against beetles, whereas diazinon is used to control ants and cockroaches; both are formulated as sprays and dusts. Their toxicity is about 60 mg/kg for rats.

Azinphos methyl. Marketed under the trade name Guthion, azinphos methyl is used on vegetables.

Malathion. Slightly soluble in water (145 ppm), malathion is effective against mites and is widely used around the home because of its very low toxicity to mammals (2000 mg/kg for rats). The toxicity in mammals is low because malathion is readily degraded by carboxyl esterases, enzymes that are lacking in insects. This chemical was used in summer and fall of 1981 to spray about 800 square miles of several suburban counties around San Jose, CA, in an effort to contain an outbreak of the Mediterranean fruit fly. It was applied at a rate of 12 ounces per acre, which corresponds to about a dozen 500 micron drops per square foot. This amount is sufficient to guarantee a more than 90% kill of insects if malathion is combined with Staley's sex hormone, which attracts the insects and ensures exposure. Once it has been sprayed, malathion decomposes within five days and becomes harmless. The chosen exposure level was less than a quarter of that routinely used in agricultural areas, and only a fraction of the indoor levels produced if widely available household insecticide sprays are used. Despite the lower than normal exposure level, protests by a toxicology professor and several political groups influenced the city council to ban any aerial spraying. When the federal government overruled the local government the infected area had grown to over 800 square miles. Eventually more than 50,000 gal of malathion were sprayed from helicopters and airplanes. This compares with 10 gal which would have been sufficient to contain the infestation in the affected neighborhoods when the spraying had first been planned.

TEPP. bis-*O,O*-diethylphosphoric anhydride, the first organophosphate insecticide, persists for only one day. Thus, it can be used on crops just before harvesting.

Dichlorvos. 2,2-Dichlorovinyl dimethyl phosphate is widely used indoors. Its toxicity is fairly high but it rapidly metabolizes in mammals. This compound is used in flea collars for dogs and cats and in many veterinary preparations. In the 1960s it was sold as an indoor fly killer in the form of impregnated resin strips from which it vaporized at a steady rate. (The trade name Vapona was used in household products.) Use of the strips has been deemphasized because some people are hypersensitive to them, and others questioned the prudence of life with a pesticide vapor in the habitat. Currently, the U.S. EPA investigates oncogenic, mutagenic, teratogenic, fetotoxic, and neurotoxic effects (Repace, 1982).

Naled; Ethion; Fenthion. The trade name of naled is Dibron; it is closely related to dichlorvos, and is used in greenhouses and mushroom caves. Its far less soluble relative ethion is used in sprays to control mites. Fenthion is used to control fruit flies and cattle parasites.

Disulfoton. One of a group of organophosphorus compounds that enter the systemic system of plants and are thus effective against leaf-eating insects, *O,O*-diethyl-*S*-[2(ethylthio)ethyl]phosphorodithioate is used in the soil. It is quite toxic, as are the related compounds mevinphos, dimethoate, and dicrotophos. Crufomate and ronnel are less toxic and are applied to cattle or added to the feed to fight cattle grub.

Carbamates. The carbamates, another group of highly effective insecticides, act as inhibitors of acetylcholinesterase. This class of insecticide is in the midst of development and its long-range effects are not yet known.

Carbaryl. 1-Naphthyl-*N*-methyl carbamate is the oldest member of the carbonate group. It decays within a few days after application. Other similar compounds are carbofuran, methiocarb, metalkamate, and propoxur. The last is slightly soluble in water (0.2%) and widely used in homes to fight flies, mosquitoes, ants, and cockroaches. It has an activity of several weeks if used indoors.

In future new classes of insecticides are likely to evolve. Among the most likely to become commercial are the hormone mimics, antimetabolites, pheromones, and microbial insecticides.

5.6.5.2 *Fumigants*

The oldest and safest fumigant is burning elemental sulfur which yields sulfur dioxide. According to Homer, Ulysses used it to fumigate his home when he returned from his travels. Sulfur has been used for centuries to fumigate oak barrels and to sterilize wine. Most other fumigants are highly toxic and nonselective vapors that penetrate structures or objects. Fumigants must be carefully contained. Thus, fumigants should be used only by trained and licensed exterminators. Among the currently used fumigants are hydrogen cyanide, methylbromide, carbon disulfide, chloropierin, ethylene bromide, and aluminum phosphide. They are used for termite control, rodent control, and insect control of buildings, food and plants. In the United States some 180 million lb are used each year.

Benzene Hexachloride. Benzene hexachloride (BHC) has the formula $C_6H_6Cl_6$. Thus, it is a derivative of cyclohexane and not of benzene. The most active isomer is commonly known as lindane. This highly volatile, odorless compound has been extensively used in household applications. Because of its solubility, lindane does not seem to accumulate in the body, and is excreted more rapidly than other chlorocarbons.

5.6.5.3 *Rodenticides*

Rodenticide is the term generally used to denote poisons for small animals that are perceived to be or experienced as pests around the house, even though many such animals are not rodents. Control of mice, rats, squirrels, skunks, gophers, groundhogs, and even rabbits or larger animals can be achieved by chemicals. However, most of these chemicals are also poisonous to the human species and house pets. Thus, the use of rodenticides is restricted by law almost everywhere.

Use and storage of rodenticides around the home are extremely hazardous because there are dozens of ways in which they can contaminate food, clothing, or places to which children or pets have access.

Two types of poison are commercially available: anticoagulants and direct poisons. Among the coagulants, warfarin and diphacinone are frequently used. Both are solids, and so do not contaminate the air. Among the direct poisons, strychnine and red squill are widely used. Both are of botanical origin and both are solids that do not normally enter the air. One of the most toxic rodenticides is sodium fluoroacetate, a white powder that does not normally occur in air. Fenthion, mentioned earlier as an insecticide, is used by pest control professionals for pigeon control around public buildings.

5.6.5.4 Herbicides

Herbicides are widely used in agriculture and around the home garden. They are also used to kill weeds in the crawl space under buildings. 4,6-Dinitro-O-cresol (DNOC), introduced in 1935, was the first highly selective herbicide. It was used for killing broadleaf weeds in grasses.

Phenoxyalkanoic acids, such as picloram (agent white), were introduced after World War II. They are not used indoors. Benzoic acid herbicides, uracils, dinitroanilines, and triazines are all used in large quantities but are also unlikely to enter indoor air. The chloroacetamines have low toxicity for mammals but can cause skin and eye irritation. Thiocarbamates are low in toxicity and metabolize readily. This was proven by the fact that ^{14}C-labeled thiocarbamates cause rats to exhale $^{14}CO_2$.

Diquat; Paraquat. Both diquat and paraquat are bipyridylium. They act as nonselective contact poisons and have been used to destroy existing vegetation as a basis for "zero tillage" agriculture. Several human fatalities have been observed from ingestion of paraquat liquid. It causes fibrosis of the lungs. At normal levels it is not toxic, but it may cause irritation to the skin. In a recent government program paraquat has been sprayed on marijuana crops to discourage their illegal sale.

2,4-Dichlorophenoxyacetic Acid (2,4-D). Early recognized as highly toxic to broad-leaved plants but not perceptibly toxic to mammals, 2,4-D has increased in use from 1400 lb in 1950 to 40 million lb in 1962, when it was introduced as a defoliant of jungle vegetation and crops in the Vietnam War. No adverse health effects have been recorded to result from standard application levels.

2,4,5-Trichlorophenoxyacetic Acid (2,4,5-T). Successfully used as a herbicide for many years, commercial grades of 2,4,5-T were then found to contain about 10^{-3} wt. % of 2,3,7,8-tetrachloro-dibenzo-*p*-dioxin (TCDD). Therefore, its use was restricted in 1971 and has been essentially halted by the EPA since 1979.

*2,3,7,8 Tetrachloro-dibenzo-*p*-dioxin (TCDD).* TCDD, or dioxin, was never used by itself as a herbicide but it occurs as an intermediate and inadvertent impurity in 2,4,5-T. The lethal dose is 0.6 mg/kg for the guinea pig, which makes it one of the most toxic substances known. It also causes skin eruptions. Laboratory animals exposed to it have experienced fetal death and teratogenic effects. The women in Alsea, OR, manifested a three-fold increase in miscarriages after a national forest nearby was sprayed with TCDD in 1970.

In an arena in eastern Missouri, 57 horses died after the turf was sprayed with oil heavily contaminated with TCDD in 1972; and in July 1976 an explosion in a pesticide plant in Seveso, Italy, released 2 kg of TCDD over an area of 5 km × 700 m (Figure 7.16); 1100 animals died and 37 cases of acute chloracne were reported among children. The interpretation of potential teratogenic and carcinogenic effects is not yet conclusive. The follow-up study is far from complete (Pocchiari et al., 1979).

Agent Orange. A 50:50 mixture of 2,4-D and 2,4,5-T, agent orange is named for the color coding of the 55-gal drums in which it was shipped to Vietnam. Over the period from 1962 to 1971, 11 million gal of agent orange were sprayed, along with an additional 7 million gal of agents purple (a mixture of the butyl esters of 2,4-D and 2,4,5-T), green and pink (isobutyl esters of 2,4,5-T), and white (picloram) and blue (cacodylic acid), over an area of 3.6 million acres. The total herbicide use corresponded to about 400 lb of dioxin in the form of impurities of the 2,4,5-T component. The agent was sprayed undiluted at a rate of 3 gal/acre, corresponding to 1 g/m^2. Core areas were sprayed up to four times. Over the entire period, some 4.2 million American soldiers came into contact with the herbicide; those who loaded the herbicide onto the spraying planes were exposed to very heavy doses, and some suffered acute chloracne. Tests of the body fat of veterans showed that many retained a body burden of up to 60 ppt of TCDD. The follow-up study of this large-scale exposure is likely to take many years (Meyers, 1981; USVA, 1982).

5.6.5.5 *Fungicides*

Fungicides are widely used in the indoor environment. They are employed to treat timber, wood products, and structural parts of buildings. These compounds can be applied directly, but are often incorporated into paints and varnishes as well.

Elemental Sulfur. Elemental sulfur is the oldest and most widely used agricultural fungicide. Reduced sulfur is phytotoxic to plants (Meyer, 1977). Elemental sulfur acts as a latent fungicide. It is insoluble in water and only slowly oxidizes in soil to sulfate. It is not toxic to man or mammals. Sulfur is currently not widely used around buildings.

Copper Sulfate. Copper sulfate has been used as a phytotoxic wood preservative for 200 years. Its solubility is 32%, and it has a toxicity (LD_{50}) of 300 mg/kg to rats. Copper has recently been replaced by dithiocarbamates.

Mercury. Both oxidation states of mercury have been used in inorganic fungicides. The lower oxidation state is less toxic to mammals. Organomercurial compounds became available around 1915. These compounds have a reduced phytotoxicity but retain their full fungicidal and bactericidal power. Mercury compounds are very persistent; they accumulate in living tissues and even 0.5 ppm is sufficient to cause permanent liver damage.

Many mercurials are volatile and irritating to the respiratory tract. Mercury-based fungicides are no longer used in North America because they can enter the food cycle and have caused serious episodes of illness and death.

Organotin Compounds. Since 1950 organotin compounds have been used to treat textiles. Triphenyl tin acetate, fentin acetate, and triphenyl tin hydroxide are the best established compounds. Their toxicity is not yet sufficiently known.

Phenols. Pentachlorophenol (PCP) has been widely used to preserve wood. Many formulations designed for both professional and home use are available for painting or spraying onto wood. PCP is long lasting in wood but degrades rapidly in sunlight or in the soil.

PCP is toxic to humans. Furthermore, commercial PCP contains traces of TCDD (tetrachloro-dibenzo-*p*-dioxin) and of other dioxins, which are highly toxic. An accident with PCP in connection with a laundry detergent caused two deaths in St. Louis in 1969 (Barthel et al., 1969). The metabolism of PCP has been repeatedly investigated, and several analytical methods are available to establish the presence and concentration of PCP (Mes, 1982).

This chemical has been found in the air in log cabins in Kentucky in concentrations up to 0.4 $\mu g/m^3$ of air. The logs had been treated with a 5% PCP solution in 1975. In 1980 the blood serum of 30 log cabin dwellers had a geometric mean PCP concentration seven times that of control individuals living in conventional homes (Falk, 1980).

The more effective and more toxic 4,6-dinitro-*o*-cresol (DNOC) is not used for wood treatment.

Dithiocarbamates. Two groups of dithiocarbamates are used. One group comprises ethylene-*bis*-dithiocarbamates, as incorporated into Nabam, Zineb, Macreb, Metimar, and similar products. The mammal toxicity is low but they produce ethylenethiourea (ETU), which can cause goiter. The other group of dithiocarbamates includes Ferbam, Ziram, and Thiram. Their toxicity is low.

Captan. The phthalimide captan is degraded in the plant cell to thiophosgene, which inhibits enzymes. Its toxicity to mammals is very low. The phthalimides also include folpet and captafol. The latter has acute LD_{50} of 5000 mg/kg but it can cause skin sensitization.

Benomyl. Benomyl is one of the modern fungicides that act systemically.

5.7 Ionizing Radiation

The human body is constantly exposed to cosmic, terrestrial, and internal radiation. Cosmic radiation, including neutrons, constitutes 50 mrem/year and accounts for 40% of the annual body burden; terrestrial radiation contributes around 40%; and internal irradiation, including ^{40}K, ^{226}Ra, ^{228}Ra,

^{210}Pb, ^{14}C, and ^{222}Rn absorbed into the bloodstream accounts for the remaining 20%.

The earth's surface continuously receives ionizing radiation from the sun and other sources. Radiation enters the body by three pathways: inhalation of radioactive dust and gases, surface radiation by gamma and beta rays, and ingestion of radioactive material in liquid and solid food. The focus of this book is on airborne radioactivity from gases and suspended particulates which upon inhalation may increase the internal radiative body burden. Airborne radiation can be due to natural sources, or due to nuclear explosions.

5.7.1 COSMIC RADIATION.

Cosmic radiation is highly penetrating, with energies of up to 10^{19} eV. It penetrates all building materials and enters any building except deeply buried shelters. Thus, cosmic radiation cannot be controlled in the normal indoor environment. When cosmic radiation hits the earth's atmosphere, it consists of 79% protons; 20% helium nuclei; 0.8% carbon, nitrogen, and oxygen nuclei; and 0.2% heavier nuclei. In the atmosphere this radiation collides with oxygen and nitrogen and produces electrons, neutrons, mesons, and x rays. Only about 0.05% of the primary electrons penetrate to sea level, but secondary electrons make up 20% of sea-level radiation, with mesons and neutrons accounting for almost all the rest. Cosmic radiation at sea level depends on the latitude. It ranges from about 25 mrad/year at the equator to about 70 mrad at the poles. At 40° north latitude it is about 50 mrad/year. The intensity doubles with every 5000 ft of altitude in the atmosphere. Passengers and crews of jumbo jets at an altitude of 10 km experience about 100-fold increased radiation exposure, about 1 mrem/hr, while SST occupants experience about 1.6 mrem/hr (Gesell and Prichard, 1975), but these levels are still very low and harmless. Martell (1977, 1981) pointed out that radiation can form Aitken particles and condensation nuclei that settle to the ground and form respirable dust.

5.7.2 TERRESTRIAL RADIATION

The nature and intensity of ground level radiation vary over a wide range. They depend on the local geological structure and soil. Every isotope of every element with an atomic number greater than 83 (antimony) is radioactive. Of the about 70 radioactive substances found on earth, radon 222 gas is the most respirable radioactive source. Radon may contribute as much as 30 mrad of the total dose (George, 1980). This noble gas has a half-life of 3.8 days, and is a progeny of radium 226 (Figure 5.4). The most important decay series are those of ^{238}U and ^{232}Th. The composition of radiation in air depends on whether the original substance travels with its daughters, or whether they get separated

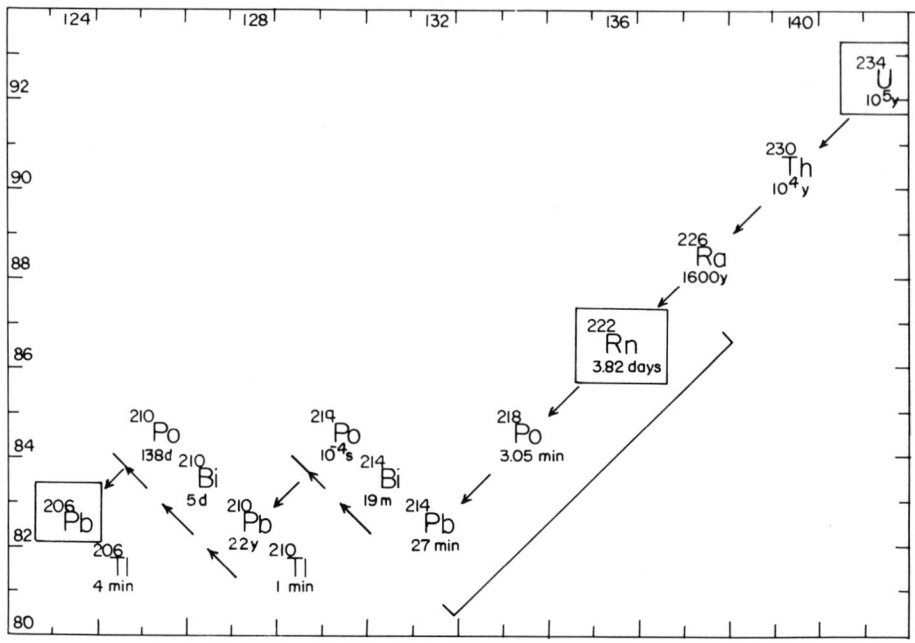

Figure 5.4. Decay series of the uranium isotope.

along the way. Since radium 226 is relatively mobile, its decay series is often observed separately from uranium 238. Its daughter radon 222, originally called "radium emanation," is yet more mobile, since it is a gas and readily separates from its mother, yielding its own independent decay series:

$$^{222}\text{Rn} \rightarrow {}^{218}\text{Po} \rightarrow {}^{214}\text{Pb} \rightarrow {}^{214}\text{Bi} \rightarrow {}^{214}\text{Po} \rightarrow {}^{210}\text{Pb} \qquad (5.1)$$

The steps are all shorter than four days down to lead 210, which has a half-life of 22.3 years. As a result, radon gas rapidly forms Po, Pb, and Bi (Figure 5.5). The radiation from a radon sample changes qualitatively as well as quantitatively as the decay products accumulate. For example, a 3-min-old sample contains 83% Po, 17% Pb, and 0.5% Bi; after 10 min the composition is 54% Po, 42% Pb, and 4% Bi, and after 3 hrs the equilibrium mixture contains 10% Po, 51% Pb, and 28% Bi. The common measure for radioactivity is the "working level," (wl), defined as any combination of the short-lived radon daughters in 1 liter of air at STP that will result in an emission of 1.3×10^5 MeV of alpha-particle energy (Table 5.4). The radiation from a fresh radon source, measured in working levels, increases in the manner shown in Figure 5.6 until equilibrium

5.7 Ionizing Radiation

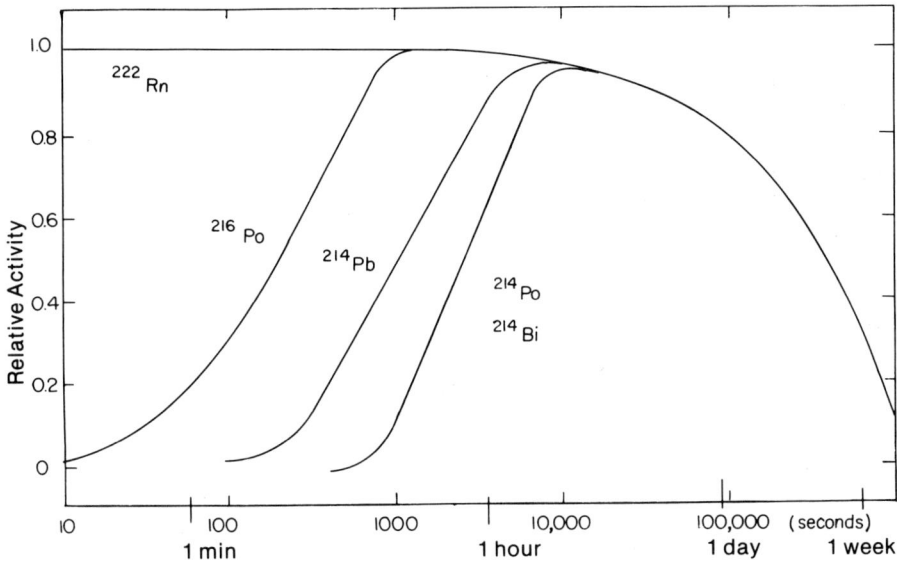

Figure 5.5. Formation of radon daughters from gaseous ^{222}Rn. (after Collé, 1980).

Table 5.4

Definition of Working Level (WL) Units[a]

Nucleide	α-Particle Energy (MeV)	Half-life	Number of Atoms per 100 pCi	Ultimate α-Particle Energy per Atom (MeV)	Total Ultimate α-Particle Energy[b] per 100 pCi (MeV)	Total α-Particle Energy (%)
^{222}Rn	5.490	3.8235 days	1.763(10^6)	Excluded	—	—
^{218}Po	6.003	3.05 months	9.768(10^2)	6.003 +7.687	0.1337	10.4
^{214}Pb	0	26.8 months	8.583(10^3)	7.687	0.6598	51.4
^{214}Bi	0	19.9 months	6.374(10^3)	7.687	0.4900	38.2
^{214}Po	7.687	1.643(10^{-4}) sec	8.770(10^{-4})	7.687	0.0000	0.0
Rn and all daughters:					1.2835	100.0

[a] From Collé (1980).
[b] Value in table is to be multiplied by 10^{-5}.

has been reached. An excellent review of the physics of radon and its progeny has been provided by Collé (1980).

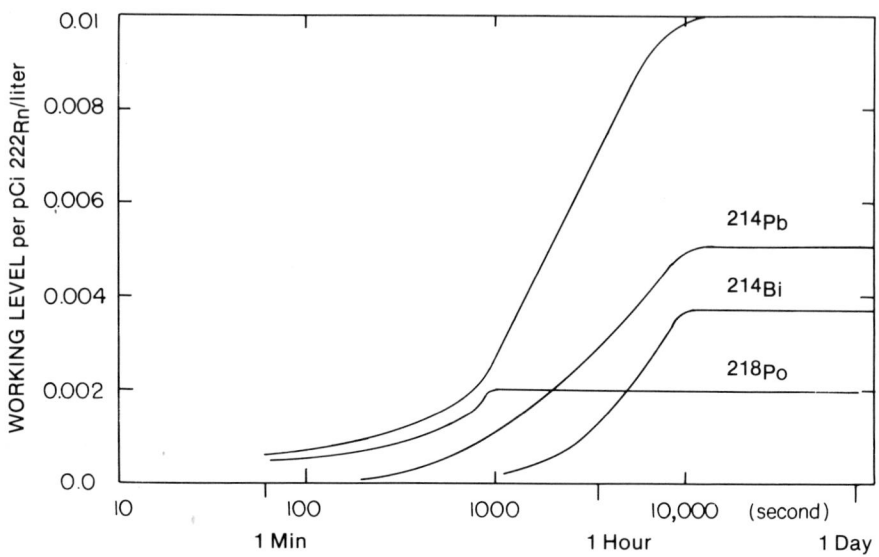

Figure 5.6. Growth of the working level for initially pure ^{222}Rn (after Collé, 1980).

The main sources of radon are coal, which may contain up to 0.1% uranium; natural gas, which has an average activity of 37 pCi/liter at the well head; liquid petroleum gas, which may have up to 20 pCi/liter; uranium milling wastes, with 10–30 pCi/liter; and phosphate mining and reclaimed land with comparable levels. Normal soil contains about 0.1–3 pCi/liter. Phosphate fertilizer contains 1–60 pCi/g (Collé, 1980; Gesell and Prichard, 1975). When open strip-mining pits were first reclaimed in the early 1960s and converted into attractive parks and housing developments, often at great expense, it was not realized that doing so could create a potential radiation hazard in unventilated basements.

The radon concentration depends on the composition of the soil. Its release increases during periods of low atmospheric pressure (Pohl, 1975). Anderson (1954) observed 25 years ago that radon air levels were proportionate to particulate levels. If radon decays, the resulting solid polonium 216 immediately converts to a lead-212 ion, which because of its charge is rapidly absorbed on atmospheric dust; the latter, depending on size, may serve as a respirable "hot" particle. Normal ambient air contains about 5×10^{-14} Ci/liter. One cubic foot of air yields enough radon and its daughters to produce about 200 alpha decays per minute. The energy spectrum is characteristic and can be unambiguously identified.

Radon is a gas nine times heavier than air; it remains close to the ground, emanating and diffusing into the stagnant air of wells, caves, or the crawl space under houses. In open spaces radon is mixed with atmospheric air and flushed into the troposphere, from which the daughter products settle randomly.

High indoor radon levels have been identified in Florida and in Montana, in homes located on uranium rich soil, and in some parts of New England where groundwater contains high radon concentrations. This was accidentally discovered several years ago by a student testing a Geiger counter that he had received as a Christmas present. Radon levels in 228 deep wells in Maine contain up to 50 pCi/liter, and average 5×10^{-3} nCi/m³, with a maximum of 9×10^{-1} nCi/m³ (Hess et al., 1979). These levels are up to 100-fold higher than the federal standards (Section 10.2). Hess estimated that more than 80% of all deep wells had radon concentrations that were above safe levels, but more than 95% of the radon is evaporated when the water is aerated and boiled. A significant effort has been made to identify spots with excessive radiation throughout North America, but a systematic survey has not yet been undertaken. Some representative radon air levels are shown in Figure 5.7.

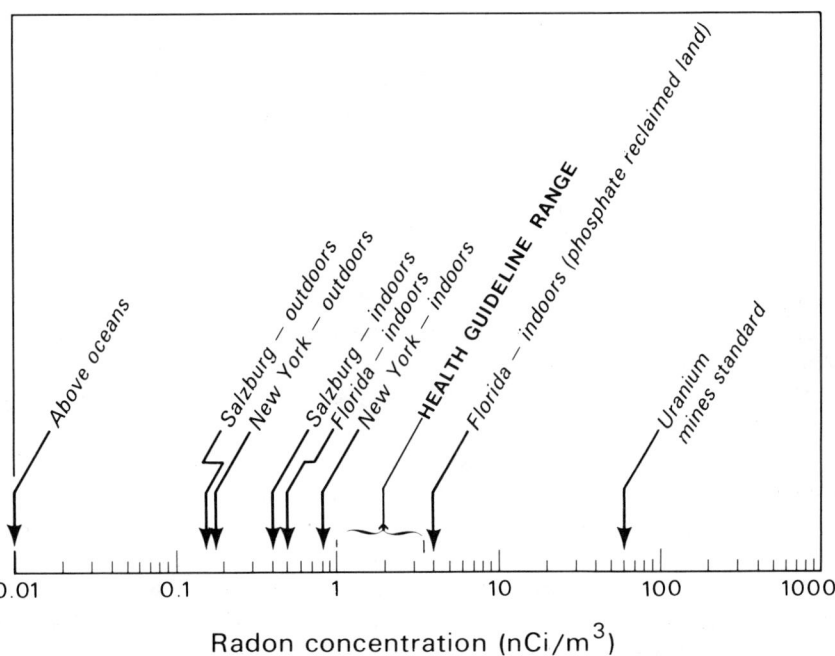

Figure 5.7. Radon concentrations in air in 10 locations, and current health guidelines (Hollowell et al., 1980).

The geometric mean of the annual average of 21 residences within 30 miles of Manhattan were followed over 2 years and very carefully evaluated by George (1980). The radon concentration in the basements averaged 1.7 pCi/liter; on the first floor the value was 0.83 nCi/m^3, four times higher than ambient air in the same place. Similar differences have been found in other places (George, 1980; Rundo, 1980; May, 1980). The mean particle size was about 0.12 μm, yielding an occupant dose of 50–420 mrad.

Sweden was the first country where high indoor Rn levels from building materials were recognized (Hultqvist, 1956). High soil levels and indoor levels have been observed in Uranium City, a city of 3000 on Lake Attabaska, Saskatchewan. Typical high values ranged up to 1 rad/year of gamma radiation, 111 pCi/liter of Rn, and 0.3 working level of Rn daughters. Control measures reduced the values to 24 pCi/liter or 0.03 working level (AECB, Canada, 1978). Very high levels are also found on monazite sand in Brazil, where street levels of 0.2 mrad/hr were measured in the city of Guarapair in the State of Espirito Santo. In the village of Minas Gerais levels of up to 12 rad/year were found. In Kerala State in southern India more than 100,000 natives live and sleep on the bare sand, which contains 0.1% thorium, yielding an annual exposure ranging from 0.2 to more than 2 rad/year. These levels compare with an annual whole-body exposure limit of 0.5 rad/year and a lifetime exposure limit to the gonads of 5 rad set by the U.S. government in 1960. The latter value corresponds to 0.17 mrad/year, based on a 30-year reproduction life cycle. Only in 1960 was it realized that closer to home some 600 of the tens of thousand homes in the Grand Junction and Durango areas of Colorado that contain uranium mill tailings in the walls also emitted radiation at a level that now requires many millions of dollars of restoration work to reduce to a safe amount (Chapter 10).

The radiation level in hot springs and spas (such as Bad Gastein, Austria, and Radium Hot Spring, British Columbia, Canada), which advertised their radiation at the turn of the century as a health bonus, is usually above 10^{-10} Ci/ml, while the groundwater in the vicinity is at the normal level of 10^{-16} Ci/ml. The radium level at the bottom of oceans is about 1000 times higher than at the surface.

The main source of indoor radon is uranium rock. Natural uranium 238 is found almost in every substance; even photographic film contains 0.2–1 ppm by weight (Eisenbud, 1963). Limestone, shale, and sandstone contain about 1.2 ppm but igneous rock contains an average of 4 ppm. Phosphate rock contains between 20 and 120 ppm, bituminous shale between 50 and 80 ppm. The corresponding radium 226 levels are a factor of 1 million lower, but this is enough to yield measurably increased Rn levels, and with modern recovery methods enough to produce reactor fuel.

5.7.3 NUCLEAR EXPLOSIONS

The public's awareness of atmospheric radiation began in the summer of 1945 when the first nuclear bomb was used in Japan to end World War II. It peaked in 1963 when the second temporary moratorium on atmospheric nuclear testing was reached between the United States and the Soviet Union. Over 350 devices were exploded during this period, and radiation levels in the air worldwide increased far above the natural level. Extensive field experience has shown that the danger is not primarily by inhalation (Eisenbud, 1963) but by long-term exposure to ^{90}Sr, mainly through the ingestion of contaminated food.

A nuclear explosion produces almost 200 isotopes of 35 elements. About 15% of the energy that is released in a burst of 10^{-8} sec produces ionizing radiation. The major atmospheric products are 1 Ci/megaton of tritium, ^{3}H, with a half-life of 12.3 years; 3.4×10^4 Ci/megaton of ^{14}C, with a half-life of 5000 years; and 59 Ci/megaton of ^{39}Ar, with a half-life of 260 years. The fallout contains about 3.6×10^3 Ci/megaton of ^{239}Pu, with a half-life of 24,000 years, and 10^7 Ci/megaton of ^{90}Sr, with a half-life of 28 years.

During the 3-year moratorium between 1958 and 1961 it was established that ^{90}Sr spread via the atmosphere in a belt circling the earth from 80° north latitude to 50° south latitude with the center over 25°–45° north latitude. Strontium fallout on the ground was directly proportionate to the rainfall. Strontium 90 was found worldwide in cow milk and reached levels of up to 20 pCi/liter during active atmospheric testing. This corrsponds to about 70 mCi/square mile, or about 0.1 mCi/acre of grassland and meadows. It is unlikely that much of this fallout was resuspended as respirable dust. The same is true for cesium 137 and plutonium 239, which are readily absorbed by soil. Iodine 131 is more volatile and constitutes a more serious hazard but its main impact is also on the milk supply.

The radiative yield of a nuclear explosion depends on the power of the device, the altitude and latitude at which it is set off, the amount of soil and water that are involved in the immediate neutron flux, and many other factors, especially the local meteorological conditions at the place where the fallout occurs.

Libby estimated that a 1-megaton bomb produces about 3.2×10^{26} carbon atoms—about 500 moles or 6 kg. Thus, over the 20-year test period man-made carbon 14 accounted for about 1200 tons, which entered the carbon cycle for an average lifetime of 8000 years and accumulates in living tissues and enters the food cycle. The current whole-body dose is estimated to about 1 mrad/year. Indoor radiation levels are not significantly smaller than ambient levels.

The highest fallout ever recorded in the United States was after a strong local rainstorm at Troy, NY, in April 1953, 2000 miles away from and two days after a relatively small explosion in Nevada.

5.8 Odors and Stale Air

Except for the scents of flowers and a few perfumes, people perceive most odors as unpleasant. Even food odors are regarded as undesirable except in the immediate vicinity of the dining table at the appropriate mealtime. As explained in earlier sections, the human metabolism can be monitored by CO_2 production. If carbon dioxide exceeds 0.1%, body odors become noticeable, as Pettenkofer (1873) discovered in 1867. In 1937 Yaglou started a series of landmark experiments in which he showed that with a ventilation rate of up to 30 ft^3/min, the odor in a room was proportional to the number of people present and inversely proportional to the ventilation rate. He found that a satisfactory ventilation rate for suppressing the odor of sedentary people was about 70 P/R, where P is the number of people and R the ventilation rate in cubic feet per hour. For children and active adults a higher rate is needed. Since then ventilation rates have been determined on the basis of odor rejection. Recent work by ASHRAE and others (Duffee et al., 1980; Duffee and Jann, 1981; Cain, 1981) points toward a return to greatly reduced ventilation rates. Since the performance of ventilation systems is dependent on maintenance and other human factors, this regressive move is bound to lead to serious odor problems in coming years.

From Linnaeus (1756) to Swaardemaker (1895) and Amoore (1970), authors have classified odors in groups. A typical seven-class comparison would involve categories such as etheral, camphorous, musky, minty, pungent, and putrid. The strength of odors varies not only with individual sensitivity, but also among different molecular groups. Table 5.5 shows that 500,000 times more formic acid has to be used to simulate the same odor intensity sensation as dimethyl sulfide.

Table 5.5

Matching Odor Standard Set[a]

Category	Concentration (ppm)	Substance
Etheral	800	Ethylene chloride
Camphorous	10	1,8-Cineole
Musky	1	Pentadecanolacetone
Floral	300	Phenylmethylethyl carbinol
Minty	6	Menthone
Pungent	50,000	Formic acid
Putrid	0.1	Dimethyldisulfide

[a]From Amoore (1970).

5.8 Odors and Stale Air

The correlation of smell with molecular structure (Amoore, 1970; MacLeod, 1980) is still a riddle. Furthermore, the sensation of smell is not always proportionate to odor concentration, and for a given system the perception of odor decreases during exposure to a constant level of a soure. Furthermore, a smell that is attractive to one person may repel another. However, a recent study by the National Research Council (1979) on odors arrived at the conclusion that odors do not affect health adversely (Chapter 8).

The malodorous metabolic components of sweat are normally about 0.03% urea and 0.001% butyric acid. Staphylococci-infected feet may cause a high skin hydrogen ion concentration (pH) and lead to bromidrosis, which is accompanied by penetrating odor. Diabetic patients excrete acetone; cystitis is characterized by ammonia odors; and stale urine smells of urinoid. Certain foods and several diseases impart strong characteristic odors to urine and feces. Thus, asparagus produces methyl mercaptan. Bacterial action in the intestine produces skatole and indole, which are also components of the flatus.

Food and cooking odors can be adsorbed on carpets, wallpaper, furniture, and clothing, and impart lasting odors that may be modified by partial oxidation or interaction with the fabric. Boiled meat releases dimethylsulfide $(CH_3)_2S$, boiled cabbage dimethyldisulfide, $(CH_3)_2S_2$. Both are formed from cystein. Similar breakdown products from the sulfur in proteins are also responsible for the characteristic odor of cheese and the characteristic smells of pork, fowl, and venison. Fresh butter smells from diacetyl, $(CH_3-CO)_2$, while cheese derives its odor from $CH_3-CO-CHOH-CH_3$, acetyl methyl carbinol. Some ripe cheeses may contain several parts per thousand of the latter. Cheese also may smell from beta-methyl mercaptopropionaldehyde, which at high concentration is reminiscent of raw pumpkin.

Sulfur compounds are important sources of odor. They form a very important component of many fragrances, but are also responsible for the rotten-egg smell of hydrogen sulfide. Raw onions smell from diallyl sulfide, $(C_3H_5-S)_2$, and garlic from allyl propyl disulfide. The close relationship between pleasant and repulsive smells is also demonstrated by chocolate, which contains traces of indole, identified earlier as a fecal odor. The smell of roasted coffee is a very complex blend. Depending on the roasting technique, the relative concentration of the components can be influenced to produce highly characteristic mixes. Burnt coffee beans smell (pungently) from allyl aldehyde, CH_2CHCHO, formed by the reduction of caffeine.

In industrial areas indoor air quality can be degraded by outdoor odors. Fish processors in coastal areas and meat processors release sulfur compounds from degraded albumin and amines as well as from some aldehydes. The osmogenic mixture is often released with steam and includes trimethylamine and hydrogen sulfide, as well as methyl and other mercaptans. Deodorization can be achieved by oxidation.

Tallow consists of palmitates, stearate, glycerol, and oleates, which upon heating form acrolein.

The odor from oil refineries is primarily an economic problem. Evaporation from storage tanks at terminals and gas stations includes the full mix of volatile hydrocarbons (Section 5.6.2). The refinery trains emit both amines and sulfur compounds, but sometimes measurable quantities of aromatics, including benzene, as well.

Paper mills may release methanol, methyl mercaptan, propanone, dimethyl sulfide, 2-methylfuran, alpha-pinene, 2-methyl furan, and ethylene. It is very difficult to fully deodorize effluents, which under certain weather conditions can be detected seveal hundred miles downwind.

Chemical solvent processors release an entire spectrum of hydrocarbons, chlorocarbons, alcohols, acetates, and ketones.

A very difficult problem can arise with the low but persistent odor in municipal water supplies, which may contain excess chlorine or other components, such as the very persistent hydrogen sulfide smell found some years ago in Santa Barbara, CA, and other communities.

If stored in large quantities, even fruit can smell offensive. The odor of ripening bananas can be suppressed with an inert gas mixed with 0.2 mg/m^3 of ozone.

Table 5.6

Olfactory Threshold and Maximum Allowable Concentration (MAC) for 12 Chemicals with High Threshold[a]

Vapor	Olfactory Threshold	MAC[b] (mg/m^3)	Ratio
Acrolein	35	0.25	140
Camphor	100	2	50
Dioxane	620	360	1.7
Hydrogen selenide	10	0.2	50
Methanol	7800	260	30
Methylformate	5000	250	20
Methylene glycol	190	80	2.4
Ozone (latest proposal)	0.2	0.05	4
Gasoline	3300	200	16.5
Sulfur dioxide	79	13	6.1
Trichlorethylene	1350	520	2.6

[a]From Summer (1971).
[b]Maximum allowable concentration in the United Kingdom.

6. Air Monitoring and Analysis

The tools for monitoring and measuring pollutants are in a state of rapid evolution. Instrumental techniques for the chemical analysis of many air impurities are now five or six orders of magnitude more sensitive than they were five years ago. For example, it is now possible to analyze chemically and measure 10^{-10} g or less of pesticides, halocarbons, or metals, and 10^{-12} parts per trillion (ppt) of such substances in body tissue or breath. (See Appendix I for units). Furthermore, much of the equipment used for these analyses is intrinsically more reliable than ever; the data can be read on digital electronic meters seconds after the air has been sampled, and so days of tedious laboratory wet chemistry and calculations can be avoided. Other equipment has been simplified to self-contained pocket-sized units on which chemical concentrations are instantly displayed electronically. This makes it possible to keep a running total of field measurements of, say, carbon monoxide or sulfur dioxide while simultaneously reviewing the time averages, standard deviations, and other trends. On the other hand, the measurement of seemingly trivial chemicals such as CO_2 and of some conceptually simple quantities—such as humidity, which influences every other physical, chemical, and biological property of air—remains tedious and subject to large systematic and instrumental errors.

Finally, air quality involves measurement of comfort and odor, two quantities that are not yet accessible to instrumental measurement and must be determined from the purely personal judgments of panels of observers who must use lengthy questionnaires to develop data by consensus.

6.1 Standard Methods

This book is not an analytical treatise; it is assumed that interested readers are familiar with recent textbooks or will consult such sources.

Although much past and current research on indoor air quality has been excellent, the available data must be considered exploratory because there are no recognized or validated standard methods or procedures for monitoring and analyzing residential and nonoccupational indoor air. Current techniques are either research techniques or standard methods developed for other uses. They derive from methods that were developed for monitoring ambient outdoor air, for source control, or for enforcing standards developed for monitoring much higher, occupational levels of work-related contaminants. The methods that come closest to those suitable for normal indoor air are those developed for confined atmospheres in space labs or submarines (USAF, 1978, USN, 1964).

It will be several years before standard indoor reference methods are developed because the process involves round-robin testing of techniques by people in the academic community, industry, and government. Both the EPA and NIOSH have developed exemplary state-of-the-art procedures for devising, reviewing, evaluating, and improving protocols, methods, and measurements, and for the quality assurance of analytical methods based on procedures established by the National Academy.

6.2 Equipment for Monitoring Physical Properties of Air

6.2.1 THERMAL COMFORT MEASUREMENTS

Personal comfort depends on the transfer rate of heat and moisture, that is a combination of temperature, humidity, and air movement (Section 3.1). About five years ago direct-reading comfort meters became commercially available; they display all three factors in an integrated manner. A Danish instrument consists of an ellipsoidal sensor that simulates the human body. An electronic calculator can be used to apply connections for clothing and activity level. A small commercial table-top analyzer with an electronic evaluator has been described by Marx and Schlüter (1980). However, these instruments are not yet sufficiently advanced to replace panels of human observers.

6.2.2 TEMPERATURE

The federally mandated temperature control program in effect from 1979 to February 1981 in the United States amply demonstrated the difficulty of monitoring temperature. Inside walls, air, and exterior walls are always at different temperatures. In winter and during direct insolation, especially in summer, the differences can easily be 30°F (Figure 4.4). Furthermore, the temperature of the air changes between heating and cooling cycles (Figures 4.7 and 4.8, pp. 76-82), and on the floor is usually 1–10°F lower than that at the breathing level. These temperature differences cause conductive, convective, and radiative gradients, all of which contribute to the observed, as well as perceived, pollutant levels.

Temperature measurements can be made by a variety of methods. The most accurate device is a gas thermometer calibrated against the melting point of a series of standard materials. The best absolute and relative readings are obtained if a calibration curve is drawn over a sufficient range so that the room temperature regions can be derived by interpolation. Chemists measure narrow temperature intervals with a metastatic Beckmann mercury thermometer, which can be read to within ±0.01°C once it has been set. Depending on their material age, and quality, thermocouples can be used to read temperatures to within ±0.1 to ±15°C. Resistance thermometers can read to within ±0.02 to ±5°C.

Glass-stem thermometers can detect temperature differences quite accurately, but laboratory thermometers frequently are not calibrated well and may be off by up to 2°C, a difference that is noticeable to many occupants of thermostated rooms. Conventional room thermometers are not reliable, and do not read radiative heat. Furthermore, a quarter-inch glass thermometer takes 4 min to equilibrate at an airflow rate of 20 ft/min. (The principle of radiant temperature is discussed in Section 3.2.2, p. 41.)

6.2.3 HUMIDITY

Measurements of humidity are difficult and not yet reliable. The best method would be to capture moisture from a measured volume in a cryogenic trap and measure the weight difference of the trapped air, but this is not convenient. All other methods depend on instruments that need frequent recalibration. The gravimetric piezoelectric meter equipped with a moisture absorbant yields values that are accurate to within ±0.1 to ±2%, depending on its construction and maintenance. The electrolytic hygrometer absorbs moisture, and the electric current needed to electrolyze the water is measured; the accuracy is ±5%. The Pope and the Dunmore hygrometers use an ion-exchange resin and lithium chloride, respectively, to absorb moisture and measure change in the electrical impedance of the probe; their accuracy is ±3%. The dew point hygrometer measures the temperature of the condensation point to within ±0.2 to ±1°C. Readings must be corrected for barometric pressure. The cheapest, but also least reliable and least accurate, instruments are based on dimensional changes, that is swelling, of human hair, nylon, dacron, wood, or other organic materials. Carbon sensors combine the hygroscopic effect with measurement of the electrical resistance.

The psychrometer is equipped with two temperature sensor bulbs. One is covered with a wet wick or sock, and room air is moved relative to the device at a rate of 700 ft/min (4m/sec) until both thermometers reach equilibrium. Psychrometric tables are used to compute the humidity.

6.2.4 AIR VELOCITY

Measurements of air velocity are always based on pressure measurements. The absolute air pressure can be determined with a mercury macrometer (which has an intrinsic accuracy of about ±0.02 Torr), a diaphragm gage (±0.05 Torr), or a pressure transducer (±0.1 to ±0.5%). Differential pressures can be sensed with a draft gage to within ±0.01 Torr, a macrometer to within ±0.1 Torr, a micromanometer to within ±0.001 Torr, or a Bourdon gage to within ±0.1%.

The monitoring of indoor heating, ventilation, and air-conditioning parameters is regularly reviewed by ASHRAE, which has sponsored basic as well as applied research for almost a hundred years and participates in drafting private as well as government regulations. The state of the art is summarized in a series of handbooks by ASHRAE dealing with fundamentals as well as equipment. Many books cover heating and ventilation engineering and design (McQuiston and Parker, 1977); systematic research on heating, infiltration, and ventilation has been conducted by several research groups in the U.S. and abroad. Research for DOE has been conducted by Hollowell and co-workers at Berkeley since 1975. Airflow metering has been reviewed by Ower and Parkhurst (1977), Coblentz and Achenback (1963), and Davies and Moore (1964).

Hospital ventilation is a well-established field (Cain and Hollowell, 1981, and Luciano, 1977).

6.3 Measurement of Air Pollutants

The many excellent reviews of air monitoring and analysis procedures could fill a library. During the last 10 years, the emphasis has shifted from wet-chemistry laboratory techniques to real-time, computer interfaced instrumentation based on the physicochemical properties of the contaminant. Only a short and arbitrary selection of references can be listed here. Quality assurance is the subject of several EPA and NIOSH manuals; calibration methods and accuracy have been reviewed by Belanger (1978) and LaFleur (1976), respectively.

6.3.1 AMBIENT AIR

Ambient air quality monitoring sites are listed in an EPA directory (1972). Source testing was described by Meinke and Scribner (1967), Cooper and Rossano (1971), Brenchley et al. (1973), and Maxwell et al. (1977). Analysis of air pollution has been summarized in books by Leithe (1971), a series of volumes by the Instrument Society of America (Scales, 1972), books by Butcher and Charlson (1972) and Leadbetter and Corn (1972), a Noyes data book (1972), and books by the Mitre Corporation (1973), Sittig (1974), and Stevens and Herget (1974). Instrumental methods have been reviewed by Willard et al. (1974) and in a series of volumes by the Lawrence Berkeley Laboratory (1973; 1975; 1976); they have been treated by Wang (1975, 1976) and in the perennial classic by Stern (1976), as well as by Noll and Davis (1976), Noll and Miller (1977), Perry and Young (1977), and Cheremisinoff and Morresi (1978). Remote sensing has been described by Gjessing (1978).

Official manuals have been prepared by many private and governmental organizations, among them NASA (1969), the American Public Health Association (APHA, 1977), the EPA (1974), the World Meteorological Organization (WMO, 1974), the Lawrence Berkeley Lab for the U.S. Department of Energy and the National Science Foundation (1973; 1975; 1976), the American Society for Testing and Materials (ASTM, 1976, 1978), the World Health Organization (Katz, 1969), the American Conference of Governmental Industrial Hygienists (ACGIH, 1978), NIOSH (1977, 1978, 1980), the International Council on Scientific Unions (1975), and the Air Pollution Control Association (APCA, 1975). Excellent reviews have been provided by the World Health Organization (Katz, 1969), the World Meteorological Association (Pack, 1977), and Katz (1980).

Methods for individual pollutants are included in monographs published by the National Academy of Sciences and the National Research Council, and official government procedures are published in handbooks and updated in the *Federal Register*. Standard AFGL trace gas frequencies for atmospheric infrared analysis of ambient NO, SO_2, NH_3, formaldehyde, and several inorganic acids have been updated by Rothman et al. (1981).

6.3.2 INDOOR AIR

As mentioned above all current indoor air testing methods derive from experience and efforts relating to confined environments and work environments. The confined environments of spacecraft and submarines have been analyzed by standard military methods derived from the ACGIH, NASA, U.S. Navy and Air Force, and U.S. Federal Aviation Administration procedures and technology. Valuable monitoring methods were devised as a result of the extended dive of Piccard's *Ben Franklin* (Abeles, 1970). A portable unit was described by Bartels and Crump (1977); exhaled air analysis was reported by Nefedov et al. (1969).

Industrial air measurement techniques, protocols, and procedures have been developed by ASHRAE (Dravnieks, 1977), ACGIH, and NIOSH. NIOSH has an extensive program for analytical procedures that follows a well-established evaluation procedure and includes backup data on sampling that provide extensive validation for all recommended methods.

Although analytical and sampling methods for the work environment are, by and large, reliable and mature, they were specifically developed for measuring the upper range at which toxic substances are tolerable at work for a healthy adult group exposed for five 8-hour days a week. Thus, they are optimized for reliability at levels 10 times higher than those considered safe for a general indoor population, which always includes children, the elderly, and people whose health is poor, all of whom spend most of their time indoors without the daily 16-hour recovery periods available to workers.

Procedures devised specifically for offices and homes have been explored by several government contractors that have acquired extensive experience in this field. Much of this invaluable work still lies dormant in reports prepared for the U.S. government and industry (see, e.g., reports by Moschandreas, 1975–1981 (for AGA, DOE, EPA, EPRI, HUD); Jarke, 1979 (for ASHRAE); Yocom, 1971–1981 (for EPA); Walcott, 1978–1982 (for HUD); Andrews, 1982 (for CPSC); Pickrell, 1982 (for CPSC); Long, Frank, and Schutte, 1978–1980 (for DOE), Ferris, 1979 (for EPA) and the extensive work of the late Craig Hollowell and co-workers 1976–1982 (for DOE).

6.3.3 MONITORING DEVICES

6.3.3.1 *Field Monitoring Laboratories*

The capability of field monitoring laboratories has been described by Moschandreas et al. (1977-1981) and Hollowell et al. (1978–1982), and many other government and private contractors. Table 6.1 shows the current capability of a typical field laboratory. A large-scale six-city study has been carried out by Ferris et al. (1979) and Spengler et al. (1978–1982) at Harvard for EPA. This work has made valuable contributions to field monitoring equipment. A field audit strategy for residences has been discussed by Traynor (1981).

Table 6.1

Capability of a Typical Mobile Field Monitoring Laboratory[a]

Pollutant	Sampling Rate (min)	Sampling Period (hr)	Analytical Method	Limit of Detection
Total suspended particulates	100	24	Filtration-gravimetric	0.1 $\mu g/m^3$
Respirable suspended particulates (3.5 μg)	50	24	Dichotomous-gravimetric	0.1 $\mu g/m^3$
Organic vapors	0.2	24	Charcoal absorption–gas chromotography	ppb as CH_4
Aliphatic aldehydes	0.5	4	Bubbler-MBTH	1.5 $\mu g/m^3$
Ammonia	0.5	1	Bubbler-phenate	5 $\mu g/m^3$
Sulfates	100	24	Filtration-methyl-thymol blue	0.5 $\mu g/m^3$
Nitrates	100	24	Filtration-brucine	0.1 $\mu g/m^3$
Lead	100	24	Filtration-atomic absorption	0.005 $\mu g/m^3$
Elemental analysis atomic nos. 16–35 and no. 82	1	Continuous	Streaker sampler–PIXE[b]	ppb to ppt

[a] After Moschandreas (1980).
[b] Proton-induced x-ray emission.

A variety of field tests has been developed and used for threshold level measurement. These devices, such as the Gaeke tubing and Palmes samplers, are still popular. They simply indicate whether or not a preset concentration has been reached; laymen have occasionally been misled, taking a negative test to indicate that no toxic substances are present. If the pollutant's toxicity is cumulative, this assumption is dangerous.

Field sampling for laboratory analysis still relies heavily on the proven polyethylene "grab bag" method (Schuette, 1967). With this procedure particulates are collected in portable cyclones or on filters, and organic vapors are increasingly collected on organic polymer foams that are specifically absorbant for certain classes of materials. Among the best tested materials are Tenax (Jarke, 1979; Zweidinger, 1981; Pellizzari et al., 1981; Wallace et al., 1981) and polyurethane (Simon and Bidleman, 1979).

6.3.3.2 *Personal Monitoring Devices*

The first personal monitoring equipment was probably the candle carried by vintners intent on checking the viability of air in storage and fermentation vaults. Today a range of portable instruments for occupational work exposure is available (Lee and Mage, 1979; Mage and Wallace, 1980; Wallace and Ott, 1982); many such instruments were developed by the U.S. Bureau of Mines and Mine Safety and by NIOSH. Two types of devices are used: (1) passive devices, which absorb chemicals by diffusion through some type of membrane or a pack of small fiber tubes; and (2) active devices, which use an air pump to force air through a filter or a solution or into a bag. The first passive devices were film badges used for detecting radiation exposure. Some devices of both types contain a chemical reagent that allows analysis in the field; others require evaluation in a laboratory. Some recent commercial monitors are listed in Table 6.2. Feigley (1982) compared high and low exposure responses of three different badge types.

6.4 Analytical Methods

6.4.1 AMBIENT AIR GASES

Standard methods for pollutants for which air quality criteria have been established by the EPA follow a rigidly and thoroughly established procedure that is regularly updated by publication in the *Federal Register* and the Federal Code of Regulations. The EPA has also determined which instrumental methods are suitable as equivalents to aqueous chemistry.

Sulfur Dioxide. The concentration of SO_2 is measured according to the Gaeke method: SO_2 is absorbed in a tetrachloromercurate solution to form a complex that reacts with pararosaniline and formaldehyde to yield a pink solution for which the absorption coefficient is well known. Ozone and nitrogen

Table 6.2

Personal Monitors for Use at Ambient Concentration[a]

Pollutant	Manufacturer	Method	Range	Time
Respirable particles	TSI (Thermosystems Inc.)	Piezo-balance	10–1000 $\mu g/m^3$	2 min
	GCA Corp.	Optical scatter	1–10^5 $\mu g/m^3$	Instant
	Harvard Pub. Health	Cyclone/pump	–	hrs
	NBS (Nat. Bur. Stand.)	Impactor/pump	–	hrs
Carbon monoxide	General Electric Co.	Electrochem./pump	1–1000 ppm	Instant
	Energetics Science Inc.	Electrochem./pump	1–100 ppm	Instant
	Energetics Science Inc.	Diffusion	ppm	Instant
	Interscan	Electrochem./pump	1 ppm	min/hrs
Nitrogen dioxide	West-Reiszner	Membrane	1 ppb	hrs
	MDA-Palmes	Membrane	10 ppb	hrs
	NBS (Nat. Bur. Stand.)	Membrane	0.5 ppb	hrs
	Toyo-Roshi Inc.	Fiber membrane	1 ppb	hrs
Radon, radiation	Union Carbide Inc.	Thermoluminescence	0–10^6 Rem	hrs
	Terradex Inc.	Track-etch badge	0.001 WL	hrs
Formaldehyde	DuPont Co, Pro-Tek	Membrane/badge	0.3 ppm	day
	3M-Corporation	Filter/membrane	0.3 ppm	day
	Oakridge Nat. Lab.	Mol. sieve/membrane	0.02 ppm	day
	LBL-Air Qual. Res. Group	Dry filter	0.5 ppm	day
	LBL-Air Qual. Res. Group	Dry filter/pump	0.01 ppm	8 min
Vinyl chloride	Reiszner Corp.	Membrane	5–5000 ppb	hrs
Organic vapors	Research Triangle Inst.	Tenax/pump	ppb-ppt	hrs
	Monsanto Res. Corp.	Sorbent/pump	ppb-ppt	hrs
Pesticides, PCPs	Oakridge Nat. Lab	Memrane	ppb	hrs

[a] From Mage and Wallace (1980), Wallace (1980), and Wallace and Ott (1982).

oxides do not interfere. The sensitivity is 0.01 ppm in 30 liters of air (i.e., 25 $\mu g/m^3$). Sulfate is discussed in section 6.4.4. on particulates.

Several equivalent methods for sulfate determination have been developed. Published in a series of NIOSH volumes, each method has a code number assigned to it; P+CAM 146 describes an impinger collector with subsequent titration for concentrations between 10 ppb and 10 ppm. P+CAM 160 uses the impinger in connection with a colorimetric method, and covers the range from 3 ppb to 5 ppm for samples collected at 0.2 liter/min for 25 min; P+CAM 204 uses a molecular sieve to collect the sample of mass spectroscopy. The sample is collected for up to 8 hours at a rate of 0.2 liter/min, and the method covers the range of 2–265 mg/m^3. P+CAM 160 describes a passive dosimeter for collecting samples at 20 ppb to 6 ppm over a period of seven days. This method uses diffusion through a fiber membrane, which makes mechanical pumping unnecessary. Continuous field monitoring of sulfur dioxide uses flame photometry with a detection limit of 5 ppb. With methylene blue, sulfate concentrations as low as 0.5 $\mu g/m^3$ can be determined.

Carbon Monoxide. Infrared spectroscopy is used to measure CO. The sensitivity is 0.1–50 ppm. Water vapor and temperature must be controlled to eliminate potential interference. Several companies are marketing pocket meters with which CO can be measured within seconds. Three passive personal monitors are available; two have a color indicator, and the third uses an alarm that indicates when the threshold limit value (TLV) has been reached. Field monitoring for this gas uses infrared detection with a limit of 0.5 ppm.

Ozone. Concentrations of ozone in the range between 0.005 and 1 ppm are measured by recording the chemiluminescence reaction with ethylene in a photocell set at 430 nm (McKee et al., 1975). The detection limit is 5 ppb. An impinger–colorimetric method covering the range 0.1–0.4 mg/m^3 in 45 min is available, but its accuracy has not been established. P+CAM 108 employs a bubbler–spectrometer combination, and records from 10 ppb to 10 ppm within 30 min. P+CAM 231 covers a solid solvent–spectrometry technique. Six different passive monitors are available. All operate in the ppm range; all employ diffusion. The DuPont badge uses ultraviolet (uv) absorption; MDA uses electrochemistry; and MDA and DACO use TEA-coated screens to measure concentration.

Nitrogen Oxides. Nitrogen oxides are collected in sodium hydroxide and converted to nitrite. The latter is determined spectrophotometrically by reaction with naphthylethylenediamine hydrochloride, which absorbs light at 540 nm. The ASTM-D-1607 method has been tested by an interlaboratory study (Foster and Beatty, 1974). The sensitivity is from 0.01 to 0.4 ppm (Ellis, 1976). Nitrogen oxides can also be analyzed with the help of Palmes tubes (Palmes and Gunnison, 1973; Palmes et al., 1976). Nitrates (discussed in Section 6.4.3) are measured with brucine, in particulates, with a detection limit of 0.1 μg/m^3. Field monitoring uses chemiluminescent detection with a limit of 5 ppb.

6.4.2 OTHER AIR COMPONENTS

The intersociety committee (ISC) comprising representatives from the Air Pollution Control Association, the American Chemical Society, the American Conference of Governmental Industrial Hygienists, the American Industrial Hygiene Association, the American Institute of Chemical Engineers, the American Public Health Association, the Society of Mechanical Engineers, the American Public Works Association, the American Society of Civil Engineers, the Society of Automobile Engineers, the Instrument Society of America, the Health Physics Society, and the Association of Official Analytical Chemists has published a manual containing details of 57 methods for hydrocarbons, carbon monoxide, halogen compounds, metals, nitrogen compounds, oxidants, sulfur compounds, particulates, and radioactive compounds.

The American Society for Testing and Materials (ASTM) and the American National Standards Institute (ANSI) have published 26 methods for ambient air and six for the workplace.

Carbon Dioxide. Carbon dioxide is collected in grab bags and analyzed by gas chromatography. The sensitivity is low. Eight hours are necessary to establish concentrations in the range from 500 ppm. Infrared is used for continuous field monitoring at levels of 25 ppm and above.

Ammonia. At least five well-tested methods can be used to measure ammonia concentrations. Passive monitors capable of detecting 0.02–60 ppm employ fiber diffusion and spectrophotometry. The Drager method responds to 24–120 ppm at a flow rate of 1 liter/min within 15 min. P+CAM 205 describes a portable impinger–calorimeter combination that responds to 1 ppm in 15 min. A bubbler–phenate combination provides a detection limit of 5 $\mu g/m^3$.

Mercury. Gaseous mercury can be monitored with badges. Two such monitors use diffusion through a gold film for detecting 2–250 $\mu g/m^3$ by flameless atomic spectrophotometry.

Lead. Lead is measured with the help of atomic absorption spectroscopy. The detection limit is 5 pg/m^3.

Hydrogen Sulfide. H_2S can be measured at the 1-ppm level with diffusion badges.

6.4.3 ORGANIC VAPORS

Laboratory analysis of organic vapors is now largely achieved by means of a combination of gas chromatography with subsequent mass spectroscopy (McFadden, 1973; Gray, 1974; Gudzinowicz et al., 1976; Debeir and Judd, 1979). The EPA and National Institutes of Health have provided an extensive data base for the computer analysis of mass spectra (Heller and Milne, 1978). Mass spectrographs and gas chromatographs are now included in submarines as well as spacecraft for continuous monitoring of air quality (Ruecker and Shaver, 1974; Scott and Stuart, 1980). The analysis of expired air for detecting disease was pioneered by Muysers and Smidt (1958).

Gas chromatography has been highly developed as an independent tool (Hachenberg, 1973; Thompson, 1977; Walker et al., 1977; Grob, 1977; Jennings, 1978; Zlatkis, 1978). Indoor air has been analyzed chromatographically by Muchtarova and Dimov (1978). Liquid chromatography is also very useful (Horvath, 1980). Trace analysis of indoor air has been discussed by Meinke and Scribner (1967), Yocom (1974), and Oikawa (1977).

Hydrocarbons. Only a few hydrocarbons are considered here, because of their importance. A systematic description of methods for other hydrocarbons and organic substances is beyond the scope of this book.

Methane can be measured with a flame ionization detector (FID) with a detection limit of 50 ppb. Heavier hydrocarbons can be separated and measured by means of gas chromatography (Harrison et al., 1977; Budde and Eichelberger, 1979). Petroleum distillates are tested with charcoal tubes–gas chromatography, or polymethane foam or Tenax foam sample tubes.

Field monitoring stations use flame ionization, with a detection limit of 50 ppb, to determine total hydrocarbons. Total aliphatic aldehydes are measured with impingers and MBTH to yield a dye for uv photometry.

Formaldehyde. Formaldehyde is difficult to monitor because of its unique ability at room temperature to transform reversibly from gaseous aldehyde into aqueous methylene glycol or into solid paraformaldehyde (Walker, 1964; Meyer, 1979). Formaldehyde can exist in any of these three phases simultaneously, and each can behave like an independent chemical. Thus, an indoor air concentration measurement provides no indication whether the vapor is due to a puff of gas or to a pool of several pounds. Furthermore, formaldehyde air concentration depends on humidity of the air as well as construction materials because the gas dissolves quantitatively in water, only to revaporize when the water vaporizes.

The standard field procedure consists of bubbling 30–60 liters of air through a series of two microimpingers at a rate of 1 liter/min. Water or sulfite is used to collect the aldehyde, and the sample is later combined with a standard solution of chromotropic acid, pararosaniline, or acetyl acetone to produce a colored dye, which can be used to determine the concentration photometrically. NIOSH has validated protocols for several methods, including polarography. The most widely used technique is still the chromotropic acid method, which has a detection limit of about 0.5 ppm, although other methods may be more sensitive once protocols have been developed. These methods have been reviewed and studied by Miksch (1980a,b), NRC (1980), Sawicki and Sawicki (1978). Dasgupta et al. (1980) have carefully examined the Schiff reaction of pararosaniline with sulfite. The sulfite method, ASTM-D-2194, is useful for larger concentrations. Recently, a sulfite-based personal monitoring badge (DuPont's Pro-Tek), and two passive monitors (Geisling, 1981 and 3M-Corp.) have become available.

Formaldehyde Resins. The term formaldehyde resin is a misnomer, because resorcinol, phenol, and urea react with formaldehyde in the resin and convert it into chemically different species such as methylols, methylene, ether, or acetal groups. Well-cured phenolic resins are water resistant and contain less formaldehyde than ambient air. Amino resins produced by reaction with urea also contain only traces of unreacted formaldehyde, but these resins are not always carefully formulated, applied, or cured, and are vulnerable to moisture exposure, even after cure (Meyer, 1979a,b).

In some applications, such as field-manufactured insulation foam, and in some hardwood veneers the resin is intentionally not immediately fully cured, so as to impart to the product certain mechanical properties. The result is that some amino resin products may initially release formaldehyde at a level depending on the degree of cure, temperature, or humidity (Schulze, 1975; CPSC, 1980; Pressler, 1981).

The major products produced with amino resins are interior plywood, particleboard, and foam insulation; for exterior plywood phenolic adhesives are used (Meyer, 1979). The best way to prevent indoor air problems is to test the material, its design, and its application before use by having a pre-installation review by experts. Standard material tests already exist in Europe (Sundin, 1978, 1980, 1982; FESYP, 1978) and Japan (JIS, 1974), and tests are currently being developed by several North American organizations (Meyer, 1982).

Formaldehyde is also incorporated into finished textile materials. The standard method of the American Association of the Textile and Cotton Councils (AATCC, 1978) is well accepted and uses a principle similar to the jar method used for particleboard in the proposed National Particleboard Association–Hardwood Plywood Manufacturers Association test (NPA-HPMA, 1981).

Polyaromatic Hydrocarbons (PAH, BaP). Fused polynuclear hydrocarbons are not very volatile and are usually absorbed on particulates in indoor air, and thus are usually collected jointly with particulates on filters and separated by benzene extraction followed by high-pressure liquid chromatography (HPLC). Depending on the resolution of the instrument, a group of aromatics is measured as a scum, or individual compounds are resolved (Pellizzari, 1978). Thus, the term benzo-a-pyrene (BaP) does not always refer to this compound alone, but is often used as a synonym for polyaromatic hydrocarbons (PAH). Schuetzle et al. (1981) analyzed several hundred compounds in diesel exhaust and discovered that the mutagenic activity is not always in the PAH fraction alone. The high-resolution spectrum of such a mixture is shown in figure 6.1.

P + CAM 183 describes a membrane filter suitable for sample collection in the range from 2 ppm to 1 g/m^3 over a period of 4 hours. Oak Ridge National Laboratory has developed a solvent-covered filter for fluorescent analysis. Some eight methods have been developed, and it is hoped that one will be suitable for indoor use.

Chlorinated Hydrocarbons. Chlorinated and polychlorinated hydrocarbons are determined by chromatography (Zimmerli and Zimmermann, 1979; Budde and Eichelberger, 1979). Vinyl chloride is collected on charcoal. Gas chromatography is suitable in the range from 0.008 to 5 mg/m^3 with a sampling period of 1 hour. The procedure for sampling polyvinyl chloride has been reported by Elfers and Richter (1975). The analysis of DDT is described in Section 5.6.5. With electron capture detectors, 10^{-12} g of DDT, other pesticides, and Agent Orange can be detected in the air and in body tissues. Sampling of polychlorinated biphenyls (PCBs) in blood, urine, body tissue, and clothing was first described by Barthel et al. (1969). A similar method is now widely used (Wallace et al., 1981). This material contains a possible 209 congeners and is also detected by electron capture response (Cairns, 1981). Volatile chlorocarbons were reviewed by the National Academy of Sciences (NRC, 1978). Table

6.4 Analytical Methods

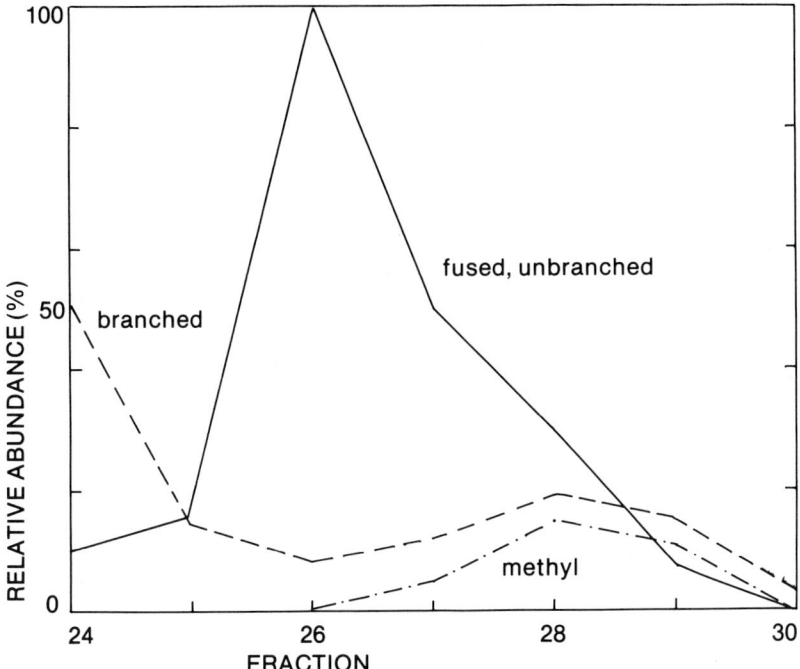

Figure 6.1. Chromatogram of three PAH mass fractions in diesel exhaust (after Schuetzle et al., 1981).

6.3 lists the estimated detection limits for 40 chlorocarbon solvents (Pellizzari and Bunch, 1979) and pesticides. Analytical methods have been described by Erickson (1980) and Krost (1982).

Nitrosamines. These potential carcinogens are collected on charcoal, Tenax polymers, in a cryogenic trap, or on potassium hydroxide. Ambient air contains 1 mg/m^3, but indoor levels of from 0.5 to 2 μg/m^3 (found in kitchens and automobiles) can be measured by thermal energy analyzers, which can detect 10–1000 ppt within 60 min.

Miscellany. Pesticide sampling has been described by Compton, Bazydlo, and Zweig (1972), and the analysis of drugs by Pesez and Bartos (1974), and McEwen, 1976.

6.4.4 PARTICULATES

Because it is very difficult to make standard particle-size control samples with which to calibrate equipment, particulate and aerosol measurements are difficult.

Table 6.3

Detection Limits for 40 Chlorocarbon Solvents and Pesticides[a]

Medium, and Compound	Estimated Detection Limit[b] (ng/m^3)	(ppt)
In air		
Vinyl bromide	250	57
Bromoform	0.340	0.03
Bromodichloromethane	1.300	0.22
Dibromochloromethane	0.667	0.07
1-Bromo-2-chloroethane	1.00	0.67
Allyl bromide	5.00	1.04
1-Bromopropane	5.200	1.06
1-Chloro-3-bromopropane	0.150	0.01
1-Chloro-2,3-dibromopropane	0.100	<0.01
1,1-Dibromo-2-chloropropane	0.100	<0.01
1,2-Dibromoethane	0.530	0.07
1,3-Dibromopropane	0.100	0.01
Epichlorohydrin (1-chloro-2,3-epoxypropane)	9.600	2.50
Epibromohydrin (1-bromo-2,3-epoxypropane)	0.300	0.05
Bromobenzene	0.100	0.02
Methyl bromide	500	135
Methyl chloride	2000	1000
Vinyl chloride	800	333
Methylene chloride	700	200
Chloroform	200	420
Carbon tetrachloride	250	400
1,2-Dichloroethane	32	8.15
1,1,1-Trichloroethane	66	12.45
Tetrachloroethylene	2.5	0.38
Trichloroethylene	10	1.92
1-Chloro-2-methylpropene	62	21.5
3-Chloro-2-methylpropene	62	21.5
3-Chloro-2-butene	83	28.8
Allyl chloride	83	28.8
4-Chloro-1-butene	38	13.2
1-Chloro-2-butene	13	4.5

[a] From Pellizzari and Bunch (1979).
[b] Limits are calculated on the basis of the breakthrough volume for 2.2 g of Tenax GC (at 70°F).
[c] Nanograms per microliter.

Table 6.3 (Continued)
Detection Limits for 40 Chlorocarbon Solvents and Pesticides[a]

Medium, and Compound	Estimated Detection Limit[b] (ng/m³)	(ppt)
Chlorobenzene	2.10	0.47
o-Dichlorobenzene	1.00	0.06
m-Dichlorobenzene	0.75	0.01
Benzylchloride	0.65	0.01
In solvent		
γ-BHC (lindane)	5–10[c]	
Heptachlor	10–20[c]	
Chlordane	25–50[c]	
p,p'-DDE	5–10[c]	
2-Chlorobiphenyl	~2.5[c]	
Hexachlorobiphenyl	25–50[c]	
Decachlorobiphenyl	150[c]	

[a] From Pellizzari and Bunch (1979).
[b] Limits are calculated on the basis of the breakthrough volume for 2.2 g of Tenax GC (at 70°F).
[c] Nanograms per microliter.

Ambient particulate concentrations are measured regularly at several thousand stations across the United States. These stations were set up over the last 15 years as part of four separate programs. Most of the stations originally belonged to the National Air Surveillance Network (NASN); six stations made up the Continuous Air Monitoring Program (CAMP), established in 1967, and about a dozen belonged to the stations of the Community Health Environmental Surveillance System (CHESS). Eighteen stations were added in 1969 for the Community Health Air Monitor Program (CHAMP). The purpose of these stations is to derive ambient air data and provide an early warning system in case of high concentrations.

The current ambient sampling method measures total suspended particulates (TSP) covering the entire range from 0.002 to 500 μm. Collection is least efficient for small particles, and the weight of large particles overwhelms the weight fraction of respirable particles below 5 μm. Adequate equipment for measuring concentrations of small particles is only slowly becoming available. The basic measurement principles involve weighing, condensation, electrical, charging, nucleation, acoustical sampling, light scattering, diffusion, and crystal sensing.

Mass Distribution. Inertial impactors grade particles by controlling the airstream velocity at deflecting surfaces. Large particles have greater momentum, and cannot follow the airstream around sharp corners. Instead, they continue on a straight path and are collected on the container wall while small particles exit with the air. The dichotomous sampler is an impactor that separates air into two fractions. Cascade impactors use a sequence of jets to grade progressively smaller sizes. The Anderson impactor uses six cascades to separate seven size fractions simultaneously. Impactors cannot grade particles smaller than 0.25 μm. Dichotomous and cascade impactors can recognize ~ 1 $\mu g/m^3$.

ANSI/ASTM test D-3365-77 measures the particle size distribution in a liquid medium by means of an electronic computer.

Chemical Analysis. The nationwide network analyzes particulates for metals, ammonium, sulfate, nitrate, and benzene-soluble organic content. The analysis protocol was developed to provide reliability for large-scale use. Since validating a protocol is a time-consuming procedure, the currently used methods are not the latest research methods available (Perera, 1979; Mitchell and Midgett, 1975; Lin, 1975; Maddalon and Quinlivan, 1977). The metal analysis includes antimony, arsenic, beryllium, bismuth, cadmium, chromium, cobalt, copper, iron, lead, manganese, molybdenum, nickel, tin, titanium, vanadium, and zinc; its accuracy is better than ±10% (Perera, 1979).

The benzene-soluble fraction is removed to determine BaP. The latter is used as an index for the entire cluster of polycyclic aromatics. Many samples have to be accumulated for this procedure, because the method is not extremely sensitive.

The current analytical methods for sulfate (ASTM-D-2010) lump together all sulfur oxyanion species (Foster, Beatty, and Howes, 1974). The nitrate analysis is notoriously difficult, as is all nitrogen analysis (Spicer and Schumacher, 1977).

Instrumental Methods. Recent research methods for particulates provide for highly sophisticated and accurate evaluation. For example, electron microscopy combined with x-ray spectroscopy now makes it possible to determine the elemental composition of different components in heterogeneous dust in buildings (Weschler, 1978), all *in situ;* and x-ray fluorescence spectroscopy (Dzubay, 1977), neutron activation analysis, and proton-induced x-ray emission (PIXE) also yield an elemental analysis. The crystal habitat can be established by x-ray and electron diffraction, as well as by electron microscopy, which is especially helpful with asbestos (ACGIH, 1980). Molecular analysis can be achieved with the help of electron spectroscopy for chemical analysis (ESCA). Infrared (ir) and Raman spectroscopy serve the same purpose by analyzing molecular vibrations. Ultraviolet (uv) and fluorescence are useful for observing electronic excitation of certain molecules; and liquid chromatography and mass spectroscopy help separate mixtures, especially organic components such as soot and tar.

The problem with these instrumental techniques is the cost of the equipment, which can easily reach $500,000 or more. Thus, sample analysis must be conducted in a central laboratory by skilled workers. The California Air Resources Board and the St. Louis Regional Air Pollution Study have each locally pooled sample analysis, and plan to add use of instrumental methods on different-sized fractions of such particulates as sulfates (Appel et al., 1977) in order to correlate bulk properties, such as nephelometric response, to chemical constituents. The California study also measures material in a field laboratory equipped with a chromatograph for gaseous products.

Recently a tabletop real-time aerodynamic particle sizer has become commercially available that, if it performs as promised, could revolutionize sample interpretation because of its ease of application. It is based on the piezobalance technique and continuously records and displays the concentration of particles with sizes of 0.5–15 μm in concentrations ranging from 5 μg/m^3 to 10 mg/m^3. Another new real-time aerodynamic particle sizer uses a laser velocimeter and can be used for particles with sizes from 0.5 to 10 μm. An accelerating nozzle is used to grade for sizes, since the optical methods are responsive not to aerodynamic particle size but to physical size.

The problem in sample dusts is the same as with microbes: If the sampling device does not provide isokinetic flow at the entrance point, the sample may well have a different composition from that of the source.

Indoor concentrations of particulates change rapidly and vary over the entire room because particles are constantly settling (Foster et al., 1974) and resuspending. If what is wanted is an indication of the actual dust exposure in a room, the people in it have to be equipped with personal monitors that collect air at the breathing level. Such devices can operate with filter–pump combinations or with any other apparatus that can be made compact enough to be worn by a person going about daily chores. Turner (Turner et al., 1979) describes a portable particulate monitor that uses a 1-μm pore filter and a 1-cm nylon cyclone with a membrane pump. This instrument was derived from a noisier pump developed for mine monitoring; its sampling is very reliable (Caplan et al., 1977). A hundred such samplers are currently in use in a six-city study (Spengler et al., 1979).

Repace et al. (1980) prefer a piezoelectric respirable-particle sampler that integrates samples over 2.5 min over the range from 10 to 1000 μg/m^3. Another such sampler, based on optical scattering, provides 10-fold better sensitivity, and thus provides readings between 1 and 200 μg/m^3. It reads instantaneously but is costlier. Both instruments cost several thousand dollars.

The general field of particle sampling, sizing, and grading has been reviewed by Perera (1979), Foster et al. (1974), Cadle (1975), and Lundgren et al. (1979).

6.4.5 MICROBIAL SAMPLING

The traditional method of sampling microbes is to collect them on moist agar gel on glass or other sticky surfaces and to obtain samples by gravitational settling (Miquel, 1890). A thorough review of suction collectors and their functioning has been provided by May (1973). Sticking coefficients, related to the stop distances, can be computed from Stokes's law, but according to May the problem of aerosol sampling has not been solved. The chief difficulty is the estimation of the true concentration in flowing coarse aerosols. The problems are the same as with inorganic dusts except that microbes are not spherical; hence, the aerodynamic and the physical diameters may differ by an order of magnitude.

Ideally, sampling should be conducted under isokinetic conditions so that the airstream entering a sampling device maintains the same speed as the bulk ambient component. This is necessary to prevent fractionation. Unfortunately, air at the ground level is turbulent, and isokinetic sampling is implicitly difficult to achieve.

Microbes can be collected with electrostatic precipitators coated with agar solution, or with cyclones, cascade centripeters, the Anderson sampler (which consists of a six-stage cascade impactor), or liquid impingers (May, 1973; Sayer et al., 1972). Whatever the method, much of the catch will lose its viability before it is analyzed. The best identification of a microbe depends on its nature. Fungi can be recognized visually under a microscope, or can be cultured, but care is necessary to prevent contamination with molds. Serological tests or skin tests are very sensitive. Bacteria may be identified under the microscope or can be microcultivated in either a fluid or a solid medium. Sample enrichment may be necessary with membrane samplers, or by incubation in animals. Viruses, which are hard to analyze because they are so small, are best identified by cultivation in fluids or solids. Furthermore, airborne bacteriophage particles can be used as models in virus studies.

Mixtures of indoor air samples are often difficult to analyze because they contain soot, ashes, and cernospheres formed by the combustion of oil or tar. They always contain desquamated fragments of skin, which must be identified by their shape because there is no satisfactory method for staining keratins. Protein, cellulose, and synthetic textile fibers can be stained by the Pressley method. Pollen grains can be identified under the microscope by comparison with diagnostic tables.

6.4.6 RADIOACTIVITY AND RADON

Equipment for measuring radiation has dramatically improved. In 1960 it took a truckload of apparatus to inventory radiation levels in mines: Today better data can be obtained with a small, light, economical track-etch film badge that can be mass-produced, machine evaluated.

Radiation measurement can be achieved by several different instantaneous methods, by continuous readout methods, or by time integration. The main problem with measuring ionizing radiation is that indoor levels may oscillate by an order of magnitude within a few hours (Knoll, 1979; Eichholz and Poston, 1979).

Real-Time Radon Measurement. The classical method of measuring radon levels in air uses a scintillation flash in which a given gas volume comes into contact with a silver-activated zinc sulfide coating that, upon irradiation by alpha particles, emits light. The light is registered on a photocathode. The sensitivity of this method is about 1 pCi/liter. Long exposure makes it possible to extend the sensitivity to 0.1 pCi/liter (Fugas, 1981; George, 1980).

Radon and its daughters can be separated by the two-filter method (Thomas, 1968): Air is drawn at a known rate through two filters held a known distance apart, and the first filter collects all the dust, including the radon daughters; only radon gas passes through this filter. From the distance between the filters and the flow rate, it is possible to compute how much of the radon decomposes before the gas reaches the second filter. The procedure takes 15 min, and a sensitivity of 0.1 pCi/l can be obtained.

The U.S. Department of Energy uses a pulse ionization technique to measure radon in grab bag samples to levels of 0.01 pCi/l (George and Breslin, 1980).

Continuous Radon Meters. A continuous-flow flash scintillation counter can observe 0.5 pCi/liter, but the data must be corrected to subtract emission from radon daughters that accumulate in the meter. Such meters are not truly portable. A two-filter monitor, sensitive to 0.01 pCi/liter, can be used if a correction is made and the accumulation of radon daughters is subtracted (Breslin, 1980). Other methods use diffusion meters, in which air is sucked through a foam barrier into a scintillation counting cavity.

Integrating Meters. Time averaging can be achieved by the slow collection of air samples. A 40-liter Mylar bag equipped with a 10-ml/min air pump allows recordings of 0.01 pCi/liter over a 2-day period. The U.S. Department of Energy has developed an electrostatic diffusion monitor that employs a thermoluminescent lithium fluoride chip to absorb and store alpha-particle energy for the field measurements. No electric current is necessary. The sensitivity is 0.03 pCi/liter^{-1} week^{-1}.

The classic mine measurements were made by the Kusnetz method using a portable air filter and a portable alpha-particle counter. The working level is computed from the radiation measurement made after a 90-min equilibration of the filter to ensure an equilibrium radon mix (Figures 5.4–5.6, pp. 128–130). The method can be used to levels of 5×10^{-4} WL. The Tsiroglou method consists of measuring the filter three times at prescribed intervals so that the mix of radon daughters can be calculated. Several variations of these methods are in use. Alpha-particle filters can be used to make the method time independent and more reliable. A thermoluminescent detector–air filter combination with an air rate of 100 ml/min can be used to sample radiation over a week and obtains a sensitivity of 5×10^{-4} WL.

Since 1980 wide use has been made of track-etch film. Originally, cellulose nitrate was used as a substrate to record alpha radiation on a film that was subsequently developed by etching in an alkaline solution. The number of pits per square millimeter is used to determine the number of particles. The accuracy is about 0.01 ± 50% WL. A great many film badges (the films are polyethylene) are currently used for the large-scale screening of sites (Alter, see Hartley, 1981). The new film material is far less sensitive to humidity and temperature than former ones. Membrane filters are used to screen particulate daughters as desired. Evaluation is now automated.

Various fully automated microprocessor-controlled portable working-level instruments are currently being developed independently by various people, including groups at the Argonne, Oak Ridge, and Lawrence Berkeley laboratories of the U.S. Department of Energy. These devices can take samples of predetermined length at predetermined intervals; some can also discriminate between different emitters.

The field sampling of radon gas is achieved with evacuated cans whose capacity is 1–100 liters, displacement cans, grab bags, or charcoal cartridges from personal respirators. Water bottles allow sampling of wells to levels of 30 pCi/liter.

Several methods for the laboratory testing of buiding materials have been reported. The quickest one consists of equilibrating the sample in a closed air volume. The gas is then collected on a glass filter and trapped with liquid nitrogen, and the concentrated sample measured on a scintillation counter (Ingersoll, Stitt, and Zapalac, 1981). The method is reproducible to within ±5%.

6.4.7. BODY BURDEN

Each contaminant has a different fate once it has been inspired, Chapter 8. Sulfur dioxide is absorbed on the moist membranes of the respiratory tract and subsequently is probably washed out, ingested, and excreted; but its exact fate is not well known. It does not seem to leave a measurable body burden. Carbon monoxide deeply penetrates the lung where it enters the bloodstream and can be measured as hemoglobin complex. Ozone reacts, and the resultant products are absorbed in mucus and blood; the metabolism of inspired nitrogen oxides is not yet known. Organic vapors enter the bloodstream, from which they can recirculate to the lung. Chlorocarbons are soluble in fat and thus accumulate in the lipoid tissue and form a cumulative body burden that can persist for many years (Anderson, 1980). The average body burden accumulated by North Americans as a result of nonoccupational exposure has been reviewed by the Oak Ridge National Laboratory for the EPA (1978). The bulk of formaldehyde is absorbed in the mouth and nose and ingested. Respirable dusts are absorbed deeper in the respiratory tract. Soluble salts of metals and other soluble inorganics from particulates either enter the bloodstream or are ingested. Insoluble materials are expelled or concentrate in lymph nodes (Section 8.1).

Once pollutants are absorbed in the body they can be recognized and analyzed by standard biochemical procedures, including liquid chromatography

(Horvath, 1980), thin layer chromatography (Touchstone and Rogers, 1980), and electroanalysis (including polarography, which is now suitable for mercury, cadmium, pesticides, and carcinogens in body fluids; Smyth, 1980), and by chemical analysis of the saliva, urine, glandular secretions, nerve cell fluids, glucose, pH and hemoglobin levels of blood, as well as by analysis of the lipids, proteins, and enzymes, and of the renal, liver, gastric, intestinal, and pancreatic fluids, the hormones, nucleotides, neurogenic amines, and every other part of the body. Furthermore, microbiological assay can be used for qualitative and quantitative analysis (Hewitt, 1977).

The interpretation of such data requires great skill, but it has recently been recognized that the dose of several substances can be readily correlated to exposure levels. Furthermore, several organic hydrocarbons, including benzene, are expired at readily measurable levels for substantial periods of time, which makes it possible to measure both the exposure level and the retention level with the same type of method and equipment. Wallace et al. (1981) are currently undertaking landmark studies in this field.

6.5 Interpretation of Data

There are many intrinsic problems in the field of monitoring and measuring indoor air. Air is not a homogeneous mixture. Air diffusion is of the order of 1.8×10^{-5} m^2/sec. Therefore, mixing depends on convection and turbulence, which are irregular and not predictable. This is specially true at the breathing level, where particulate, pressure, temperature, humidity, and chemical gradients constantly change. Thus, indoor air is characterized by large intrinsic concentration changes. As the speed and accuracy of measurements increase, the interpretation of field data will become even more difficult and will increasingly involve statistical methods. In other fields of science such problems can be overcome by performing work in a laboratory under controlled conditions. However, in this case, research in the lab differs inherently from what happens in real homes where occupants contribute moisture and heat and interact with pollutants.

Numerical values of pollutant concentrations are meaningless unless conditions at the sampling site are carefully characterized and validated with an explicitly stated and fully explained method. Thus, a zero reading on an air sampling tube using an insensitive method may correspond to several parts per million, a concentration well within the range of other instruments or perceivable to the nose of occupants. On the other hand, the discovery of a signal on, say, a highly sensitive gas chromatograph-mass spectrometer instrument for some organic chemical does not implicitly mean that the chemical constitutes a new life-threatening element in the indoor air environment, or that the observed material is responsible for discomfort encountered in the location where it has just been detected. The effect of different instrument sensitivity is especially important now, when the gap between the sensitivity of little-tested research equipment and that of standard methods in some fields is still wide.

It is very important to know whether an observation was made once or how many times it was replicated; whether the reading was compared with a control and a blank; and whether it could be replicated with different equipment and different methods. Furthermore, it is essential that all other pollutants be listed, and that the measurement conditions be described in great detail so that the reader can understand what other factors might have influenced the perceived effect. Indoor air always contains many chemicals, and thus measurements are always made in the presence of ample oxygen, moisture, carbon dioxide, and many organic compounds. Furthermore, the temperature and pressure and air velocity influence the measurement.

Finally, it must be realized that a host of imponderables can influence single readings in a real-life environment. To give just one example, the formaldehyde level in a room depends not only on the emission source of the gas, but also on the absolute temperature of the air, the inside wall, and the outside wall; the permeability of the paint; the humidity in the air; moisture absorbed on the walls; the air velocity; whether the air is filtered through a moist filter; the age of the building materials; the temperature and moisture history of the building envelope and the time profile of the corresponding gradients; and on the occupant density and the sequence of human activities.

If samples are not analyzed on the spot, the stability of the contaminant in its state of accumulation must be carefully reviewed because pollutants are often reactive, and microbes do not remain viable forever. The EPA has developed special procedures for detecting and preventing significant deterioration of samples (Fuerst, Scaringelli, and Margeson, 1976).

The preparation of samples and controls is a subject in itself. For example, The National Bureau of Standards prepares standard particulates and soots for comparison purposes. Methods for preparing standard samples are described in the specialty literature. The unusual properties of formaldehyde and similar gases preclude commercial preparation of standards for these gases. The analysis of such chemicals remains an art.

7. Exposure Levels

This chapter reviews the currently still very limited knowledge of the indoor concentration of pollutants and the resultant exposure levels. The total exposure of a human being to a pollutant is determined by the concentration of the contaminant and the duration of the exposure. The first section of this chapter considers how and where we spend our time; the second reviews the concentration of pollutants at various places; the third estimates the range of exposures we encounter; and the fourth explores the correlation between exposure and dose.

7.1 Human Activity Patterns

The functioning of industrialized nations requires a work rhythm and lifestyle that are possible only within a narrow climatic range. In most parts of the world these climatic conditions do not prevail, and so much of the population there can function productively only in an urban indoor environment. In fact, we all spend far more time indoors, and in an artificial habitat, than we tend to realize.

Table 7.1 shows how people working outside the home and homemakers spend each day (Szalai, 1972; Chapin, 1974; Ott, 1981). According to Ott's interpretation, people in all countries spend more than 20 hr/day under a roof breathing confined air, and an average urban person spends less than 2 hr outdoors. In fact, Ott found that on an average day an average American housewife spends only 20 min, and an average person working outside the home only 40 min, outdoors.

A more detailed breakdown indicates that the average person working outside the home in 44 U.S. cities spends about 14 hr each day inside the home. Averaged over the full week, about 6 hr are spent at work, 1½ hr in transit, and only 12 min in the yard or garden. The average woman employed outside the home has no time to spend in her yard. Time spent shopping or in other business or public buildings amounts to just about 1 hr each day, thus, indoor air quality is very important for personal and public health. The long indoor exposure times indicate that special attention must be given to those low-level pollutants that cause chronic discomfort by additive action of the agent. An example of such an agent is radioactive dust, which is accumulated and re-

Table 7.1

Time Budget of People Working Outside the Home and of Homemakers in 12 Countries[a]

	Average Time Expenditure (hr/day) of					
	Persons Working Outside the Home			Homemakers		
Country	Indoors	Outdoors	In transit	Indoors	Outdoors	In transit
Belgium	21.6	0.9	1.5	23.2	0.4	0.4
Bulgaria (Kazanlik)	21.0	0.9	2.1	22.1	1.5	0.4
Czechoslovakia (Olomouc)	21.3	1.1	1.6	23.2	0.5	0.4
France (six cities)	22.0	0.5	1.5	23.3	0.2	0.5
West Germany (100 districts)	20.4	1.9	1.7	22.2	1.2	0.6
East Germany (Hoyerswerda)	21.6	0.7	1.7	23.2	0.5	0.3
Hungary (Cyor)	20.4	1.6	2.0	21.5	2.3	0.2
Peru (Lima-Callao)	20.8	0.7	2.5	22.9	0.7	0.4
Poland (Torun)	21.9	0.4	1.7	23.3	0.2	0.5
United States (44 cities)	21.7	0.7	1.6	22.8	0.4	0.8
Soviet Union (Pskov)	21.0	1.0	2.0	22.6	0.7	0.7
Yugoslavia (Kragujevac)	21.4	0.8	1.8	22.4	0.9	0.7

[a] Derived from data originally collected by Szalai (1972); data are weighted to ensure equality of days of the week and number of eligible respondents per household (after Ott, 1981).

tained by the body. However, acute irritation by such agents as formaldehyde, pesticides, and paint solvents at low levels must also be considered. Fourteen hours is almost twice the length of a normal workday and sufficient time to build up two or more times the work dose or body burden of most pollutants. Furthermore, the recovery period, shortened from 16 hours for work exposure to 8 hours for home exposure, is less sufficient for reducing to threshold levels the body burden or any other residual impact. Thus, anyone who suffers daily respiratory discomfort in his home is more vulnerable and susceptible to secondary effects, such as invasion by infectious agents or generally reduced functions.

Figure 7.1 shows the time spent at eight activities over a 24-hour period. Clearly, most daily activities coincide with diurnal cycles. Therefore, we expect to find a systematic trend in the correlation of weather with activities. For example, most people spend the night, when outdoor temperatures drop toward the daily minimum and relative humidity approaches its peak, in their bedrooms. Thus, the formation of dew and its effect always coincide with residence in the bedroom, and air exchange rates should be adjusted to take this into consideration. On the other hand, the nightly temperature minimum always coincides with the lowest occupancy period of office buildings. In summer this usually means that all ventilation has been turned off to conserve energy just at the best time for purging hot stuffy buildings with fresh air. In winter the situation reverses: Offices as well as bedrooms should be purged with warm dry air at midday in order to evaporate condensation. Thus, a proper air quality control strategy has to take into account occupancy hours of

7.1 Human Activity Patterns

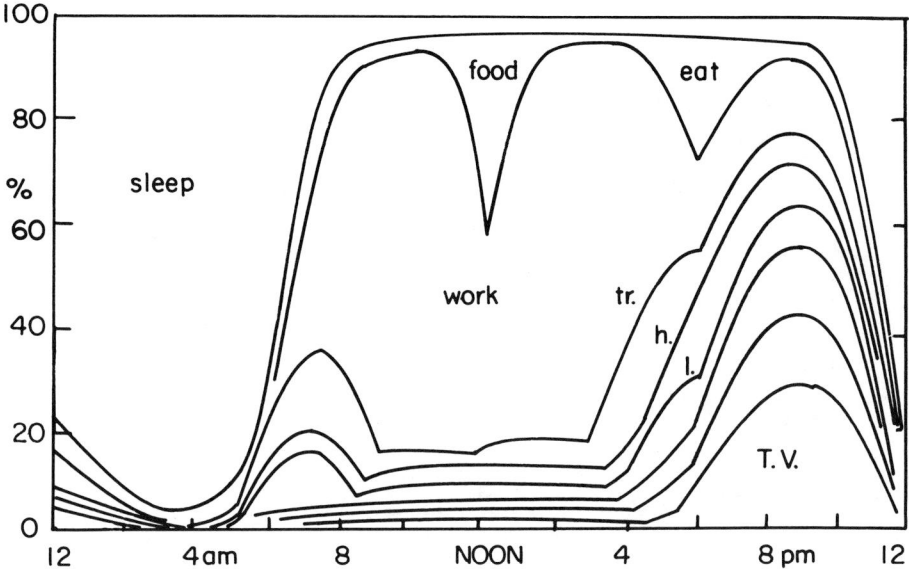

Figure 7.1. Pattern over 24 hours for eight activities: average of 44 U.S. cities (Szalai, 1972).

various types of buildings. Obviously, public buildings and offices, which are occupied during only about 40 daytime hours a week, need a different air management plan than homes, and even in the home the bedroom and living room require a different strategy from other rooms.

An extreme situation exists in churches and public theaters, which are used for only a few hours each week. In contrast, city center arcades and domed sports arenas have become so large that it is sometimes more efficient to maintain a comfortable temperature all year round, regardless of occupancy.

In the early part of this century, many studies directly correlated industrial job productivity with indoor air quality, and it was shown that thermal comfort could be related to employee attitude. The same is surely true for indoor comfort in the home.

Unfortunately, indoor activities constitute the main source of indoor pollution. Since indoor air is confined and allows pollutants to accumulate, the indoor habitat is a sensitive indicator of self-control. Probably no example demonstrates this fact better than smoking. Wherever a smoker goes a cloud of fumes advertises his presence, and when he departs he leaves a trail of odors that can linger for hours or even days (Figure 7.2). Currently, about 40% of the U.S. population smokes. Figure 7.3 shows the trend in cigarette consumption over the last 30 years.

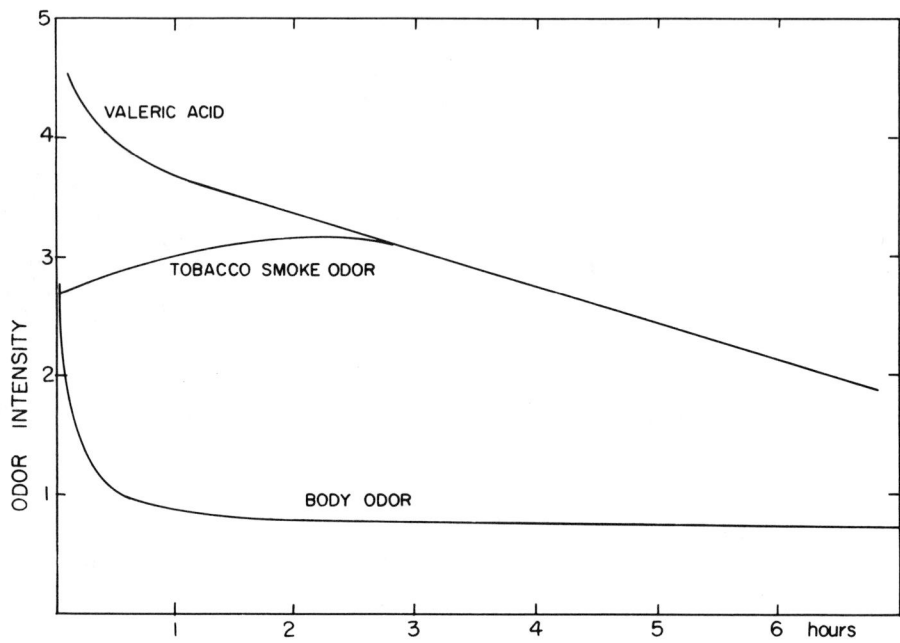

Figure 7.2. Decay of tobacco smoke odor in test chamber (Yaglou, 1937).

Work-related activities contribute heavily to indoor pollution. In fact, cleansers, cleaning fluids, copy machines, and dust from paper, clothing, and other textiles contribute the bulk of contaminants in almost every office. At home, cooking, cleaning, washing, and hobbies are a steady source of pollution. Half of all North American homes use natural gas as a fuel. Table 7.2 shows the regional prevalence of gas and oil as home fuel. Water from food and carbon monoxide and nitric oxide from natural gas-fueled kitchen stoves account for a large fraction of the load of these pollutants, as shown later in Figure 7.9 and in Table 7.3. The composition of combustion gases from a wood stove is listed in Table 7.4.

7.2 Pollutant Concentrations

A survey of research before 1973 of indoor pollution concentrations was prepared by the EPA (Benson et al., 1972; Henderson et al., 1973). The U.S. Department of Housing and Urban Development and the EPA commissioned an update of literature up to 1979 (HUD, 1979).

7.2 Pollutant Concentrations

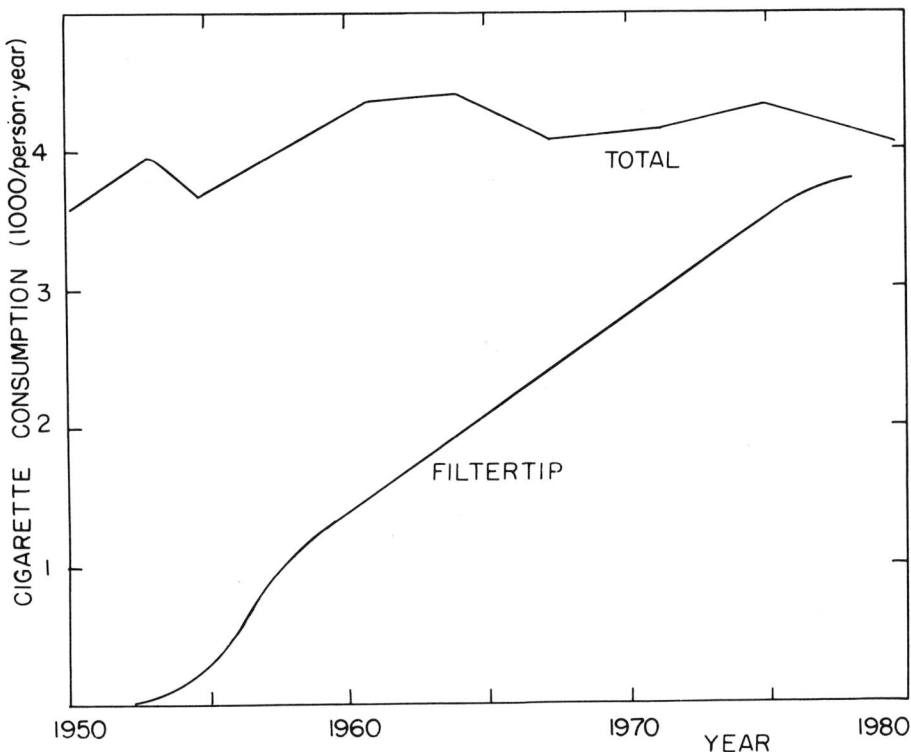

Figure 7.3. Trend in tobacco (cigarette) smoking from 1950 to 1980 (U.S. Surgeon General, 1979).

The concentration of pollutants depends on the size and nature of the source, the size of the building, the ventilation rate, and many other factors.

In rooms with solid or liquid pollutants the air concentration always increases with temperature, because the vapor pressure above solids and liquids is a thermodynamic function. The concentrations reported in this section are not standardized, and not all are equally meaningful. Some represent single exploratory measurements; others are averages derived from periodic measurements with field laboratories. The exposure at breathing level experienced by people living or working in the area can differ substantially, depending on the air mixing and temperature gradient within rooms and within the building.

By definition, air pollutants always appear in the company of nitrogen, oxygen, water, and carbon dioxide. Both oxygen and water can react with many contaminants or influence their concentration.

Table 7.2

Regional Distribution of Home Fuel Use[a]

Region	Distribution of Fuel Use (%)			
	Gas	Oil	Electricity	Coal and Wood
Northeast				
New England	20	76	3	1
Middle Atlantic	46	45	3	6
North central				
East north central	71	23	2	4
West north central	76	20	2	2
South				
South Atlantic	41	39	13	7
East south central	60	4	20	16
West south central	93	—	4	3
West				
Rocky Mountain	81	8	5	6
Pacific	76	12	10	2

[a] After Keyes (1978).

Table 7.3

Emission Rate of Gas Stoves and Space Heaters at Breathing Level

	Emission Rate (μg/kcal)			
	Gas Stove[a]		Space Heater[b]	
Pollutant	Average	Range	Low flame	High flame
CO	890	720–1090	632	319
CO_2	209,000	196,000–217,000	200,000	200,000
NO	31	21–47	76	135
NO_2	85	69–100	46	44
SO_2	0.8	0.6–0.9	—	—
CH_2O	7.1	3.6–10.5	—	—
kcal/hr	2500		2800	6200

[a] From Hollowell et al. (1980).
[b] From Cote (1974).

Table 7.4

Emission from Residential Wood Stoves[a]

Pollutant	Emission (lb/cord)	
	Range	Average
Particulates	3–93	30.3
SO_x	0.5–1.5	0.7
NO_x	0.7–2.6	1.6
Hydrocarbons	1–146	41.6
Carbon monoxide	300–1220	598.3
Polycyclic organic materials	0.6–1.22	0.9
Formaldehyde	0.3–1	0.8
Acetaldehyde	0.1–0.3	0.4
Phenols	0.3–8	3.3
Acetic acid	5–48	21.1
Aluminum	—	1.3
Calcium	—	10.2
Magnesium	—	2.0
Silicon	—	1.6
Manganese	—	1.6
Phosphorus	—	1.0
Potassium	—	3.6

[a] After Duncan (1980).

7.2.1 HUMIDITY

Water is not normally considered a pollutant, but moisture is an excellent solvent for many pollutants and thus influences their concentration; moreover, moisture acts as an important ingredient in the corrosive action of many pollutants. The moisture content of saturated air changes from 5 g/m³ at 32°F to 50 g/m³ at 100°F. Thus, the water content of air changes whenever the temperature changes, and moisture constitutes a major driving force for the migration of many pollutants. Furthermore, moisture is a prerequisite for the growth and survival of many microbes and for the release of formaldehyde from building materials.

The atmospheric water vapor concentration is listed in Eq. 3.9–3.11 and recorded in Figure 3.7. Extremes of high and low humidity are unpleasant, Chapter 3.2.1. An extremely dry climate is encountered in the cabin of long distance aircraft, where the relative humidity can drop to 5%, that is lower than in the driest deserts.

The human body releases about 40–80 ml water per hour via respiration and perspiration. In a closed, crowded room or motor vehicle this rapidly causes stuffiness. If the outdoor air is very dry, as on a cold winter day, or in an aircraft, metabolic water is an important factor in keeping humidity at tolerable levels.

7.2.2 CARBON DIOXIDE

Carbon dioxide levels of occupied rooms are higher than outdoor levels, because the human metabolism produces about 5 ml/min. Typical indoor levels range from 50 to 1200 ppm. Average levels are 100–500 ppm. The recommended ASHRAE 8-hour level is 500 ppm. Typical levels for a San Francisco office are shown in Figure 7.4 (Hollowell et al., 1979). Biological CO_2 from fermentation can be an odorless, hence unperceived, danger in farming areas and wine-producing areas because CO_2 is heavier than air and accumulates in basements, vaults, and wells and can cause suffocation. CO_2 is chemically inert; it does not react with building materials, and must be removed by ventilation.

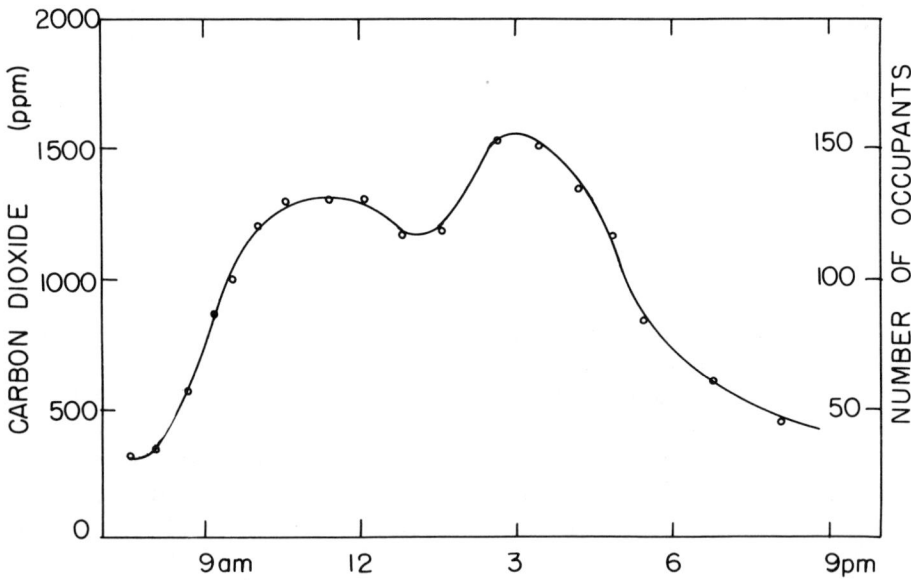

Figure 7.4. Diurnal concentration of carbon dioxide in a San Francisco office building (Hollowell et al., 1979); airflow 6.4 m³/hr (3.8 ft³/min) per person at maximum occupancy.

7.2.3 SULFUR DIOXIDE

Sulfur dioxide (SO_2) is a reactive gas that can be adequately identified and measured with currently available techniques. It fully mixes with air at all temperatures. SO_2 dissolves in water and is absorbed by and reacts with particleboard (Anderson, Lundquist, and Molhave, 1975; Meyer, 1979, 1980), wallpaper (Spedding and Rowlands, 1970), gypsum board (Wilson, 1968), and many furnishings. However, because of its reactivity, its concentration will decrease when it comes in contact with certain substances. SO_2 can react with wood adhesives and wall surfaces; it dissolves in water adsorbed on surfaces, and it oxidizes, especially on active particulate surfaces, to form sulfate particulates, which are often respirable and are known to cause asthma and other health problems. Thus, SO_2 gas is in equilibrium with sulfate particulates in air. The ratio changes as a complex function of time, humidity, particulates, sunshine, and many other factors.

Outdoor SO_2 levels have been thoroughly studied for several decades, and diurnal and seasonal variations have been carefully analyzed (EPA, 1974).

Natural SO_2 levels are very low, probably below 0.1 ppm, except around volcanos (Meyer, 1977). SO_2 levels are continually recorded by several thousand measuring stations.

Sulfur dioxide stems mainly from coal- and oil-burning electric power plants, and its concentration is related to the sulfur content of the coal and the total amount of coal consumed. Since energy consumption is related to the size of a community, there is a direct correlation between the population density of an area and the SO_2 air levels; see Figure 7.5a (Warner, 1979). Unfortunately, peak energy consumption (i.e., SO_2 peak concentrations) can coincide with hot or cold air inversions, which can substantially enhance SO_2 levels. The dispersion of SO_2 is well understood, and is determined by mixing of flue gases with ambient airstreams (Section 3.4.2). During the last 20 years, SO_2 levels in most cities have dropped to about 10% of former values. This reduction is the direct result of environmental legislation and cooperation by the power industry. Typical average values are 8 $\mu g/m^3$ in Portage, WI; 4 $\mu g/m^3$ in Topeka, KA; 36 $\mu g/m^3$ in Kingston, TN; 15 $\mu g/m^3$ in Watertown, MA; 115 $\mu g/m^3$ in St. Louis; and 110 $\mu g/m^3$ in Steubenville. The National Air Pollution Quality standard is currently 80 $\mu g/m^3$. Since the sources are well known, and subject to EPA permits, it is now possible to predict average SO_2 levels for the next 10 years.

Indoor–outdoor ratios have been measured for several decades. If outdoor levels exceed 30 ppb (Yocom et al., 1977), indoor levels are lower than outdoor levels and the indoor/outdoor ratio gradually decreases to about 0.6 (Gentilizza, 1977). Among the classical studies are Andersen's (1972, 1975) research in Denmark, Yocom's study in Hartford (1969), Wilson's measurements in 1968 in London, Grafe's in Hamburg (1966), Miura's in Japan (1965), Biersteker's in 1964 in Rotterdam, Phair's in Cincinnati in 1957 and 1964, and

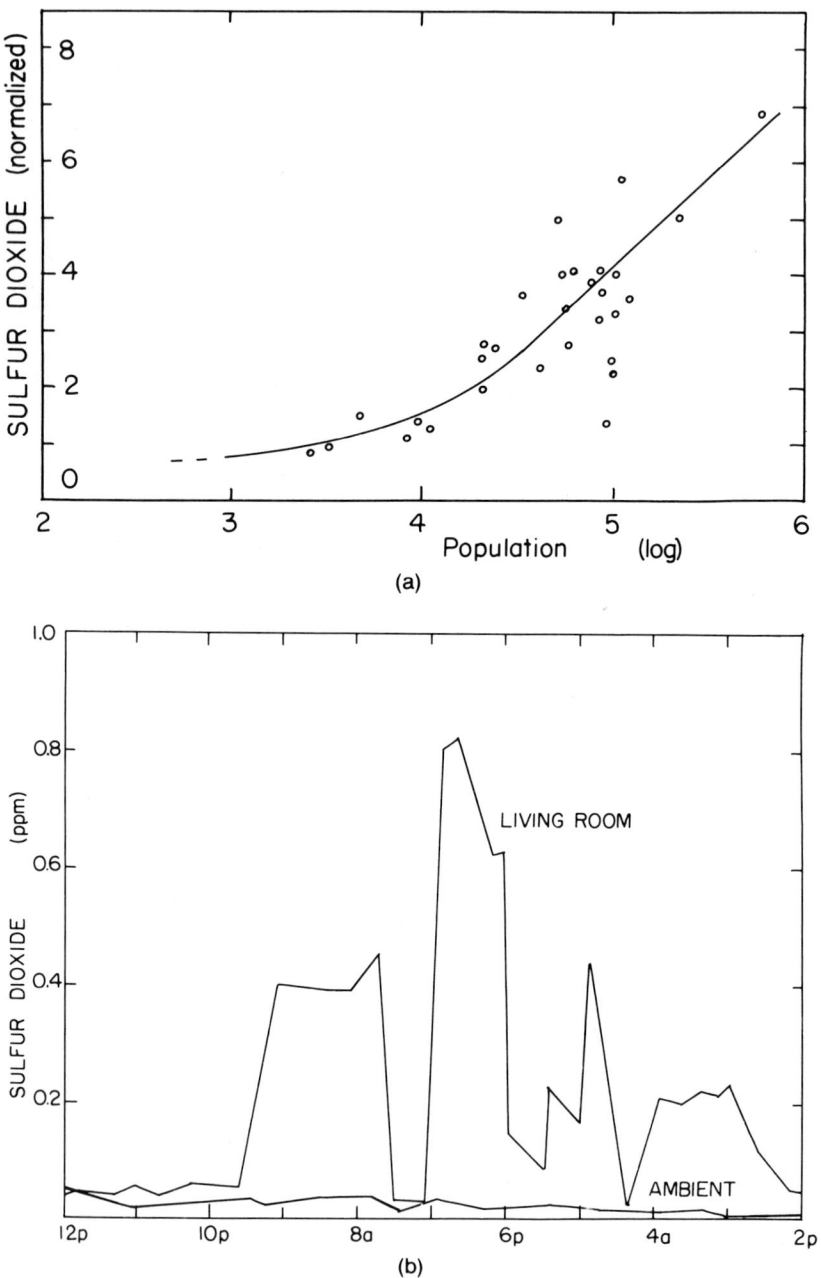

Figure 7.5. Sulfur dioxide concentration (a) as a function of population density (Warner, 1979) and (b) in a coal-heated home in New England (after Yocom, 1971).

the Russian work by Kraglikova in Moscow in 1954. SO_2 measurements are integrated in a current EPA six-city study (Spengler et al., 1979).

Moschandreas (1978) observed values in the range from 3 to 29 ppb in a Pittsburgh mobile home, 0 to 110 ppb in an apartment house in the same city, and 2 to 114 ppb in Baltimore; the lowest level, 1.4–2.5 ppb, was found in Washington. Sullivan (1979) observed 12 sites in Los Angeles and found average values of 7–10 ppb in that area, with individual variations of the same order at each location.

Sulfur dioxide does not currently constitute a serious indoor problem in the United States, except in the few areas where local coal heating and heating by high-sulfur crudes is customary. However, the situation could change if coal and high-sulfur residual oils were reintroduced into the heating market. Yocom (1971) observed in Hartford, CT, that SO_2 levels can reach 0.4 ppm at night and 0.8 ppm during the day in homes where coal is used as fuel (Figure 7.5b). Coal burning in residential fireplaces was responsible for the smog and smoke nuisance that blighted many parts of England for more than 500 years. With the production of oil and gas in the North Sea, the SO_2 problem has been solved.

High SO_2 levels were common around metal smelters in the western United States, and bothered the residents of Tacoma, Shelby, Salt Lake City and Butte for many decades (EPA, 1974, 1980). The situation in Trail, British Columbia, proved in 1935 that SO_2 abatement can be economically acceptable (Robinson, 1974).

7.2.4 CARBON MONOXIDE

Carbon monoxide (CO) is a chemically fairly stable gas that can be rapidly and accurately measured with pocket-sized digital meters. It mixes with air at all temperatures, is biologically 25–27 times more active than oxygen in binding to blood hemoglobin, and causes acute poisoning above about 20 ppm (see Figure 8.12); yet low CO levels are noted even in unpolluted blood and breath. Background levels are about 3–5 ppm in Los Angeles (Sullivan, 1979) and Osaka City (Kaneko et al., 1977). The most important urban source is automobile traffic, which causes concentrations up to 90 ppm along arterials, at intersections, and at traffic lights, especially during commuter rush hours (Ott, 1978–1980), as well as in parking garages (Flachsbart and Ott, 1981). Emission control has drastically reduced CO emission from passenger cars. Diesel engines emit about 1.6 g of CO per brake horse power hour (Sheehy, 1980). Carbon monoxide from city traffic and parking garages drifts into residences, retail stores, and businesses. In tall buildings CO can be swept up to the twentieth floor or higher by elevators and by open doors in emergency staircases in such a predictable manner that CO profiles can be drawn. Figure 7.6 shows the extent to which CO spreading via an open fire exit can affect an entire building. Similar stack effects are observed on the outside of high-rise buildings that are

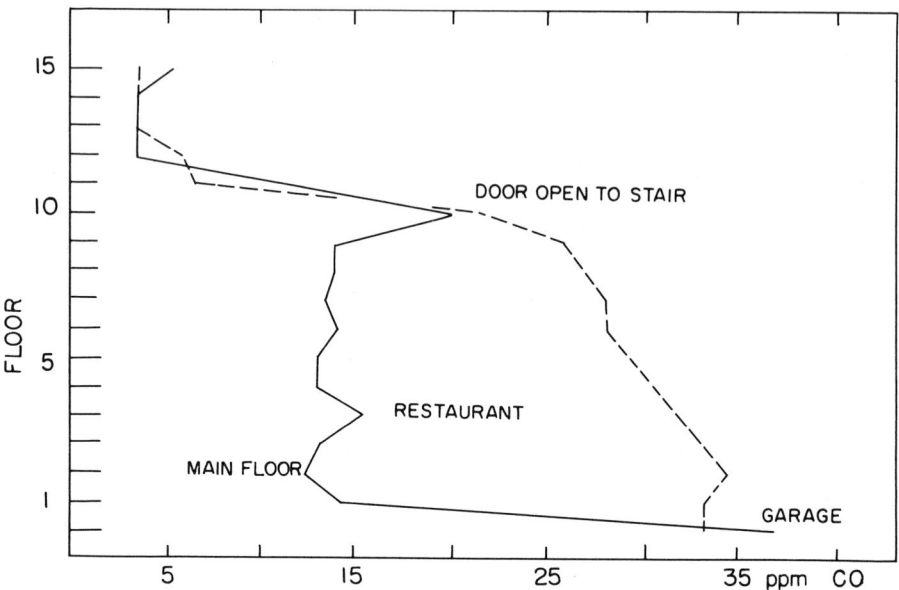

Figure 7.6. Vertical carbon monoxide profile in a tall office building in Palo Alto (after Ott, 1980).

located over traffic arteries in New York (General Electric Co., 1972) where CO may be picked up by window air conditioners. Normally, however, the CO concentration decreases in closed tall buildings approximately exponentially with height (Giles et al., 1974).

Indoor CO levels depend strongly on such indoor sources as gas stoves, heaters, fireplaces, and people who smoke. The Institute for Gas Technology measured CO in 57 kitchens during cooking and found a 1-hr average value of 7.3 ± 6.7 ppm. One of the values was above 39 ppm and one was 22 ppm because of badly adjusted appliances (Elkins et al., 1974). Yocom (Yocom, et al., 1977) estimates that a gas range produces between 1 and 3 mg of CO per hour of operation. Hollowell (1976, 1977) observed similar levels. Ott (1980), Wallace and Repace (1979) have made careful field studies and Ott (1980a) and co-workers have developed careful exposure profiles for CO (Figure 7.7). Typical exposures range from 5 to 50 ppm. In public arenas they are 9–30 ppm (Elliott, 1975), depending on the presence of smokers.

Sebben (Sebben, et al., 1977) found average CO levels of 14 ppm in nightclubs, with values as high as 40 ppm in some bars. Fischer (Fischer et al., 1978) and Weber (Weber, et al., 1979) established that above a threshold of 3 ppm of CO combined with 150 ppb of NO healthy adults show increased eye blink-

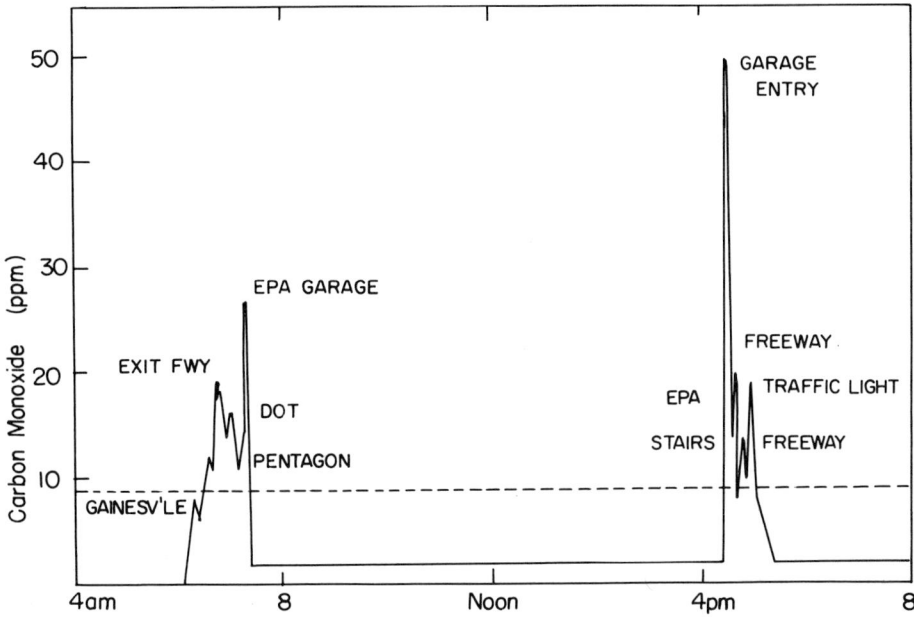

Figure 7.7. Carbon monoxide profile as a function of location and time as measured by personal monitors (from Repace, Ott, and Wallace, 1980).

ing and other signs of discomfort. Substantial seasonal variations have been observed (Yocom, 1971), with superimposed diurnal variations. In a kitchen the CO concentration can increase by a factor of 15 above ambient levels, reaching 13,000 pg/ml during cooking on a gas stove (Wade, Cote, and Yocom, 1975). Unvented space heaters (Lao, 1982), and woodstoves (EPA, 1981) can produce high CO-levels. Sullivan (1979) measured CO for 12 homes in Los Angeles and observed levels averaging from 3 to 7 ppm with ambient values of 0.3–1.2 ppm. The distribution within residences was measured by Moschandreas (1978) and Yocom (1971). The distribution is strongly time dependent (Figure 7.8). The superposition of the local variation with the time-dependent activities (Figure 7.7) shows why the computation of CO exposure levels for individuals requires careful analysis and a multifactor computer program (Ott, 1980).

In most buildings CO from the garage constitutes the strongest source. Depending on weather conditions and building design, significant quantities of CO may be drawn or swept into buildings, where the gas then slowly spreads until it is removed by ventilation. The observed concentration depends on air mixing by recirculation, people movement, and other mixing by thermal or

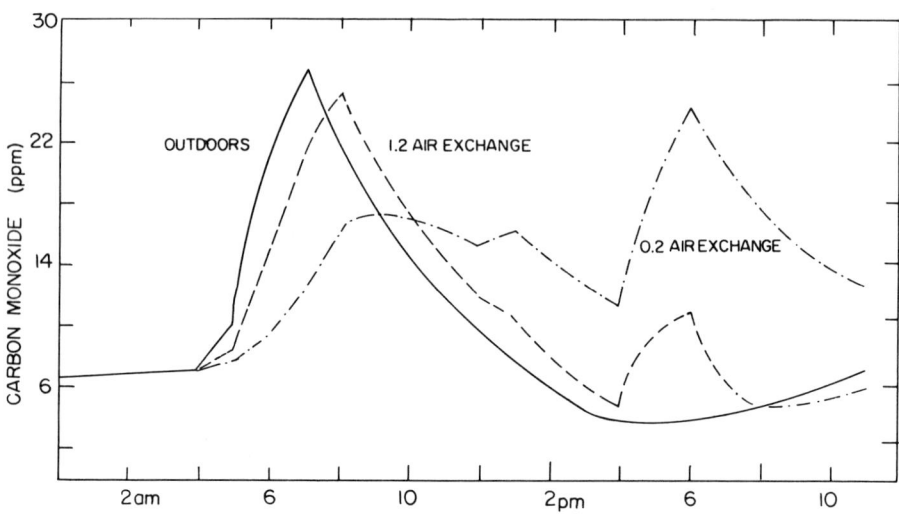

Figure 7.8. Diurnal CO concentration profile as a function of ventilation rate (Moschandreas, 1978).

pressure gradients. Thanks to the work of EPA researchers, CO is well on the way to being the best understood indoor air contaminant. Carbon monoxide concentrations can be predicted from Ott's model and can be modeled with his computer program.

In rooms with smokers the CO level can increase to the point where acute poisoning symptoms are possible; that is, levels may reach peaks of 40–80 ppm (HEW, 1979) if several smokers crowd into an unventilated room (Satish, 1975). In a room with a ventilation rate of 1 air change per hour, one heavy smoker can cause transient CO concentrations of up to 30 ppm. Nonsmokers share concentrations of up to 10 ppm in offices (Repace, 1980), residences and hospitals (Allan, 1982).

Levels of CO are high whenever internal combustion engines are used in an enclosed space. An example of this type of situation is an ice skating rink. Johnson and Breysse (1975) measured CO levels of 45–250 ppm at the breathing level above ice, depending on whether the ice-cleaning engine was in operation. Similar values have since been confirmed by Spengler et al. (1978).

Other indoor/outdoor studies have been reported by HEW (1970), Yocom (1971, 1977), Spengler et al. (1978, 1979), Allen (1978), Moschandreas (1978), and Smith (1978). White and Kragulski (1972) and Barrett et al. (1973) studied the effect of indoor combustion equipment; Ingersoll (1971) measured the effect of soil as a sink on CO; Godin et al. (1972) generalized urban exposures; and Kahn et al. (1975) identified sources within St. Louis. Joshi and Wanner

(1975) found CO lower indoors and Yocom (1976) carefully reviewed his extensive indoor/outdoor work as a function of location, infiltration, season, and meteorology. Koepsel (1977) found that ozone reduces CO, a fact that might be relevant in smoggy areas such as Los Angeles. In 1977 the NAS (1977) published a thorough review of CO and its effects.

Yocom (1971, 1977) found that the indoor/outdoor ratio of CO is always higher in office buildings during the day than at night, and higher in residences during the night than during the day. That is, CO reflects human activity patterns and follows man. He found that CO is highest in fall and winter and lowest in summer.

The Federal Aviation Administration (FAA, 1971) measured an average CO level of 3 ppm in military and commercial aircraft, with a peak at 5 ppm. Levels at airports range from 5.4 to 6.2 indoors and from 10.5 to 11.4 ppm outdoors. The baggage handling areas register 11-ppm levels if gasoline-powered tractors are used (Bellin and Spengler, 1980).

The CO level inside automobiles (Jaeger, 1981; Ott, 1981) depends on the source strength, ventilation rate, and residence time:

$$C_{(car)} = \tau \frac{dC_2(t)}{dt} + C_2(t) \tag{7.1}$$

where τ is the time constant, which depends on the window position and the car's speed; a typical value is 4.5 min for a closed car with only one window at the driver's side one-third open (Ott and Willits, 1981). In a first approximation the indoor level is similar to the outdoor level, which varies greatly along the way (Petersen and Sabersky, 1975). Typical levels range from 6 ppm for low traffic points to 16 ppm at intersections with 30 cars/min (Willits and Ott, 1981; Switzer, 1981). Of the observed values, 10% are above 20 ppm, 5% above 25 ppm, and 1% above 35 ppm (Cortese and Spengler, 1975).

Much higher CO levels are observed in stalled traffic where the exhaust from one car may feed the air intake of the following vehicle. In one case in St. Louis, Brice (1966) observed 77 ppm. In one underground garage Rueden and Langer (1976) found 140 ppm, and in old cars with defective exhaust systems CO levels in the vehicle can become very high, sometimes high enough to cause acute toxic effects, and in some recent accidents (Yabroff et al., 1969) even death. Koch (1976) observed levels of 35 ppm in school buses when he investigated the cause of hospitalization of several schoolchildren.

7.2.5 NITROGEN OXIDES

Nitrogen forms several oxides. Ambient air pollutant components are NO and NO_2; both are gases that are fully miscible with air. Both contain nitrogen in unstable oxidation states, and both are active redox reagents, can interconvert, and can act as catalysts for oxygen transfer. Both gases are irritating to the lung and eye. Nitrogen oxides react quite rapidly in the indoor environment. Cote (1974) found a half-life of 1.8 hr for NO and 0.6 hr for NO_2. Cote (1974), Hollowell (1980), and others found the NO/NO_2 ratio to be almost always greater than 1 and often than 2.

Moschandreas (1978) observed a range of 0–300 ppb for NO, with a maximum of 470 ppb, while NO_2 levels were close to ambient in the five cities studied. Part of the NO_x is oxidized to nitrate particulates, which accounted for 1–6 $\mu g/m^3$ in air. No_x levels have been reviewed by Goldstein (1981).

The nitrogen oxides observed indoors can arise from sources outside or inside the home. Outside NO_x results from the combustion of nitrogen-containing heterocyclic compounds in coal-fired power plants and from reaction of air in internal combustion engines. Automobile engines are now equipped with exhaust control devices, but Diesel engines still emit approximately 4 gm of NO_x per brake horsepower-hour (Sheehy, 1980). The primary indoor sources are gas stoves (Eaton et al., 1972). In Los Angeles, Derham et al. (1974) found NO_x levels between 0.2 and 0.6 ppm, depending on time of day and weather conditions. Sullivan (1979) observed average levels of 0.09–0.2 ppm in 12 locations in Los Angeles in 1979. He observed strong daily peaks of up to 0.4 ppm of NO during the daily cycles, with low values of 0.05 ppm. The ratio of NO to NO_2 varied from 0.3 to 1.2.

Speizer et al. (1980) compared 24-hr average NO_2 levels in recreation rooms in homes in six cities. Figure 7.9 shows daily indoor and outdoor levels of nitric oxide for offices and residences with electric and gas heat (Moschandreas, 1980). Kaneko et al. (1979) found ratios between 0.3 and 8 for NO_2 indoor/outdoor ratios, and between 0.3 and 33 for NO. Spengler (Spengler et al., 1979) compared homes with gas-supplied and electric kitchens and found a direct correlation between homes in which electric and gas cooking units were used in the kitchen. In each city NO_2 averaged 25–400% more in homes with gas appliances than in the electric homes. In four cities the indoor levels of NO_2 were clearly higher than outdoors when gas utilities were present, and lower when electric ranges were present. The geometric mean was 49 $\mu g/m^3$ (25 ppb) outdoors, 41.4 (20.7 ppb) for the recreation room in homes with electric kitchens, and 54 (27 ppb) for the recreation room in homes with gas kitchens. The NO_2 levels in the kitchen 1 m from the stove were 200 $\mu g/m^3$ (0.1 ppm) when the pilot light alone was on and 500–1200 $\mu g/m^3$ (0.25–0.6 ppm) while the stove was in use. Earlier, Elkins et al. (1974) had found a 24-hr mean concentration of 40 ppb in 37 electric kitchens and 74 ppb for 121 kitchens with gas in Chicago. In Columbus OH, levels were 54 ppb in 69 gas kitchens and 18 ppb in 50 electric homes, with outdoor levels at 29 ppb. Summer levels were 29 ± 2.4 ppb for 51 gas kitchens and 9 ± 1.1 ppb for 12 electric kitchens, while winter levels were 31 ± 3 ppb for 70 gas kitchens and 4 ± 1.7 ppb in 24 electric kitchens. In rooms remote from the kitchen Elkins found 7 ± 1.6 ppb in gas homes. Daily variations of NO_x in gas kitchens have been reported by Yocom (1975), Hollowell (1979a), and Speizer et al. (1980). Yocom et al. (1977) estimated NO production at

7.2 Pollutant Concentrations

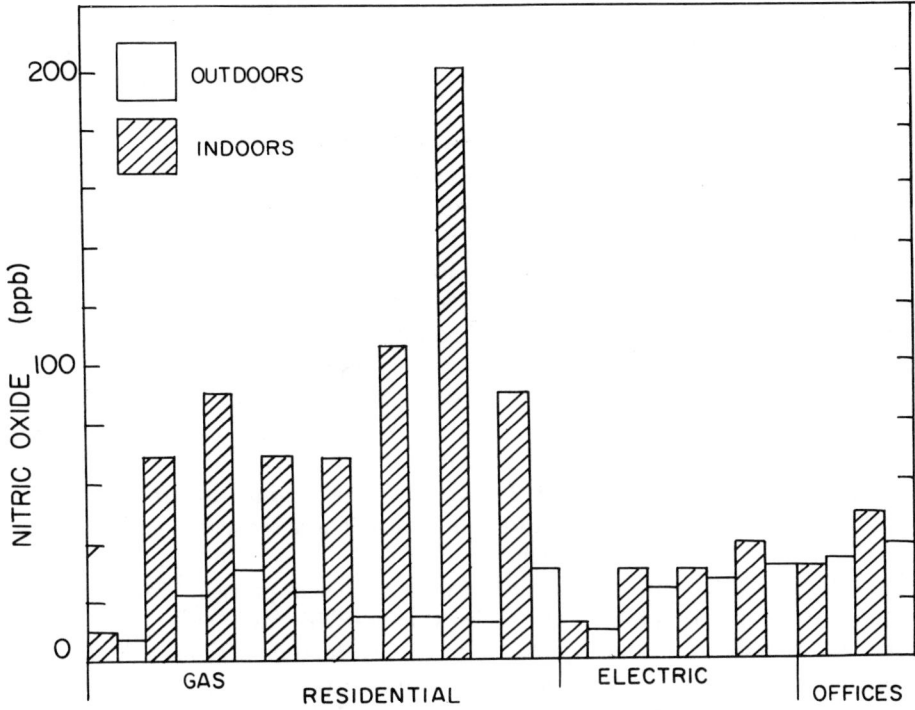

Figure 7.9. Indoor and outdoor levels of nitrogen oxides for residences and offices with gas and electric heating (Moschandreas, 1980) in the Boston area (95% confidence limit).

150–1500 mg/hr and NO_2 at 100–700 mg/hr during operation of the stove. Macriss (1977) explained the influence of ventilation.

7.2.6 OZONE

Ozone (O_3) is a reactive gas that is fully miscible with air. Outdoors it is produced by solar irradiation of polluted air. Indoors ozone is produced by uv light fixtures and by electrostatic office copying machines. Although ozone is beneficial as a sterilizing agent against indoor fungi, bacteria, and viruses, even at low concentrations it irritates the lung and eye tissue and it is responsible for photochemical smog. Ozone occurs jointly with peroxyacetyl nitrate (PAN) and is an indicator of other air pollutants (Dmitriev et al., 1979). The EPA has published an extensive review of research on ozone done before 1971.

Shair and Heitner (1974) showed that indoor ozone in Los Angeles is largely due to outdoor sources. Accordingly, the diurnal concentration follows the outdoor concentration with a time lag of 1–3 hr, and increases from 7 A.M. until noon. It drops to the night level around 9 P.M. The concentration range is between 0.1 and 0.3 ppm, depending on the weather conditions. Average values measured in 12 areas of Los Angeles ranged from 1 to 7 ppb in 1979 (Sullivan, 1979). Indoor O_3 levels are 50% or more below ambient levels because O_3 reacts rapidly indoors or decomposes by surface catalysis on aluminum, steel, and other building materials. The rate of decomposition of ozone increases fivefold when the air humidity increases from 5 to 55%. At 5% RH the half-life is 10 hr; at 85% RH it is 10 min (Shair, 1974). Shair and Heitner (1974) gave catalysis rates for various materials and found wool and cotton three times more effective than plywood and nylon, and 20 times more effective than metals or glass. However, metal oxides are also effective.

Ozone production by photocopying machines is significant. A freshly serviced machine produces about 4 µg/copy. After extended use the production can increase to up to 131 µg/copy (Allen et al., 1978) with an average of about 40 µg/copy (Selway et al., 1980). This results in breathing-level ozone concentrations of about 4–300 µg/m³ for the operator. There seems to be a significant trend between different makes and models.

If the U.S. Department of Energy pursues a recent proposal for promoting mercury-enhanced indoor light bulbs that can be mounted in traditional sockets, indoor ozone levels would increase dramatically. This might stamp out indoor microbes, but could cause distress to asthmatics and those with sensitive eyes. Current indoor and outdoor O_3 levels for an energy-efficient research home in Iowa are shown in Figure 7.10.

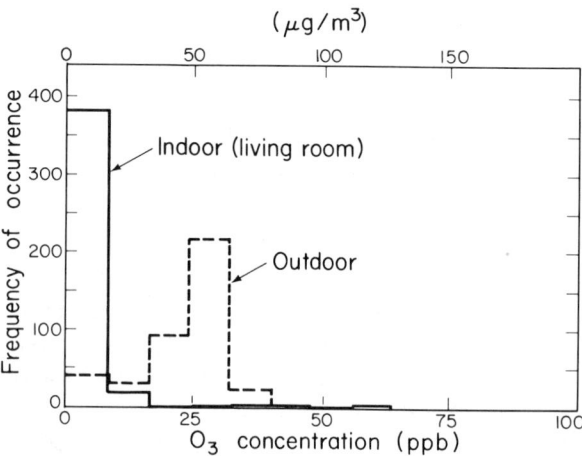

Figure 7.10. Indoor and outdoor levels of ozone (O_3) for the Iowa State University energy research house (Hollowell, 1980).

Ozone levels in automobiles are equivalent to ambient levels. Ozone occurs simultaneously with nitric oxide and carbon monoxide and forms a synergistic mixture. Hospitals have been studied by Allen (1981).

A special situation exists in aircraft because jumbo jets and supersonic aircraft travel at an altitude of 10–15 km, which is close to the earth's ozone layer. At these altitudes, ozone concentrations change regionally and seasonally but can reach up to 0.30 ppm. Many people can smell ozone at such concentrations, which are sufficient to rapidly degrade rubber hoses and similarly sensitive materials. In 1970 the FAA estimated cabin ozone levels to be below 0.02 ppm. In long-distance aircraft ozone coincides with very low humidity; this causes irritation for people with contact lenses or respiratory problems (Higgins et al., 1979, 1980). Ozone can be removed by catalytic decomposition. This is now mandatory (14CFR25.832), as is the provision for nonsmoking sections to reduce synergism with smoking (45FR22763); see Section 10.5.

7.2.7 PARTICULATES AND DUST

Even in remote areas with very clean air we daily breathe about 200 μg of particulates, ranging from harmless sea spray to potentially toxic radioactive dust. Particulates occur in a wide spectrum of sizes and shapes (Figure 5.1). Until recently it was very difficult to measure the size distribution of particulates, and current methods are not yet adequate for quick and reliable field measurements of rapidly changing concentrations. The measurement of microbes is especially difficult and tedious, p. 153.

Figure 7.11 shows the particulate size distribution above a gas stove in a kitchen and during a pollution episode. Most dusts contain two or more distinct size fractions. Table 7.3 shows the chemical analysis of all gases above a stove. Figure 7.11 shows the chemical analysis of two different smog samples. This figure shows a typical range of variation for the constituents. Nitric oxide varies by a factor of five, ammonia by a factor of two. Since the different weight fractions settle at different speeds, and since none of the respirable particulates form stable homogeneous mixtures with air, their concentration changes constantly. Furthermore, they fractionate in air pressure gradients around and inside houses, where dust settles and accumulates, and from where they are resuspended when the wind direction changes or local turbulence develops, as when a person walks close by (Sansone and Slein, 1978; Sehmel, 1980).

Only particulates in the range from 0.1 to 1.5 μm reach the lung, and particulates larger than about 50 μm settle too fast to reach the nose or mouth (Chapters 3 and 8). Particulates can contain almost any type of chemical. Most dusts consist of mixtures, and almost all dusts are coated with water films. The humidity depends on the fraction of soluble salts in the particles. Biological particulates and aerosols change their size as they travel through humidity gradients in the air, and thus change their behavior. Obviously, there is no true

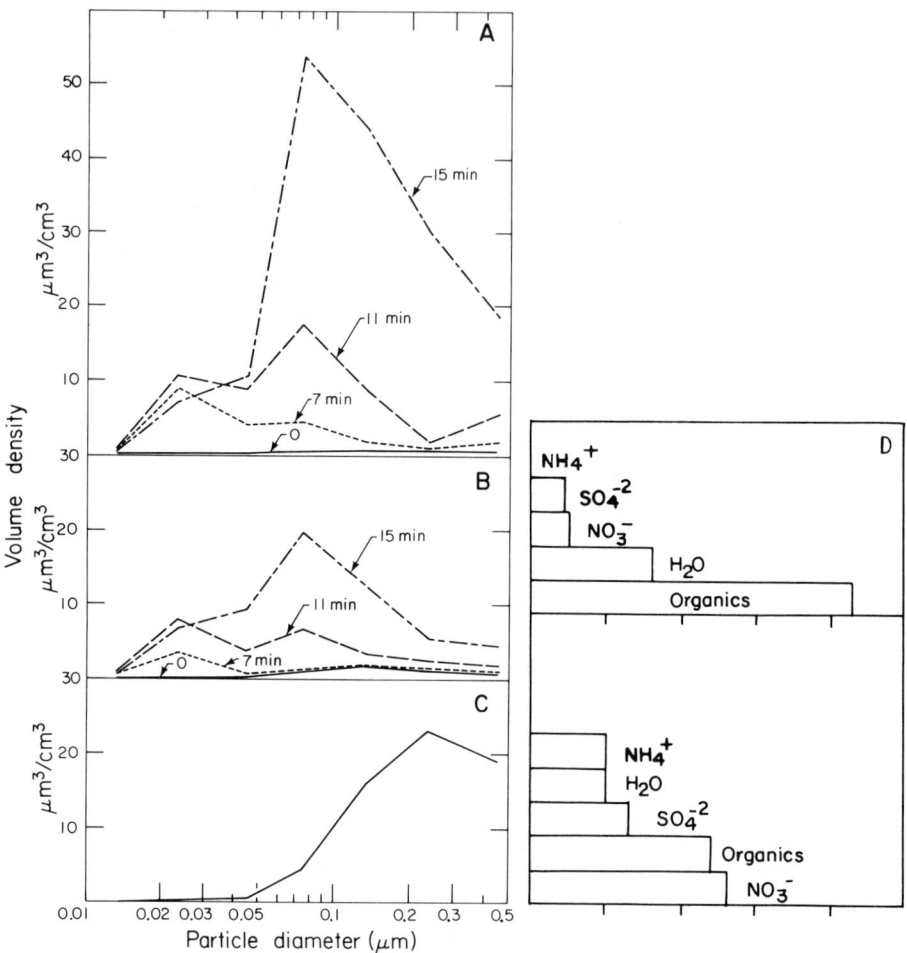

Figure 7.11. Size distribution of smog and kitchen stove particulates. Volume density and distribution of particulate emissions smaller than 0.5 μm in diameter. (A,B) Two burners with water-filled cooking pots at 7, 11, and 15 min after turning the burners on; (C) particle concentration distribution for a San Francisco Bay area air pollution episode; (D) organic fraction in two cities in Southern California (after Hollowell, 1980).

equilibrium concentration of particulates in air, and it is not generally possible to predict their concentration unless they are prepared in a laboratory directly in front of the mouth.

In the United States ambient particulate levels are measured by a network of 2700 stations that report regularly to the EPA. About half these stations record values exceeding 75 mg/m^3, and some exceed the national standard for a total 24-hr limit of 260 μg/m^3. Human activities account for about 200–400 $\times$ 10^6 tons/year of particles with a radius of less than 20 μm. About half are sulfate; about one-sixth is nitrate, and two-sixths are organics. Man-made emission is most densely concentrated in the eastern United States. Natural sources emit about 800–2200 $\times$ 10^6 tons/year of particles in the same size range. Four sources—natural sulfate, soil and rock debris, vegetation, and forest fires—each contribute about 200 $\times$ 10^6 tons/year. Volcanic dust accounts for 25–150 $\times$ 10^6 tons/year, but in 1980 the eruption of Mount St. Helens in Washington produced over 8000 $\times$ 10^6 tons of dust, one ton for each human being living on earth, in one six hour period.

7.2.7.1 *Particle Resuspension*

Dry dust travels with the wind until it is trapped by moisture. In ambient air the suspension rate of soil and dust depends on the local geography, humidity, wind speed, and the nature of the surface crust. Dust transfer occurs on a local as well as a regional scale: In residences and commercial buildings, local eddy currents stirred up by the motion of people are sufficient to resuspend batches of dust. In arid open areas soil moves by three distinct mechanisms: saltation, surface creep, and suspension. The average North American wind erosion rate is estimated to be about 2.4 cm per 1000 years. Agricultural soil losses can reach up to 340 tons/acre each year. The resuspension rate is about 3 $\times$ 10^{-3} per suspended fraction per hour. Soil losses in vegetated areas are up to 3 m per 1000 years.

Saltation involves particles of 0.1–0.5 mm. They bounce along the air-surface interface, hit the soil periodically at an angle of 6–12°, and grind up dust and pollutants as they move along. This causes secondary resuspension and avalanching. Surface creep affects particles with sizes of 0.5–1 mm. They slide and roll with the wind. Particles 0.1 mm and smaller rise vertically with the wind and are carried in the air. Saltation is by far the most efficient method for the transport of the soil mass.

Transport phenomena depend on a very large number of variables. Sehmel (1980) lists 24 meteorological, 14 topographic, and 19 surface-dependent variables; 21 particle–soil interaction factors; 8 soil property factors; and 10 factors determined by the pollutant. The total mass transport increases with more than the third power of the windspeed. The vertical concentration profile depends on local geography and wind speed, but always exhibits a maximum at a height of a few meters. The pollutant concentration above the ground is defined as the "airborne concentration half-life," a phrase that is easily confused with, but has no relation to, the residence time of suspended particulates. Measurements of the airborne concentration half-life have been made for many radioactive substances. The values range from 1–2 months for plutonium in the Nevada desert to a few years in rocky flats and in some Russian locations. In other areas, for example Hanford, WA, the settling is apparent at least 1000 times slower because of lower wind speed and higher humidity.

In all locations seasonal resuspension variations are very large. In spring, for example, the rates in Hanford range from 5×10^{-13} to 2×10^{-11}, depending on wind speed. In summer and fall they range from 4×10^{-11} to 2×10^{-8}.

Road traffic increases resuspension to a rate of from 10^{-5} to 10^{-2}, depending on the speed of the disturbance and the road roughness (Figure 7.12). The rate for pedestrians is from 10^{-5} to 10^{-3} (Sehmel, 1980) for particles with an average size of 8 μm. Indoor resuspension rates have not yet been reported, but they are probably similar.

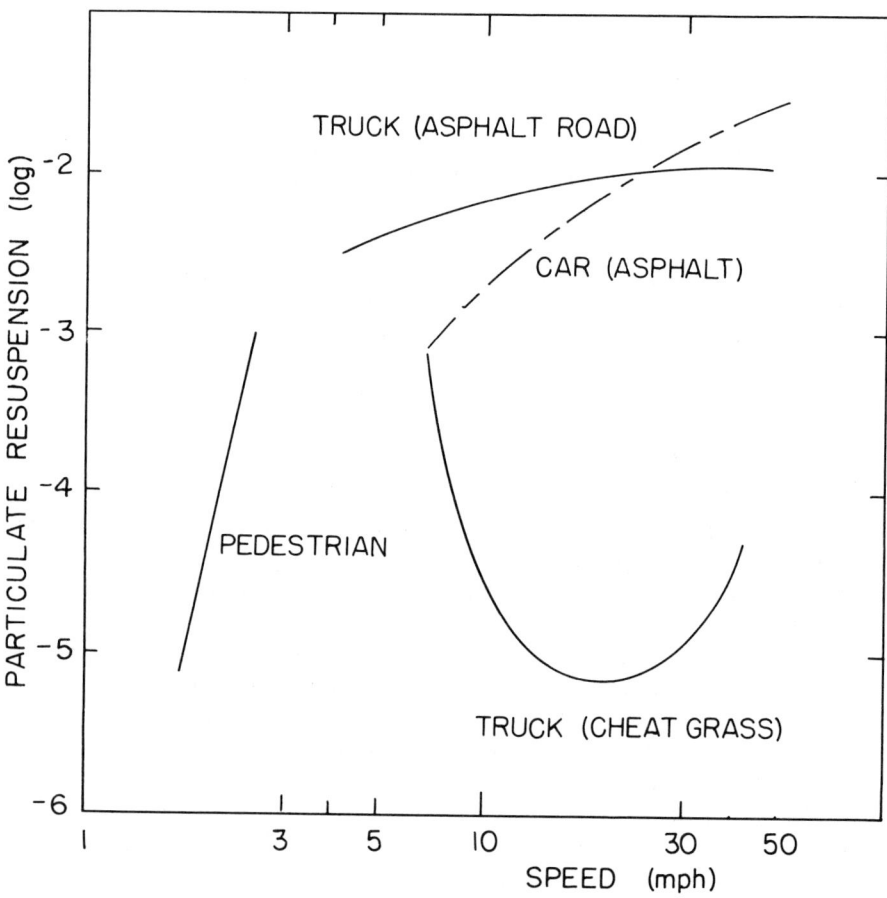

Figure 7.12. Particulate resuspension by automobile on asphalt road, truck on asphalt, truck on grass, and pesticides suspended by pedestrian (Fanger, 1979; Sehmel, 1980).

7.2.7.2 Indoor Particulate Levels

The overall concentration of indoor and outdoor dust is frequently comparable, but the composition of indoor dust differs from that in the ambient air. Ambient air pollutants are typically lower. For example, respirable lead is reduced from 0.65 µg/m³ outdoors to 0.60 µg/m³ indoors in New York City (Halpern, 1978). On the other hand, cigarette smoke, which is negligible outdoors, reaches very high levels indoors because each cigarette releases about 10^{12} particulates in volumes of 10-50 m³. Likewise, resuspended house dust can create very high local indoor levels.

Dockery and co-workers (1981) measured respirable sulfate in 68 homes in six cities over a period of a year and found that indoor sulfate levels are always about 70%. In fully air-conditioned homes the level is only 50% of outdoor levels, but as expected, such homes retain particulates longer and have higher resuspension rates. Dockery developed an indoor–outdoor model, based on mass conservation, that yields an average indoor concentration of

$$\bar{C}_i = PC_o + (1/q)\bar{S} \tag{7.2}$$

where $\bar{C}_i$ is the average concentration, P the penetration rate of the pollutant (as a percentage), C_o the outside level, q the flow rate in volume units, and $\bar{S}$ the interior generation rate. This equation is valid at ventilation rates of about 1.5 air changes per hour, at a settling rate of 0.5 an hour for respirable sulfate. Indoor sulfate stems from smoking and gas stoves.

Dockery found the mean concentration of respirable particles to be 32 µg/m³ and that of sulfate to be 5 µg/m³. The highest sulfate level was 7.6, in Steubenville, and the lowest 2.7 µg/m³, in Topeka and Portage; the highest respirable particle concentration was 47 µg/m³, in St. Louis, and the lowest 19.6 µg/m³, in Portage. The largest effect in individual homes could be correlated to smoke from cigarettes. In homes without smokers the values were 15 ± 2 µg/m³ for respirable particles and 0.4 ± 2 µg/m³ for SO_4^{-2}. One smoker per home increased the average to 35 ± 8 µg/m³ for respirable particulates and 0.8 ± 3 µg/m³ for SO_4^{-2}, and more than one smoker yielded values of 50 ± 15 µg/m³ for respirable particulates and 1.8 ± 0.6 µg/m³ for SO_4^{-2}. The difference for all other factors was overshadowed by the presence of smokers. Homes with electric kitchens had 22 ± 5 µg/m³ for respirable particulates and 0.6 ± 2 µg/m³ for sulfate as compared to 24 ± 6 for respirable particulates and 0.9 ± 3 µg/m³ for sulfate in homes with gas heaters and kitchen ranges. Full air conditioning increased the respirable particulate concentration from 21 ± 2 µg/m³ to 28 ± 8 µg/m³ and sulfate from 0.6 ± 2 to 1.1 ± 4 µg/m³. The difference between oil and gas heating fuel was not pronounced. The author also reports the effect of air conditioning, heating system, and storm windows, as well as kitchen ventilation methods, on the indoor–outdoor exchange rate, but for the purposes of this book the absolute indoor concentrations are more significant.

Moschandreas (1978) reports indoor sulfate values of from 2.5 $\mu g/m^3$ in Chicago to 15 $\mu g/m^3$ in Pittsburgh at outdoor levels of from 5 in Chicago to 17 $\mu g/m^3$ in Pittsburgh. He observed that in the latter city outdoor peak concentrations developed over periods of about three days with variations of from 4 to 16 ppb, while the corresponding indoor peaks were substantially blunted to 1.5–9 ppb. These values correlate well with the extensive and thorough work of Yocom (1971), who studied a public library, four private homes, an office building, and the city hall in Hartford, CN, over all seasons over the various diurnal and other cycles. He found indoor–outdoor ratios that ranged from 0.2 to 1.2 $\mu g/m^3$, with the lowest value in winter and the highest in summer and the highest outdoor levels and variation during the day, probably as a result of vehicular traffic. He differentiated between suspended particulates and soluble particulates as well as soiling particulates. Neal et al. (1978) measured indoor particulates in three urban hospitals in Chicago over a 6-month period and found values of about 30 $\mu g/m^3$ in two intensive care stations. The 48-hour integrated indoor–outdoor values at two locations were 32/63 and 17/68 $\mu g/m^3$ in summer, and 14/61 and 50/77 $\mu g/m^3$ in winter, with an annual average of 21/73 and 39/71 $\mu g/m^3$. These values are lower than those obtained by Thompson, who had observed peak indoor–outdoor values of 60/210 $\mu g/m^3$ in the intensive care units.

Lefcoe and Inculet (1975) and Benarie, Chuong, and Nonat (1977) recognized early that indoor particulate levels were directly correlated to the number of occupants in a building and that they increased during rush hours and dropped at night and during business holidays, independently of outside levels. Halpern (1978) also found that indoor levels could not be directly correlated to outdoor levels.

Large daytime and local variations were also observed by Repace et al.,(1980; see Figure 7.13) and by Wallace (1980), who carried piezobalances as real-time personal monitors throughout the day and painstakingly recorded data every fifteen minutes.

Yocom (1969) studied the removal efficiency of air filters, electronic air filters, activated charcoal filters, air washers, mist eliminators, air cooling coils, heating coils, and humidifiers for particulates (Chapter 9), and found filters and electronic devices capable of removing particulates by an order of magnitude over the entire observed range from 15 to 112 $\mu g/m^3$.

7.2.7.3 Smoking

Repace and Lowrey (1980) studied respirable particulate levels in nonsmoking areas and found indoor levels of 24–55 $\mu g/m^3$ in 16 places in Washington, DC. The lowest and highest values were in private residences, with intermediate values of 30 $\mu g/m^3$ in a church during Sunday services and at the Library of Congress, with outdoor levels at 40 $\mu g/m^3$. In contrast, the levels in similar places, with 3–10% of the occupants smoking, ranged from 90

7.2 Pollutant Concentrations

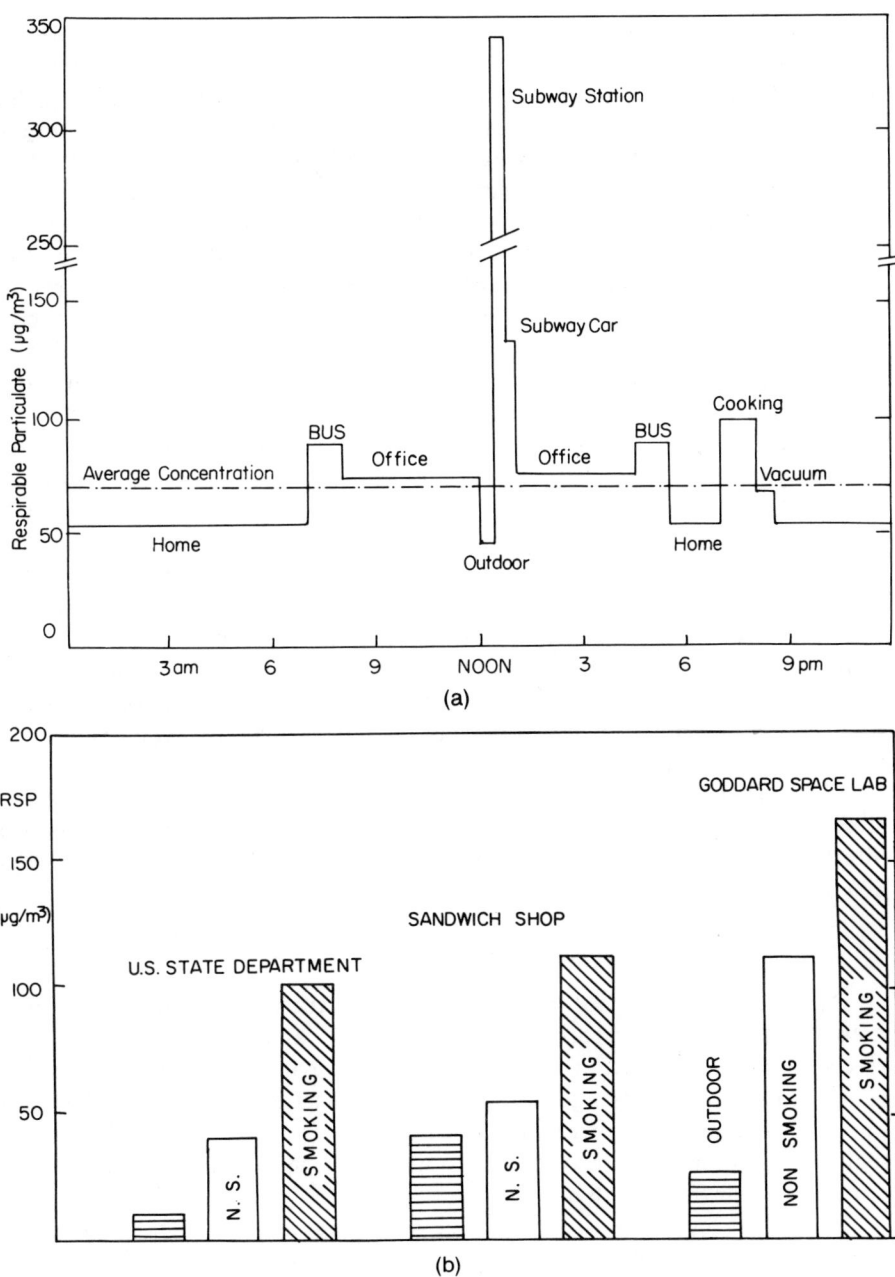

Figure 7.13. Particulate time and space profile, (a) during 24 hr day (after Ott, 1980), and (b) in three public places (Repace, 1982).

to 700 µg/m³, with the lower values in restaurants and the higher levels in a lodge hall, at bingo games, and at a cocktail party. In a 22-m³ unventilated room with natural air mixing he observed a buildup from 30 to 300 µg/m³ within 5 min after the lighting of a cigarette. The upper decay curve in Figure 7.14a represents natural air mixing in the wood-paneled den; the lower curve was obtained in a repeat experiment by mixing the air with a fan.

The effect of occupancy and ventilation rate on tobacco smoke particulate concentration is shown in Figure 7.14b. It is readily seen that six smokers at a party can easily produce air levels of 900 µg/m³, even at an air change rate of 1.5 per hour. This concentration corresponds to the air pollution emergency level. The air levels for commercial premises and aircraft are computed from ventilation rates corresponding to current practice, which is generous compared to the reduced rates recently adopted by ASHRAE (Chapter 10).

Many dozen smoking studies have been made since the U.S. Surgeon General focused his attention on smoking (1964) and its effect on nonsmokers (1979). Passive smoking was studied by Harke et al. (1972); Miyazaki et al. (1977) included incense and mosquito-repellent sticks in his smoking study, and Sebben et al. (1977), Brunnemann (1978), Fischer et al. (1978), and Weber et al. (1979) all studied smoking indoors. The last two groups developed an annoyance index for healthy people. (The levels of pollution found in different locations where smokers congregate are summarized in Table 7.5.) Tobacco smoke contains up to 10 µg/m³ of nicotine, as well as measurable amounts of cadmium, BaP, and other toxic agents (Section 5.3). Concentration measurements have been compared by Sterling (1982).

Elliott (1975) measured particulates in arenas of various sizes and found levels of 150 µg/m³ in an arena seating 11,000 in which smoking was prohibited, whereas an arena for 2000 where smoking was permitted had levels of 620 µg/m³. Levels of CO and BaP were likewise related.

7.2.7.4 *In Transit*

Particulate levels in automobiles depend on traffic conditions. The main sources of particulates are resuspended dust and exhaust fumes from preceding automobiles. Exposure levels for total particulates are estimated to be 60–200 µg/m³ in passenger cars, 70–150 in diesel buses, and 80–190 in subway cars (Repace and Lowrey, 1980). The relatively lower levels in buses conceal the fact that buses have no emission control as yet and emit about four times more particulates than cars.

Particulate levels are estimated to be 40–120 µg/m³ in airplanes where 39% of the passengers smoke (FAA, 1971), but these data were obtained by insensitive methods that might have underestimated respirable particulate levels by a factor of two.

High particulate levels are found in subway stations, which are characterized by levels of between 200 and 500 µg/m³, even in the new and still very

7.2 Pollutant Concentrations

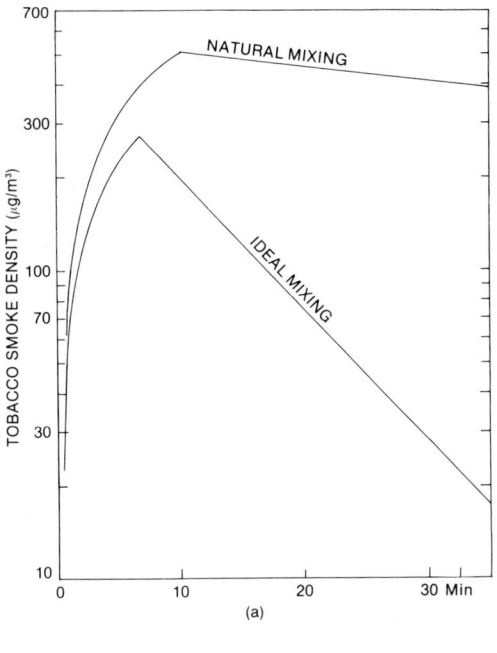

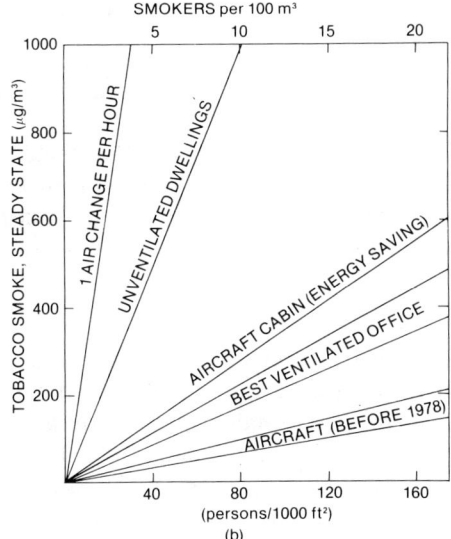

Figure 7.14. Respirable particulate concentration as a function of smoking, (a) Decay curve in a ventilated and unventilated room; (b) tobacco smoke particulate concentration as a function of occupancy and ventilation rate (after Repace and Lowrey, 1980).

Table 7.5

Nicotine, Particulate, and Carbon Monoxide Levels
in Smoking Sections of Public Places[a]

Location	Constituent	Level
	Nicotine	
Commuter train		0.0049 mg/m^3
Commuter bus		0.0063 mg/m^3
Bus waiting room		0.001 mg/m^3
Airline waiting room		0.0031 mg/m^3
Restaurant		0.0052 mg/m^3
Cocktail lounge		0.0103 mg/m^3
Student lounge		0.0028 mg/m^3
	Particles	
House		48 × 10^6 per cubic foot
	CO	
Offices		2.7 ppm
Nightclubs		13.4 ppm (6.5–41.9)
Restaurants		8–28 ppm
Bus		7.3 ppm (6–14)
Conference room		
8 air changes/hr		8 ppm
6 air changes/hr		10 ppm

[a] Report of the U.S. Surgeon General (HEW, 1979), and Report on Indoor Pollutants (NRC, 1981). See also Sterling (1982).

clean-looking systems such as the Washington, DC, metro and the San Francisco Bay Area Rapid Transit, or BART (Repace and Lowrey, 1980). Figure 7.15 shows the range of particulates observed in transit. Even moderate smoking overwhelms all indoor locations.

7.2.7.5 Asbestos

Asbestos consists of distinct needle-like fibers, of which those that are 0.1–5 μm in size are readily respired and absorbed in the lung, and migrate to the lymph ganglions where they remain permanently as latent irritants. The fibers can be readily recognized and counted under an electron microscope. Concentrations of up to 50 particles/ml and more can be found around and above natural deposits of asbestos. An ideal and fire-proof insulating material, asbestos was extensively incorporated into buildings and building materials in almost all construction until about 1960, when it was realized that asbestos products release respirable fibres that, airborne, threaten not only construction workers at the time of installation but also occupants of the building for its entire lifetime. Asbestos from air-duct insulation has been found in

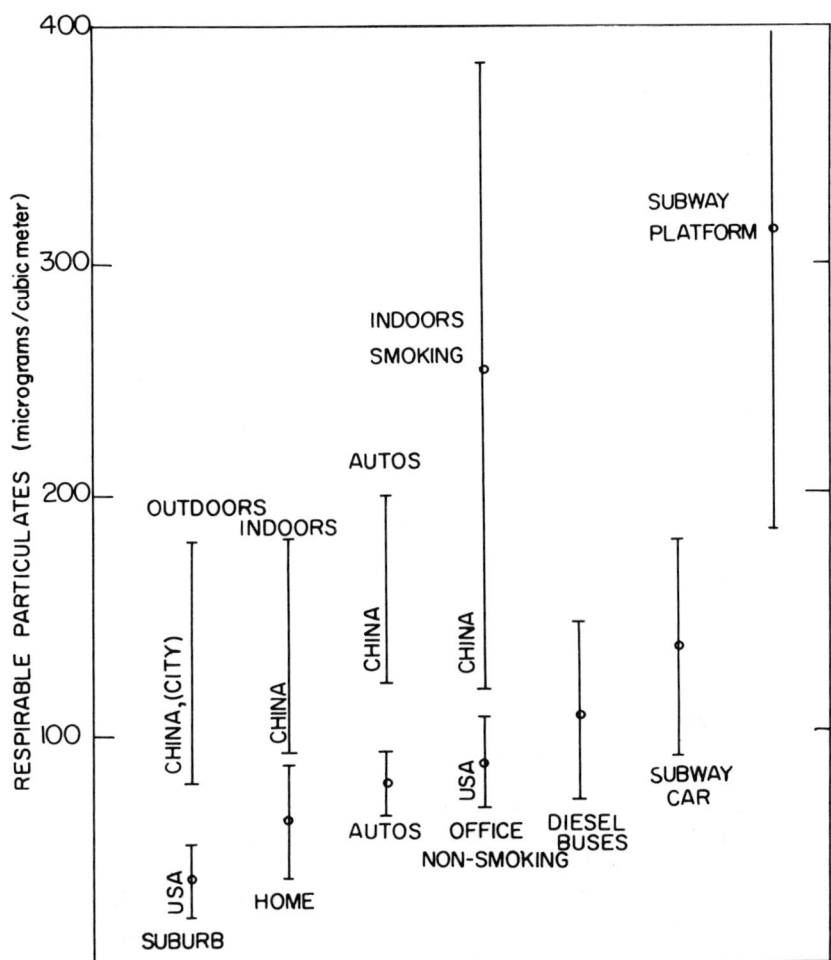

Figure 7.15. Particulate exposure range at seven locations (after Repace and Lowrey, 1980).

many school buildings built during the 1960s, and the Environmental Protection Agency had to initiate a special program (EPA, 1977–1980) to safely remove asbestos from places where its concentration exceeds the NIOSH safe level of 5 particles/ml. By way of example, 40 out of 250 schools in Maryland contain a total of 36,000 ft^2 of friable asbestos that should be removed. It is not yet known whether fiberglass dust is also dangerous (Leadbetter and Corn, 1972). Typical asbestos levels are shown in Table 7.6. The long term health effect is shown in Figure 8.17.

Table 7.6

Airborne Asbestos Concentrations in 11 Locations
Determined by Electron Microscopy Count

Location and Activity	Mean (ng/m^3)	n	Range (ng/m^3)
Urban outdoors			
48 U.S. cities	<10	187	—
New York City	17	22	2–65
New Jersey schools	14	3	3–30
Indoors[a]			
Office buildings	79	3	40–110
New York City schools	99	5	9–135
Massachusetts schools	151	5	38–260
New Jersey schools	217	27	9–1950
Office buildings	2.5–200	116	0–800
Custodial activity			
New Jersey apartment	296	1	—
Connecticut schools	643	2	186–1100
New Jersey school	1950	1	—

[a] Friable asbestos from structural surfaces.

7.2.7.6 House Dust

The composition of house dust changes depending on location, season, the nature of the building, its external and internal environment, and the indoor activities. House dust is not homogeneous, and thus the different components cannot be analyzed with equal accuracy. (The chemical analysis of house dust is described in Chapter 5.) Along freeways, house dust contains lead, as it does when paint that contains lead has been used. Dust around aluminum plants contains fluorides (NRC, 1971). It is controversial whether certain elements, such as potassium or bromine, may be used to determine the indoor/outdoor ratio of house dust. If house dust circulates through a heater, its composition (Joshi and Wanner, 1975) and the pH (Satish and Wanner, 1975) change. Desaedeleer and Winchester (1975) measured the concentration of trace metals in exhaled breath to determine the retention ratio by the lung. Cotton dust is known to be an allergen to certain groups of people (Hammad and Corn, 1971). Certain detergents, manufactured mainly during the middle 1960s, contained enzyme dust that was highly allergenic (Fulwiler, Abbott, and Darcy, 1972). Carbon fibers have been investigated by Harvey (1979).

General observations on house dust have been reportd since Pettenkofer correlated tuberculosis with stale indoor air in 1870. Jennison (1974) used high-speed photography to observe the spread of aerosols from coughs and sneezes. These activities release clouds of small droplets at high speed; the droplets evaporate during the process and a fraction converts to dry airborne particulates that can carry microbes as long as they are viable.

The dust of house mites, which is vigorously disseminated during bed-making, remains allergenic long after the death of the parasites. Vacuum cleaning reduces the dust concentration on the floor, but contributes heavily to the resuspension of small particulates (Annis and Annis, 1973). Frequent showering increases the shedding of skin scales, and thus the concentration of indoor bacteria. Research on house dust has been reported by Goldwater, Manoharan, and Jacobs (1961), Lefcoe and Inculet (1975), and many others.

Combustion plumes have been described and vividly photographed by Marsch (1947). Their fly ash content has been analyzed by Davison et al. (1974), and the chemistry of aerosol and dust has been discussed by Spedding (1974). Wilson (1968) measured (in his lab under unspecified conditions) the half-life of air retention of smoke as 2–5 hours, of sulfur dioxide as 40–60 min, and of hydrogen chloride as 7 min.

Pesticides. House dust contains many pesticides. DDT, for example, was found in house dust worldwide in 1975 (Davies, 1975). In July 1976, the release of 2 kg of TCDD during a factory accident in Seveso led to fumigation of a large area inhabited by 350,000 people. The resultant pesticide concentration is shown in Figure 7.16 (Doll, 1979); health effects so far have been less adverse

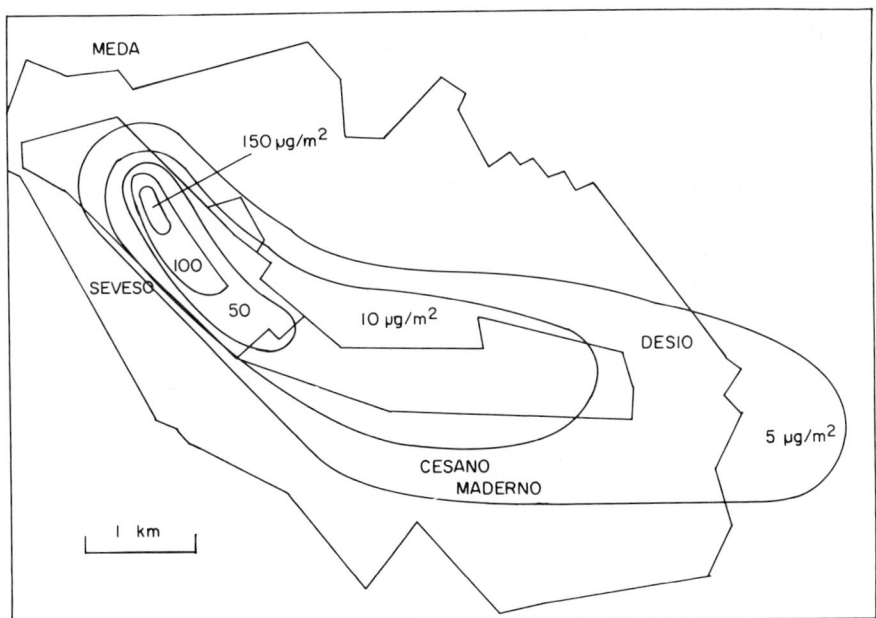

Figure 7.16. Dioxin concentration in the Seveso region (after Doll, 1979).

than anticipated. The indoor level of these agents is not predictable because their vapor pressure changes at ambient temperature and shifts the ratio of the vapor-to-particulate transport rate. At the beginning of the winter heating season, pesticides that have accumulated with dust at pressure gradients around the basement furnace may vaporize or become resuspended with other dust and enter living areas in spurts at substantial concentrations. They are likely to deposit preferentially on moist or cold sheltered surfaces, again at concentrations far above what would be assumed to be the average. They also accumulate on air filters, from which they can become resuspended upon changes in temperature, humidity, or pressure, or after the filter has outlasted its usefulness (Air Filtering System Designing Committee, 1967).

Many aerosol sprays release aromatic organic particulates (NRC, 1972). The particle size in aerosols frequently makes them respirable (Vos and Thomson, 1974).

7.2.7.7 *Microbial Dust*

Microorganisms are indigenous to man (Rosebury, 1962). Several human activities produce biologically active dust. Certain wood panels and the cutting, planing, and sanding of redwood, oak, cedar, and other species provoke allergic reactions in some people and can cause cancer in those exposed on the job (WHO, 1981).

A peculiar problem is posed by the high-speed, water-cooled air turbines used in modern dentistry (Miller, Buron, and Spore, 1963), which carry potentially powerful biological and inorganic agents. Concentrations of microbes are discussed in Section 5.3. Spore counts can range from 10 to 20,000 per cubic meter. Fungus spores on damp walls can reach 360,000 per cubic meter (Gregory, 1973). The speed of their settling depends on their aerodynamic size. The indoor/outdoor ratio of microbial particulates may be either larger or smaller than 1. The indoor concentrations of pollen are lower and depend on the proximity of the source, the aerodynamics of the species, the climate and meteorology, and of course, on whether a building is air conditioned or not. Yocom (1971) found indoor/outdoor ratios of 0.2/300 for *Cladosporium*. The indoor/outdoor concentration ratio always decreases linearly with increasing outdoor concentration. The bacteria and fungi are greatly influenced by housekeeping and the cleanliness of humidifiers and air filters (Gregory, 1973; Edmonds, 1979), which can become infested with microbes if they are not properly maintained (Sterling, 1981; Pennington et al., 1966; Grieble et al., 1970; Rosenzweig, 1970; Scott and Jacobson, 1970; Hodges et al., 1974; Crowley, 1978; Rylander et al., 1978; Broome and Fraser, 1979.) Figure 7.17a shows a plot of their relative size against their settling speed. Figure 7.17b shows typical settling speeds, expressed by Eq. (7.3). Ishido et al. (1956) and Seisaburo et al. (1959) found high bacteria counts indoors in Japan, as did Winslow and Browne (1914) in schools and offices in New York, but more re-

7.2 Pollutant Concentrations

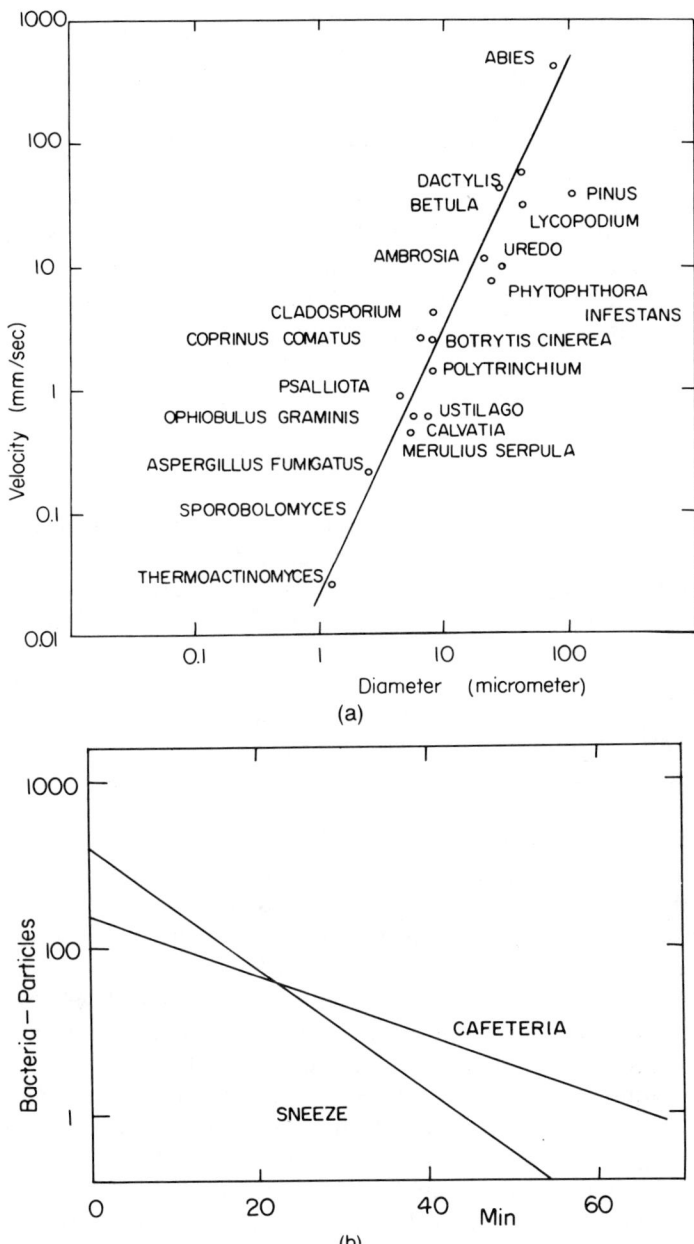

Figure 7.17. Settling of microbes: (a) terminal velocity of 20 spores and pollen; (b) settling of bacteria-carrying particles in a military canteen and after a sneeze (after Gregory and Monteith, 1967; Lidwell, 1979).

cent work by Spiegelmann et al. (1963) in Philadelphia and by others indicates that indoor bacteria counts are lower in many locations. Obviously, this has to do with indoor temperature and humidity of the building envelope. Drastic energy conservation measures are likely to foster increased indoor microbial growth, as the history of tuberculosis (Section 2.1) shows.

Airborne infections have always been feared in crowded places, such as schools, navy vessels (Wright et al., 1968), and military barracks (von Pettenkofer, 1873; Buchbinder et al., 1945). Hospital air is an excellent testing ground for microbes and their transport mechanisms (Fincher, 1969). If the pressure gradient between rooms is controlled and filters in the air conditioners are well maintained, as is done in hospital operating theaters, it is possible to keep the microbe count close to zero. The rate of removal of microbes by sedimentation, ventilation, and death is approximately exponential:

$$N = N_o \exp(-BT) \qquad (7.3)$$

where N is the microbial count and B represents a constant incorporating all three factors. At night after everybody has settled down, B values are about 6, and during the day they are 4 (Gregory, 1973). The location of bacterial aggregations has been carefully studied by Daws (1967), who found that heating and ventilation contribute to resuspension. He computed the effect of the isothermal shear stress against flat surfaces and of the convection about heat sources as well as the effect of moving occupants.

While 98% of outdoor hay fever spores can be removed readily from a clean home, bed-making may inject into the air as many as 4×10^5 skin scales per cubic meter, each of which is capable of carrying many microbes and mites (Gregory, 1973) while they settle to the ground at an average speed of only 2 mm/sec (5 in./min). Hospital air in Paris had bacterial counts of 10^4 per cubic meter in 1883 (Miquel), but by 1948 these values had decreased to 10^2 per cubic meter. However, with closed windows, mold counts increased in wards. Unfortunately, one tuberculosis spore per 5×10^2 m^3 is sufficient to cause infection (Gregory, 1973). Allergens may be responded to at similar levels by sensitive people. For example, one ambrosia (i.e., ragweed) pollen spore may induce rhinitis in some people (Gregory, 1973; Harrington, 1965).

Today, clean-room technology makes it possible to produce microbe-free air in virus labs, space shuttles, electronics factories, and the like, but only if people are kept out of these rooms or are detoxified with masks and clothing.

Appropriate balancing of ventilation requirements with energy conservation is necessary to determine the best possible ventilation rate (Russenberger, Scholz, and Wanner, 1974).

General aspects of aerobiology have been discussed by I.H. Silver (1970). Daws (1967) photographed smoke movement in rooms. Williams (1967) evaluated the spread of pathogenic microbes related to pulmonary tuberculosis and anthrax; of staphylococci and streptococci, which cause throat infections;

of meningococci and pneumococci; and of the microorganisms that cause plague and other airborne diseases that need to be checked in hospitals, where 1–10 microbes/m³ are sufficient to cause havoc. Rati and Ramalingam (1979) compared indoor and outdoor levels of *Aspergillus* around chicken sheds. Tyrell (1967) followed the response of the nasal passage to microbes, and the resuspension of parasitically infested dust and found up to 2×10^6 spores/ft² on a plate held under the chin of a sneezer when the nasal count was estimated to be 8×10^8. Thus, a person can discharge 25% of the entire microbial count in the septum in one sneeze. He also followed the decay curves of microbes ejected by volunteers infected with Coxsackievirus in a study initiated by Buchland (1964). The very complex subject of aerobiology has been analyzed in excellent books by Gregory (1973; Gregory and Monteith, 1967) and Edmonds (1979), wo found that microbes, like other particulates, can travel as plumes for many hundreds of miles. Thus, microbial concentrations are not reliably predictable, and freak infections are possible anytime that microbes are present. This message was brought home 20 years ago when 13 people in Oakland, CA, contracted Q fever in their homes from microbes released from one imported sheepskin processed several blocks upwind from their residences (Wellock, 1960).

7.2.7.8 *Radioactive Dust and Radon*

Radioactive sources are described in Section 5.7. Radiation levels in basements and inside homes in areas with high soil activity can exceed safe work levels for healthy adults (Douglas, 1977). Detailed inventories are not yet available and exposure assessments are only tentative (Section 7.2). A summary of currently measured levels is given in Table 7.7. The radon flux depends on atmospheric pressure (Duwe, 1979; Pohl–Rühling, 1979; Harley, 1981).

It has long been known that ambient air contains traces of radioactive nuclei (Junge, 1963; Eisenbud, 1963) and that uranium miners are exposed to both radon gas and radon progeny particulates (Blanchard, 1969). Stranden calculated in 1977 that in Sweden outdoor levels averaged 7.3 mrad/hr, but indoor levels averaged 48 mrad/year as opposed to about 44 mrad/year outdoors. Kusuda, Hund, and McNall (1979) proposed that ventilation rates should be adjusted to maintain appropriate radiation levels. The potential for increased radon levels in passive solar homes was explored by Rogozen (1980) and Desrosiers and Farber (1980), and indoor levels were modeled by Kusuda, Silberstein, and McNall (1981). The radon concentration depends on the source strength and infiltration. Table 7.8 shows Rn concentrations in seven building materials. The correlation between source strength and indoor levels is shown in Figure 7.18. The frequency distribution of Rn in 87 English and 15 American homes is shown in Figure 7.19. The infiltration path is shown in Figure 7.20. Radon concentrations vary significantly within homes. Very little is yet known about general radon trends in homes, except for the few cases where abnormally high levels were accidentally discovered.

Table 7.7

Radon Concentration in Ten U.S. Locations[a]

Location	Radon-222 Concentration (nCi/m³)	Progeny PAEC Concentration (WL)	Number of Residences	Type of Measurement	Comments
Normal areas					
Tennessee	—	0.008 (0.0008–0.03)	15	Grab bag	Shale area; concrete construction
Houston	0.07 (0.005–0.2)	(Up to 0.002)	7	Grab bag and ventilation	Single family; 1–6 air changes/hr
New York–New Jersey	0.8 (0.3–3.1)	0.004 (0.002–0.013)	21	Several integrated measurements over year	17 single-family, 3 multiple-family, 1 apartment building
Illinois	0.3–33	—	22	Grab bag	Wood frame construction, unpaved crawl spaces
San Francisco area	0.4–0.8	—	26	Grab bag and ventilation	0.02–1.0 air changes/hr
United States–Canada	0.6–22	—	17	Grab bag and ventilation	Energy-efficient houses; 0.04–1.0 air changes/hr
Problem areas					
Grand Junction, CO	—	0.006	29	Integrated year round	Geometric mean of control homes used in remedial action program
Florida	—	0.004	28	Integrated year round	Controls on unmineralized soils
	—	0.014	133	Integrated year round	Houses on reclaimed phosphate lands
Butte, MT	—	0.02	56	Integrated year round	Intensive mining area
Anaconda, MT	—	0.013	16	Integrated year round	Intensive mining area

[a] From NRC (1981).

7.2 Pollutant Concentrations

Table 7.8
Radiation Level in Seven Building Materials

Material	Concentration (pCi/g)			Comments
	^{238}U Series	^{232}Th Series	^{40}K	
Concrete	0.29–1.32	0.28–1.58	6.6–9.8	General average
Concrete	1.4	1.5	21	Atlanta area
Brick	1.8	1.8	17	Atlanta area
Tile	1.9	1.1	8	Atlanta area
Concrete	0.2–1.0	0.2–1.0	5–12	Nine metropolitan areas (preliminary values)
Solar rock bed	1.5	1.4	25	New Mexico (preliminary values)
Concrete	0.9–2.0	0.8–2.3	9–19	European measurements

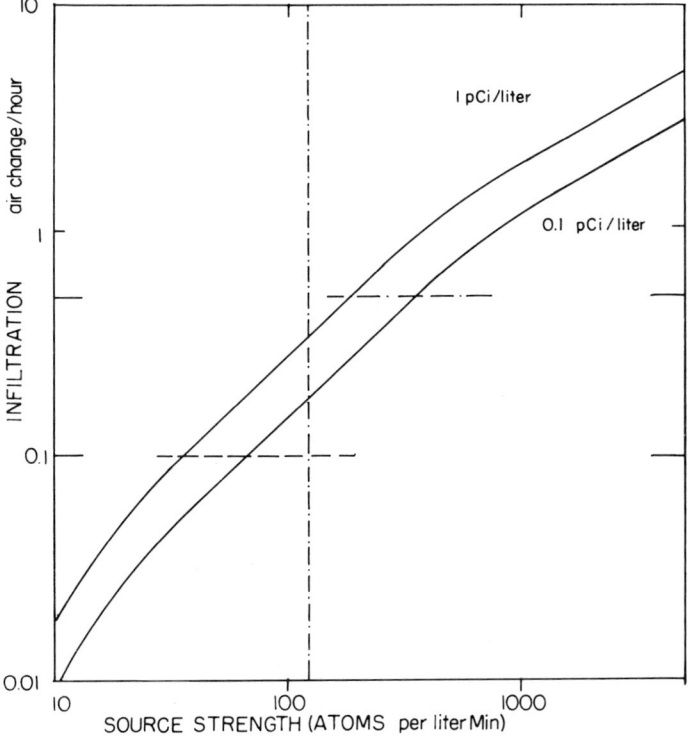

Figure 7.18. Radiation level as a function of infiltration and source strength.

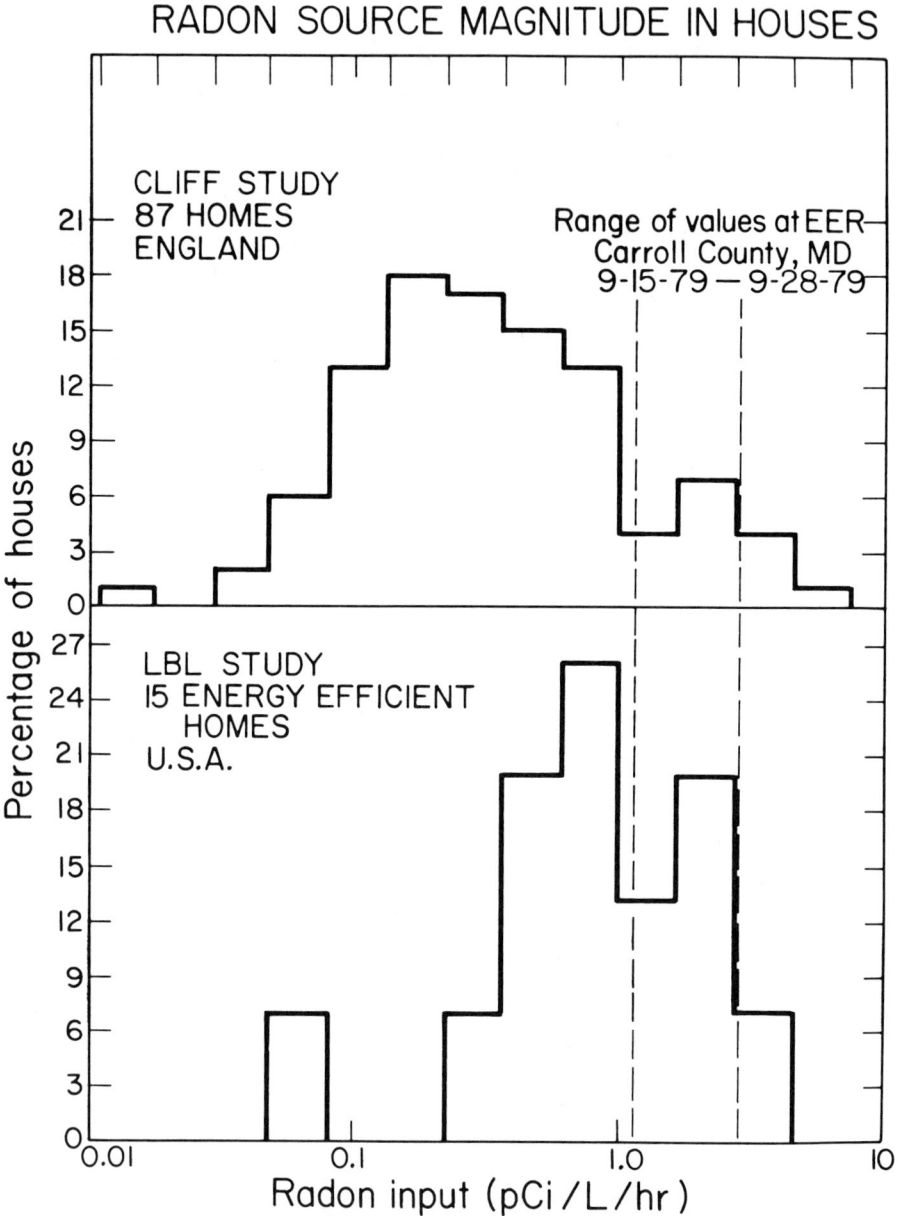

Figure 7.19. Radon concentration frequency distribution for 87 English and 15 U.S. homes (Hollowell, 1981).

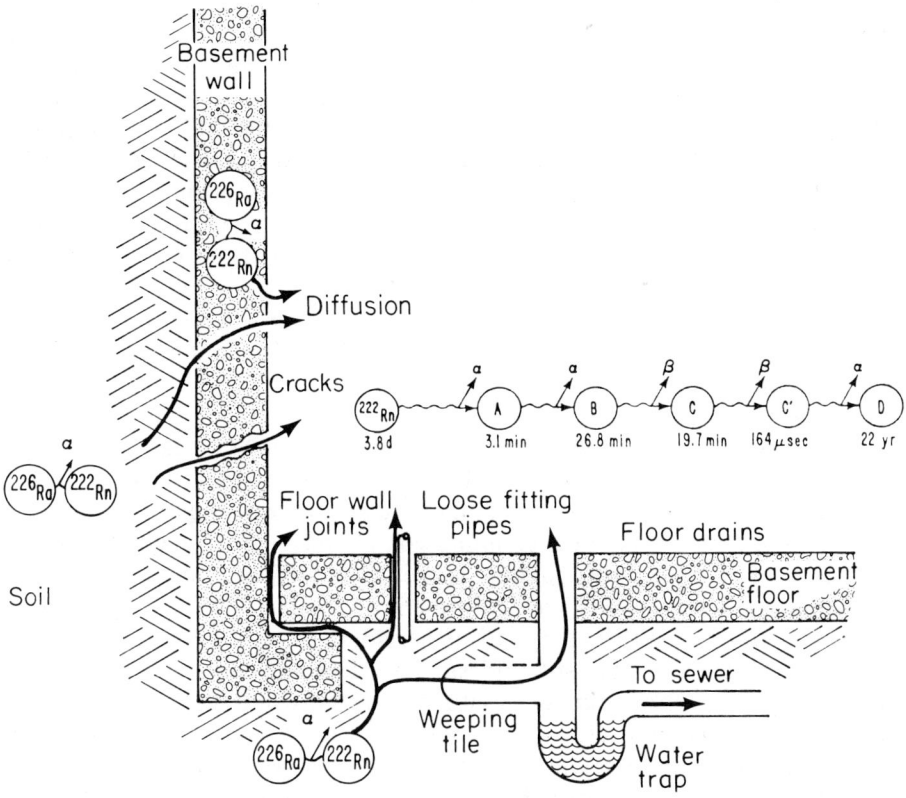

Figure 7.20. Entry of radon into buildings (after Hollowell et al., 1980).

7.2.8 ORGANIC VAPORS

Each year more than 200×10^6 tons of organic material are released into the air by vegetation (NRC, 1977), but much of this material promptly enters local biological cycles and does not reach urban areas in significant quantities. The main source of urban organic air pollutants is the manufacture and use of products derived from oil and coal. The EPA has compiled 26 reports that provide industrial process profiles especially prepared for environmental use. The chapters on the plastic and resin industries and on solvents vividly reflect the variety of compounds and of mechanisms by which organic vapors can reach ambient air. Use of the compounds has increased exponentially over the last 40 years.

The concentration of airborne contaminants depends on the physical and chemical states of the material. Gases mix readily and usually have a short residence time. Solids and liquids release contaminants more slowly, but are persistent, often after being dormant on cold surfaces, or in winter. Until recently little was known about the concentrations and the fate of airborne and solid organic wastes. On occasion this lack of awareness culminated in the disposal of unwanted chemicals in unmarked containers buried in foot-deep soil, as at Love Canal, NY. Currently, much work is under way (Pellizzari et al., 1981; Wallace, 1981) to characterize the air concentration of organic vapors; at the same time, significant efforts are being made to control, and thereby reduce, these levels. A careful analysis of trends has been reported for the Great Lakes area (Eisenreich et al., 1981). Ironically, these reductions will affect primarily ambient outdoor sources, while indoor organic sources due to seepage from soil or release from construction materials and human activities will, at least initially, stay high.

More than 250 volatile organic compounds have been found in quantities of more than 1 ppb in indoor air of urban residences in American cities (Jarke, 1979) and in offices (Schmidt et al., 1980); almost every class of organic chemicals except acids, which are soluble and adhere tenaciously to surfaces, is represented. Among the most common compounds are alkane hydrocarbons with up to 16 carbon atoms, but air has been found to carry potentially toxic or carcinogenic compounds such as dichlorobenzene, tetrachloroethylene, benzene, carbon disulfide, butylnitrile, acrylonitrile, acetonitrile, and ketene, as well as the less toxic but still undesirable xylene, toluene, and acetone. The main sources for the latter are household solvents and cleaners (Table 7.9) and construction adhesives. The indoor concentrations are due to ambient outdoor air as well as indoor sources. However, indoor concentrations can exceed outdoor concentrations substantially (Figure 7.21), even in a 20-year-old office building (Jarke, 1979, Schmidt et al. 1980). In NASA's *Skylab 4* mission 300 organic compounds, with concentrations between 1 ppb and 8 ppm (Thomas, 1971), were found in the spacecraft.

7.2.8.1 *Aerosol Propellants*

The invention of economical and reliable mass-produced needle valves in the 1950s made possible the development of throwaway spray cans. Each use of a product in a pressurized can releases aerosols, particulates, solvent, and propellant, most of which mixes with indoor air. The following quantities of propellant are released per average application: 1 g for deodorants, 5–7 g for hair spray, 0.5 g for shaving cream, 6–12 g for air freshener, 8 g for disinfectant, 8 g for furniture polish, 4–8 g for dusting spray, and 20–25 g for oven cleaners. This is about 1–10 liters of gas, which in a standard-sized room amounts to about 400 ppm of propellant (Bridbord et al., 1975). During application the concentration locally is far higher; in fact, if the propellant is flammable, it is sufficient to cause a fire or explosion if a combustion source is present.

Table 7.9
Active Ingredients in Household Solvents and Cleaners

Compound	Use	Indoor Air Level (ppm)	(mg/m^3)	Exposure–Response Data
Methylene chloride	Paint remover	200	720	180–200 ppm per day results in increment of 4.5% COHb. Narcosis above 4000 ppm. Brain damage reported above 500 ppm.
Perchloroethylene	Cleaning fluid	100	670	100 ppm for 7 hr produced narcotic effect, headache, mild eye and respiratory irritation.
Naphtha	Paint thinner	200	900	Multiple hydrocarbon content (TLV based in nonane, xylene)
Benzene, toluene, xylol, xylene	Lacquer and cement solvents	10	30	Myelotoxicant; skin absorption can produce chronic poisoning. Nausea, narcosis above 200 ppm. Neurological and vascular effects reported in workers. May contain 6–15% ethyl benzene.
Methanol	Shellac and varnish thinner	200	260	No worker injury at average 160–780 ppm. Repeated exposure above 3000 ppm may result in cumulative body concentration.
Acetone	Lacquer solvent	1000	2400	Eye and nose irritation above 2500 ppm
Ethyl ether	Lacquer, wax, plastic solvent	400	1200	Nasal irritation above 200 ppm. No worker injury demonstrated from regular exposure of 500–1000 ppm. Chronic effects noted after long-term exposure.
Sodium hydroxide	Drain cleaner	—	2	Upper respiratory irritant
Ammonia	Window cleaner	25	18	Rat ciliary motion stopped at 3 ppm. Eye and nasal irritation at 20–25 ppm.
Turpentine	Painting	100	560	Respiratory irritation at 75 ppm; nausea above 750 ppm.

In the 1960s the fluorocarbons were extensively used as self-pressurized propellants because they are inert, stable, and not very toxic. Only later was it recognized that these compounds can be contaminated by traces of toxic

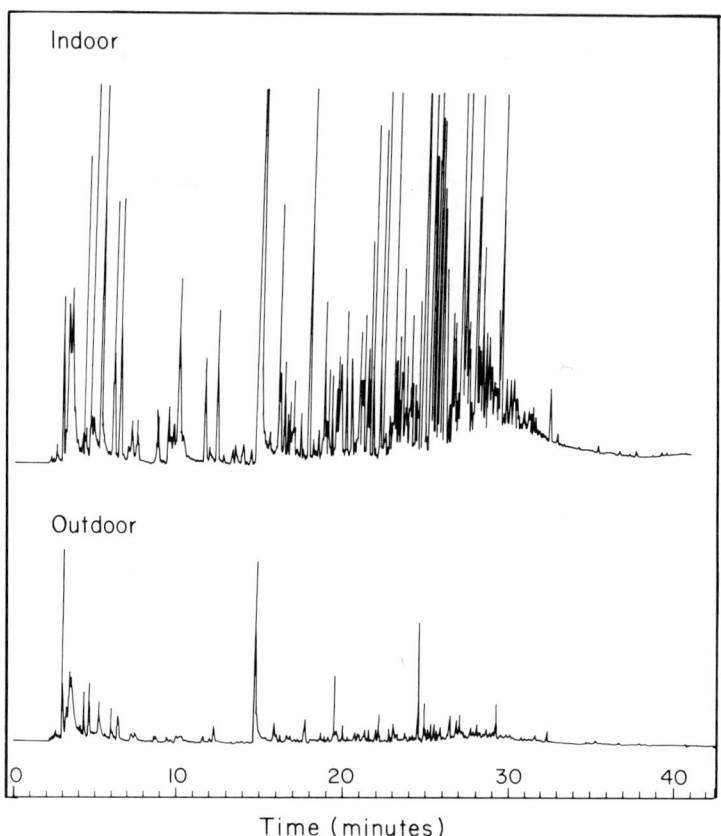

Figure 7.21. Trace organics inside and outside an office building at the Lawrence Berkeley National Laboratory (Miksch et al., 1980a).

impurities, and that the former, because of their unusual stability, end up in the earth's atmosphere, where solar radiation can initiate free radical reactions that act as ozone scavengers. Had the release of fluorocarbons continued for another 20 years at the 1960 rate, these reactions could have irreversibly altered the climate of the entire planet. Manufacture of these compounds is now controlled by the EPA (1979) and their use as consumer spray propellants has been curtailed. Currently, flammable hydrocarbons are used as propellants, but many consumer products are no longer sold as sprays, because experience has shown that some of the most effective spray components are caustic, toxic, or potentially carcinogenic.

7.2 Pollutant Concentrations

Volatile chlorocarbons have been used and studied for several decades. They are quite stable, except against photochemical reactions in sunlight. They can be analyzed with very sensitive analytical techniques, such as gas chromatography combined with electron capture, mass spectroscopy, or infrared spectroscopy. The first two methods allow detection at a concentration of 10^{-12} (i.e., 10^{-6} ppm) for CCl_4, $CHCl_3$, and CH_2Cl_2. For CH_3Cl, 10^{-3} ppm is necessary. The current atmospheric background level of carbon tetrachloride, CCl_4, is about 0.120–0.14 ppb. It was extensively used as a fumigant during World War II, but its use was discontinued when it was realized that it is highly toxic. The background concentration of chloroform, $CHCl_3$, is about 20–40 ppb. Dichloromethylene, CH_2Cl_2, has reached an ambient concentration of 0.5–1.5 ppb because about 80% of the annual production of 200,000 tons is released into the air after it is used as a solvent. Water contains up to 1.4×10^{-4} µg/liter of chlorocarbon, which can be reevaporated.

The National Academy of Sciences studied chloromethanes in 1978 and recommended caution with these compounds, which are suspected teratogens and are widely used in many applications and places, even beauty parlors.

Fluorocarbons and chlorocarbons are not found in any advanced form of life, and only very few bacteria and fungi incorporate halogens (Zimmerli, 1977). An exception is chlortetracycline, a highly beneficial antibiotic. Organobromine compounds are found in *Asparagopsis taxiformis,* a tasty edible seaweed that is harvested in Hawaii and is apparently nontoxic. Iodine is incorporated in thyroxine, an amino acid that is essential to mammalian life. Otherwise, organohalogens have been virtually excluded from biosynthesis. These compounds are chemically unsuitable for storage of metabolic energy and therefore do not compete with lipids and carbohydrates for energy storage, but they are chemically soluble in lipids and thus accumulate in such tissues, where they form an additive body burden of about 1–50 ppt that persists for many years (Section 5.6.3).

The chlorides, bromides, and iodides of methyl, butyl, propyl, and isopropyl alkylating compounds are known causes of pulmonary adenomas in mice (Kraybill, 1977).

7.2.8.2 Formaldehyde

Formaldehyde is a highly water soluble gas that can adsorb on wood and other hygroscopic surfaces, from which it can re-vaporize after a dormant period. Ambient formaldehyde levels range from 0.2 ppm on Pacific islands to 2 ppb in tropospheric air. In urban areas the air level depends on industrial activities and weather conditions. In 1960 Los Angeles air contained from 5 to 40 ppb, depending on the smog level and time of day. During a very severe inversion in the fall of 1979 the weekly average value was 30.8 ppb, but normal formaldehyde levels are now about 5 ppb. In the industrial agglomeration of

Camden, NJ, the daily average reached between 3.8 and 7.1 ppm, and the daily peak between 7 and 14 ppm, during the five summer months in 1974 (EPA, 1980).

Tap water contains less than 0.01 ppb. Acid rain may contain up to 0.5 ppm under extreme conditions. Industrial effluents with 1.5 ppm have been reported.

Occupational levels are 0.05–4 ppm, with extreme levels, up to 15 ppm, in biological research labs, autopsy rooms, mortuaries, and in a few cases in the textile, urea-formaldehyde foam resin, and forest products industries. (See also Tables 7.10; 7.11; 8.8–8.10, and Figure 7.27.)

Food may be packaged in formaldehyde-containing materials, but formaldehyde may not be used as a food additive in the United States. However, fish and processed turkey contain formaldehyde in concentrations up to 1 ppm as decomposition products. Paraformaldehyde is used by maple sap collectors in tapholes to prevent fermentation. Finished maple syrup may contain up to 2 ppm of formaldehyde. Slow-release fertilizer does not leave formaldehyde on produce, or the soil, even though formaldehyde has been reported in some air samples taken above soil.

The half-life of formaldehyde in ambient air in full sunshine is about 75 min; therefore, atmospheric formaldehyde levels can be maintained only if the fraction that is oxidized in the atmosphere is continuously replenished. In aerobic water, 50 μg/liter of formaldehyde is oxidized within 30 hours. In soil it reacts promptly. Some fungi and bacteria can utilize formaldehyde as a carbon source.

In 1961 Breysse (Breysse, 1977; Carbone 1979), investigating local complaints about pungent stale air in mobile homes, measured formaldehyde concentrations there of up to 1 ppm and in some cases even higher. This work led to the discovery that continued exposure to formaldehyde, even at a moderate level of 0.1–0.5 ppm, can be irritating to many people. This fact surprised the chemical industry, because formaldehyde has been used in much higher concentrations in hospitals as an antiseptic for almost 100 years, and it is used in anatomy and biology labs as well as by undertakers to preserve specimens and embalm corpses with few complaints, except by the approximately 3% of the workforce who react with skin rashes to direct exposure.

Formaldehyde is used to manufacture adhesive resins. Each year the forest products industry uses about 500,000 tons of formaldehyde-based resin to manufacture plywood and particleboard which are important staples for building construction, cabinet work and furniture. Particleboard invented in the current form by Fahrni in 1947, is in many respects superior to wood, from the wastes of which it can be made. It is manufactured in large panels or any desired shape by pressing wood chips with 7 wt. % urea-formaldehyde resin in a hot press. Fresh panels have a pronounced formaldehyde odor that dissipates from well-made products within a few weeks (Meyer, 1979). Unfortunately, urea-based resins are not yet very water resistant; upon extended exposure to moisture they gradually hydrolyze and in the process may release traces of formaldehyde (Myers, 1979, 1981; Myers and Nagaoka, 1980; Roffael, 1976-1980). The same can happen with urea-formaldehyde insulation foams (Long et al., 1979-1981), which were invented by Curs in 1930, make excellent thermal insulation, are economical (Allan et al., 1980), and resist fire far better than any other organic insulation or foam (Meyer, 1979).

It is estimated that more than 4 million mobile homes and more than 50 million homes and office buildings contain urea-formaldehyde bonded wood products, and that 200,000 to 600,000 homes in the U.S. contain foam insulation.

Formaldehyde levels are always highest in new buildings (Rutkowska, 1981; Guberskii, 1981), in which particleboard, because of its otherwise attractive properties, is extensively used. School buildings have led to well documented complaints in Canada, Germany (Deimel, 1976), and Russia (Meyer, 1979): School buildings are especially problematic because of their high occupant density and the high metabolic rate of children, which combine to provide a hot, humid atmosphere that conduces formaldehyde release (Section 4.2.3). In extreme cases formaldehyde in poorly ventilated rooms can reach 1 ppm or more. This causes acute discomfort to the eyes of students and teachers. It also can cause asthma in sensitive children or teachers. In 1974 Andersen and coworkers measured 16 Danish homes and observed an average level of 0.64 mg/m^3 with a range from 80 ppb to 2240 ppb. Table 7.10 shows formaldehyde levels observed in nine studies by contractors for state and federal agencies in the United States.

A recent comparison of consumer products (Pickrell, Griffis, and Hobbs, 1982) observed that new urea-formaldehyde bonded interior grade plywood with an emission rate of up to 34 mg/m^2 day and particleboard with a rate of up to 25 mg/m^2 day were the strongest emitters. Emission rates for some other materials are listed in Table 7.11. The recorded values are based on only a few specimens and on a small number of measurements. Therefore, they represent a range, but not average values, for the entire class of materials. Some formaldehyde can also stem from cooking, smoking, fireplaces, and heaters.

It has recently been demonstrated that carpet shampoos can leave enough formaldehyde residues to cause health complaints (Kreiss, 1981). Indoor formaldehyde levels measured in Wisconsin are shown in Figure 7.22a. The median values are 0.47 ppm for mobile homes, 0.35 ppm for all homes, and range from 20 ppb to 4 ppm (Anderson, 1980, 1981; Dally, 1979–1981; Eckmann, 1981; Hanrahan, 1981; Woodbury, 1978); in Minnesota, they averaged 0.58 ppm with a range of 0.1-3 ppm (Oatman, 1979). Gary (1979) observed levels of up to 3 ppm in new homes and 0.2 ppm in ten year-old homes. Godish (1981) studied homes in Indiana. Homes without formaldehyde-containing materials average 40 ppb (Stolwjik, 1979). In California

Table 7.10

Formaldehyde Concentrations Observed in 9 Studies[a]

Sampling Site	Concentration[b] (ppm)		Method of Analysis
	Range	Mean	
Danish residences	1.8 (peak)	—	Unspecified
Netherlands residences built without formaldehyde-releasing materials	0.08 (peak)	0.03	Unspecified
Residences in Denmark, Netherlands and Federal Republic of Germany	2.3 (peak)	0.4	Unspecified
Two mobile homes in Pittsburgh, PA	0.1–0.8	0.36	BTH bubblers
Sample residence in Pittsburgh, PA	0.5 (peak)	0.15	MBTH bubblers
Mobile homes registering complaints in state of Washington	0–1.77	0.1–0.44	Chromotropic acid (single impinger)
Mobile homes registering complaints in Minnesota	0–3.0	0.4	Chromotropic acid (30-min sample)
Mobile homes registering complaints in Wisconsin	0.02–4.2	0.88	Chromotropic acid
Public buildings and energy-efficient homes (occupied and unoccupied)	0–0.21	—	Pararosaniline and chromotropic acid
	0–0.23	—	MBTH bubblers

[a] Reprinted from Indoor Air Report, NRC (1981), see also CPSC (1982) Rutkowska (1981) and Guberskii (1981).

Fanning (1979) and Chin-I-Lin et al. (1979) found levels of 0.2 ppm in trailers with 1 air exchange per hour. In Washington in 600 mobile homes an average indoor formaldehyde level of 0.5 ppm was measured (Breysse, 1977; Carbone, 1978). Measurements in Pittsburgh made for the U.S. Department of Housing and Urban Development (HUD) yielded an average of 0.43 ppm (Moschandreas, 1978). An interesting study of 104 mobile homes whose occupants had no complaints was made by the EPA in Wisconsin, where a possible correlation between formaldehyde concentration and clinical complaints is being investigated (Dally et al., 1981). Average levels of about 0.4 ppm have been found. Another important study, sponsored by HUD is investigating the correlation between formaldehyde release from construction materials and indoor air levels in new mobile homes (Walcott, 1982). Long and his co-workers Frank and Shutte in Iowa have conducted thorough and careful studies on formaldehyde release, as well as air levels, observing levels of 0.3–1 ppm, depending on age and season (Long et al., 1979). The purpose of this work, sponsored by the

Table 7.11

Approximate Formaldehyde Emission Rates for 28 Construction Materials and Consumer Products

Material	Release Rate (mg/m² day)	Reference
UF-foam insulation	1–50	a,b
Plywood		
(UF-bonded)	1–34	a
(phenolic)	0–0.05	b
Hardwood paneling (UF)	1–34	b
Particleboard		
standard (UF)	2–34	b
low emission (UF)	0.5–3	b
phenolic	0–0.001	b
Fiberglass ceiling panel	2.8	a
Lady's dresses	1.2	a
Men's shirts	1.0	a
100% cotton drapery fabric	0.2–0.70	a
Paper plates and cups	0.70	a
Fiberglass insulation	0.45	a
Rigid round air duct	0.40	a
Blackface insulation sheathing	0.35	a
Paper plates and cups	0.33	a
Latex backed fabric	0.19	a
Rigid round fiberglass duct	0.15	a
Foambacked carpet	0.12	a
77% rayon/ 23% cotton drapery	0.10	a
3½ inch fiberglass	0.090	a
Plywood	0.055	a
Blend fabric	0.050	a
Foambacked carpet	0.020	a
Nylon upholstery fabric	0.018	a
Olefin upholstery fabric	0.005	a

[a]Pickrell et al., 1982
[b]Meyer, 1979, 1980, see also Spedding, 1980, Lehmann, 1982 and Sundin, 1982.

U.S. Department of Energy (USDOE), was to establish UFFI standards for both the ASTM (Rossiter, 1980) and the USDOE (1980). Correlating work was performed by Osborne (1980) and Hawthorne (1981) for the Consumer Product Safety Commission, Table 7.11 and Figure 7.22a.

The EPA (1981) estimates that about 2,400,000 people in mobile homes are exposed to an average of up to 400 ppb, and another 2 million to average levels

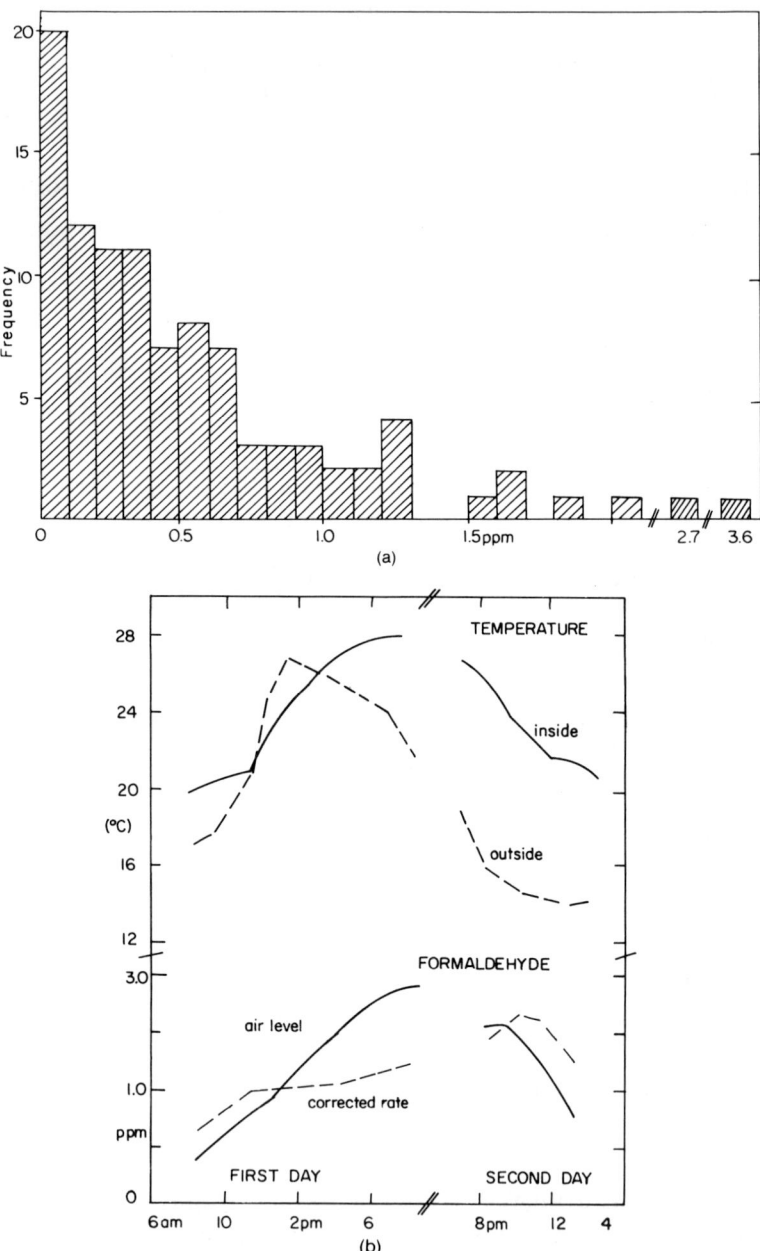

Figure 7.22. (a) Formaldehyde concentration in 150 homes and mobile homes (after Dally, 1982). (b) Diurnal formaldehyde levels in a Wisconsin mobile home as a function of outside temperature (after Myers, 1982).

of up to 350 ppb, of formaldehyde from urea-formaldehyde foam insulation. This is hundred times more than ambient formaldehyde levels in rural areas.

Material tests indicate that formaldehyde release from forest products is strongly temperature and humidity dependent (Sundin, 1978, 1982; Meyer, 1979; Myers, 1980; Roffael, 1980). Indoor air levels also depend on these parameters, and thus correlate strongly with occupancy, weather, season, and many other factors. Furthermore, formaldehyde dissolves on all moist surfaces and revaporizes when the building envelope dries; this process can lead to constantly changing air levels in a home with forced air heating, where each heating cycle might be followed by an influx of formaldehyde. Therefore, measurements taken at different times rarely coincide.

Figure 7.22b shows the indoor formaldehyde level measured in an unoccupied mobile home on a sales lot in Wisconsin (Myers, 1982). The dashed curve shows the values predicted by the best current model for particleboard emission. The wall cavity levels, not shown in the figure, follow a parallel trend. Figure 7.23 shows the correlation between air temperature and humidity and the indoor air concentration for urea-formaldehyde insulation foam. Table 7.12 summarizes formaldehyde concentrations in a 4" deep cabinet-like compartment attached to wall cavities which were fitted with foam by eight commercial contractors in spring of 1980. This table shows that levels of 5–24 ppm could still be measured during simulated hot weather spells 18 months after installation. Obviously, in a well ventilated room the levels would be drastically lower. In fact, in properly placed installations supervised by the UF-foam insulation quality assurance association in Europe, levels above 0.5 ppm have been reported in only a very few cases during the last twenty years. However, much basic research and field work remains necessary to develop a reliable relationship between building material properties and indoor exposure levels (Meyer and Nunlist, 1981; Tomita, 1980).

7.2.8.3 *Hydrocarbons, Halocarbons, and Pesticides*

Indoor air contains hydrocarbons with 1–16 carbon atoms (Jarke, 1979), that is, essentially all the gaseous or vaporizable hydrocarbons. In 12 locations in Los Angeles total hydrocarbon levels averaged 1.7–5 ppm in 1979 (Sullivan, 1979). Moschandreas (1978) found indoor levels between the detection limit and 8 ppm, with an average below 3.5 ppm, in a five-city study; the indoor/outdoor ratio was always above 1. Peaks were reached when housekeeping was in progress. Hollowell observed typically 300 ppb of *n*-octane, 150 ppb of *n*-nonane, 25 ppb of benzene, 75 ppb of toluene, and 150 ppm of xylene. The latter is occasionally very high in hospitals, where it is used as a cleaning fluid (NIOSH, 1978; Verschueren, 1977). Ambient air quality criteria were established by the EPA in 1970. An inventory of organic pollutants was prepared by the National Academy of Sciences in 1976; California air levels were recorded by Mayrsohn et al. (1976; 1977). In the vicinity of some oil refineries, benzene

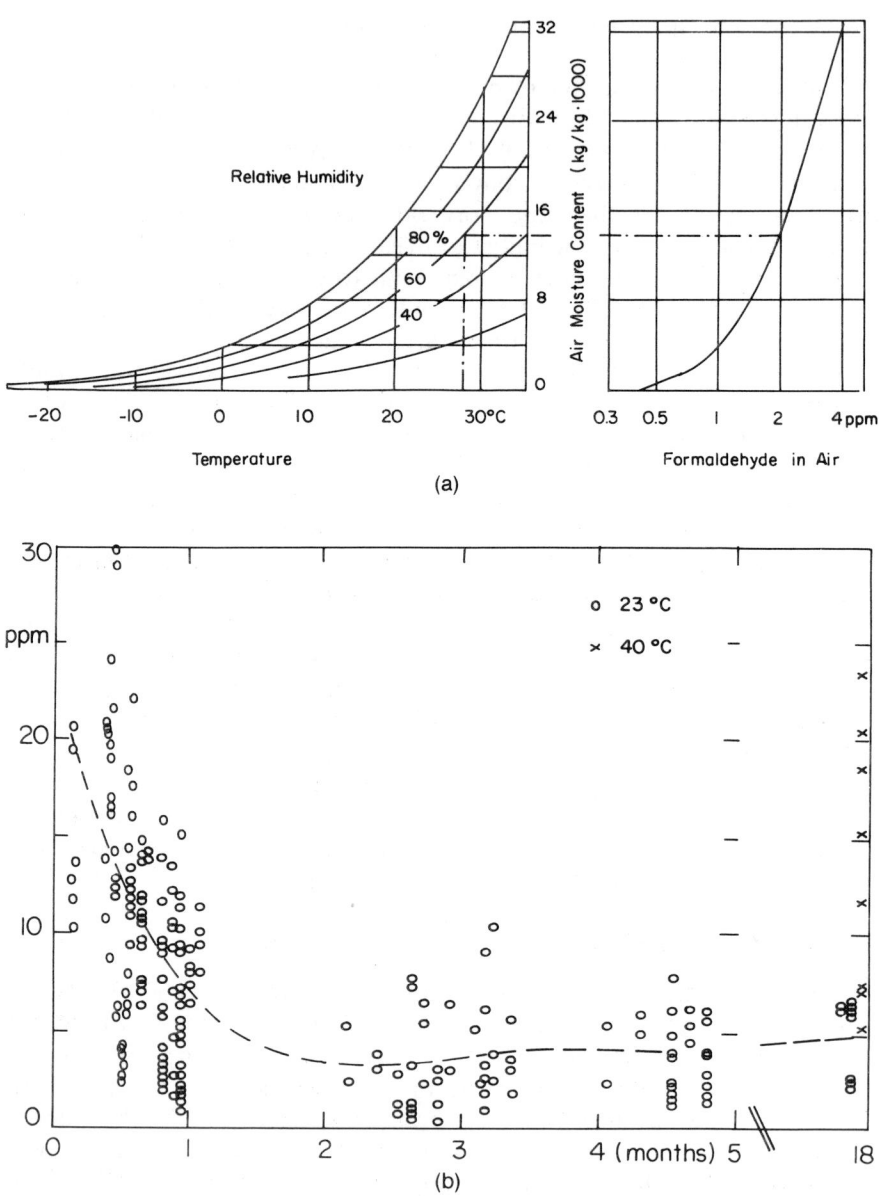

Figure 7.23. (a) Formaldehyde air concentration as a function of temperature and moisture in UFFI Cavity (after Bowen, Shirtliffe, and Chown, 1981); (b) formaldehyde air concentration in wall cabinet as a function of age, temperature, and humidity for eight commercial UF foams (after Osborne, 1980; Hawthorne, Gammage, Mathews, et al., 1981; Pressler, 1981).

Table 7.12

Formaldehyde Concentration for UFFI Placed in a Test Wall Section by Contractors for Nine Commercial Firms[a]

Product	10 days[b]	3 months[b]	16 months[c]		18 months[d]		
			Cavity	Cabinet	23°C	33°C	40°C
1	2.2	1.3	6.8	2.2	0.4	2.9	5.5
2	2.5	1.0	5.8	2.6	0.6	2.8	7.3
3	10.2	4.2	8.3	6.0	1.8	11.9	15.4
4	1.2	2.1	5.2	2.4	0.8	3.2	6.9
5	3.1	5.4	17.8	6.4	0.6	5.7	18.8
6	11.8	—	9.1	6.4	2.1	12.2	23.7
7	5.4	3.4	12.3	6.6	0.9	5.7	11.7
8	8.0	5.8	10.1	6.6	1.8	9.5	20.6
9	3.4	2.3	8.4	6.2	—	—	—

[a] At 23°C and 50% relative humidity.
[b] From Osborne (1980).
[c] From Hawthorne and Gammage (1981).
[d] From Pressler (1981).

and other hydrocarbon concentrations of from 1 to 5 g/m³ have been noted, and can be so high as to be detectable in the breath of nonoccupational residents (Wallace et al., 1981).

Incompletely burned hydrocarbons from automobiles are a major source of Los Angeles smog. The strength of this source has been dramatically reduced by automobile emission control by catalytic devices mandated by the EPA during the last decade. A major source is now exhaust from trucks and other diesel engines that are not yet effectively converted.

For a long time, one of the most toxic halocarbons in wide use was the vinyl chloride monomer (VCM), used to produce polyvinyl chloride, one of the most durable polymers known. The EPA has recently set a target level of zero for this substance in ambient air because of its toxic effect on workers in the polyvinyl chloride (PVC) industry. The monomer boils at −20°C and travels up to 5 miles with ambient air. Wallace et al. (1981) has observed chloroform, trichloroethylene, tetrachloroethylene, vinylidene chloride, and similarly volatile halocarbons in concentrations of 0.3–10 μg/m³ in air and in the breath of college students (see Figure 8.14).

Automobile upholstery can release more than 147 different organic compounds, including aniline, biphenyl, dibromoethane, dichlorobenzene, maleic anhydride, naphthalene, and other potentially toxic substances (Zweidinger, 1977; Figure 7.24). The most prevailing toxic agent is the vinyl chloride monomer, which stems from the extensive use of polyvinyl chloride in car inte-

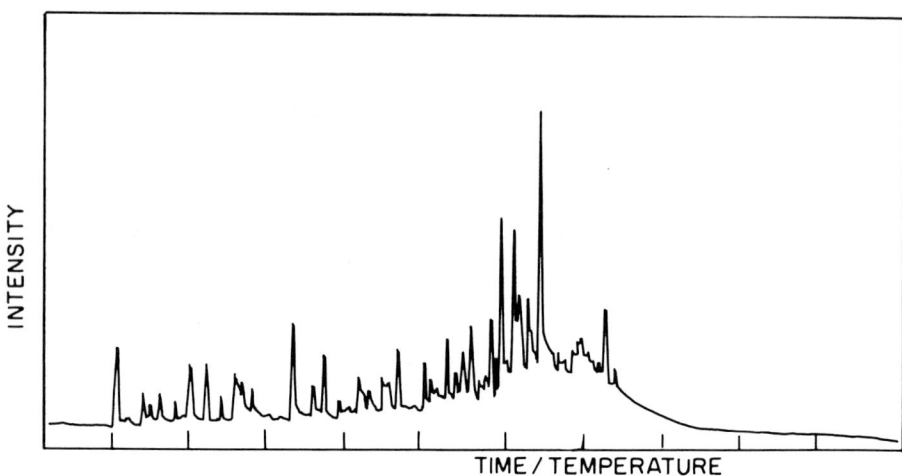

Figure 7.24. Organic components from automobile upholstery (after Zweidinger, 1977).

riors. Levels of up to 7 ± 2 ppb of the VCM were observed in subcompact cars, with typical levels averaging 2–3 ppb at 45°C. The levels decrease below the measurement threshold of 1–2 ppb after about 3 months.

Pesticides (Section 5.6.5) are used in 90% of all North American homes (Savage, 1982). Since most of them have low vapor pressures, in the range of 10^{-3}–10^{-7} Torr, they can be very persistent. Vapor pressures for seven pesticides

Table 7.13

Vapor Pressure and Airborne Concentration of Seven Pesticides during Four Days Following Indoor Application[a]

Insecticide	Concentration (μg/m^3) on day				Vapor pressure (10^{-6} Torr; 20°C)
	0	1	2	3	
Acephate (1.0%)	1.3	2.9	0.5	0.3	1.7
Bendiocarb (0.5%)	7.7	1.3	—	—	5
Carbaryl (5.0%)	1.3	0.2	0.1	0.01	5000
Chlorpyrifos (0.5%)	1.1	1.1	0.8	0.3	19
Diazinon (1.0%)	1.6	0.6	0.5	0.4	140
Fenitrothion (1.0%)	3.3	1.1	0.8	0.5	6
Propoxur (1.1%)	15.4	2.7	1.8	0.7	3

[a] After Wright (1981).

are listed in Table 7.13, together with indoor air concentrations during the three days following their application (Wright, 1981). Care must be taken in estimating exposure levels from vapor pressure measurements and intermittent air analysts alone (Arthur, Cain, and Barrentine, 1976) because the thermodynamic vapor pressure might belie the size of the latent pool in the form of dust that can be resuspended any time (Savage, 1975) by the slightest turbulence. In fact, this innocuous, but pervasive, presence is an intrinsic quality of most pesticides (McEwen, 1979). Very extensive studies of DDT have been made all over the world for several decades; these are reviewed in Section 5.6.5. In 1966 DDT levels ranged from 0.1 ppb in rural areas to 8 ppm in urban areas. The malathion level was 0.43 ppm in the urban United States and methyl parathion and toxaphene levels were up to 2.5 ppm (Arthur et al., 1976).

An example are the No-Pest insecticide strips, which saturate air for weeks with powerful agents (Collins and Decries, 1973). Indoor levels have been estimated by Wright (1981), but it is quite easy for a person to inhale particulate pesticides and experience a very high local concentration around the specks absorbed on the internal membranes. Emission from mosquito sticks was studied by Miyazaki et al. (1977).

A unique information basis has been acquired for TCDD as a result of the use of Agent Orange as a defoliant in the Vietnam War: 4.2 million U.S. soldiers were exposed to more than ten times agricultural levels of the defoliant, which was contaminated with 1 ppm (wt. %) of TCCD, a very toxic impurity (Section 5.6). Several follow-up studies are under way (Meyers, 1981). Another unique exposure arose from the accidental discharge of 2 kg of dioxin—containing intermediates from a pesticide plant in Seveso, Italy. Figure 7.16 shows the concentration distribution that was measured in the fields as well as computed from distribution models (Doll, 1979). A very intense international effort was made to remove the dioxin from the soil.

A poorly charted but very extensive source of pesticide and by-product emissions is the estimated 10,000 hazardous waste disposal sites where buried containers lie decaying. Even with the several billion dollars allocated for a special bill passed by Congress in December 1980, a thorough cleanup of all sites seems improbable, and in coming decades many homeowners throughout the United States are liable to experience the effects of discarded chemicals that may be hazardous and might well contaminate indoor air, at least in the basement, as has already happened in the well-publicized area of Love Canal, NY, and elsewhere.

The same situation holds for halocarbon wastes. Thus, measurement of these substances is usually effected by absorption or a filter (Zimmerli and Zimmerman, 1979) or an organic foam (Jarke, 1979), charcoal, or a zeolite resin, p. 148.

Pentachlorophenol (PCP) and biphenyl (MacLeod, 1979; Zimmerli and Zimmerman, 1979; Bridbord, 1977) are no longer used for indoor application and have been phased out of industrial applications, with a 1985 target date for total elimination. However, many houses still contain capacitors with polychlorinated biphenyls (PCBs) as dielectric liquid, an application in which they have unsurpassed electrical properties. MacLeod (1980) observed outdoor PCB background levels of 4 ng/m³ and indoor levels between 40 and 580 ng/m³. All levels above 200 ng/m³ were in the kitchen and those above 400 ng/m³ were usually in rooms where PCB-filled fluorescent light ballasts had been used or were still in use. She determined the decay curve after removal of the source from a contaminated kitchen and found a half-life of about 2 weeks. The level dropped exponentially from an initial level of 10 g/m³ to 1.3 in the first month to 0.5 after two months and to 0.25 ± 0.05 μg/m³ for the ensuing 3- to 13-month period.

Benzo-α-pyrene (BaP). The aromatic compound benzo-α-pyrene is known to be carcinogenic; it is produced as a byproduct during combustion of coal as well as tobacco. Ambient levels in urban areas were at an all-time high in 1959 and ranged from 0.5 ng/m³ in Bismarck, ND, San Jose, CA, Shreveport, LA, and Salt Lake City, UT, to 31 ng/m³ in Chattanooga, TN, 39 in Charlotte, NC, 45 in Richmond, VA, 54 in St. Louis, MO, and 61 ng/m³ in Altoona, PA. All values are averaged for January–March 1959 (Wilson et al., 1980). Fluoroanthene, pyrene, benzoanthracene, chrysene, perylene, and similar polynuclear aromatics occur in similar concentrations. The national average was about 6 ng/m³ in 1959 and had dropped to less than 1 ng/m³ by 1975.

Elliott (1975) measured BaP in sports arenas in the midwest and found it only in places where smokers were present. The indoor levels were between 7.1 and 13 ng/m³, depending on the number of smokers. On the basis of established estimates that an average smoker inhales about 8 ng of BaP per cigarette, he computes that in an arena with an attendance of 11,800, where he measured a level of 12.5 ng/m³, each nonsmoker would inspire 67 ng (i.e., the BaP equivalent of 8 cigarettes) during the two-hour event.

7.2.9 ODORS

Almost every indoor area exudes some type of odor which is at least weakly noticed when one enters (Figure 1.8). Most odors are due to organic vapors. However, the odor intensity is not proportionate to the concentration of osmogenic substances (Tables 5.5 and 5.6).

7.3 Exposure Level Models

Indoor exposure levels cannot yet be reliably predicted because exposure models depend on accurate pollutant profiles that reflect local and temporal variations of all pollutants. At present, the measured indoor and outdoor air pollution concentrations are sporadic values, or average values at best, and do not represent the actual breathing-level dose that a person inhales. This gap

7.3 Exposure Level Models

will soon be filled because convenient and reliable personal monitoring equipment is now available for several pollutants, p. 144. However, data for each chemical can only be correctly interpreted if a complete chemical analysis of potential synergens is available. Any theoretical model depends on information for at least five parameters: human activities, pollutant strength, indoor characteristics, structural characteristics, and monitoring constraints.

Human Activities. Four cases can be envisioned: A structure may be (a) unoccupied, in which case the contaminants are independent of the human presence; (b) heavily occupied, and so the indoor air is saturated with human pollutants; (c) lightly occupied, so human activities have to be factored into the pollution situation jointly with other equally important factors; and (d) occupied by a population that changes significantly over a given period.

An analysis of common building use shows that apart from hospitals, very few places are evenly occupied day and night. In almost all commercial, public, and residential structures, cases (a) and (b) or cases (a) and (c) alternate in diurnal cycles. Examples of extreme peak uses are airplanes, subways, commuter trains, buses, churches, sports arenas, theaters, classrooms, and nightclubs. In private homes, vacation homes, and military barracks the greatest use occurs at night. In office buildings combination (a) and (c) prevails; that is, the buildings are empty 16 hours a day and lightly occupied 8 hours a day, with 48-hour weekend intervals.

In the past, two approaches were used to bridge the diurnal span. In some tall buildings and domed sports arenas the ventilation and heating systems were run at full load day and night. In older buildings heating and ventilation were run only during occupancy and the preceding hour. Today the empty-building cycle is increasingly adjusted to minimize energy use over the weekly span. The ideal approach differs for each building and each season. In cars, airplanes, and churches, ventilation may be shut off during off-cycles. A light intermediate cycle might be used for stadia and office buildings, in which case the time lag between extremes becomes a crucial factor.

Pollutant Strength. Each pollutant has different characteristics and, usually, different sources. Factors are often additive. Ambient pollutants such as particulates, sulfur dioxide, and nitric oxide occur at low levels but are pervasive, and stem from a steady source that is perceived as infinitely large compared to the indoor environment. Similar characteristics are exhibited by some indoor air pollutants, including formaldehyde from building materials, radon from the ground or building envelope, nitric oxide from pilot lights in ranges, carbon dioxide from occupants during business hours, body odors, and ozone from copy machines. Levels of these pollutants can be computed quite reasonably from models discussed below.

A large group of pollutants occurs with periodic concentration variations. Examples are carbon monoxide, moisture, and nitric oxide from cooking; particulates; nitric oxides; radon; and lead released during heating cycles of forced air heaters or during air conditioner cycles. Often, periodic cycles of different duration overlap, which can result in amplification or diminution, depending on chance or on intentional building management. Examples of overlap might be particulates in a living room from outdoor dust infiltration during a storm; dust from a heating duct; smoke from the kitchen, a fireplace, and a person smoking; plus, possibly, dust resuspended by people walking in the room. To make the case extreme, there might also be dust from a vacuum cleaner and an aerosol used in conjunction with a hobby, and insecticide dust from a family pet.

It is unlikely that particulate levels in an office, a home, or a vehicle are ever due to a single source. Since particulates constantly settle and are resuspended by turbulence, the determination of real-time levels of particulates at the breathing level can only be determined experimentally with a personal monitor. On the other hand, average concentrations can probably be reasonably calculated.

The same is true for point-source pollutants, which are released continuously or in puffs. For example, smoke from a diesel truck at an intersection constantly changes concentration. One person might cross the street and breathe fresh air all the way; another, trapped between stalled cars or standing behind an exhaust pipe spewing smoke, might breathe the exhaust gases full strength. The real-time exposure levels vary, but the two pedestrians would receive similar integrated exposures if both crossed the same intersection every day. The crucial question in the analysis of this case depends on the dose and the response. If prolonged additive exposure leads to bronchial problems, average values count. If the level crosses the acute toxic threshold, real-time exposure is more important.

Indoor Characteristics. Humidity, pressure, and temperature have an important influence on all indoor exposure. For one thing, changes of each of these parameters might activate or deactivate some pollutants, such as fungi, bacteria, formaldehyde from building materials, radon from soil, and dust from floors. Second, the differentials, and especially the changes of differential gradients between indoor envelope and indoor air, might activate or conceal potential contaminant traps, such as cold windows, walls, or floors, which draw moisture that subsequently absorbs certain particulates and gases and then introduces secondary sources and sinks. Finally, each of these factors will influence interaction among pollutants. For example, in the extreme case, ambient sulfate particulates might provoke aerosol formation in a moist cool room, both in winter and in an improperly air-conditioned room in summer; or moisture might cause surface condensation of gases or particulates and in the process change both their size and their toxicity, and thus alter both their dose and their effect on health.

Structural Characteristics. Building size, room size, heat transfer rate through walls, floors, ceilings, and partitions, windows, and surface materials

7.3 Exposure Level Models

all directly influence levels, residence time, and the general fate of pollutants. This subject has been thoroughly explored by air-conditioning engineers and has been thoroughly reevaluated from the viewpoint of energy conservation. Further factors are ventilation ducts, window use, and other environmental factors of the building (Hittman, 1974). This topic is the subject of an entire U.S. Department of Energy program (Hollowell et al., 1976, 1978, 1979, 1980).

7.3.1 EXPOSURE MODELS

Whether the source and sink of pollutants is mechanical, chemical, or biological, all indoor models use as their basis the conceptually simple mass balance equations.

Stirred Setting Chambers. Thoroughly mixed biological, organic, and inorganic particulates settle in quiet air at a predictable rate:

$$N(t)/N(o) = \exp(-kt) \tag{7.4}$$

where $N(t)$ is the concentration at time t and $N(o)$ at the time of entry. The constant k is called Stokes's velocity. In a room 3 m high, half of the particles that are 1 μm in diameter settle to the floor in 20 hours, 75% in 40 hours and only 4% in the first hour.

The same equation applies for the residence time of well-mixed gases in a room that is ventilated at a constant rate. In this case it is customary to define $1/k$ as the ventilation rate, since it represents the number of air changes per hour (ach). Of course, this model applies only for one-shot injection of pollutants.

The Biological Warfare Model. Also called a dosage model (though the term dosage has changed and should no longer be used in this connection), the biological warfare model includes several additional sources and sinks:

$$dN = (1 - k) \, rx(o) \, dt - aN \, dt + q \, dt - rx(i) \, dt \tag{7.5}$$

where N is the mass; k the fraction removed by filtration; r the ventilation rate; $x(o)$ and $x(i)$ the outside and inside concentrations, respectively; a the removal or settling rate; and q the internal generation rate. This equation can be used to compute the indoor concentration change, $dx(i)/dt$; therefore, it is used to derive the infiltration rates of buildings from the observed decay rate of such tracers as sulfur hexafluoride. This model was originally developed by Calder in 1957.

The Well-Mixed Tank. The well-mixed tank model considers a combination of three factors: a steady-state indoor source, a sink, and ventilation:

$$dN = V\,dC - G\,dt - (C - Ci)\,Qi\,dt - CEQr\,dt \qquad (7.6)$$

where N is the total quantity of the pollutant, V the volume, C the concentration at time t, Ci the infiltrating concentration, Qi the volume rate of infiltration, E the vapor removal rate, G the pollution generation rate, and Qr the volume rate of the filtering device. A mixing factor is usually added to compensate for slow indoor air mixing. Typical correction values lie between 0.3 and 0.1.

This model was developed by Turk (1974) for osmogenic contaminants; it is extremely useful for any steady-state indoor source (such as formaldehyde from building materials, pesticides from dusts, and wood preservatives from the structure) and for correlating indoor to outdoor air.

The equilibrium concentration after infinite time is $C = (CiQi + G)/(Qi + EQr)$; if outdoor air is pure, the equilibrium concentration is $C = G/(Qi + EQr)$; if the indoor source subsides, $C = CiQi/(Qi + EQr)$; if the filter removes all pollutants on each pass, $C = (CiQi + G)/(Qi + Qr)$; if the outdoor air is pure and the filter 100% efficient, $C = G/(Qi + Qr)$; and in a tightly closed room

$$C = C_o \exp\left(\frac{-EQrt}{v}\right) - \frac{G}{EQr}\left[1 - \exp\left(\frac{-EQvt}{V}\right)\right] \qquad (7.7)$$

The Black Box Model. In the black box model, which was developed by Hunt at the National Bureau of Standards (NBS), outdoor air is filtered jointly with recirculated air. The response time is added explicitly to the well-mixed tank:

$$dQ/dt = ab(1 - E) + G - QEr - Qb/V \qquad (7.8)$$

where the first term consists of concentration a and volume b of outside air, G is the indoor generation rate, E the efficiency of the filter, r the recirculation rate, V the indoor volume, and Q the indoor quantity of pollutant. A similar model has been used as the basis for residential modeling by Moschandreas (1978–1981). The parameters suffice to give good agreement with observations for several contaminants. Observed and calculated levels correspond well for methane and carbon dioxide; for carbon monoxide and nonmethane hydrocarbons the model works adequately; it does not work well for sulfur dioxide, ozone, or the nitric oxides, especially at the lower threshold.

The Linear Dynamic Model. In the linear dynamic model three well-mixed airstreams operate simultaneously: infiltration and exfiltration, filtered makeup air, and an exhaust stream of which part is recirculated and separately filtered. This model necessitates six mass components and 17 parameters, but it represents a common ventilation situation and is therefore of practical importance.

Smoke Ventilation Models. The movement of indoor air in large buildings has been investigated by fire researchers, who need to predict the flow of smoke within buildings. Three well-known models are the National Research Council (NRC) of Canada model, the NBS air-movement simulation program, and the residential infiltration program, all of which are written for computers. The NRC program correlates infiltration to the wind velocity and direction and to temperature, and the stack effect is evaluated at each floor. The program is designed for 75°F and simplifies internal friction. The air flow is computed as a function of pressure at each point by linear approximation. The NBS program, conceived in 1975, can accommodate 25 separate air-handling units in a building with 10 rooms on each of 100 floors, and incorporates provisions for windows, ducts, elevator shafts, and 20 separate temperature settings. This program uses a nested reiterative procedure. The residential infiltration program computes the stagnation pressure from the wind velocity and direction. The inside pressure and air flow are derived by iteration. A multicompartment model has also been discussed by Ozkaynah (1982).

7.3.2 FORMALDEHYDE RELEASE MODELS

Formaldehyde release from particleboard, plywood, and urea-formaldehyde foam depends on temperature, humidity, the ratio of board surface to air volume, and ventilation. Andersen (1974) was the first to develop a model:

$$E = (RT + S) \frac{aH + b}{L + (nc/x)} \qquad (7.9)$$

where E is the observed indoor concentration; $a, b, c, S,$ and R are constants that represent surface finish and other factors; T is the air temperature; H the humidity; n the infiltration rate; and x the load factor in cubic meters of particleboard per cubic meter of air. He was able to represent observed values between 0.4 and 0.75 $\mu g/m^3$ in 12 Danish homes by this model.

Several parallel efforts are now under way by particleboard and plywood manufacturers and their trade organizations to develop a reliable model to correlate formaldehyde release rates from the board surface at standard conditions to the approximate range of indoor air levels at various temperatures and humidities.

Myers (1980–present) has carefully tested many forest product systems and found that a modified equation (Berge, 1980) represented the air levels quite reliably:

$$C\infty = k \cdot \alpha \cdot C_s(n + k\alpha) \qquad (7.10)$$

where C is the equilibrium air concentration, C_s the formaldehyde concentration at the panel surface, k the mass transfer coefficient in m/hr, α the loading factor (m^2/m^3), and n the ventilation frequency in air changes per hr.

The data have been checked by correlation of various emission measurement methods with the levels observed in full-scale air chambers. So far, however, no consensus has been reached on standard methods to be used in the United States for measuring the emission rate C_s (Lehmann, 1982).

Japan has had a mandatory standard for both plywood and particleboard since 1974, and the European Federation of Particle Board Manufacturers uses two parallel methods: a perforator method to determine the total free formaldehyde content, and a release rate method (Hoetjer, 1981; Meyer, 1979; Meyer and Nunlist, 1981; Sundin, 1982).

7.4 Twenty-four-Hour Total Exposure

The pollutant dose that an individual receives depends on the local exposure level with which that person comes in contact, the duration of the contact, and the absorption rate. For most air pollutants the lung constitutes a very efficient filter and the dose is close to the total integrated exposure. This section deals with the pollutant concentration at the point of contact (i.e., the exposure level). The absorption process and the dose are described in Chapter 8.

In earlier pollution studies it was assumed that, to a first approximation, each of these factors could be taken to be constant, and that the average level observed for a standard location and a standard person could be used to estimate the total dose and total health effect. The assumptions were that the variation of pollutant concentrations was negligible and that air was well mixed. While these assumptions might hold for emission from large industrial sources, we now know that these assumptions are incorrect for indoor air, and that pollution concentrations can differ several fold within a city, within a house, and even within different parts of a room; and that the concentration at any point may change as a function of time. EPA is currently conducting the third stage of a total exposure assessment methodology (TEAM) study involving 369 residents of New Jersey to develop experimental and theoretical experience.

7.4.1 AMBIENT POLLUTION EXPOSURE

Fugas (1975) was the first to compute integrated exposures, which he called weighted weekly exposure (Table 7.14). He also recognized the potential synergism and estimated total exposure to sulfur dioxide, lead, and manganese.

Nitrogen Oxide exposure calculations were attempted by Maeda (1981), and Quackenboss (1981). Both also calculated doses.

Particulate exposure levels have been measured and calculated by Fugas (1981) and Spengler (1981).

Table 7.14

Exposure Level for Sulfur Dioxide, Lead, and Manganese[a]

Type of Exposure	Hours per Week	SO$_2$		Pb		Mn	
		C	C × t	C	C × t	C	C × t
Home	110	89	9790	2.5	275	0.04	4.4
Work	42	8	336	0.3	12.6	0.02	0.84
Street 1	10	600	6000	6.0	60	0.80	8.0
Street 2	4	180	720	3.5	14	0.12	0.48
Countryside	2	25	50	0.1	0.2	0.01	0.02
Total	168		16896		361.8		13.74
Weighted weekly exposure			101		2.2		0.08

[a] After Fugas (1975). The concentration (C) is expressed in units of $\mu g/m^3$ and the integrated exposure ($C \times t$) in units $\mu g\ hr^{-1}\ m^{-3}$.

7.4.2 CARBON MONOXIDE

A more detailed mathematical and experimental approach was undertaken by Ott (1981), Flachsbart (1981), Ziskind (1981), and by Atherton (1981). In the case of carbon monoxide, a working person might experience a concentration of x ppm for 8 hours each day; she might then enter her car in the basement garage, where CO is usually at a high concentration, y, and travel for 1 hour by car to her home. Along the freeway the concentration, $z(t)$, changes as a function of time. It will increase at intersections (Ott, 1979), especially if she has to stop for a red light; it will decrease as she travels along the parkway, and so on. At home she encounters a lower CO level, $h(t)$, for the next 12 hours, with peaks whenever the heater switches on. The CO concentration will peak in the kitchen while dinner is being prepared; it will be lowest at night and then increase after breakfast when the car engine warms up in the garage. The commuter then repeats the preceding evening's trip in reverse, this time with other local CO concentrations.

The total concentration would be

$$E = \int Cx(t)x\ dt_w + \int Cy(t)x\ dt_c + \int Cz(t)x\ dt_h + \int Ci(t)x\ dt_i \qquad (7.11)$$

where t_w is the time at work, t_c the commuting time, t_h the time at home, and t_i the time at other places and activities. From this quantity we can calculate for each activity an average exposure

$$E = \frac{1}{t_a} \int_0^a Ci(t)\, dt \qquad (7.12)$$

and the "standardized exposure"

$$E = \frac{1}{t_s} \int_0^a Ci(t)\, dt \qquad (7.13)$$

where t_s corresponds to the time period used for a given environmental standard, say 8 hours for the TLV, or 30 min for peak loads, and so on. The exposure levels observed by samples taken by a person as she goes about her daily activities clearly will differ from the value calculated by taking the pollutant levels at a nearby weather station and multiplying them by the total time. A numerical total exposure analysis has not yet been performed for CO.

In field experiments it is obviously highly desirable to repeat the exposure experiments over several days in order to obtain consecutive average times. Furthermore, it is important to break activity time blocks into small enough segments to observe short peaks. Suitable procedures have been explained in several recent papers by Ott (1980a) and Repace and Lowrey (1980) and co-workers at the EPA. A similar procedure was prepared by Moschandreas (1978), who presented three-dimensional block charts (Figure 7.25) with a segment for each activity.

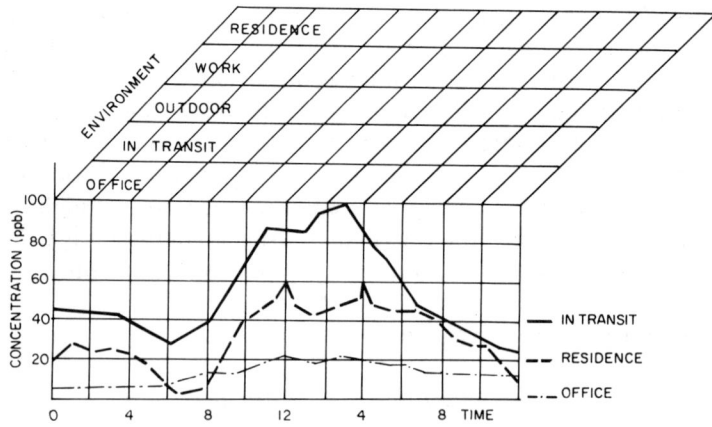

Figure 7.25. Diurnal variation of ozone in three environments (Moschandreas, 1978).

7.4 Twenty-four-Hour Total Exposure

Such calculations can now be readily made and generalized by means of compact electronic field meters that can be equipped with computer memories, and of printout and evaluation programs. Generalized exposures can also be derived from previous measurements and previous studies by computer synthesis or modeling. Ott (1980) used a Monte Carlo simulation to combine 24-hour pollution profiles for each microenvironment with the activity pattern, represented as minute-by-minute random variables.

$$E_i = \sum_{j=1}^{m} C_j t_{ij} \tag{7.14}$$

where m is the number of microenvironments. A sample printout is shown in Figure 7.26 (Ott, 1980).

It is not yet economically feasible to use this model for all pollutants of interest. Furthermore, for many pollutants the measuring methods are not yet sensitive enough to permit the collection of large-scale data in the field. In any case, a thorough analysis of all possible sources must be made before modeling becomes reliable.

7.4.3 FORMALDEHYDE

Formaldehyde must still be measured by a slow and tedious wet method. A very thorough draft total exposure analysis was prepared by R. Dailey of EPA for the Interagency Regulatory Liaison Group (IRLG, 1981). The predominant indoor source, other than building materials, is smoking, which can bring indoor levels to 0.2 or 0.4 ppm in crowded places such as bars or nightclubs. Gas stoves also contribute to indoor formaldehyde levels (Section 7.2).

Formaldehyde is an intermediate in outdoor combustion. Typical concentrations range from 2.4 to 60 mg/m^3 in natural gas combustion products, and from 140 to 1300 wt. ppm in fuel oil. Mobile sources contribute an estimated 223,000 million tons (mtons) each year, with 188,000 due to diesel trucks and buses and 31,000 tons due to jet engines. Photochemical oxidation of hydrocarbon probably adds another 20 mtons/year. The relative importance of the various sources has changed drastically over the last 10 years because the compulsory installation of catalytic devices has reduced automobile emissions, the dominant factor prior to 1975, to approximately 2300 tons/year.

Formaldehyde, an important industrial chemical, is currently produced in 52 plants in the United States. The annual U.S. production is 3 million tons. Less than 0.08 wt. % is inadvertently emitted with the tail gas at these plants; most of the formaldehyde is used captively. About 60% is used to make formaldehyde resins. About 0.8 mtons is used for urea-formaldehyde resins (Meyer, 1979), 0.5 mtons for phenolic resins, and 0.1 mtons for melamine. About 12% of formaldehyde is used to produce hexamethylene tetramine and

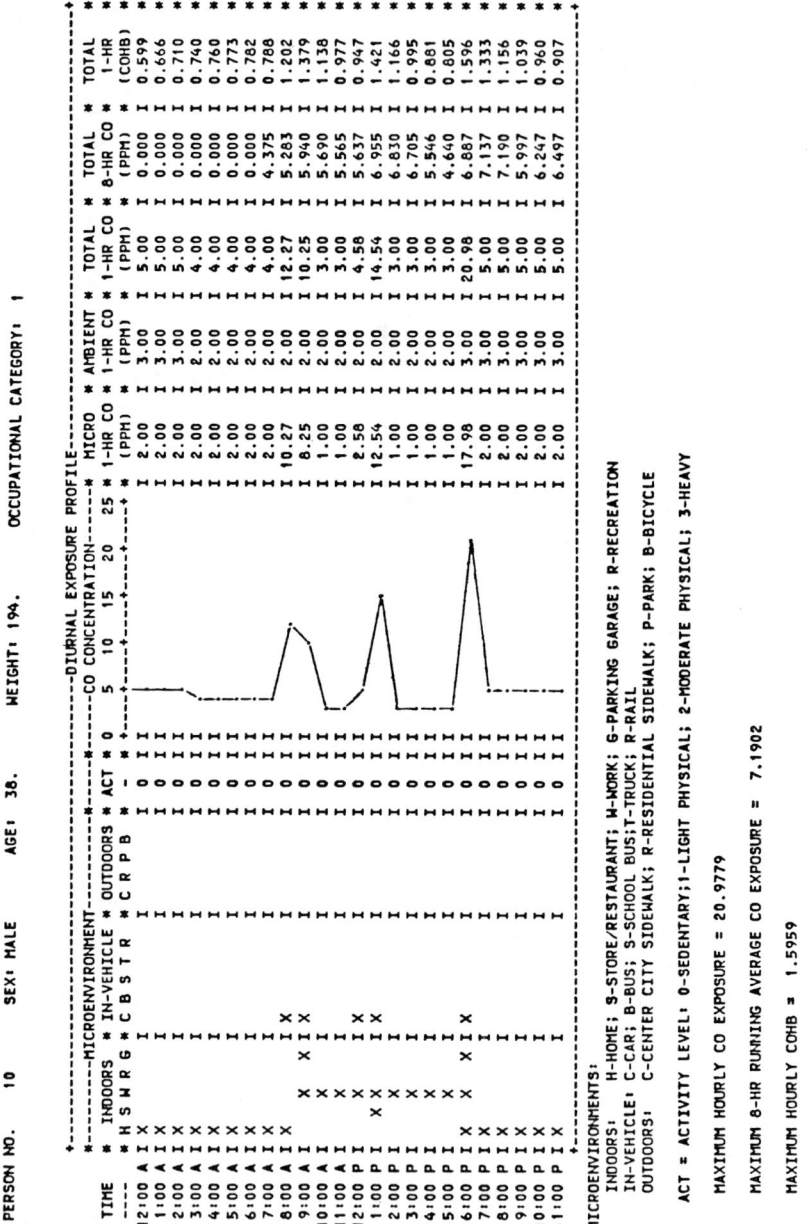

Figure 7.26. Hourly diurnal exposure profile for commuter exposure to carbon monoxide (Ott, 1980).

pentaerythritol. About 0.6 mtons of urea-formaldehyde resin are used to produce particleboard and indoor grade plywood. Particleboard contains 7 wt. % urea-formaldehyde resin, plywood 2 wt. %. Phenolic resin is used to make exterior grade plywood. Melamine resins are used extensively in Europe for wood bonding, but in the United States their main use is still for moldings. Urea-formaldehyde resins and melamine-formaldehyde resins are also used in the paper industry for sizing. During the last 10 years about 75,000 tons of urea-formaldehyde resins have been used for preparing urea-formaldehyde insulation foam.

Phenol resins are prepared in 116 plants in the United States. About 30% of the annual output is used for bonding plywood and specialized particleboard grades; the rest is distributed over about a dozen applications ranging from the electronics industry to home appliances.

Formaldehyde is also used to prepare 1,4-butanediol, acetal resins, diisocyanate, pyridine chemicals, trimethylol propane, and nitroparaffin derivatives. An increasing amount is used to prepare slow-release fertilizers (Meyer, 1977), which hydrolyze without releasing formaldehyde, since it is chemically altered in the granulating process and in the soil. Small quantities of formaldehyde are used in chemical labs as reagents.

Formaldehyde is used in several (the EPA lists 98) professional and consumer products; among the latter are shampoos (0.05%), foot powders (1%), antiperspirants and bubble baths (0.6%), and even mouthwash. About 1-2% of the population have skin allergies to formaldehyde (Bardana, 1980) and cannot use these compounds.

In the past, large quantities of formaldehyde were used in hospitals and by doctors as antiseptics, since formaldehyde attacks bacteria, fungi, molds, and yeast by reacting with DNA. At present, 50 ppm is used to stabilize influenza virus vaccines and 1 ppm in tetanus toxoids and for stabilizing pollen and venoms without interfering with their ability to produce antigens *in vivo*.

Formaldehyde reacts with protein in hair and leather and can be used for fumigating moldy furs, leather goods, and shoes. In the United States large quantities of formaldehyde are used for embalming corpses and preserving scientific specimens. Recent studies of formaldehyde levels in autopsy and embalming rooms have shown concentrations averaging about 0.7 ppm, but peaks of up to 5 or even 8 ppm have been observed in almost all locations because large open trays of formaldehyde solution are kept at hand while autopsies are in progress.

Formaldehyde is also used extensively in the textile industry to seize cotton dyes, impart crease resistance, reduce shrinkage, and make wool water resistant. New fabrics release up to 14 ppm of formaldehyde during storage for several months, but it disappears upon laundering.

Exposure to formaldehyde is possible in each of these applications at the workplace and wherever formaldehyde is released from products. Occupational levels are limited by OSHA to 3 ppm, and the recommended NIOSH value is 1.2 ppm. Most workplaces are now well below the limit. It is estimated that the 12,000 people employed at formaldehyde plants are exposed to an average of less than 1 ppm, even though the total flux is 1 mtons/year. Formaldehyde emission at the plants is estimated to be about 4000 tons in the urea-formaldehyde resin industry, 3.9 tons in the phenolic resin industry, 1.5 tons in the production of pentaerythritol, 1100 tons in the production of acetal, and only 1100 tons in formaldehyde manufacturing plants. Individual exposure varies greatly according to industry; in the particleboard industry, for example, almost all emission stems from the hot press room, whereas the wood-chip preparation train is free of formaldehyde.

Obviously, everybody is exposed most of the time to formaldehyde. The total exposure depends on the microenvironmental levels with which individuals come in contact. Inspection of Table 7.10 shows that people who live in manufactured homes and in homes with gas utilities will have the highest home exposure. This population segment comprises occupants of some 70 million homes where levels are between 0.1 and 1 ppm. The highest occupational levels are probably experienced by the 12,000 employees of formaldehyde production-related industries; the 23,000 workers in the textile industry; the 20,000 in the furniture industry; the 20,000 foundry workers; the 50,000 in the funeral services industry; the 480,000 in health and hospital services (especially pathology workers); and finally, some 1000 people in university biology and autopsy labs and the 500,000 students who enter these labs each year. The highest work exposures are restricted to a few workers, probably 1% or so, in each of these fields, who work at and occasionally above the OSHA limit. A total of perhaps 5 million other people have daily contact with formaldehyde-containing products. (This figure includes high school students enrolled in biology classes.) In addition the entire U.S. population is exposed to formaldehyde in consumer products used at home (EPA, 1981). These include treated cotton in underwear, shirts, blouses, pants, dresses, pillow covers, and linens, as well as cosmetics and pharmaceuticals. Levels are not known and vary according to the age and use of the product. Sensitized individuals experience formaldehyde levels on new pillow cases that are high enough to trigger severe asthma attacks. This is not surprising if one compares the formaldehyde release rates for cotton and particleboard and plywood (Figure 7.27), but such levels are clearly related to the proximity of the textiles because the formaldehyde reservoir on textiles is only on the order of milligrams, whereas building materials contain several kilograms of resin (Meyer, 1981).

Exposure in transit depends on ambient air levels, which may be between 0.4 and 200 ppb, depending on location and traffic density.

Total Exposure. The integrated quantity of formaldehyde with which a person comes in contact is the same as equation 7.14:

7.4 Twenty-four-Hour Total Exposure

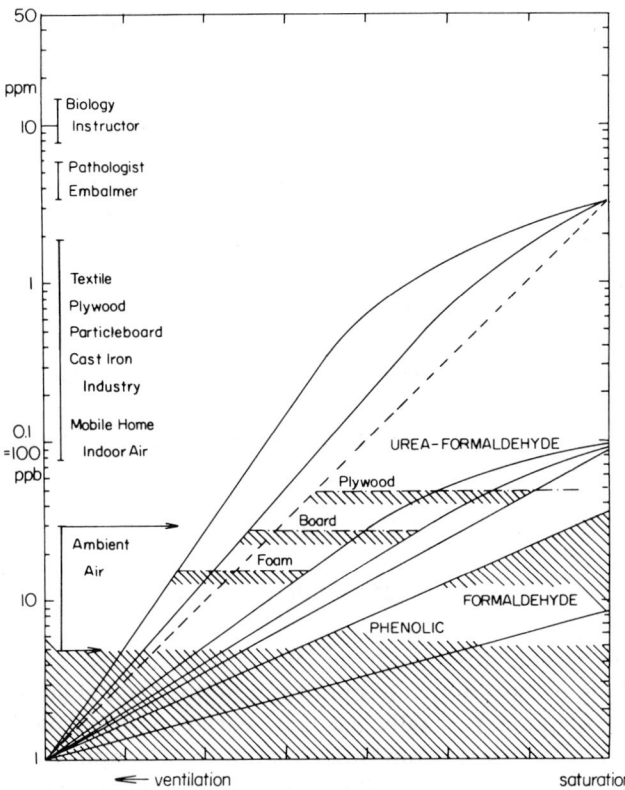

Figure 7.27. Formaldehyde levels in various environments and formaldehyde release rates for UF foam, hardwood plywood, particleboard, and crease-resistant cotton (after Meyer, 1981; Pickrell, 1982).

$$E_i = \sum_{j=1}^{m} C_j t_{ij} \tag{7.15}$$

but in the absence of personal exposure measurements, average values are used. For simplicity's sake the IRLG report assumes only three activities:

$$E_i = C_{work} \cdot t_8 + C_{home} \cdot t_{12} + C_{ambient} \cdot t_{24} \tag{7.16}$$

Exposures were based on 25 weeks with 40 work hours; 15 hours × 365 days of exposure to residential air, and 24 hours × 365 days of ambient air. The lowest conceivable exposure would be encountered in rural air with no other exposure:

$$E_{\text{lowest}} = 0 + 0 + 35 = 35 \text{ ppm} \cdot \text{hr} \cdot \text{year}^{-1} \tag{7.17}$$

An infant in an average clean urban mobile home might be exposed to

$$E_{\text{infant}} = 0 + 3500 + 350 = 3850 \text{ ppm} \cdot \text{hr year}^{-1} \tag{7.18}$$

while a textile worker might experience 0.6 ppm at home and 4 ppm at work:

$$E_{\text{FPW}} = 4000 + 3285 + 350 = 7600 \text{ ppm} \cdot \text{hr year}^{-1} \tag{7.19}$$

An autopsy instructor could conceivably experience 2 ppm at home and 14 at work:

$$E_{\text{biol.}} = = 14{,}000 + 11{,}354 + 350$$
$$= 25{,}000 \text{ ppm} \cdot \text{hr year}^{-1} \tag{7.20}$$

An exposure level of 1 ppm at home corresponds to a work exposure of 5.5 ppm for working parents and 7.30 ppm for a mother and child who spend all but 4 hours a day indoors. Finally, a 4-ppm work exposure would correspond to 0.55 ppm and the NIOSH limit of 1.2 ppm to 0.16 ppm for a mother or child at home, or to 0.73 ppm and 0.22 ppm for a working person who spends 15 hours at home.

These computations assume that the effect of formaldehyde is additive and constant for all concentrations. This might not be correct because the working person has a 16-hour recovery period for each 8-hour exposure at work; furthermore, the dose differs significantly for adults and children.

Formaldehyde Dose. Animal studies with labeled carbon (CIIT, 1980) indicate that formaldehyde is quantitatively absorbed in the turbulent airstream in the nasal passage where it dissolves in mucus. Thus, it is assumed that all formaldehyde is absorbed. The total formaldehyde dose is obtained by multiplying the exposure level by the inspired air volume, which is approximately 1 m³/hr. The highest level, for biology instructors, would be, for example

$$E \times V = 23{,}000 \text{ ppm hr year}^{-1} \tag{7.21}$$

This amounts to 1 mole, or 30 g of formaldehyde per year. This is an hourly average exposure of 3.1 ppm. A hard worker would inhale up to 10 mg per work shift. A summary of the integrated formaldehyde exposure for several population segments is shown in Tables 8.8–8.10. Even though it is now clear that 1 ppm of formaldehyde at home adds up to a longer exposure than 10 ppm at work, it is not yet established how the health effects compare.

7.4.4 RADON

No official analysis for radon has yet been completed. Observed radon levels are listed in Table 7.7. The highest levels were found in 56 homes in Butte, Montana, with average values of 0.02 working levels (WL), the lowest, in Tennessee, were in 15 homes with 0.008 WL. The integrated exposure for the first would be clearly above current safe EPA levels (Section 10.2) if the occupants were to spend 24 hours at home. The associated health risk prediction ranges from 0 to 4000 cancer cases over a lifetime, as compared with 75,000 from smoking (Nero, 1981). A comparison of the exposure levels from cosmic and terrestrial radiation is provided in Table 7.15. It is clearly too early to judge the public health hazard by radon, but it is clear that remedial action should be taken in several homes to prevent a long range, cumulative hazard to smokers or others with pre-existing tendencies to lung cancer. Such action is currently prepared by HUD and EPA.

7.4.5 ASBESTOS

The EPA completed a thorough exposure study for asbestos in 1979 that led it to ban the use of asbestos, because a direct correlation between asbestosis and the fiber count in air at the workplace had been established. This link could be demonstrated, Figure 8.16, because it is a very toxic agent, causes characteristic symptoms, and because the asbestos fibers can be readily observed, both in air and in the lung lymph nodes of the victims, where they accompany the victim until, and beyond, death.

Table 7.15

Summary of Average Dose of Natural Radiation in the United States[a]

Radiation Source	Average Dose Equivalent Rates				
	Gonads	Lung	Bone Surfaces	Marrow	Gastrointestinal Tract
Cosmic (90% of ambient)	28	28	28	28	28
Cosmogenic radionuclides	0.7	0.7	0.8	0.7	0.7
External terrestrial (60%)	26	26	26	26	26
Inhaled radionuclides[b]	—	100	—	—	—
Radionuclides in the body	27	24	60	24	24
Total	80	180	120	80	80

[a] From National Council for Radiation Protection (1961).
[b] Lung only; doses to other organs are included in "Radionuclides in the body."

8. Health

Indoor air must be fit for consumption by anyone, including infants, pregnant women, and individuals with hypersensitivities, allergies, or idiosyncratic responses to specific air contaminants because any large population always includes people who are unusually sensitive to airborne pollutants. Thus, indoor air quality requirements for the home differ significantly from those for the workplace, where people are generally alert and fit, where they spend only 40 hours a week, and where the infirm and allergic can avoid exposure by changing jobs.

High sensitivity can be due to developmental factors, such as age or seasonal and circadian rhythms, or to underdeveloped metabolic pathways, immune responses, and irradiation. It is enhanced by synergism upon continued exposure to high humidity or temperature, or to high levels of dust or microbial agents. Secondary effects of sensitization or aging caused by abuse or the deterioration of specific organs or general health are important factors, as are nutritional deficiencies, which might involve psychological factors as well as vitamin and trace mineral deficiencies. Another factor is socioeconomic stress. In a recent conference, the U.S. Department of Health, Education and Welfare listed some 60 environmental factors that affect health in the indoor environment (Hinkle and Loring, 1977); see Table 8.1.

Among the internal health factors, genetic disorders (such as blood serum disorders, red blood cell disorders, homeostatic disorders, and immunological disorders) are as important as acquired disabilities of the heart, lung, kidney, liver, and all other organs. The sensitivity of predisposed groups can theoretically be determined on the basis of chemical, biological, or physiological principles, but such are not yet fully known. Thus, information must be derived from exposure studies of selected groups, or by analysis of general population responses to ambient exposure. The analysis of such data is incredibly complex because the different factors are all interrelated and operate at the same time. Usually, the difference in average exposure levels is smaller than the fluctuations in pollutant concentrations; all groups are exposed to mixtures of agents; and in each case the mixture is somewhat different. Furthermore, the analysis of epidemiological data for low exposure levels presents tremendous statistical challenges—for example, because of the mobility of the observed people.

Tobacco smoking presents a good example (Section 8.4). Smoke is a complex chemical mixture, each component of which affects health differently.

Table 8.1
Variables in the Relationship between the Residential Environment, Health, and Well-Being[a]

	Conditioning Variables in the Person		
Demographic	Cultural–Subcultural	Stable Personality	
Age, sex, race	Religion, religiosity	Ego strength	
Marital status	Ethnic/racial origin	Flexible/rigid	
Education	Extended family kin contact	Coping styles	
Occupation	Values about deviant behavior	Self-identity	
Income		Habit patterns, needs	
Stage of life cycle			

Objective Environment	Subjective (Perceived) Environment	Behavioral Reactions	Mediating Processes	Outcome Variables
Physical environment			Biochemical and physiological	Disease states
Characteristics of dwelling unit	Perceived crowding and privacy	Daily patterns of living	Blood pressure	Illness and sick-role behavior
Distance to facilities	Perceived convenience to facilities	Behavioral adaptions	Pulse rate	Social deviance
Characteristics of neighborhood	Perceived pollution in neighborhood	Coping behaviors	Lipids	Social effectiveness and competence
		Interpersonal behavior	Glucose	Addictions
Social environment		Leisure-time activities	Clotting time	
Availability of relatives, friends	Perceived closeness to kin, friends		Plasma cortisol	
Demographic similarity to others in neighborhood	Perceived similarity to neighbors		Affective	
Presence of crime, addiction in neighborhood	Perceived dangers in neighborhood		Anxiety/tension	
			Depression	
			Anger/irritation	
			Morale	
			Satisfaction–complaints	

[a] From Hinkle and Loring (1977).

Moreover, smokers are a diverse group, and the health effects of smoking are indistinguishable from those of many other diseases. Therefore, it took a quarter of a century of intensive coordinated research to enable the U.S. Surgeon General to state flatly that "the old claims that no link between smoking and cancer was 'proven' ... [are] utterly vacuous now."

The EPA undertook a similar careful analysis of the effects of sulfur dioxide and particulate exposure on health in six selected communities. This Community Health Environmental Surveillance Study (CHESS) (EPA, 1974, and 1980) has led to the discovery of several trends. For example, it was found that children living in certain regions suffered increased incidence of prolonged bouts of croup; the elderly were found to have an increased incidence of respiratory distress; and asthmatics had a greater number of attacks. The increased incidence of health problems occurred in areas where the average exposure level of sulfur dioxide, particulate sulfate, and other particulates was higher than average. The same effect has been observed in areas where photochemical oxidants, ozone, and nitric oxides are abundant. The status of this study is described in Section 8.5.

An important problem in deciphering health effects is that we are living in a period of rapid transitions. Often our lifestyle changes quicker than it can be analyzed. Furthermore, we face new experiences for which we are ill prepared. For example, clothing made from new synthetic fibers differs in moisture absorbance and thermal properties from clothes made of the traditional cotton or wool, but the garments are so similar that we have to read the labels to determine the climate for which the fabric is suitable. Similar changes occur within our homes, where building materials such as particleboard and fiberglass are used in new applications, and ventilation, illumination, and increased thermal insulation are stepwise altered. The stress is increased when changes are covert.

8.1 Dose–Response Relationship

Exposure levels of many indoor air pollutants can be estimated to within 5 or 10 percent from the case studies and average concentrations summarized in Sections 7.2–7.4. The dose is obtained by adding the weighted exposures or by integrating the exposure levels over 24 hours and 365 days. The study of the dose–response relationship is the subject of pharmacology and biostatistics.

The response to a pollutant depends on the mechanism of its action. In the simplest case, the response is directly proportional to its concentration:

$$R = k[C] \tag{8.1}$$

where R is the response, C the concentration, and k a constant. This is the case if a pollutant is a mild irritant and metabolizes rapidly, as with low concentra-

8.1 Dose-Response Relationship

tions of sulfur dioxide and similar gases. However, the response and the value of the constant differ among individuals and for each exposure, according to statistical laws [Eq. (8.6)]. Furthermore, at higher levels the initial response often triggers secondary acute or chronic problems, such as pulmonary edema. These secondary effects can be cumulative: The body defense against allergens, for example, overwhelms the initial effect and lasts far beyond the irritation, and the response becomes extremely complex.

If the pollutant accumulates in the body, the response is

$$R = k[C] \cdot t^n. \tag{8.2}$$

This is the case with asbestos and cigarette smoke. This correlation is called Haber's law. Often the exponent n is not 1. Furthermore, it is characteristic for cumulative exposure that the response increases with time. Thus, the cumulative effect might be linear up to a threshold where secondary effects enhance the response. In the case of asbestos the cumulative effect is connected with the body burden. The body burden (BB) is determined by the relative rates of absorption A, storage S, elimination L, and biotransformation T:

$$BB = A + S - L - T. \tag{8.3}$$

Each factor depends on physicochemical and biological factors. The body burden is highest for fat-soluble organics, organic bases, and weak acids. Asbestos, chlorocarbons, and radioactive fluorides also may be stored for life. In contrast, water-soluble compounds are rapidly eliminated. The EPA has reviewed the body burden of the average North American (Holleman, 1980) and compiled reports of over 100 organic substances and metals at measurable levels. Some gaseous materials, such as chloroform, equilibrate within hours via breath. Others, such as DDT and lead, are retained for several years.

At the cellular level, the response is determined by the local concentration and the gradient at the interface:

$$R = k \frac{d[c]}{dt} \tag{8.4}$$

where t is time and c is the local concentration. This fact explains why doctors dread all chemicals that accumulate and form a body burden that constitutes a driving force in transferring the chemical to other sites. An example is the transfer of chlorocarbons from adipose tissue in pregnant women through the placenta.

For soluble materials the effective local concentration is smaller than the concentration of the dose. The concentration at the effective site depends on the dose, the rate of metabolism, and the rate of absorbance:

$$[c] = x[C] - T \qquad (8.5)$$

where x is the rate of absorbance and T the metabolic removal rate. An effective drug is one for which x is close to 1, so that all of the therapeutic agent can reach its target and the dose can be low. With air pollutants the situation is the same. The most dangerous pollutants are those that are readily absorbed and effectively transmitted to their target. Unfortunately, the respiratory tract is designed to be efficient in transmitting air pollutants.

Biotransformation can render air pollutants either harmless or toxic. For example, BaP only becomes viciously carcinogenic by transformation. Formaldehyde, on the other hand, is converted in the liver to formic acid, which is a normal metabolic by-product and harmless at a reasonable concentration.

The response of an individual to a given dose cannot be quantitatively predicted because it depends on many extraneous factors, such as general health and diet, that vary constantly and that are rarely fully known. This is true even for homogeneous groups such as soldiers or sailors (Hers, et al. 1969; Voors, et al. 1968), who share the same diet and activities and belong to the same age group. In fact, even if people could be selected and kept under controlled conditions, like laboratory animals, a statistical spread in responses would still be expected.

Figure 8.1a shows a normal distribution curve of the formula

$$Y = [s^{-1}(2\pi)^{1/2}] \exp\{-1/2[(X-\mu)^2/S]\} \qquad (8.6)$$

where S is the standard deviation and μ the mean of the distribution. If such a curve is drawn for a response that is absent in a normal unexposed population, the plotting is relatively simple, especially if the response is a tumor or an infection for which the onset is a simple all-or-nothing proposition. Three ranges are always observed: (a) from 0 to the first point at which a tumor has been observed (point A in Figure 8.1b); (b) from the first to 100%, where individual response varies; and (c) from 100% on, where all guinea pigs have responded. It is common to represent data in a cumulative curve (Figure 8.1b) where the 50% response corresponds to the mean.

The relative dose might be plotted on a weight scale, or on a dimensionless scale of weight of dose per weight of host, such as milligrams per kilogram or parts per million (ppm), a weight per surface area, or any other suitable scale. The practical problem with the curve in Figure 8.1b is that a large number of observations is needed to determine the best slope of the dose–response curve.

If the curve is experimentally derived, it is much easier to determine the approximate location of the 50%, or mean, value, than to establish point A, since the latter shifts with increasing number of experiments. As far as exposure to air pollutants is concerned, we clearly do not want to tolerate more than a very small effect in a very large population. Thus, we have to determine the shape of the curve close to point A (Figure 8.1b). If we want to determine

8.1 Dose-Response Relationship

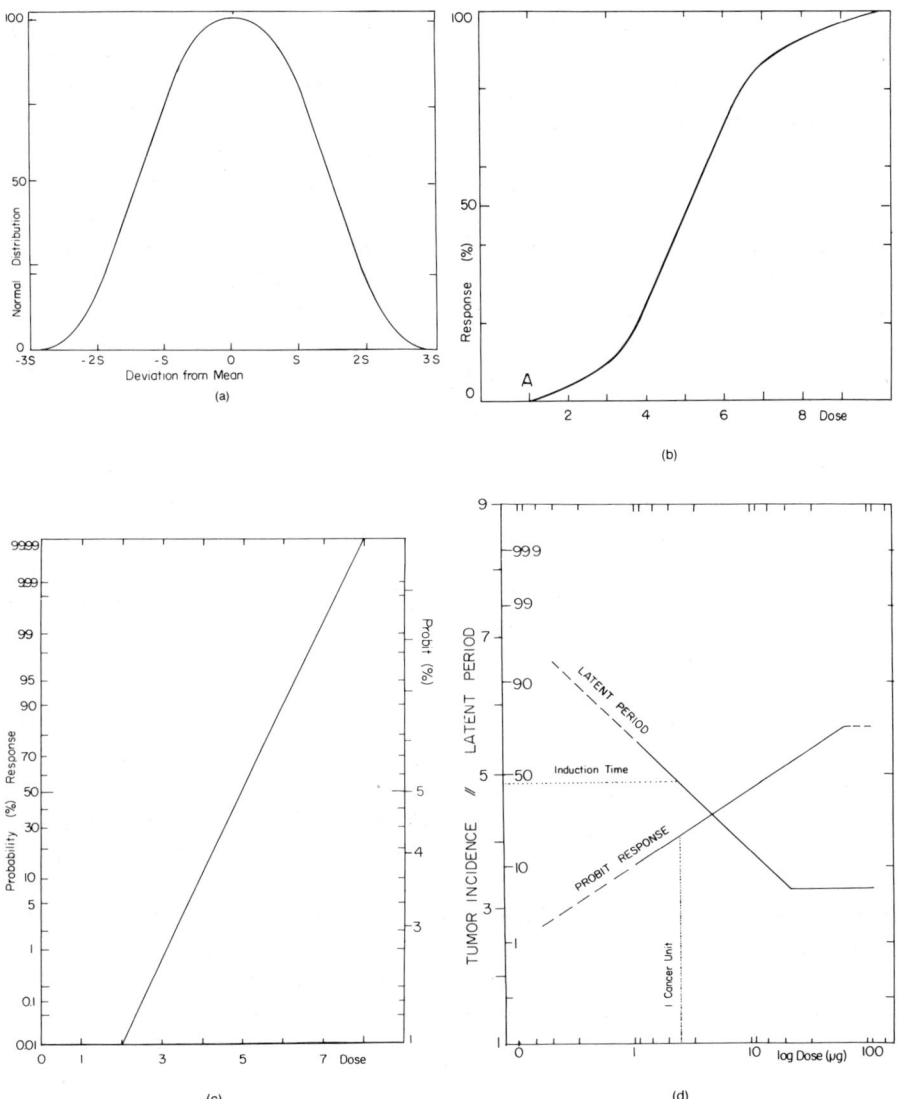

Figure 8.1. Dose–response relationship. (a) Normal distribution curve; (b) cumulative distribution curve; (c) probit transformation; (d) probit response curve and latent-period response curve (from Bryan, 1940).

the point at which only 1% of the population is affected and if we want to have a 50% chance to observe this effect, the laws of statistics require the study of 135 subjects for a significant answer. For any irritating or toxic substance, this type of reliability is insufficient. Therefore, the number of subjects needs to be increased. With expensive animal experiments that might go on for many years a compromise has to be found between the "megamouse" experiment and the reliability of point A. Often, the compromise consists in deriving the curve by extrapolation from the slope at a higher point; but if only very few experiments are available, it becomes difficult to establish the slope of the curve in Figure 8.1b.

There are other complicating factors. It may become necessary to determine the effect of an air pollutant superimposed on a background effect in a "normal" population. Statistical analysis indicates that if the background incidence of a disease is 1%, and we want to have a 50% chance to observe the dose necessary for doubling the effect, we would have to study more than 775 subjects. If different doses cause different clinical symptoms, the situation becomes challenging indeed. Such a case is represented by the effect of diethylnitrosamine on rats (Schmähl, 1980): Above 7 mg/kg it causes kidney cancer, below that dose it causes liver carcinoma, and below 1 mg/kg it causes carcinoma of the esophagus. If multiple effects are expected, as is normally the case, each exposed and control subject must be dissected and its body cavity and nervous system inspected. A histological examination using eight step slides yields about 80 slides per animal. Thus, a five-dose evaluation with 100 animals entails some 48,000 slides, enough to tie up a large histological institute for a full year.

Such problems can be overcome by mathematically transforming the standard distribution in such a manner that the sigmoid curve is stretched into a straight line. This is done by distorting the ordinate scale as is shown in Figure 8.1c so that the normal deviation becomes standardized. Furthermore, the axis is shifted by 5 units to avoid negative values. This probit transformation makes it must easier to plot weighted curves. Special log dose versus probit paper is available to establish linear probit regression lines from which TD_{50} values can be read accurately.

In the case of carcinogenic agents we see that the time response to dose is also linear. The latent-period dose–response curves follow equations of the type

$$t = a - b(X - \overline{X}) . \tag{8.7}$$

From this we can deduce the mean tumor-free latent period, which follows a hyperbolic function. Unfortunately, the induction time of single cancer cells is not measurable *in vivo,* and the clinical induction time is a large multiple of the molecular transformation time. In fact, the clinical induction time is not necessarily proportionate to the molecular transformation time because it depends strongly on the manifestation of a tumor. Several grams might be necessary before a wildly metastatic tumor is recognized, whereas 10 mg might suffice for a well-contained skin growth. In any case, more than half of the time of the tumor growth period may be symptom free (Schmähl, 1980). Figure 8.2 shows that 10^6 cancer cells are necessary to produce 1g of tissue (Gullino, 1977; Schmähl, 1976). Many complications can occur if doses are applied repeatedly. The proper placing of dose–response curves is a demanding science because of the need to avoid overdoses or saturation doses and other secondary effects.

8.2 Toxicity

Whether a substance is toxic depends on its dose. Almost any substance is toxic if taken in excess; even oxygen causes severe respiratory irritation, and 10 years ago a man died from electrolytic cell imbalance after rapidly drinking 17 liters (four gallons) of plain water (Schmähl, 1976). Repeated chemical irritation by caustic or acidic agents such as HCL or KOH has been repeatedly shown to produce squamous cell carcinoma (Narat, 1925; Schmähl, 1980).

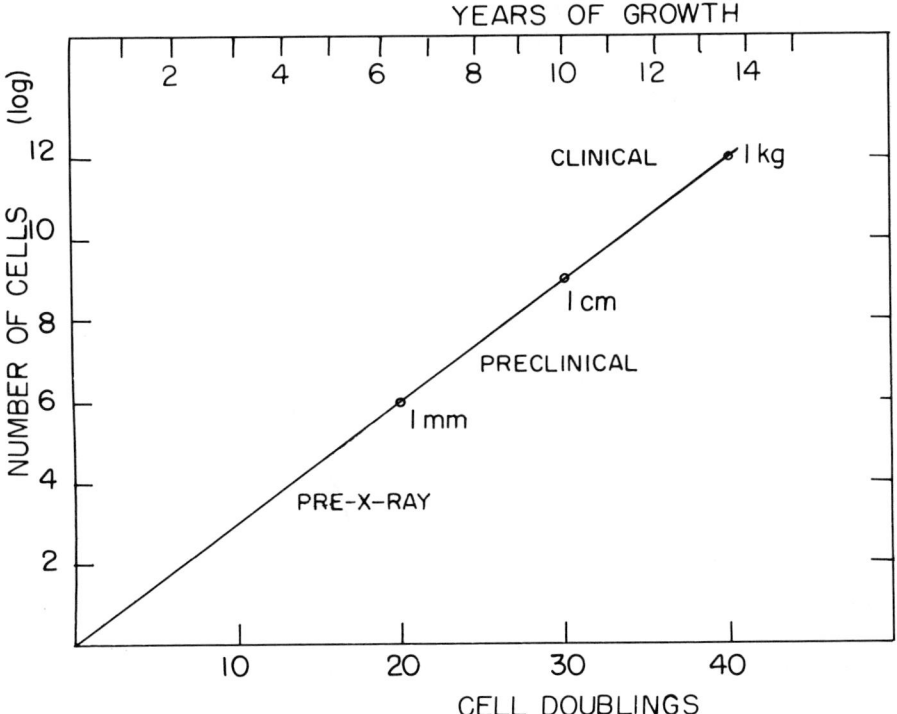

Figure 8.2. Number of tumor cells as a function of doubling time of cancer cells (Gullino, 1977).

Most chemicals have more than one effect, depending on concentration. Arsenic and selenium are typical for the behavior of many metals. They are important trace elements. If the daily dose is below the threshold, a deficiency will develop and will cause severe disease. As the dose is increased above the necessary level, the body eliminates excess metal. If the dose increases beyond the body's ability to eliminate or metabolize the chemicals, they will cause acute or chronic poisoning. The same type of response holds for many other chemicals, including all vitamins except vitamin C. In pharmacology the distance between the 95% point of the curative dose–response curve (CD_{95}) and the 5% point of the toxic or lethal dose–response curve (LD_5) (Figure 8.3), the so-called Ehrlich index, is used to characterize the safety margin of a drug. For old medicines, such as mercury, which was used against syphilis until after World War II, the margin is small and the two curves are so close that an individual might die before the drug became fully effective. For anesthetics such as chloroform and sleeping pills such as barbiturates the curves are better separated but the range is still narrow. Fortunately, the safe range for oxygen is large enough to accommodate man at most altitudes found on the earth's surface. (It ranges from 100 Torr to 480 Torr, which corresponds to 12 to 60 vol % at standard pressure (Figure 1.1b, p. 2 and Section 3.2.1)).

Typical dose–response curves for low exposure levels are indicated in Figure 8.4. Carbon dioxide clearly fits curve (c), because air in the lung always

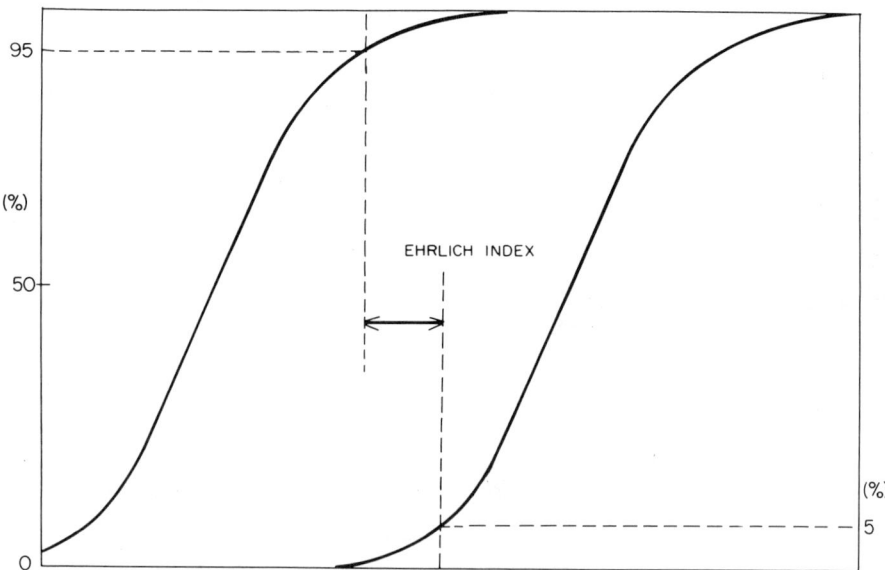

Figure 8.3. Ehrlich index.

8.2 Toxicity

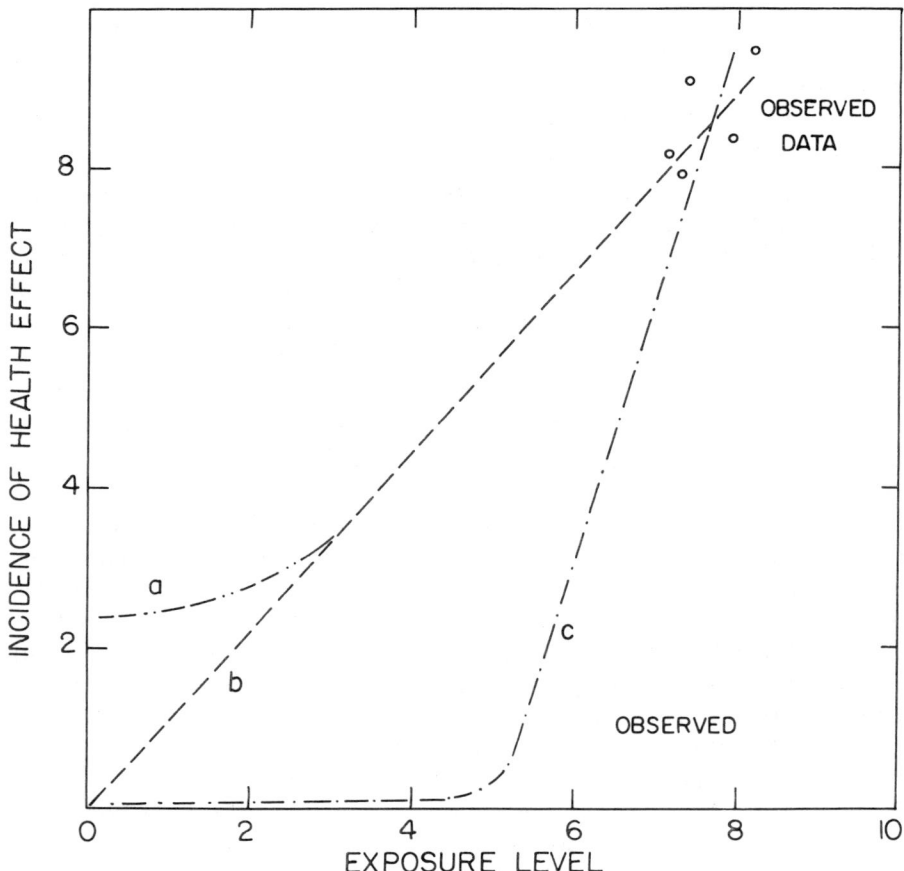

Figure 8.4. Dose–response relationships at very low levels of exposure.

contains at least 5.6%, or a partial pressure of 40 Torr (Figure 3.3, p. 37). Above 6% CO_2 rapidly becomes toxic and causes narcosis. In contrast, certain bacteria, for example the tubercle bacillus, fit curve (a), because experiment has shown that one microbe in 50 m³ is sufficient to cause the dreaded disease.

Most air pollutants are quite harmless at low concentration, and therefore, as explained above, the curve cannot be easily measured experimentally, and a dispute inevitably arises whether point A (Figure 8.1b) is at zero (Mantel and Schneidermann, 1975; Port, et al. 1976; Schmähl, 1980), and the curve is a straight line (Figure 8.4, curve b). The answer depends on the individual agent and the group at risk. We would like to assume that those chemicals that are natural impurities in air, such as CO, CO_2, NH_3, formaldehyde, and the metals, follow curve (c) in Figure 8.4 and that they are harmless at natural levels. This is the case for most. The best known exceptions are asbestos and radon, which occur in natural geological hot spots in amounts that exceed safe levels; both are carcinogenic.

From a theoretical viewpoint, Truhaut (1977) summarized the arguments for and against the carcinogenic threshold theories as follows.

Arguments for a No-Threshold Theory. (1) Inasmuch as cancer is induced by modification of DNA, any one carcinogenic molecule may suffice to initiate modification of the cell replication; (2) for some agents, such as diacrylnitrosamines, even minute doses appear to be additive; (3) for croton oil and dimethylbenzamine (DMBA), a single contact is sufficient; (4) for asbestos, response can set in decades after a single dose; and (5) cancer may result from mutation in a somatic cell.

Arguments for a Threshold Theory. (1) Inasmuch as the human life span is limited, there is for each individual a time threshold for contracting cancer; (2) the probability that an airborne agent will reach its target organ is likely less than 100%, especially if the organism has a metabolic defense for deactivation or detoxification; (3) in some microbial systems DNA repair mechanisms for radiation damage have been observed *in vitro,* this type of defense may also exist *in vivo* in mammals. Biochemical research on the cause of hypersensitivity to ultraviolet light-induced skin cancer may support this argument. (4) The first step of irritating agents consists of mutation. The human body has the ability to host many mutative cells without the secondary, clonal multiplication. (5) Antigens may reject tumors. (6) The decline of risk in people who discontinue smoking may be interpreted as due to a threshold. (7) In some experiments with DMBA secondary exposure resulted in reduced response. (8) Some secondary carcinogens act only after preliminary changes, such as hyperplasia or cirrhosis, have been observed. (9) Some carcinogens, such as selenium and arsenic, are trace elements; that is, trace quantities of these substances are not only harmless, but are necessary for survival.

A simplistic answer is not possible because the term cancer denotes a variety of malignant tumors that can be produced by different mechanisms, and as long as the mechanism is not known, any discussion will remain speculative.

Combination Effects. One of the characteristic aspects of air pollution is that air pollutants rarely occur alone, and that exposure to several occurs simultaneously or sequentially. A study of the combination effect requires knowledge of the regression line of each of the agents. It has been shown that the combination of subthreshold level doses can lead to five different types of responses:

1. Finney predicted in 1952 that the effect of multiple agents should be additive:

$$\frac{1}{LD_{50}^{(total)}} = \frac{P_A}{LD_{50}(A)} + \frac{P_B}{LD_{50}(B)} + \frac{P_C}{LD_{50}(C)} + \ldots \quad (8.8)$$

This behavior has been established for a combination of lead, cadmium, and arsenic; cadmium and mercury; or zinc and aluminum (Calabrese, 1978).

2. Some agents act "indifferently" that is independent of each other. Thus, the oberved effect is due to the strongest individual component. It has been claimed that SO_2 and NO_2, or SO_2, NO_2, and CO, are indifferent, as are SO_2, H_2SO_4, and fly ash; but it is likely that the particulate concentration is above saturation level, and that in reality the observed effects are always due to a combination.

3. Antagonism is observed for hydrogen sulfide and ozone or other chemicals that neutralize each other chemically or negate each other's biological effects. Disulfide and sulfide may retard radiation damage. Several reports indicate that particulates, even silicates, antagonize nitrous oxide, aldehydes, and other agents. This may be due to the increased mucus production induced initially by particulates. However, such antagonistic effects may merely hide delayed new reactions, or long-range synergism.

4. For some agents, the development of tolerance has been indicated. Ozone, for example, leads to tolerance after several weeks of applications in athletes (Calabrese, 1978; Hackney, 1974); ozone also is reported to induce cross-tolerance to cumene, nitric oxide, and even radiation.

5. Many toxic substances act synergistically: Beryllium and fluoride cause exaggerated pulmonary lesions; organic mercury enhances the effect of parathion. Lead enhances the effect of benzopyrene, ethanol, oxygen, ozone, cadmium, and zinc. A further example of synergism is formaldehyde, which above 0.7 ppm enhances the action of 0.04 μm of sodium chloride mist in guinea pigs (Amdur, 1959) and of mineral oil, glycerin, and glycol, as well as sodium chloride and amorphous silica (LaBelle, 1955); BaP enhances diethylnitrosamine, SO_2, N-methyl-n-nitrosourea dodecane phenol, ferric oxide, benzene hexachloride, and radiation.

Radiation, like many other agents that can breach the body's natural defenses and open pathways for exposure, must be assumed to act synergistically. Synergism has been proven for the combination of tobacco smoke with many other irritant or toxic substances. Also well established is the effect of tobacco smoke with radon on uranium miners (Lundin, Lloyd, Smith, Archer, and Holaday, 1969), with asbestos on construction workers, with dust, and with many other agents. In fact, tobacco smoke itself is a synergistic mixture of particulates, smoke, organic carcinogens, and carbon monoxide (La Belle, 1955).

Finally, heat and humidity can act synergistically on chemicals. Experiments with mercury chloride indicate that air movement can enhance the hyperthermia caused by the former combination (Thaxton, Yonushonis, and Baughman, 1975).

In some cases synergism might be a clinical effect only, while the microscopic effect might be a threshold effect by one agent, followed by an additive, or even an indifferent, action of the secondary agent.

Toxicity Testing. The only reliable approach for establishing the toxicity of any chemical is full-scale human testing under real-life conditions. In the past, drugs were tested on human volunteers, often prisoners, without preceding animal screening. This practice was discontinued after World War II. However, many pollutants were introduced into the environment before the Toxic Substance Act (Chapter 10.4) and are now so abundant that the only test left is the absence test, in which an airborne polutant is prohibited until it has been established whether the public health significantly improves or not. This was the approach used by the EPA when it eliminated SO_2 by substituting low-sulfur oil and gas for coal. The same approach is now used with smoking, but the concept of "unnecessary risk" is not yet popular. The conventional approach is to determine acute toxicity in laboratory experiments before a hazardous substance is banned (de Bruin, 1976).

The current standard procedure (NRC, 1977a) for testing toxicity and drugs is a stepwise method that starts with *in vitro* experiments on yeasts and bacteria and proceeds to animals. There is currently much public and political debate on the value of rat and mouse experiments for the determination of effects on humans. Animal data depend on the dose, form of application, diet, sequencing, and many other factors. Furthermore, different strains of a given species might yield different results. Since the main purpose of animal experiments is the screening of toxic effects, not only mammals but even vertebrates are sometimes useful because they have certain basic phylogenetic similarities in their response. The problem of the best choice of species is complex. For one, the natural lifetime of the animal is crucial. Then, the resistance of the species to respiratory cross-infection is important. Healthy animals are vital in order to keep manageable the number of animals necessary to yield significant differences between the control and the test group. However, such healthy animals do not represent the spread found in a normal human population, where the infirm also deserve protection. Many other practical considerations, such as the availability of animals, space, and funds for keepers and veterinarians, are involved. Other factors affecting the selection are the absorption, distribution, storage, metabolism, and excretion mechanisms for various agents in relation to the target organ.

Interspecies extrapolation must be corrected for differences in each of these factors, as well as in body size. However, it is vital that interspecies extrapolation be held to a minimum by testing different species of animals. In some cases, doses act on the basis of relative body weight, whereas in others the dosage per surface area is more decisive. Mechlorethamine provides an example where the dose response is very closely related to surface area: The human response range is 11.3–15 mg/m^2; other responses are 11.1 mg/m^2 for the rat, 10.9 mg/m^2 for the hamster, and 9.6 mg/m^2 for the mouse. Substantial literature is available on this subject (Calabrese, 1978). However, multispecies testing remains the safest approach.

Occupational Threshold. If toxic properties are not anticipated, they usually become apparent first at the workplace, where relatively high concentrations act on a well-defined group of generally healthy people. The need for fresh air in the workplace was recognized as soon as underground mining started. The effect of temperature and humidity on productivity was fully recognized in textile mills, where hot moist air is needed for "conditioning" the print cloth. Smoke fumes irritated workers 4000 years ago. Coal dust was first identified in 1831 in the lungs of miners. Dr. Birmingham published a study in 1900 which showed that ganister, a highly friable silicate rock used to line metallurgical ovens, caused an annual death rate of 4.2% among ganister miners, 18% among ganister grinders, and 2% among ganister brickmakers (Birmingham, 1900). This mortality rate was three times higher than the normal 4.3% for respiratory disease in the mining industry at that time; "2-3 years of work bowled most of the men over," and death (usually from tuberculosis) occurred 1-2 years after symptoms were detected. In 1912 the first tables of safe toxic exposure levels were published, with 20 recognized acute toxins. Today detailed threshold levels are known for many hundreds of chemicals (NIOSH, 1974; ACGIH, 1976; EPA, 1974-1981; Decoufle et al., 1974; Section 10.2).

8.3 The Respiratory System

The human respiratory system can be divided into three compartments: The nasopharyngeal compartment, or upper respiratory tract; the tracheobronchial tract; and the pulmonary compartment. The first part comprises the anterior nares, anterior pharynx, posterior pharynx, larynx and epiglottis (Figure 8.5). Only some 3% or less of the air inspired passes through the nose. The sensation of smell arises from two independent sources. The first, the olfactory epithelium has a surface of less than a square inch. This epithelium contains several million bipolar receptor cells, which lead to the olfactory bulb in the brain. The second, the trigeminal nerve system, extends over a larger part of the nasal cavity. Its main reaction is to initiate sneezing and other protective reflexes (Sewall, 1897).

The second compartment includes the entire bronchial tree. The two comprise the entire epithelial area, which is ciliated and covered with mucus, which stems from the goblet cells and secretory glands. It forms the anatomical dead space of the respiratory tract. The pulmonary compartment includes the respiratory bronchioles, alveolar ducts, atria, alveoli, and alveolar sacs. This functional, or exchange, space contains a moist epithelium that is not ciliated.

A healthy sedentary adult pumps about 10,000 liters (2600 gal) of air through this system per day. This air includes 1-20 mg of respirable particulates. The vital capacity of the lung is about 4 liters (1 gal). The active

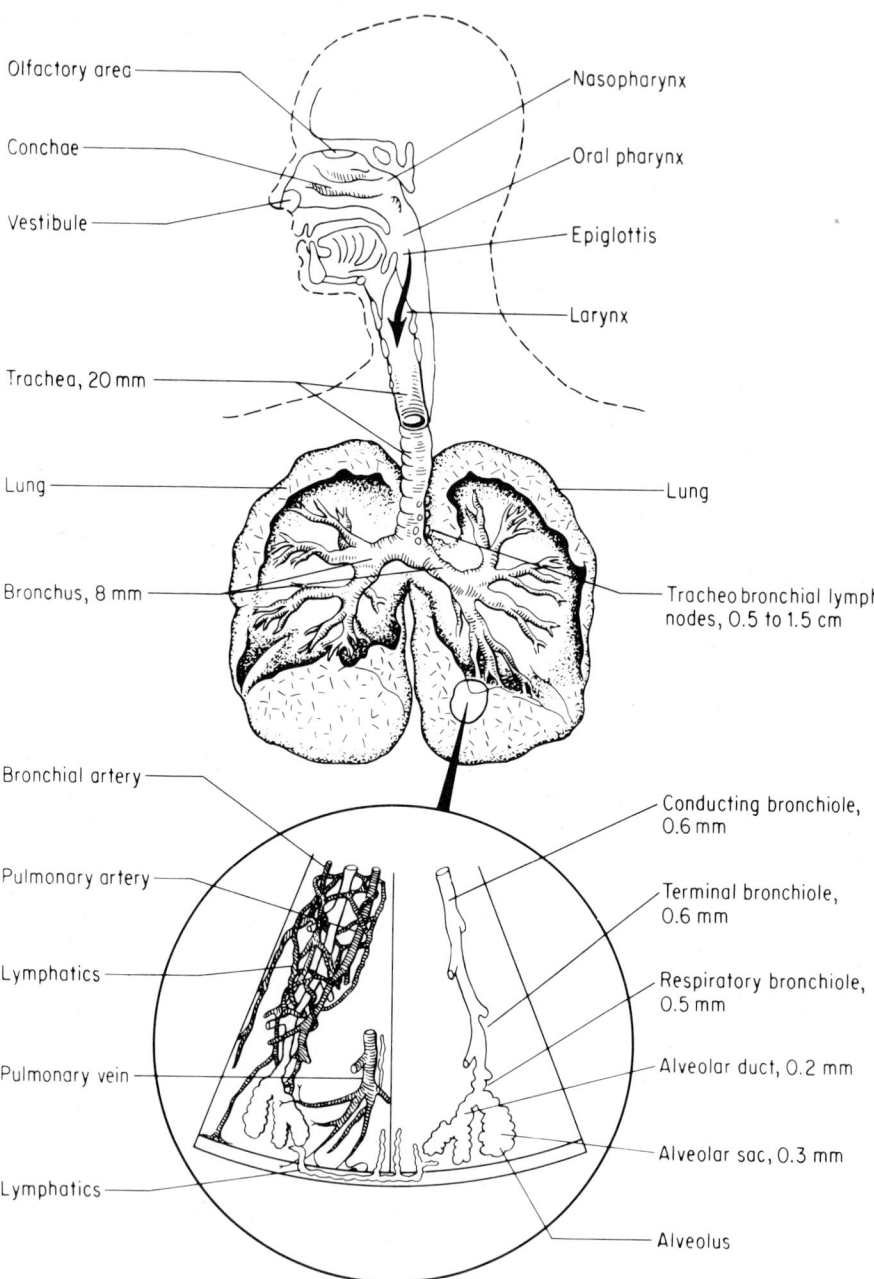

Figure 8.5. The human respiratory system (NRC, 1961).

alveolar area is 50 m² and the upper respiratory tract, including the bronchi, measures 20 m². The principal features of the respiratory tract are shown in Figure 8.5 (NRC, 1961). The processes in the lung involve oxygen and carbon dioxide exchange at a concentration gradient determined by the ambient air (20% O_2, 0.03% CO_2, 20%–90% H_2O), and the alveolar air (inspired at 15% O_2 and 5.6% CO_2, and expired at 14% O_2, 6% CO_2 and 100% H_2O).

The normal airway resistance is 0.6–2.4 cm of H_2O per liter per second at a lung volume of 4.2 liters. The pulmonary resistance is 1.9 cm of H_2O per liter per second. The oxygen exchange in an alveolus that has a diameter of 0.1–0.2 mm can be graphically represented by a plot of blood oxygen saturation versus oxygen tension (Briscoe, 1974); see Figure 8.6. The lines correspond to

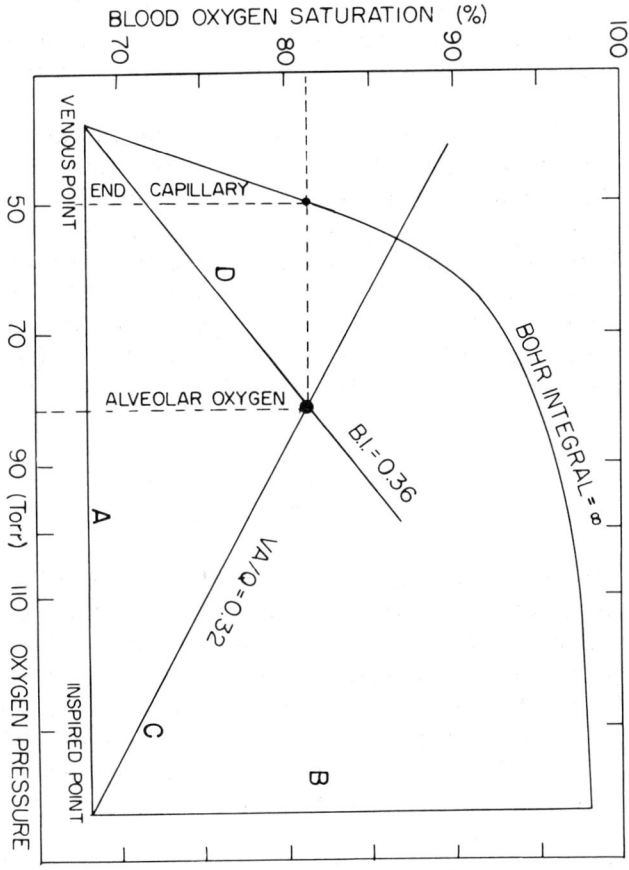

Figure 8.6. The Bohr integral (from Briscoe and King, 1974)

mixed saturation in the vein (A), oxygen partial pressure (B), alveolar ventilation (C), alveolar perfusion (D), and diffusion capacity. Lines C and D form a ventilation–perfusion isopleth. Perfusion (D) and diffusion form an isopleth that is named the Bohr integral, after Christian Bohr, who in 1909 incorrectly believed that oxygen transfer was a process actively supported by secretion. The difference between alveolar and capillary oxygen tension is

$$B = \int_{S_1}^{S_2} \frac{dS}{(P_a - P_c)} \tag{8.9}$$

where B is the Bohr integral, dS the momentary change in saturation, and P_a and P_c the alveolar and capillary oxygen partial pressures, respectively. The exchange differs among the 200 million alveoli in the fast and slow space areas of the lung. Gravity and other factors are involved in the distribution. The anatomy of the lung and its mechanism are the subject of many specialized books.

Exercise and altitude increase the pulmonary gas exchange and the tidal volume, but up to a factor of 3. The respiratory frequency is about 15 cycles/min. In a highly trained athlete the alveolar capillary exchange of O_2 and CO_2 can exceed 100 ml/sec (Whipp and Mahler, 1980). At sea level and 37°C, blood dissolves 0.3 ml of oxygen as gas and 19.5 ml as oxyhemoglobin per 100 ml of blood at pH 7.4. In venous blood (at pH 7.37) the hemoglobin saturation drops from 97% to 75%. The oxygen–carbon dioxide exchange is aided by the change of 0.03 pH units, reflected in the Bohr effect (Figure 8.6). The correlation between pH, CO_2, and HCO_3^- follows the Henderson–Hasselbach equation:

$$pH = 6.1 + \log \left[\frac{(HCO_3^-)}{(CO_2)} \right] \tag{8.10}$$

The oxygen use is reflected in the hydrolysis of adenosine triphosphate (ATP), which transduces free energy during muscle contraction. The control of O_2 consumption is a very complex process that is not yet fully understood, but during light or moderate work the oxygen uptake increases, as demonstrated by Hill in 1924, with an exponential time factor, with a rate constant of about 80 sec^{-1}. In the dog the time constant is 15 sec. The CO_2 release is determined by metabolic CO_2, by change in the bicarbonate blood buffer, and by hyperventilation. The bicarbonate buffer is responsible for the dissociation of the lactic acid, which has a pK of 3.86.

About 16 classes of respiratory diseases are distinguished (Bates and Christie, 1971). They can result in imbalance of the ventilation perfusion relationship and yield a feeling of increased respiratory effort, called dyspnea. This may lead to, but does not necessarily cause, increased breathing (tachypnea) and increased ventilation (hyperpnea), or hyperventilation; the latter helps reduce the CO_2 level in the alveoli.

Breathing would be unduly difficult if the lung did not secrete dipalmoyl lecithin, which acts as a surfactant and reduces the surface tension, which would otherwise obstruct distension of the lung. The importance of this function is demonstrated by the respiratory distress syndrome of the newborn.

8.4 Climate and Health

The diagnosis of environmental disease is an art the practice of which requires extensive experience and time. Since most environmental diseases are not immediately life threatening, this art is not glamorous and some doctors treat indoor air complaints hesitantly or even with benign neglect. As a result, many millions of people live with varying degrees of discomfort that more or less subtly affects their private lives and their professional performance. During recent years, however, a small group of medical doctors and health specialists has revived the field of clinical ecology (Dickey, 1976; Silver, 1979; Lave, 1972, 1977; Fanger, 1979; Thomson, 1979; WHO, 1979; Pfeiffer and Nikel, 1980).

8.4.1 THE INDOOR CLIMATE

The relationship between indoor comfort and professional productivity is well established (Fanger, 1979). As pointed out in Section 3.2, a lightly dressed, sedentary person feels most comfortable at a temperature between 20 and 25°C, a relative humidity between 40 and 60%, and an air velocity of 0.1–0.4 m/sec. The comfort range is shown in Figure 3.5, p. 44. If these conditions are not met, the healthy body takes instinctively defensive actions. In the indoor environment, however, if a person is seated or in a social setting, corrective action is deferred and the person easily drifts into a state of discomfort. In crowded meeting rooms high humidity and high temperature develop rapidly, producing an atmosphere of stress that enhances any unpleasant feeling. Recognition that a chronic sedentary life causes circulatory sluggishness has caused several million Americans to take up jogging or other forms of exercise. As early as 1919 climographs (Figure 8.7) correlating changes in the local death rate with humidity and temperature in the United States and France had been published. Statistics show that weather is the dominant reason for absenteeism in the military (Sale, 1972).

There is no question that the physical and chemical conditions of stale air can interfere with normal body functions in homes and offices (Fanger, 1979), as well as in spacecraft, submarines, and army tanks, but the physical, physiological, and psychological factors are so deeply intertwined with personal perception that it is almost impossible to distinguish how much of an individual's

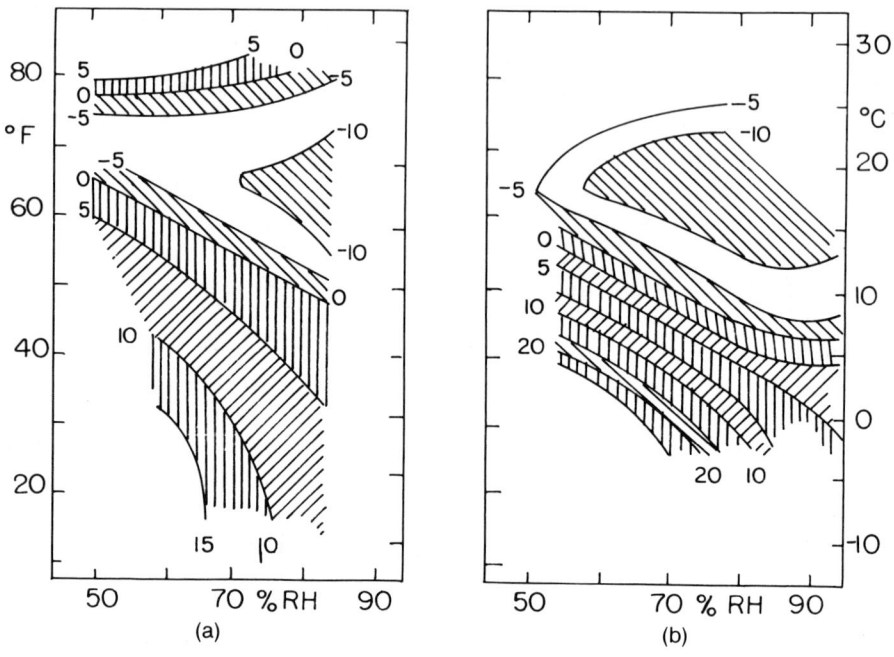

Figure 8.7. Death rates as a function of temperature and humidity in (a) the United States, 1912–1915: 921,000 deaths; and (b) southern France, 1901–1910: 838,000 deaths. (After Huntington, 1919.)

discomfort is due to climatic stress. Furthermore, unsatisfactory climatic conditions are frequently related to socioeconomic factors (Table 8.1; Biersteker, et al. 1965; Lave and Seskin, 1972; Ayers et al. 1973; Bouhuys et al. 1975; Martin, et al. 1976; Breslow and Whittemore, 1979; Spivey and Radford, 1979; Beck and Bouhuys, 1980). Finally, the clinical symptoms caused by such conditions, including fatigue, headache, sweating, respiratory discomfort, and changed pulse frequency, are not specific, and therefore cannot be conclusively related to air quality unless a more comfortable climate can be restored. The situation with other pollutants is similar because they are usually present in low but steady concentrations and the onset of symptoms is too gradual to be easily and conclusively correlated with a pollutant whose existence might not be suspected. Thus, it is virtually impossible, except under extreme conditions, to make generally valid, conclusive, and quantitative statements about the health effects of air contaminants unless extensive testing of volunteers in enclosed environmental chambers has been conducted (Colligan, 1981).

An extreme climate is encountered in commercial air travel. Contrary to expectations, the unusually crowded seating does not conduce to high humidity. On the contrary, the outdoor air is so dry at high altitudes (Figure 3.11) that the humidity in commercial aircraft is only 5-10%, far lower than in the driest deserts of the world. This causes problems for passengers and flight crew with sensitive eyes. Corneal hypoxia can cause epithelial damage and edema in lens wearers. Hard contact lenses are the worst. In a survey of 774 flight attendants, 95% reported eye discomfort (Eng, 1979). The discomfort was attributed to smoking by 92% of those polled, air conditioning by 63%, cabin lights by 56%, and light reflected from the aircraft wings by 47%.

The low humidity also causes excessively dry oral mucosa. The effect is similar to that in overheated rooms. Accordingly, flight crews report a larger than average number of sinus discomfort complaints and suffer from increased sinus infections. The latter, however, are also aggravated by the proximity to passengers, the constant cabin pressure changes, and other environmental factors, and Andersen (Andersen, Lundqvist, and Proctor, 1972; Andersen, Lundqvist, and Jensen, 1974; Gelperin, 1973) claims that low humidity, by itself, is harmless.

8.4.2 THE OUTDOOR CLIMATE

In the past doctors were amazingly knowledgeable about indoor air, and many were keen meteorologists. Hippocrates wrote a volume called *Air, Water and Places* in which he correlated the general health and well-being of people with "the variability of the weather," and the Romans sent people to the Yugoslavian seashore to convalesce.

More recently, the medical profession frequently added moral connotations to the indoor climate surrounding their clients (Ross, 1852), and did not hesitate to educate patients in housekeeping skills. Thirty years ago when doctors still made house calls, a doctor typically treated bronchitis by instructing the housewife on how to reduce drafts in the bedroom (Wells, 1955) and smoke from the coal stove, or how to keep air humidity under control. Today, the patient is not seen in the habitat that causes the complaint, and the doctor or nurse practitioner hardly has time to worry about the climate or the patient's clothing or home environment. There are many reasons for this change. One is that chemistry has made it easier to arrest or treat a disease than to eliminate the cause. Boethius analyzed the underlying problem thoroughly in his treatise *De causa rerum,* published more than 1000 years ago. Recently, however, the art of clinical ecology and the science of human biometeorology have been revived.

Important statistical data on regional distribution of disease are kept by the Center for Disease Control and published in the form of maps and tables (Hove, 1977). It is well established that the incidence of certain types of illness and disease shows large geographic variations. For example, in 1979 in Egypt the incidence of bronchitis was 172 cases per 100,000 population; the corresponding numbers were 92 in Ecuador, 68 in Ireland, 62 in England, 44 in Peru and Colombia, 40 in Germany, 14 in the United States, 15 in Canada, and about 6 in France. The strikingly high values in the English Isles are unequally distributed among the sexes. The value is 99 for men and only 35 for women. Since the incidence of bronchitis has steadily decreased in the English Isles since World War II, it has been proposed that the high bronchitis incidence might have been due to sulfur and particulate exposure from domestic coal heating. However, this type of conclusion is highly speculative for many different reasons. For one, the CHESS study showed that the diagnostic methods used to identify bronchitis by doctors in England differ from those in the United States and other countries (EPA, 1974). Furthermore, the English lifestyle differs from that of the Americans, and the socioeconomic profile of men in the three most affected countries—England, Egypt, and Ecuador—differs significantly—as do, of course, the climates. A more careful analysis of the data shows that the English Isles harbor mainly chronic bronchitis cases, whereas Egypt, Ecuador, and Colombia report predominantly acute cases, since the life expectancy is generally lower in these countries. Thus, the health effects in the British Islands simply cannot be compared with those in Egypt, Ecuador, or Colombia.

Similar problems arise if engineers or scientists from different cultures seek to define what constitutes a healthy climate. At a recent conference (Fanger, 1979), all participants complained about unhealthy winter air in their home countries. However, they differed in the analysis of the problem. The Europeans said it was too wet, whereas North American participants complained about the dry indoor air. In this case, the apparent conflict arises not so much from the definition of a healthy climate, but from the fact that some older heavy-walled European homes remain at a lower indoor temperature in winter, close to the dew point, whereas many homes in North America are poorly insulated and leaky, and can only be kept habitable in winter if large amounts of hot, and therefore very dry, air are pumped through them. Occupants of dry buildings are more likely to suffer nasal distress, whereas occupants of cold and humid buildings suffer more frequently from pneumonia.

Other effects of local climates has been reviewed by Thomson (1979). There is no question that certain climatic conditions influence personal comfort: The stimulating sea winds on the windward coasts of continents have been popular for many thousands of years and are said to have produced the hardy Danish, German, and Dutch nations. It is not exactly known what makes sea climates healthy, except for allergic patients who can evade landborne microbes there. Salty aerosols also stimulate the flow of mucus in the sinuses and the entire respiratory passage. On the other hand it has long been known that the dry, warm mountain winds cause dreaded headaches. The Roman settlers called the Swiss Föhn "faronius." Similar uncomfortable winds prevail in many other parts of the world: The bora storms through Trieste, the mistral through the Rhone valley and Monaco, the sirocco through the Strait of Gibraltar, the chinook through Alberta, the Santa Ana through California, the sharav through Israel, the thar through India, the lxokk through Malta. Dozens of other regional storms periodically plague sensitive people around the world. The main cause of discomfort, as with thunderstorms, is the accompanying transient pressure gradient that precedes the weather front and affects temperature, humidity, and the electrostatic earth potential, all of which some individuals sense for several hours before the onset of the wind. The electrostatic field can change from 50 to 400 mV/m within a few minutes.

Atmospheric pressure differences affect chronic rheumatics and people with amputated limbs. The correlation between increased pain and atmospheric pressure differences was valid in 93% of several thousand responses reported by 347 arthritic patients at Mayo Clinic (Thomson, 1979).

The effect of altitude was firmly established in the 19th century in connection with tuberculosis. The beneficial effect of altitude on general health has been reconfirmed by epidemiological studies of the Indian Army in high-altitude medical observations during the 1967 war at the Chinese border, but it is not known whether the effect is due to the cosmic radiation level, which increases threefold for every 3000 m; the strong uv radiation; the lower air density; the less polluted air; or a combination of these or other factors.

In contrast, the hazards of cold and hot climates are less mysterious. They are usually related to the change these climates impose on the heat content of the body surface as well as on the temperature of the core. The threshold for the onset of a statistically significant increase in death rates is determined by a heat gradient that can be evoked by a range of combinations of humidity and temperature, indicated in Figure 8.7. Acute exposure to strong sun in hot climates can cause brain damage due to heat accumulation; chronic sun exposure increases the probability of contracting skin cancer, at least among Caucasians. It has been reported recently that Inuit Eskimos who live in the extreme cold exhibited a 50% lower forced midexpiratory flow at age 40 than those who had moved indoors. It was suggested that chronic exposure to cold dry air causes damage to the lung tissue (Schaefer, et al. 1980).

Thomson (1979) pointed out that adventure-seekers who travel great distances to exotic places should be prepared for the impact of unaccustomed climates, which according to health statistics have a more advese effect on travelers than is commonly realized.

8.5 Particulates

This section explains some basic elements of the mechanism of dust inhalation and then describes house dust, microbes (including infective agents), allergens, and carcinogens (including radioactive dust, cigarette smoke, and other carcinogenic dusts).

8.5.1 INHALATION OF DUST

Clean ambient air at the breathing level (1.5 m above the ground) contains 20–40 $\mu g/m^3$ of dust in the form of sea spray, soluble salts, organic matter, and microbes. The lung is very efficient at filtering air, but it is also very efficient at cleaning the lung. Thus, much of the inhaled dust is harmless (Holt, 1980). Acute or chronic health problems can arise in all three segments of the respiratory tract (Figure 8.5). In the nasopharyngeal compartment insoluble particles are either transferred to the gastrointestinal tract or eliminated with sputum. Soluble dust can cause damage to the mucosa or paralysis of the cilia, or can provoke an allergic or hypersensitive response. In the tracheobronchial compartment, the same mechanisms hold for both insoluble and soluble particles, but their effect is enhanced because of the larger surface area and the more tender mucosa. Additionally, they can cause a bronchoconstrictive reflex or become the cause of local infection. Furthermore, particulates can potentiate the effect of almost any irritant gas. In the pulmonary tract, the reactions are yet more enhanced (Brain, Proctor, and Reid, 1977). Soluble matter can cause damage to the epithelium, constriction of the peripheral respiratory unit, release of proteolytic enzymes, and emphysema with alveolar constriction. Insoluble matter may react with tissue locally, remain inert and embedded in tissue, or migrate to lymph nodes, as described below.

The fate of dust depends on size. Particles between 1 and 100 μm are retained in the nasopharyngeal tract; particles between 0.01 and 10 μm are absorbed in the tracheobronchial tract or proceed into the pulmonary cavity. A large fraction of the particles smaller than 0.05 μm can be exhaled unless they dissolve or react with or on the surface. Figure 8.8 shows the screening model accepted by the National Academy of Sciences (1961; Cartwright and Nagelschmidt, 1961; Hatch and Gross, 1964). It is based on the work of the International Radiological Protection Commission (UNSCAR, 1977). The figure includes size as well as gravimetric distributions of particles derived from an analysis of log-normal distribution curves. If the tidal volume is increased by exercise, the retention in the upper respiratory tract and the bronchial tract increases over the entire range of particles. However, retention in the pulmonary tract depends on particle size. Retention of particles with a size less than 0.8 μm increases with decreasing size, whereas particles larger than 0.8 μm are increasingly filtered by the preceding section. The deposition in the nasopharyngeal tract has been experimentally studied, and the fraction of particles N retained is approximately

$$N = 0.475(-1 + \log d^2 V) \qquad (8.11)$$

where d is the diameter and V the volume inspired. Mouth breathing seems to affect retention less than might be assumed, but a detailed comparison of mouth and nose breathing in relation to particle retention has not yet been

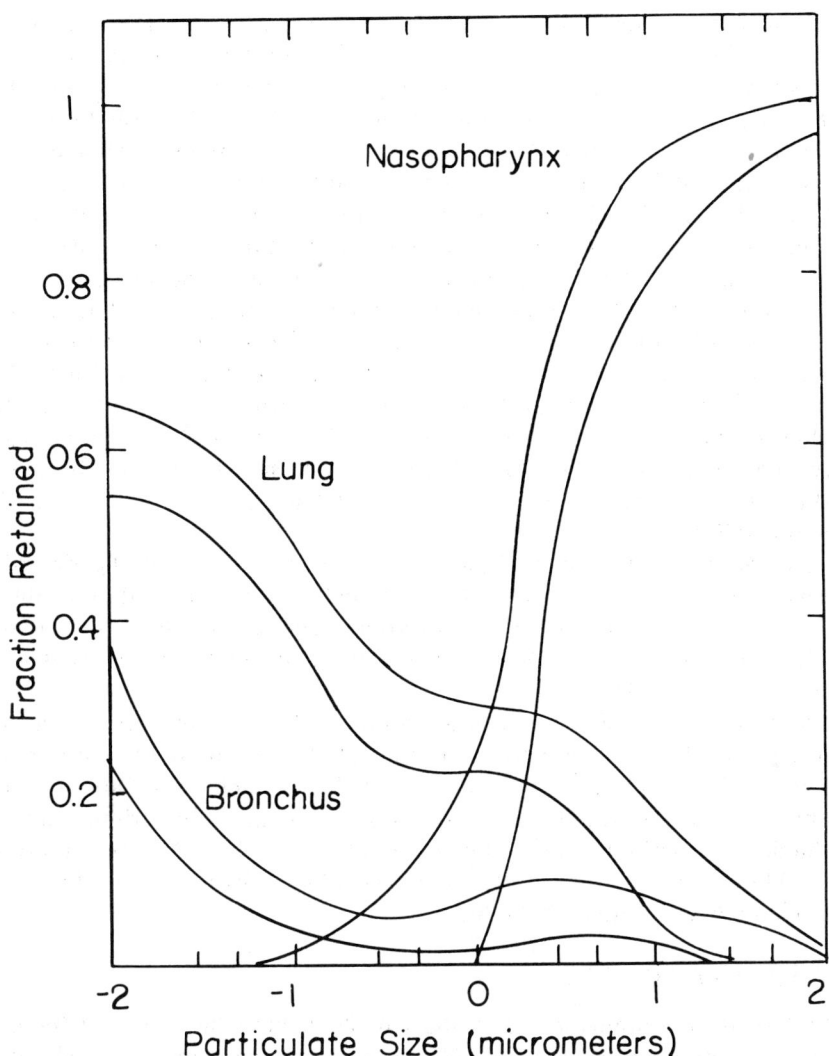

Figure 8.8. Deposition of particles in the three compartments of the respiratory tract (after NRC, 1965).

made. Earlier experiments are of doubtful value because the inspired air was directly inhaled from tubes. Hygroscopic particles, such as pollen, increase substantially in size as they travel along the tract; the density change depends on their structure.

Particle deposition in the lung is not even. It depends on aerodynamic factors, the solubility of the particle, and other factors. Particles larger than 5 μm have sufficient mass and momentum to be deposited at the junctions of the entry walls. Particles deposited on the ciliated epithelium of the bronchioles are rapidly swept by mucus to the pharynx and are swallowed (Dadaian, Yin, and Laurenzi, 1971). If the lung function is abnormal, the particle deposition will be influenced by obstructions; this can lead to amplification of damage.

Dust that reaches the alveoli is met by macrophages. These are large, freely moving cells that surround particles within hours after deposition and thus protect the alveolar membrane against puncture. The origin, concentration, and renewal of these cells is not yet understood. The macrophages slowly drift to the cilia so that the dust can be expelled. The process of macrophage transport is not yet conclusively established. Another possible transport mechanism involves penetration of the particles through the stomata into the interstitial fluid of the epithelial cell, which could carry them to the bronchiole (Davies, 1961; Leadbetter and Corn, 1972; Chamberlain, 1970; Muir, 1972; Morrow, 1970).

Radioactive studies have shown that a large fraction of the deposit is removed from the lung within a day after deposition (Radford and Martell, 1977). The rest is removed by a different process, which is measured in months. In mixed systems, the different components might be removed at different rates (Holt, 1980).

Noble gases, sea spray and other soluble alkali halides, and oxides and hydroxides of the chemical periodic groups 1a, 3a, 4a, 5a, and 6a cations clear the lung and lymph nodes within minutes or hours. The most stubbornly persistent chemicals are the insoluble silicates, Zr, Y, and Mn carbides, and the lanthanide and actinide oxides and hydroxides, as well as the fluorides of the latter. These are retained for years (Mercer, 1966). Thus, dust can lead to an appreciable body burden in the lung.

8.5.2 HOUSE DUST

Indoor dust contains the same ingredients as ambient dust, but the composition is significantly different. House dust can carry microbes, allergens, radioactive dust, asbestos, pesticides, and just about every type of inert or poisonous agent. Lead levels are between 600 and 1400 ppm (Solomon and Hartford, 1976). Dust contains a mixture of inorganic and organic, as well as biological, debris reflecting the human activities and the surroundings of the building (Chapter 5). House dust, especially in bedrooms, frequently contains dermatophagoides, mites that live off human keratin and sebum, and produce allergenic feces. House dust also contains hairs and feces of pets, fur fragments, and feather and insect debris. These substances contain proteins and polysaccharides that can be allergenic. House dust may also contain pollen. It remains a mystery how a pollen spore with a size of 30 μm can penetrate far enough into the lung to triger asthma, but it appears that some plasma and most cells lie free within the bronchial space, and thus avoid the need for bronchial wall penetration. House dust may also contain molds, with a size of 1–2 μm, which predestines them for deep penetration. Some bacteria also act as indirect allergens by stimulating the release of mediators from the host cells. A special case is the *B. subtilis* used in commercial enzyme detergents, which directly causes allergies of the skin and asthma (Rimington, Stillwell, and Maunsell, 1947).

8.5.3 MICROBES

Until 30 years ago man's principal enemy was infection. Figure 8.9 shows the mortality from the 11 worst infectious diseases of 1912–1916 (Vaughan, Vaughan, and Palmer, 1922); all except influenza have been brought under control (O'Hare, 1942).

The source, properties, and concentration of microbes are discussed in Sections 5.3 and 7.18. The human body and human activities are the main source of infective agents or vectors. Human skin scales are a very important source; each of the million scales that are shed daily carries an average of four bac-

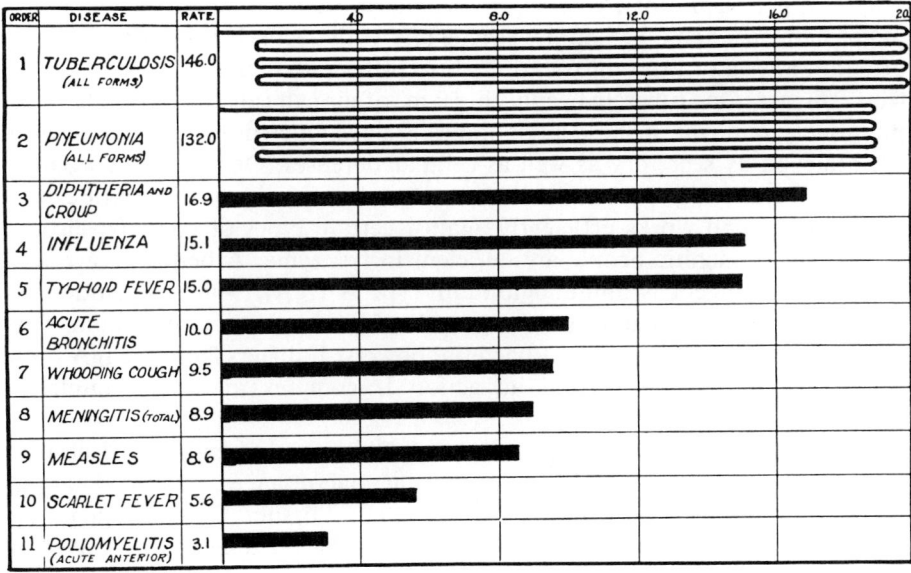

Figure 8.9. Mortality rate (annual rate per 100,000 people) from the 11 leading communicable diseases in the United States, 1912–1916 (after Vaughan, 1922).

teria, such as staphylococci (Speers et al., 1966; Sidorenko, 1967; May and Pomeroy, 1973; Dickgiesser, 1978). Air conditioners, air filters, humidifiers, spirometers, and anesthetic apparatus can be a breeding ground for *Pseudomonas aeroginosa* and *P. legionella* (NRC, 1981; Mogabgab, 1968; Hers, 1969; Weiss, 1971; Banaszak et al., 1970; Fink et al., 1971; Broome, 1979; Salvaggio, 1979; Granier et al., 1980; Rylander et al., 1978; Fumarola, 1981; Imperato, 1981). Microbes can spread by a variety of pathways. The common cold virus can survive only in wet aerosols (Wells, 1955), which are normally too large for inhalation. Fine aerosols, dry particles, and air serve as vehicles for transmitting chickenpox, smallpox, measles, and tuberculosis (Riley at al., 1962; Tyrrell, 1967; Williams, 1967). Transmitted by the same aerial pathway are the spores of *Bacillus anthracis,* which lodges on wool and causes anthrax; the rickettsia *Coxiella burnetti,* which causes Q-fever (Wellock, 1960); other microbes, which reside on cattle (Nikolov, 1977; Hendricks, et al. 1962) or sheep hides and on straw; the virus of foot-and-mouth disease (Norris, 1970); the *Francisella tularensis,* carried on rodents; the *Chlamydia,* hosted by birds (Riley, 1973); and the *Histoplasma capsulatum,* which thrives on bird droppings and causes histoplasmosis (Boyer, 1974). Viri have been discussed by Couch (1981).

All of these agents are vulnerable to dry air and sunlight, and thrive on moist house dust. Fortunately, the carrier agents are usually too large to penetrate beyond the nasolaryngeal compartment; they enter the pulmonary region only if upper respiratory, gastric, or esophageal fluids drain into the lungs, as might happen when a person lies supine and with decreased consciousness, as after anesthesia or after intoxication. Very little contamination is necessary to cause disease. One tubercle bacillus in 50 m^3 of air can cause infection because pulmonary removal of these large bodies is slow. Once in place, the microbes are attacked by secretory as well as cellular defenses. One secretory agent is lysozyme. Little is known about the source and properties of this agent except that it exists and works efficiently against certain types of bacteria; also, it seems that bronchitic people are deficient in lysozyme. A better known and better studied agent is immunoglobulin Type A (IgA), which is synthesized locally by plasma cells in the lamina propria of the respiratory epithelium. This agent has both antiviral and antibacterial powers. It differs from the IgA present in the blood serum, and its action is complex and not completely understood.

The cellular defense consists of phagocytic cells that engulf and destroy bacteria. Phagocytes occur either as alveolar macrophages, described earlier, or as leukocytes in the subepithelial connective tissue. Their origin is not known, but bone marrow has been identified as a possible source. Alveolar macrophages readily undergo aerobic glycolysis; their phagocytic ability is enhanced by oxygen. The polymorphonuclear leukocytes, called neutrophils, provide the bulk of the early defense against microbes. They are delivered via the bloodstream, and reach their highest concentration within about 5 hours after infestation. The half-life of bacteria in the lung is about 2 hours, compared to 40 min in the upper airways, or days or months for neutral dust in the lung. Virulent pathogenic agents against which the lung has no immunological resistance may multiply upon first contact (Figure 8.10).

Air contaminants may influence body defense. For example, ozone may reduce the effect of macrophages (Coltin, 1972), which are activated by oxygen. Lack of vitamin A will transform the mucus-secreting epithelium. Corticosteroid levels, influenced by stress or cold, will influence leukocyte release. Virus infections decrease resistance against bacteria (Green, 1979). In contrast, silica dust seems to increase the resistance of standard rats, which are usually chronically infected with bacteria, against an additional bacteria burden (Davies, 1961). Normally, however, dead alveolar macrophages cause edema, and exfoliation of mucus-covered epithelium will interfere with the normal defense mechanism.

Overwhelmed defense mechanisms will result in inflammation, which can cause interruption of the cilia movement in the bronchi and alter the permeability of alveoli and their supporting cells. Repair of inflammation will take 1–2 weeks, during which the surface is sensitive to secondary infection, against which the body reacts with adaptive changes that will persist for several weeks. It is not well understood whether and how such adaptive processes lead to metaplastic epithelial changes. Normal fibrosis may involve enzymatic leaching during phagocytosis, due to collapse of the antiprotease protection.

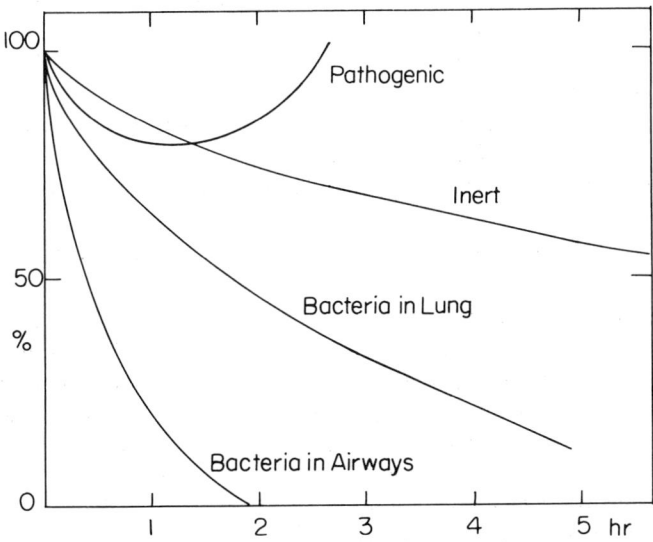

Figure 8.10. Concentration of particles in the lung following deposition (after Rylander, 1976).

8.5.4 ALLERGENS

The indoor environment is loaded with all types of potential allergens (Rimington et al., 1947; Reed, 1981), but this book cannot possibly do justice to this subject. Briefly, the field of immunology deals with all those processes by which a host recognizes and responds to foreign matter. Allergens involve immune responses. They distinctly differ from idiosyncratic hypersensitivity reactions, which are caused by the lack of detoxifying enzymes. The immune responses consist of the antigenic response, which is defined as the immune reaction between agent and antibody or the sensitized cell, independent of its effect on the host tissues (such reactions are usually studied *in vitro*); or the allergic response, which is defined as the immune reaction by the host. The latter may be protective, without or with only slight damage to the host, or hypersensitive (i.e., involving disproportionate damage to the host).

The host response depends on the physicochemical properties of the antigen; the dose, duration, and route of exposure to the antigen; and the immunological responsiveness of the host. Whether the tissue damage is minimal or lethal depends on the state of sensitization, or protective immunity, of the host, as determined by previous exposure, the nature of the antibody or lymphocyte sensitization, the molar ratio of antibody to antigen, the nature of the host tissue, the presence of secondary events in the tissue, the nature of potential pharmacological mediators liberated from the tissue, and the potential effect of corticosteroids or histamines. The most important clinical syndromes are (1) allergic rhinitis, (2) bronchial asthma, (3) alveolar infiltrates with blood eosinophilia, and (4) alveolitis.

The most common allergens are house dust, animal products, molds, bacteria, chemicals, and pollen (Green, 1979; van Assendelft, et al., 1979). The active ingredients in house dust are proteins and polysaccharides from pollen, dead mites, animal feces, hair, and other biological matter (Rimington et al., 1947). A substantial number of chemicals act as direct or indirect allergens; among these are the isocyanates, piperazine, and platinum salts. Formaldehyde is probably not allergenic, even though it causes hypersensitivity similar to type I or type IV allergens.

The tissue damage can be classified as one of four types: Type I is immediate and anaphylactic. In this reaction IgE becomes attached to mast or other cells. This causes the release of cytoplasmic granules from mast cells, which release histamine, which prompts blood vessel walls and smooth muscle fibers to produce the configurations responsible for the bronchial wheal and the flare on skin. In type II the IgM or IgG antibody reacts with antigens of the cell membrane, causing direct injury. One form of this type of response is Goodpasture's syndrome. Type III damage is delayed and the result of the formation of an antigen–antibody complex that may release enzymes. This Arthus reaction is delayed. The lung responses include fever, malaise, and leukocytosis; breathlessness without wheeze but a restrictive ventilatory effect; or increased airway resistance. Type IV, finally, does not involve antibodies at all. Instead, antigens stimulate lymphocytes to turn into sensitized cells that may release irritating factors.

In many allergic disorders, several responses occur simultaneously. Allergic alveolitis can be caused by extrinsic agents as diverse as hay, sugarcane bagasse, mushrooms, maple bark, barley, wheat flour, cork bark, redwood sawdust, pigeon dust, and pituitary powder.

Asthma is a common reaction to allergy. It occurs if the mucosal tissue fluid balance, the secretory activities, and the bronchomotor are not in harmony. Asthma occurs in about 2% of the general population. Frequently, it is initiated by only one agent, but in many cases an extended sensitivity is acquired and the threshold for its onset may be significantly reduced, thus leading to more or less chronic asthma. About 50% of childhood asthma cases will be repeated in adult life. Among those who commence allergic reactions as adults, about 80% will experience repeated attacks. About 10% of all asthmatics experience severe cases. Not all asthma is due to allergies, since asthma can be a symptom of wheeze bronchitis.

Most allergic asthma is multifactorial. Grass pollen asthma may occur simultaneously with hay fever. House dust asthma may be caused by reaction of the aldehydic groups of the carbohydrate allergen with the amino group of the lysine residues in peptide chains, in a process initiated by enzymes.

8.5.5 METALS

The metal content of dust has been described in Sections 5.2.5 and 5.4. Metals can be very poisonous, but (except in special circumstances) the bulk of the body burden stems from ingestion and not from inhalation. The seasonal content of Te, Zn, Ca, Mg, Mn, Ni, Cu, Pb, Cd, and Co in ambient air was studied by Takeushi et al. (1974).

Lead is an EPA "criteria pollutant" and has been extensively characterized (Campbell and Mergard, 1972). Rabinowitz et al. (1975) studied the balance between absorption, storage, and excretion. Lead occurs in the air of indoor rifle ranges (Fischbein et al., 1979; Trattner et al., 1976; Landrigan, 1976; Gunter et al., 1977). The aerial transport of lead and other metals has been reviewed by Winchester and Nelson (1979). Lead is also present in newsprint (Las and Blackwell, 1979), and can enter food when newspapers are burned in outdoor grills.

Very high exposure to nickel can cause cancer (NRC, 1975), but high concentrations of this element are not common in indoor air. Mercury levels have been extensively tested (Section 5.4; Foote, 1972; NRC, 1978). Concentrations of this element, which causes nerve damage (Thaxton et al., 1975), can be high in dental offices, sometimes up to 0.1 mg/m^3 (Gronka, 1970).

8.5.6 SULFATE

Sulfate particulate is the oxidation product from sulfur dioxide released from coal-fired power plants. Sulfate acts synergistically with SO_2 and sulfuric acid mist. It is likely that soluble sulfate is metabolized promptly.

8.6 Potential and Recognized Carcinogens

The group of potential and recognized carcinogenic pollutants comprises chemically dissimilar organic and inorganic dusts as well as some gases. These substances have in common that they can cause cancer. Respiratory carcinogens invade the host via the respiratory tract. This does not mean that these agents have this part of the body as their target; in fact, many of them aim selectively at other primary sites. These are, in decreasing order of incidence, the breast, colon, prostate, ovary, stomach, pancreas, cervic, blood system, bladder, uterus, and rectum, as well as the oral cavity and esophagus (Wynder and Hoffmann, 1972).

According to Webster, the word cancer denotes a malignant tumor characterized by potentially unlimited growth, with local expansion by invasion and systemic spread by metastasis. Clinically, cancer is recognized by the appearance of tumors and neoplasms (that is, noninflammatory growths that arise without obvious cause from cells or preexisting tissues and that serve no physiological function). Cancerous growth can be unambiguously and rapidly identified with the help of routine laboratory tests; the same cannot be said of the agents that cause cancer. It is currently believed that about 30–70% of all cancer is caused by environmental factors.

Carcinogenic substances have no common intrinsic chemical function, and range from asbestos to radon gas and inorganic components of tobacco smoke to organic solvents. Intensive occupational exposure has led to the recognition of carcinogenic action by several classes of chemicals. Aromatic hydrocarbons present in coal soot, coal tar, and other combustion products can cause cancer of the lung, larynx, skin, scrotum, and bladder, with an incubation period of 9–23 years. The risk was discovered among asphalt workers, coke oven workers, and chimney sweeps. Anthracene, benzo-α-pyrene (BaP) and fused aromatic rings are responsible for this. Benzene has been found to increase the risk of leukemia of the bone marrow threefold among shoemakers, rubber cement workers, and distillers, with an incubation period of 6–14 years. Alkylating agents, especially mustard gas, can increase the incidence of cancer of the larynx, lung, and bronchi up to 36-fold (as was found in mustard gas workers), with a delay time of 10–25 years. Isopropyl oil increases the chance of cancer in the nasal cavity 21-fold in production workers; the delay is more than 10 years. Vinyl chloride increases the chance of liver cancer 200-fold in plastic workers; the delay time is 20–30 years. *Bis*(chloromethyl) ether increases the risk of oat cell carcinoma, a characteristic form of lung cancer among chemical workers, 45 times; the delay time is 5 years.

Arsenic increases the chance of skin, lung, and liver cancer eightfold among smelter workers; the delay time is 10 years. Chromium exposure among acetylene, aniline dye, bleach, pottery, linoleum, and battery-assembly workers increases their susceptibility to sinus, nasal, and laryngeal cancer 40-fold; the delay time is 15-25 years. Iron oxide, especially hematite, increases lung and laryngeal cancer fivefold among miners, foundry workers, and grinders; the delay time is short. Nickel increases the incidence of sinus and nasal cancer 100-fold and that of lung cancer 10-fold in electrolysis workers; the delay time is 3-30 years.

Asbestos causes pleural and peritoneal mesothelioma, and lung cancer at a rate 2-12 times greater than that for the general population in textile, shipyard, and insulation workers. Finally, woodworkers experience nasal and sinus cancer twice as frequently as others. The delay time is 30-40 years. For leather workers, the same risk is increased 50-fold (WHO, 1980; Fraumeni, 1975 and WHO, 1981).

These agents can be well controlled in the occupational setting when the risk is recognized. It is much harder to protect the consumer, for whom the far lower exposure may be almost negligible, because the population at risk may be 100,000 times larger. The principal problem is that many of these agents, no matter how strongly they correlate to cancer, cause so few cases of disease that screening and protection of the population is prohibitive.

Laymen and many health specialists consider almost any irritant or toxic substance a potential carcinogen. There are several reasons for this. One is that some carcinogens act very slowly, with time delays of up to three decades. Thus, many widely used substances, such as asbestos, were not recognized as carcinogens for centuries because until recently the average life expectancy was shorter than the incubation period. Another reason is that the technological and economic revolutions of the last 50 years have made possible the mass production and use of such synthetic chemical compounds as pesticides, detergents, polymers, and drugs on a scale not possible in earlier times. Thus, most of the world's population is now rapidly sharing the benefit of new inventions, and thereby has transformed the surface of the world into a gigantic laboratory for sophisticated epidemiological testing. Thus, modern medicine and mass communication have made it possible to locate and identify isolated cases of rare diseases, and modern analytical chemistry is capable of identifying chemicals in concentrations of parts per trillion, so that it has become feasible to correlate, for example, individual cases of birth defects with personal exposure to dioxin impurities in agent orange during the Vietnam War (Meyers, 1981). This capability tends to overshadow the fact that many hundred million of human beings are daily exposed to dozens of indoor agents that cause non-specific, but equally devastating suffering on a far larger scale.

8.6.1 CLASSIFICATION OF CARCINOGENS

The following four-step classification has been suggested (Kraybill, 1977) until we know more about the biochemical mechanism of carcinogenic agents.

Recognized or proven carcinogens must meet certain criteria, all of which must be backed with statistically significant data: The substance has been conclusively demonstrated to cause tumors; epidemiological observations have established them as human carcinogens; they appear on a list generally accepted by scientists as agents that have demonstrated a carcinogenic response in humans and experimental animals; they evoke a response at a low dose, and a dose–response relationship is confirmed in several or many species and strains and by several laboratories. Furthermore, tumors are transplantable. Currently recognized respiratory carcinogens are benzidine, *bis*(chloromethyl) ether, chloromethyl methyl ether, vinyl chloride, and asbestos. Other recognized carcinogens are beta-naphthylamine, auramine, aminobibenzyl, and arsenic.

Suspect carcinogens are agents for which *in vitro* cell cultures show malignant transformation; an association of increased tumor incidence occurs with known exposure; and a dose–response pattern or a positive response pattern is recognized. Suspect carcinogens need no epidemiological evidence. Benzo-α-pyrene belongs to this category, as do beryllium, diazomethane, dimethyl sulfate, carbon tetrachloride, nickel, and nickel carbonyl.

Potential carcinogens have structural similarities or exhibit mutagenic similarity to known carcinogens. This group includes the urethanes.

Inadequate data for classification prevent the designation of some suspect chemicals as carcinogenic. Such chemicals should be clearly identified as suspect. NIOSH maintains a complete directory of several hundred potential carcinogens, providing detailed references (Christensen and Fairchild, 1976) and The National Toxicology Program has the mandate to publish an Annual Report on Carcinogens.

8.6.2 RADON DAUGHTERS

It is currently estimated that about 30% of the 110 mrad of natural radiation exposure that the average person receives each year (UNSCAR, 1977) is due to radon gas (George, 1980, 1981; Martell, 1977; see also Section 5.7, p. 127). Radon converts to ^{226}Rn with a half-life of 3.8 days. The resulting particulate is ionic because of its continued radioactivity. The natural background air concentration is about 5×10^{-14} Ci/liter. There is currently a substantial interest in and controversy over the effects of Rn on health because several locations have been identified where residential indoor concentrations are reaching the equivalent of the occupational work limit (Harley, 1981).

The effects of Rn gas are relatively well known because of the occupational experience of miners in uranium mines who were exposed to high radon levels before the hazard was recognized. In fact, lung cancer among uranium miners in the Schneeberg mines of Germany had already been recognized by Harting in 1879. Among those exposed to high levels were fluorspar miners in Canada, iron ore miners in Great Britain, and miners in some areas of Yugoslavia and Saxonia. Of the 2000 miners who produced uranium in Colorado during World War II, 200 died from lung and bronchial cancer. However, they had been exposed to high levels and many were heavy smokers. The cancers in the Schneeberg miners were located deep in the bronchial tree from the trachea. Sublethal exposures cause spleen and kidney lesions. Recent observations indicate that the combination of ionizing radiation with dust is more potent than radiation alone because lung cancer persists even in uranium miners whose carefully controlled work levels appear too low to explain the observed frequency (Morken, 1975, 1980). The assessment of safe limits has changed several times during the last 20 years. Natural radon levels are normally less than 0.1 rem/year. It is currently estimated that 1 rem delivered to 1 million people will cause 100 cancers over 25 years. The symptoms are the same as those produced by smoking, described earlier, except that in occupational cases they also reflect skin exposure. Some communities in the U.S., Canada and abroad have identified individual homes where basement air concentrations of Rn were similar to work exposures. In these isolated cases, the same symptoms would be expected as in occupational exposure, but with doses reflecting the length of occupancy, which might be considerable. However, in the overwhelming majority of homes Rn levels are negligible.

Contaminated clothing can cause high concentrations of dust, and cancer (Bailey, 1957). The dose distribution among different age groups has been explored by Hofmann, et al. (1978). Chromosome aberrations can be used as a dose indicator (Brandom et al., 1978).

The action of radon is similar to that of x rays and alpha radiation (Sevcova et al., 1978), which also increase the risk of cancer (Section 5.7). The cancer risk is increased by about a factor of 3-9 for patients clinically treated with strong doses of x rays. Uranium miners had 25-fold increased risks of cancer. Sunlight is held responsible for as much as 40–80% of the squamous and basal cell carcinoma of the skin in Caucasians residing in Philadelphia (see: ozone report, NRC, 1973).

8.6.3 CIGARETTE SMOKE

Cigarette smoke consists of a mixture of dust, smoke, aerosol, and vapors; see Table 8.2 (Stedman, 1968; Druckrey, et al., 1960). Each cigarette has the potential to produce 5×10^{12} particles with an average size of 0.2 μm. Each particle may contain any combination of the over 3000 chemicals that have been identified in smoke (Table 5.1). Once smoke is condensed in the lung, each of the individual pollutants starts its path toward its most compatible target. Ashes act like dust. Organic components dissolve in the tissue and penetrate the blood and lymph systems, joining other toxins that have been ingested. In the lung only a very small fraction of the effects of the individual

Table 8.2

Composition of Mainstream Smoke[a]

Material	Weight (mg/cigarette)	Total effluent (%)
Particulate matter (inc. cond. H_2O)	40.6	8.2
Nitrogen (67.2 vol %)	295.4	59.0
Oxygen (13.3 vol %)	66.8	13.4
Carbon dioxide (9.8 vol %)	68.1	13.6
Carbon monoxide (3.7 vol %)	16.2	3.2
Hydrogen (2.2 vol %)	0.7	0.1
Argon (0.8 vol %)	5.0	1.0
Methane (0.5 vol %)	1.3	0.3
Water vapor (relative humidity = 0.6)	5.8	1.2
C_2–C_6 hydrocarbons	2.5	0.5
Carbonyls	1.9	0.4
Hydrogen cyanide	0.3	0.1
Other known gaseous materials	1.0	0.2
Total	505.6	101.2
Measured total effluent	500	100

[a] From the U.S. Surgeon General (1979); 85-mm nonfilter cigarettes, 30-mm butt length, 10 puffs of 38.9-ml volume each.

smoke components is known, but the overall effect is more than the sum because the irritants enhance each other's effects and overwhelm the body defenses.

Smoke causes an increase in mitotic activity in goblet cells. It also reduces pulmonary functions (Boren, 1970; Woolf and Suero, 1971) and ciliastasis, but quantitative data on reduced clearance are still sparse. Heavy smoking increases the incidence of carcinoma of the bronchus in direct proportion to the number of cigarettes smoked (see Surgeon General, HEW, 1979 for: Brewis, 1975; Hammond, 1975; Wynder, 1974). The epidemiology of smoking is discussed in Section 7.2.

Cigarette smoking claims the lives of more than 50,000 Americans each year (HEW, 1979). The main diseases are carcinoma of the bronchus (lung cancer), oral cancer, and cancer of the larynx and esophagus. The symptoms are cough, often due to concurrent chronic bronchitis (White and Froeb, 1980), chest pain, hoarseness due to involvement of the left recurrent laryngeal nerve by hilar extension of the tumor, and eventually, dyspnea. The disease is frequently accompanied by metastatic complication, extension of the tumor into the axilla followed by distressing arm pains. Homer's syndrome, neurological disturbances, thrombophlebitis migrans, and many other complaints. The radiological features usually include fluffy or round shadows of 2 cm diameter, and may range up to collapse of a lobe or lung. Treatment consists of radical surgery, palliative surgery, or radical chemotherapy; or of palliative measures only. In the localized indolent stage, lobectomy yields up to a 29% chance for five years of survival; pneumonectomy yields a 34% survival rate (Lane, 1976). The risk of smoking to those living with the smoker is explained in the section on risk assessment (Section 8.9, p. 281).

8.6.4 OTHER ORGANICS

The following chemicals have been established and confirmed as human carcinogens: benzidine, coal tar pitch, *bis*(chloromethyl) ether, chloromethyl methyl ether, vinyl chloride, beta-naphthylamine, auramine, and 4-aminodiphenyl (Calabrese, 1978; Laskin, et al. 1970). Benzo-α-pyrene is metabolized in the human liver to a carcinogen. It occurs in polluted industrial air, around coal-burning facilities and is implicated in cancer of smokers. Ambient levels in Altoona, PA, were 61 ng/m^3 in 1959, 31 ng/m^3 in Los Angeles, and 54 in St. Louis. Fortunately, the levels have now dropped to about 1–5 ng/m^3, levels similar to those experienced by people sharing a room with smokers. Smokers themselves experience about 20 ng/m^3. The concentration varies locally and seasonally, with peaks during high industrial activity. The synergism between the BaP in ambient air and the pollutants inspired by smokers is self-evident. Americans, with better ambient air, experience only half the risk of those who live in Wales, where Doll (1979) reported 50 ng/m^3 in ambient air in 1977.

BaP levels, and thus the incidence of cancer are a sensitive indicator of pollution. Since BaP is usually absorbed on particulates, any relaxing of the particulate standard in the revised Clean Air Act of 1982 would have a direct effect on the cancer risk in industrial cities.

Several chemicals are known of which the body has the ability to build up a substantial body burden. Among them are pesticides such as DDT, polyaromatic hydrocarbons (PAH), the pentachlorophenols (PCPs) (Mes, 1982) and other fat-soluble substances, and metal-organic compounds which can react with body tissue. Table 8.3 lists some of the many dozen organics that are known to be in ambient air (Kraybill, 1977). Indoor air contains up to 300 organics (Figure 5.7; Hollowell et al., 1978; Jarke, 1980) all of which can be inhaled. Their human health effects are not yet well understood (Beall and Ulsamer, 1981). Their exposure levels are listed in Section 7.4.

8.6.5 INORGANIC CARCINOGENS

Among the inorganic carcinogens are asbestos, other natural dusts, arsenic, and other metals (Selikoff, 1981).

Table 8.3

Suspected Carcinogens in Air[a]

Chemical	Source
Benz[α]anthracene, and PAH	Combustion products and cigarette smoke
Benzo[α]pyrene	Combustion products and cigarette smoke
Dibenzo[α]pyrene	Combustion products and cigarette smoke
Benzo[b]fluoranthene	Combustion products and cigarette smoke
Benzo[j]fluoranthene	Combustion products and cigarette smoke
Indeno[1,2,3-cd]pyrene	Combustion products and cigarette smoke
Arsenic	From pesticide use
Asbestos	Particles from asbestos products
Cadmium	Mining–smelting
Chromate (hexavalent)	Mining–smelting
Aldrin	Pesticide application
Dieldrin	Pesticide application
Heptachlor	Pesticide application
Lindane	Pesticide application
Carbon tetrachloride	Industrial effluent
Vinyl chloride	Industrial effluent

[a] From Kraybill (1977) and WHO (1981). For an updated listing see: *Annual Report on Carcinogens* by the National Toxicology Program (HHS).

8.6.5.1 Pneumoconiosis

The term pneumoconiosis refers to the tissue reaction to inorganic dust that has accumulated in the lung. There has been much speculation whether this effect is due to chemical hydrolysis of the particles, but it is currently believed that the response is more likely due to decomposition products of the macrophages, possibly enzymes released from dead cells. Although it is not clear how and why macrophages die after attaching themselves to certain types of silicates, it is certain that the toxicity of dust particles depends on their structure. Silicate droplets are far less toxic than fibers.

Our limited understanding of the transport and removal processes is a major obstacle to estimating the chronic effects of irritants. Most medical authors assume that brown or grey lymph nodes and mucus are normal, but the gradual relationship between deposition in lymph nodes and miner's lung disease and similar diseases corresponds to a concentration effect that is clearly chemically significant and would contribute to any additive exposure.

Thus, the threshold for the onset of acute discomfort surely occurs at a higher body burden than that for chemical stress of the type that leads to chronic disease. But the body burden and stress on the lung seem to be long-range processes. This fact is demonstrated by the pneumoconioses, almost all of which are irreversible irritations characterized by delayed onset. However, statistics on smoking indicate that the prognosis is improved if smoking is discontinued (i.e., the irritating effect of smoke deposits decreases with time).

8.6.5.2 *Asbestosis*

Asbestos fibers lodge mainly in the bronchioles. The pathogenesis of asbestosis is similar to that of the other pneumoconioses. Fibers in the alveoli are engulfed by a macrophage and carried to the next regional lymphoid cluster where the number of nodules increases as more dust arrives (Flowers, 1974). If exposure is long or strong, an increasingly massive fibrosis will develop. Asbestos fibers can be readily recognized by the shape of needles and counted under an electron microscope. The first symptom of light exposure to asbestos is pleural thickening, followed by calcification and mesothelioma. Up to 40 years (during which dyspnea and a dry cough develop) can elapse before x rays show fibrosis. As secondary results of exposure, mesotheliomas and bronchogenic carcinomas can develop 12 or more years after the exposure. Such carcinogenic effects have a very low threshold and have been identified in family members of workers exposed to asbestos (Anderson et al., 1979; Chambre Syndicate, 1974; NRC, 1971; ACGIH, 1975). Heavy exposure causes bronchial carcinoma, especially among smokers. Synergism with smoking increases mortality 54-fold (Enterline, 1976; Shettigara and Morgan, 1975; Sawyer, 1979; Pruett and Winslow, 1980; NIOSH, 1980). There is no known remedy for asbestosis (Stanton et al., 1969).

8.6.5.3 *Silicosis*

Silicosis occurs after exposure to dusts in mines, foundries, and potteries; stonecutters, sandblasters, tunnel workers, and others who deal with dense, fine silica dust are also often affected. Particles with a size of about 1 μm are retained most thoroughly. The necessary exposure is possibly as high as 10^7 particles/m^3, and the onset of symptoms is delayed by months. The disease can lead to acute honeycomb fibrosis. The damage consists of fibrils and nodules around lymph nodes. Silicosis greatly increases the susceptibility to tuberculosis. There is disagreement whether the disease becomes stationary when exposure is discontinued, and some doctors believe that continuation of exposure will not greatly increase the gravity of the disease once it has reached the diagnostic level. There is no effective therapy for silicosis (Phibbs et al., 1971).

8.6.5.4 *Talcosis*

Hydrate magnesium silicate, talcum, is used as a filler in cosmetics, soaps, paper, and many commercial products. The fibrotic damage it causes, talcosis, is similar to asbestosis. Similar effects result from exposure to kaolin as used for pottery and clay, and to mica, fuller's earth, and other silicates.

8.6.5.5 Coal Miner's Pneumoconiosis

The sensitivity to coal dust is far lower than that to mixed silicate and coal dust; but some 50,000 coal miners have established symptoms of the disease in England alone. In the first stage, simple pneumoconiosis lung nodes up to 5 mm in diameter can be observed on x rays. Occupational doctors use a numerical scale to categorize the darkening of the lung, but it is believed that this first stage causes no direct physiological impairment or symptoms. In more advanced stages, with opacities up to 5 cm (called *progressive massive fibrosis*) breathlessness and restricted ventilation are observed; black material may be coughed up when masses of fibrous tissue rupture into the lung tissue. Individuals with the latter symptoms are vulnerable to tuberculosis. The dose response varies greatly among workers. Coal miners with rheumatoid arthritis may develop Caplan's syndrome, which may cause clubbing of fingers and toes.

8.6.5.6 Siderosis

Iron oxides and oxides of other heavy metals are readily visible on x rays because of their heavy atomic mass. Apparently exposure to such substances can lead to strikingly abnormal chest x rays in the absence of any acute symptoms. However, it has been reported that at least the oxides of cadmium and zinc can cause mesotheliomas (Furst, 1976).

8.6.5.7 Toxic Chemical Pneumonitis

Toxic chemical pneumonitis can be caused by beryllium (Kanarek, et al., 1973), cadmium (Solomon and Hartford, 1976), arsenic (Lee and Fraumeni, 1969), vanadium, manganese, bauxite fumes, and osmium.

8.6.6 FACTORS INFLUENCING CANCER

The risk for cancer is intrinsically greater for people with congenital or genetic diseases. The most endangered groups are those with chromosomal disorders, with single-gene disease of dominant or recessive tract, and those with polygenic diseases. Chronic myelocytic leukemia is an example of a genetic chromosome disorder of the type Boveri foresaw 65 years ago as a possibly precarcinogenic condition. It is currently believed that a defective 22nd chromosome, the Philadelphia chromosome (Ph'), can lead to a deteriorating "stem-cell evolution" (Kraybill, 1977). Down's syndrome and Bloom's syndrome are further examples. Diseases that demonstrate that cancer risk can be inherited are melanoma, neuroblastoma, nevoid basal cell cancer, thyroid carcinoma, and adenocarcinomatosis, for which the probability is more than doubled in certain families. There is no definitive explanation available why such is the case. However, it is well established that people with immunodeficiencies have an increased incidence.

The role of chronic irritation is hotly debated. Skin, esophagus, and lung irritation have been suspected of causing cancer, but it is not clear whether predisposition is necessary. There is little doubt that precancerous lesions similar to the better known lesions of the uterine cervix can occur in the larynx, esophagus, and bronchus. By definition these lesions constitute morphological abnormalities with microscopic characteristics of cancer, but they are confined to the epithelium. The rate of their invasive growth cannot be predicted, but their removal will eliminate the danger of cancer in that organ.

A special problem with attributing cancer to causes is the potential for metastasis. Drugs may induce cancer or promote its spread, but their use may present a smaller risk than neglect of the disease. This is true for phosphorus-32 irradiation and other radiative treatment; immunosuppressive drugs necessary to prevent rejection of transplants; hormones; and arsenic, phenacetin, or coal tar ointments.

Some dietary ingredients may enhance cancer. Aflatoxins, which may grow on peanuts and on grain, are known to cause live cancer in man. Nitrosamines are very strongly carcinogenic in animals, and episodic case studies point toward human carcinogenic potential, but probably in conjunction with polynuclear hydrocarbons. The danger of the latter when consumed along with meat grilled over charcoal fires is not established yet, even though smoke from toasters, grills, and stoves contains a respirable fraction of these agents. There is also some concern about food dyes, cyclamates, and saccharin, because the delay time between consumption and the onset of symptoms might conceal any relationship that might exist.

The geographic distribution of cancer, reflected in reports of cancer incidence on five continents (Doll, 1979; WHO, 1973), leads to a puzzling array of observations. The incidence of lung and bronchial cancer per 100,000 people is highest for men in Liverpool, England (86.6), and for Maori women (38) in New Zealand, and lowest for men and women in Ibadan, Nigeria (1.2 and 1.0). The lung cancer rate for men in Liverpool is higher than the highest incidence of breast cancer (63) or cancer of the cervix (80) among women anywhere in the world. Cancer of the larynx is highest for men in Bombay (13.8) and lowest for Caucasian men in El Paso, TX (1). The incidence for women ranks lower everywhere. Pharyngeal cancer incidence is highest for men in Bombay (16) and lowest in rural Norway (0.3), while skin cancer is highest among Hispanic women in El Paso, TX (106) and lowest in Miyaki, Japan (1.4).

The incidence of lung cancer in the United States for 1969–1971 was very high. Black males in Pittsburgh had an incidence of 99.8 per 100,000, white males 67, white women 14, black women 15, and both races and sexes combined had 39 (see Surgeon General, 1979, for: Fraumeni, 1975). The levels in Colorado were 49.6 for white men and 10.2 for white women. There is no other form of cancer for which the discrepancy between men and women is more striking. Lung cancer also shows a distinct trend in the ratio of urban to rural incidence. In this ranking, esophageal, laryngeal and nasopharyngeal cancers lead with a value of 3. Lung cancer ranks sixth with a value of 2, while the ratio for sinus cancer is 1.0. There is no drastic correlation between socioeconomic status; the well-to-do lead by a ratio of 1.2.

8.7 Gases and Vapors

Sulfur dioxide from coal combustion has been recognized as a major health hazard since the Middle Ages. Today, the EPA has established health criteria for seven ambient pollutants: carbon monoxide, nitrogen oxides, sulfur dioxide, ozone, hydrocarbons, particulates, and lead. For each of these, safe ambient air levels have been established on the basis of carefully selected and reviewed data.

8.7.1 SULFUR DIOXIDE

Careful control of coal-burning power plants, smelters, and other sources has reduced SO_2 exposure levels by a factor of 10 during the last 10 years (Section 7.2.2). If the current air quality standards are preserved, the health effects of SO_2 will be minimal, except to those residing near a source that has obtained a variance from standards.

Health criteria for SO_2 are revised every 5 years by EPA review. At a concentration above 0.12 ppm, SO_2 causes distress that is noticed within a day or less. At a level of 0.03 ppm, SO_2 is not noticed by most people, but health statistics show that it causes a significantly increased risk of respiratory damage. Recent work has shown that this effect is due to a mixture of SO_2 and sulfate particulates (Table 8.4). The current air quality criteria account for this mixture (EPA, 1980). They are based on animal studies (Hazleton Laboratories, 1972), as well as field studies of adults and children (Hasselblad, 1977). Recent work shows that asthmatics, as a group, respond at a lower threshold with bronchomotor and bronchoconstriction symptoms than healthy adults (Sheppard, 1980).

About 10 years ago the EPA undertook an ambitious field study to establish the epidemiology of ambient SO_2 levels. The study explored whether children and asthmatics, as a group, were incurring higher than average risks from SO_2 and particulate exposures. Sulfur dioxide is very suitable for such a study because (1) it can be easily and accurately measured; (2) it stems almost exclusively from point sources, to which it can be traced; and (3) as long as it is not discharged through specially designed tall stacks, it settles within a few miles of the source, yielding a well-defined exposure area. Finally, SO_2 is transmitted exclusively through the air, and once it has touched ground it rapidly converts to harmless sulfate and is not resuspended.

Table 8.4
Health Effects of Sulfur Dioxide and Sulfate[a]

Adverse Health Effect	Concentration at Which Effect Was Observed		Averaging Time
	SO_2 ($\mu g/m^3$; ppm)	Sulfates ($\mu g/m^3$)	
Increased mortality	300–400; 0.11–0.15	NA	24 hours
Aggravation of symptoms in elderly	365; 0.14	8–10	24 hours
Aggravation of asthma	180–250; 0.07–0.09	6–10	24 hours
Decreased lung function in children	220; 0.075	11	Annual mean
Increased acute lower respiratory disease in families	90–100; 0.034–0.037	9	Annual mean
Increased prevalence of chronic bronchitis	95; 0.035	14	Annual mean
Increased acute respiratory disease in families	106; 0.039	15	Annual mean
Increased respiratory disease-related illness absence in female workers	NA	13	Annual mean
Primary standard	365	—	24 hours
Primary standard	80	—	Annual mean

[a] From the U.S. Environmental Protection Agency (EPA, 1974).

This CHESS study (EPA 1974, 1980) concluded that the incidence of croup and asthma complaints among children in different cities was directly proportionate to the average ambient SO_2 concentrations in the local area, and that a significant trend existed even if the SO_2 levels differed by only a few parts per million, that is a small fraction of the local daily variations. It also found that symptoms persisted for many months after people moved to an area of lower exposure. This study was highly criticized from every angle, and the researchers as well as the EPA were accused of overinterpretation, carelessness, and even potential fraud (CHESS report, EPA 1980), but a very careful and very extensive statistical analysis of all available epidemiological data by a public interest group (Lave and Seskin, 1977) indicates that variations of perhaps 1% in the SO_2 exposure could be correlated to clinical effects with a reasonable con-

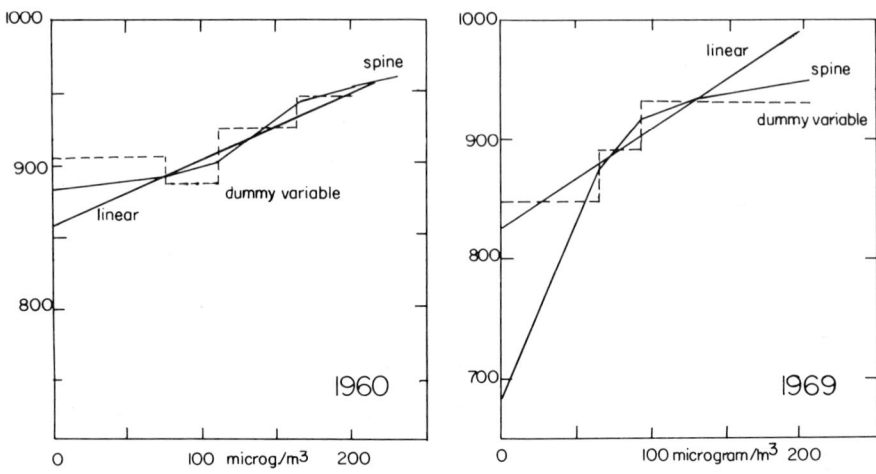

Figure 8.11. Dose–response curve for SO_2 and sulfate in ambient air (after Lave and Seskin, 1977).

fidence factor, depending on the size of the sample. Figure 8.11 shows the dose–response curve using three different mathematical models. Recent work by researchers at Harvard University (Ferris, 1978, 1980; Speizer and Spengler, 1980; Samet, 1981; Spengler, 1981) produced experimental as well as theoretical evidence vindicating much of the CHESS study.

8.7.2 NITROGEN OXIDES

A summary of health thresholds for NO_x is shown in Table 8.5. Nitric oxide is a major irritant in smog. Outdoor levels have been dramatically reduced since the EPA set ambient values of 0.05 ppm and established health criteria (Kerr et al. 1978). At a concentration above 200 ppm, NO_x can cause acute pulmonary edema and occasionally death. There have been reports for many years that NO_x might constitute a larger indoor hazard than assumed. For example, a British study among schoolchildren showed an increased number of students with reduced respiratory function and a corresponding increase in disease among those living in homes with gas stoves (Melia et al., 1977; Benes, 1981; Goldstein, 1981). A study among adult women did not find a significant effect (Mitchell, 1974). Speizer et al. (1980) studied 9280 schoolchildren in six cities and found significant correlations between home cooking, stove fuel, and illness history, as well as lung function. Parental smoking enhanced the effect. The incidence of respiratory illness among children below age two was about 20% higher in homes with gas than in those with electricity. The health hazards of nitrates and nitroso compounds have been carefully reviewed (WHO, 1977).

Table 8.5

Health Effects of Nitrogen Oxides[a]

Population Studied	Average NO_2 Concentration ($\mu g/m^3$)	Reported Effect
Japanese railroad workers	300–1130	Decrease in several measures of pulmonary function as compared to controls
Chattanooga schoolchildren, aged 7–8	150–280	Borderline decrease in lung function test
Central City vs. suburban policeman in Boston	100 vs. 80	No differences in various measures of pulmonary function
Seventh Day Adventists in Los Angeles vs. San Diego	96 vs. 43	No differences in various measures of pulmonary function
Czechoslovakian children, aged 7–12	20–70	Twofold excess in acute respiratory disease compared to unexposed group
U.S.S.R. adolescents in chemical and fertilizer plants	<10	Excess in acute respiratory disease ranging from 11 to 27%
Individuals living within 1 km of a U.S.S.R. chemical plant	580–1120	Forty-four percent increase in physician visits for respiratory, visual, nervous system, and skin problems
Families in Chattanooga, Tennessee	150–280	Excess in acute respiratory disease: 1–17% in children 9–33% in adults
Infants and children 6–9 in Chattanooga	150–280	Infants exhibited 10–58% excess of acute bronchitis; children 6–9, 39–71% excess

[a] From the U.S. Environmental Protection Agency (EPA, 1976).

8.7.3 CARBON MONOXIDE

Carbon monoxide is 250 times more efficient at hemoglobin binding than oxygen. It takes about 10 hours for the blood level of COHb to reach equilibrium (Bartlett, 1968; Stewart et al. 1969; NRC, 1969; Coburn, 1970). Woebkenberg et al. (1981) measured COHb levels in several hundred people.

In Cleveland, OH, she found average levels of 7.43 ± 0.37 in smokers and 3.21 ± 0.17 in non-smokers. In Elyria, IL, the average levels were 6.27 for smokers and 2.00 for non-smokers. Honigman (1982) observed COHb levels in joggers in Denver, CO. The dose–response curve is listed in Figure 8.12. The 1-hour national ambient air standard is 40 ppm. These levels are exceeded in rooms with heavy smokers (Section 7.2.3). Furthermore, CO can build up from combustion gases. Several cases of death in tightly closed school buses (Johnson, Moran, and Pekich, 1975a) and police cars (Savage et al., 1976) have been correlated to CO in recent years. If buildings are tightly sealed to reduce ventilation, the danger exists that infiltration from garages could lead to unsuspected and unnoticed increase in CO (Schaplowsky, et al., 1973; Jaeger, 1981). Blocked furnace flues, space heaters (Lao, 1982) and poorly maintained stoves can cause similar effects. Low-level exposure to CO can cause retinal hemorrhage (Kelley and Sophocleus, 1978).

8.7.4 OZONE

The ambient air standard for ozone is 0.12 ppm. Ozone is very reactive and reduces midrespiratory flow. Ozone is a component of smog; it is produced by uv light (Section 7.2.3) and is a natural component of the stratosphere. It has been claimed recently that man adapts to ozone (Hackney, et al., 1978), but it

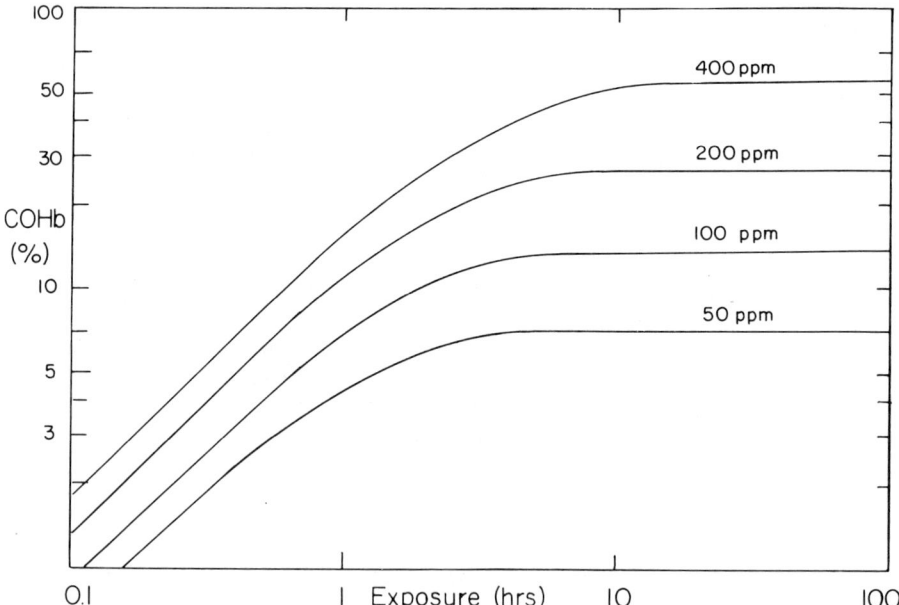

Figure 8.12. Dose–response curve for carbon monoxide (EPA, 1976).

is not clear whether the adaptation of respiratory function reflects reduced toxicity or whether subjects merely learn how to override natural defenses.

Short-term exposure to ozone at 0.3 ppm is experienced in commercial aircraft (Van Huesden and Mans, 1978). The FAA has made several studies that show noticeable discomfort but no clinical effects (Higgins, et al. 1979; 1980) except red, dry eyes (Eng, 1979). Extensive risk studies exist for California, but these data refer to smog rather than pure ozone. Smog contains nitric oxide, ozone, sulfur dioxide, hydrocarbons, and particulates. Smog levels in Los Angeles are now 10 times lower than they were 10 years ago (Section 7.2.2) because automobiles have been equipped with exhaust control devices and industrial sources have been curtailed.

8.7.5 AMMONIA

Ammonia occurs in body metabolites and in breath, but it is a strong irritant and very toxic to the eye (Smith, 1976). NIOSH limits for ammonia are 0.5 mg/m^3 for annual exposure, and a ceiling of 7 mg/m^3. The health effects have been carefully reviewed by the National Research Council (NRC, 1977).

8.7.6 HYDROCARBONS AND HALOCARBONS

Volatile organics result from solvents (Stewart and Hake, 1976; NRC, 1978), aerosol spray propellants, and home fuels. The relative toxicity of these substances has been carefully evaluated by NIOSH and is published in a continuously updated listing in the U.S. Code of Federal Regulations (29 CFR 1910.1000, table Z-1). Extensive files have been prepared by the same agency for toxic substances, for suspected carcinogens (Christensen and Luginbyhl, 1974, 1975) and for observations of health effects (NIOSH, 1976; Fairchild, 1977 and Hamilton and Hardy, 1974). A thorough study of pollutants, which can be detected in the human body, has been prepared by the Oak Ridge National Laboratory for the EPA (1978; Holleman, Ryon, and Hammons, 1980).

Hydrocarbons and halocarbons are fat soluble and thus accumulate in the adipose tissue, where they can constitute a measurable body burden. For volatile compounds, such as the fluoro- and chlorocarbon gases that are used as aerosol propellants, the lung is the main point of entry (Figure 8.13), but for pesticide dusts the distribution among air, food, and water intake is more equal. The intake of PCBs is mainly via water and food (Huff et al. 1980). Falk (1980) observed blood serum levels of 710-1750 ppb and urinary levels of 47-216 ppb in Kentucky residents living in log homes impregnated with a 5% solution of PCP. The indoor air levels were between 0.20 μg/m^3 and 0.38 μg/m^3. The recommended indoor limit for most chlorocarbons is 2 μg/m^3. The human health effects of these substances are not yet well known. In constrast, animal toxicity of pesticides is very carefully established before they are registered by the EPA. Table 8.6 shows some select animal toxicity data for some compounds.

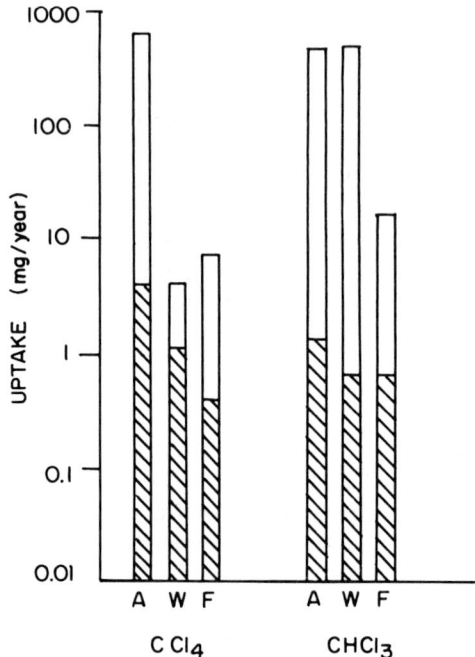

Figure 8.13. Relative intake of carbon tetrachloride and chloroform by an adult man from air and food (after NRC, 1978).

Thorough studies indicate that in 1977 the human DDT body burden had reached 8 mg/kg worldwide. Since the chemical has been banned, the level is rapidly decreasing. The health effects of DDT are minimal and no chronic human health effects have been proven. The main reason for banning DDT was damage to fish in contaminated lakes, and fear of possible teratogenic effects as a result of long-range accumulation in human adipose tissue.

A detailed study is now under way to follow possible health effects of 2,4,5-T, called Agent Orange, which has been shown to be teratogenic in animals at high exposure levels. Reports of human miscarriages are not frequent enough to be attributable to contamination, but acute chloracne is clearly caused by high exposure. Long-term effects have been discussed by Huff et al. (1980). The interest in this compound stems from the exposure of over 4 million people to high doses of it during the Vietnam War (Meyers, 1981; Section 5.7). So far 65,000 veterans have been examined (JRB Associates, 1981). Poisoning by paraquat has been described by Fitzgerald et al. (1978).

In a recent study the air levels of benzene, chloroform, vinylidene chloride, trichloroethane, tetrachloroethylene, trichloroethylene, and dichlorobenzene

Table 8.6

Toxicity of Pesticides to Animals[a]

Substance	Animal	LD$_{50}$ (mg/kg)
Lead arsenate	Rat	50
Calcium arsenate	Rat	40
Sodium fluoride	Cockroach, rat	200
Nicotine	Rat	30
Pyrethrins (chrysanthemum)	Rat	820
Sulfur	Rat	$<10^4$
DDT	Rat	200
	Cockroach	8
	Housefly	2
	Fish	0.01
Chlordane	Rat	400
Heptachlor	Rat	100
Aldrin	Rat	50
Endrin	Rat	10
Mirex	Rat	400
Kepone, chlordecone	Rat	100
Endosulfan	Rat	75
Toxaphene	Rat	100
Parathion	Rat	10
Diazinon	Rat	60
Malathion	Rat	2000
Dichlorvos	Rat	25
Disulfoton	Rat	2
Ronmel	Rat	20
Diquat	Rat	100
Paraquat	Rat	200
Copper sulfate	Rat	300
Pentachlorophenol	Rat	210
Captan	Rat	9000
2,4,5-T	Guinea pig	0.6

[a] From McEwan (1979); LD$_{50}$ means lethal dose for 50% of species tested.

were studied on two college campuses by means of personal monitors (Wallace et al., 1981; Zepinski, 1981), and compared with the level in breath (Figure 8.14). The levels are amazingly close. Figure 8.15 shows that there is a direct correlation between dose and breath levels over the entire observed range. These data indicate that the adipose tissue concentration reaches equilibrium with the air level within less than 6 hours. No adverse health effects are expected at the observed levels, but health scientists consider all compounds that cause a

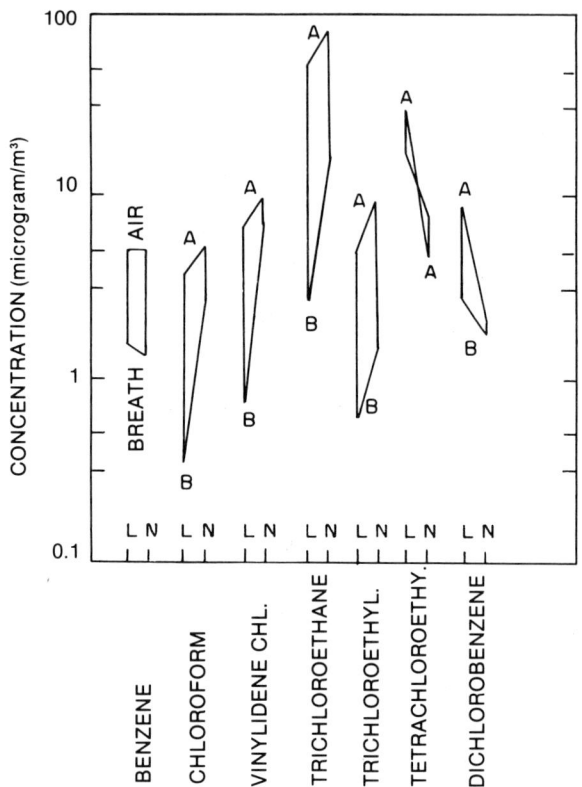

Figure 8.14. Concentration of benzene and six chlorocarbons in the air at LaMar University and the University of North Carolina, and in the exhaled breath of 17 students (Wallace, 1981 and Zepinski, 1981).

measurable body burden to be very undesirable, since the body burden prolongs the dose beyond acute exposure and in pregnant women serves as a source of exposure of the fetus through the placenta.

The toxicity of polychlorinated biphenyls (PCBs) was first fully recognized in 1964 when rice oil contaminated with PCBs from a heat exchanger poisoned people in Japan causing the so-called Yusho incident. The main danger is ingestion (Cairns, 1981).

8.7.7 FORMALDEHYDE

Currently a controversy rages over the health effects of formaldehyde because (a) formaldehyde is present at concentrations of 0.3 ppm and higher

8.7 Gases and Vapors

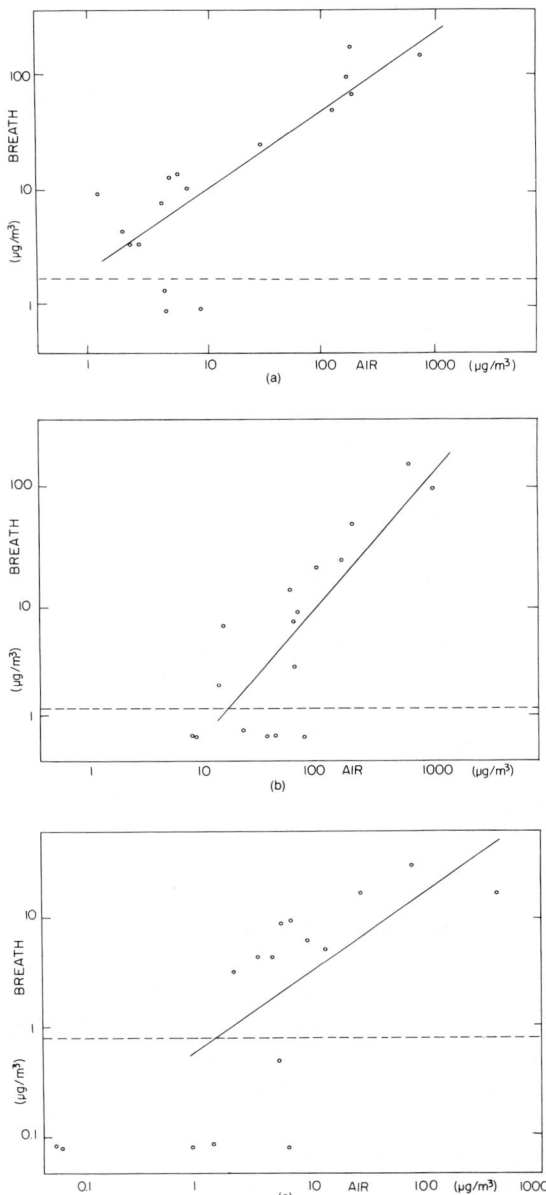

Figure 8.15. Correlation between the mean air concentration of three chlorocarbons and the breath concentration in subjects exposed for 24 hours: (a) tetrachloroethylene; (b) 1,1,1-trichloroethane; (c) vinylidene chloride (after Wallace, 1981 and Zepinski, 1981).

(i.e., at 10–100 times the ambient air level) in several million buildings in the United States (section 7.2.8.2 and 7.4.3); (b) the level is so high in several thousand homes that its pungent odor can be noticed; (c) the vapor can cause asthma, eye irritation and other effects in sensitized individuals; and (d) it has been shown that formaldehyde can produce cancer in rats.

In the animal studies three groups of 120 rats were exposed via inhalation at 2.1, 4.6, and 14.1 ppm for 6 hours a day, 5 days a week for 24 months. At 14.1 ppm 92 animals exhibited squamous cell carcinomas of the nasal turbinates, 3 had respiratory epithelial tumors, and 24 had distinct hyperplasia and metaplasia; at 5.6 ppm 3 developed tumors; at 2.1 ppm no tumors were observed (CIIT conference, 1980). A panel of experts appointed by the National Toxicology Program confirmed the data after reviewing the bioassays, and an NAS committee decided that formaldehyde was a potential human carcinogen (NRC, 1980, 1981). These findings came as a great surprise to everyone, including those who had planned the study, because formaldehyde has been used for hundreds of years in copious quantities in hospitals, and by literally millions of doctors and pathologists, who surely should have noticed tumors such as those appearing in the lower nose passage of the rats if formaldehyde caused such tumors in humans; but so far an extensive search has failed to reveal such cases. Thus, there is no reason for the general population to worry about an acute risk of cancer because indoor levels are normally 50–100 times lower than those in the rat study. Both industry and the government are now conducting extensive studies to verify the situation and evaluate the risk. The best current estimate of the potential worst risk is about 180 cases of cancer (U.S. Consumer Product Safety Commission, 1981 and 1982), as compared to 70,000 cancer deaths a year from smoking.

There is no question that formaldehyde in textiles can cause acute skin irritation and is a potent sensitizer (Fisher, et al. 1962; O'Quinn and Kennedy, 1965; Wayne, et al. 1976; Weber, et al. 1977; Bardana, 1980). About 8% of the population are vulnerable to sensitization if repeatedly exposed to 10% aqueous solution. Textile workers and about 1% of the general population are sensitive to freshly treated linen and must wash cotton shirts, clothing and bed linens before the first use. However, many embalmers, pathologists, pathology nurses, biology instructors, and forest product industry workers have inhaled air with 1–6 ppm for many years without chronic ill effects. In fact, formaldehyde exposure seems to desensitize many people. On the other hand, 0.1–3 ppm of formaldehyde can cause eye, nose, and throat irritation and aggravates chronic bronchial problems ranging from asthma to bronchitis because case studies indicate consistently higher incidences of symptoms among those exposed to formaldehyde. A large number of cohort studies is now under way. The first epidemiological studies are now available. Nonoccupational groups have been studied in New Jersey (Thun, 1980), Wisconsin (Anderson, 1981; Dally, 1981), Oregon (Williams, 1979), Washington (Breysse and Carbone, 1979), Connecticut (Giulietti, 1980), New Jersey (Marshall, 1980), New Hampshire (Hilgemeier, 1980), and Minnesota (Gary and Oatman, 1980). The results of all these studies indicate increased acute discomfort among people who live in homes with 0.1–2 ppm. Epidemiological mortality studies of occupational formaldehyde exposure have been completed by the National Cancer Institute for 5000 medical technologists, 1000 New York embalmers, 5000 embalmers in California, 1200 Ontario embalmers, and 10,000 pathologists in California without finding a significant increase in mortality.

Formaldehyde is present in maple syrup, shampoos, and some 120 other products without any noticed adverse health effects (Section 7.2.8.2 and 7.4.3). Even the human cell contains traces of formaldehyde, and white bread contains about 0.4 ppm wt. % in the crumbs and up to 2 ppm wt. % in the crust (Linko, et al. 1962). Thus, many doctors claim that the fear of formaldehyde is due to neither allergy nor hypersensitivity but to idiosyncrasy or hysteria (Berger, 1981).

For those who live in homes that contain urea-formaldehyde bonded particleboard, plywood, or urea-formaldehyde insulation foam, the question is less one of health risk than one of living quality and the quality of products installed in their homes because well-made and properly installed products release so little formaldehyde that even minimal ventilation will reduce indoor air levels to normal ambient levels (Meyer, 1979; Stone, et al. 1981). In fact, pure cured urea-formaldehyde resins contain so little formaldehyde that they can be used as surgical powder (Baumann and Schmidt, 1958). Thus, reports of toxicity of UF-based products (Harris, et al. 1981; Most, 1981, Marshall, 1980) are obviously based on faulty products that should be replaced anyway because their mechanical performance will be inferior.

Formaldehyde resins may contain other toxic malodorous or additive substances such as furfural, benzaldehyde, and phenol derivatives (Schoenberg and Mitchell, 1975). In fact, acrylo compounds (Davis, et al. 1973) and the isocyanates (Rye, 1973) used to reduce odors, may be more toxic than the resins currently used.

8.8 Ocular Pollutants

The sensitivity of the eye is self-evident. Dry air interferes with lacrimation and causes distress to contact lens wearers. Corneal hypoxia in airplanes, where the humidity can be less than 10%, can cause epithelial damage and edema (Eng, 1979). Ozone at a level of 0.3 ppm causes redness but no permanent damage (Van Huesden and Mans, 1978).

Several household agents are potentially damaging to the eye. Ammonia in vapors from cleaning fluids can cause permanent visual defects, lead to cataracts, and necessitate surgery to drain the anterior. Bathroom cleansers containing sodium silicate, lye, or carbonate are caustic and cause corneal damage. Oven cleaners contain sodium hydroxide and act like the bathroom cleansers.

Chlorocarbon solvents such as Freon–12 or 1,1,1-trichloroethylene, used as spray propellants in Scotch Guard impregnating agents, shoe polish, or similar products, cause reversible irritation. The toluene solvent in typing fluid or epoxy spray, used by teenagers in the 1960s for "glue sniffing," is harmless to the eye up to 300 ppm, and irritation is reversible up to 800 ppm. Allyl isothiocyanate, as used in Duco Cement, can cause lacrimation and, eventually, keratitis. Paradichlorobenzene in moth killer is safe up to 100 ppm; perchlorobenzene in spot removers causes mainly pain. Lysol contains EDTA, an allergen. Listerol disinfectant is quite harmless; isopropanol in silver polish causes irritation; and hypochlorite, used as a disinfectant in swimming pools or as a cleanser, is harmful in a concentration above 5%. Many other gases and particulates are harmful to the eye (Smith, 1976). Irritation by formaldehyde is reversible.

8.9 Risk Assessment

The purpose of a risk assessment is to allow comparison of risks for alternate options. Thus, medical doctors will weigh the risk of, say, future lymph cancer from radiation treatment of a tumor against the short term risk of living with a rapidly growing, uncontrolled tumor. For an older person with a painful metastatic cancer, the assessment will be different from that for a child with an operable tumor.

All of us are daily making similar assessments, say, about the merit and risk of driving an automobile, but most of us do so subconsciously and not on the basis of reasoned scientific facts. Otherwise few people of sound mind would routinely drive a car on a busy freeway, gamble at a gaming table, or smoke tobacco. However, careful risk assessment is a very important prerequisite for setting any type of rational voluntary or government standard. In the last 10 years, the U.S. Congress has increasingly requested federal agencies to present formal comprehensive risk analysis data for public scrutiny before regulatory action is taken. Currently, formal federal risk assessments exist only for asbestos and formaldehyde. Risk assessment constitutes a very complex task because it entails carefully analyzing the dose–response curve and integrating the risk for each segment over all other parameters. This type of multidimensional integration requires considered judgment and the ability to weigh the reliability of conflicting experiments (Conway, 1982).

The risk of contracting an air pollution-related injury or disease depends on the exposure level, the duration of the exposure, and many other factors involving the nature of the agent, the mode of transmission, and the nature of the response. For example, the risk of contracting measles depends not only on the number of infected people we might meet in a specific room, the amount of time we spend with them, and our previous exposure to the disease, but also on the size of the room, ventilation rate, direction of the airflow, air humidity, and temperature. Furthermore, the risk also depends on the rate of our breathing and on our general health and well-being. The latter is a complex quantity that depends on our previous health history and on a variety of periodic, diurnal, and seasonal factors. For some agents, such as sulfur dioxide, the risk remains constant over extended periods of time. For asbestos and radiation the risk increases over a period of several decades. Furthermore, for some contaminants the risk increases due to chemically and physically unrelated agents from other sources. Finally, for some pollutants there is both an acute and chronic risk.

In order to assess a risk, it must be compared with other threats. Table 8.7 lists the 10 leading causes of death. The mortality rate for infectious diseases is summarized in Figure 8.9, p. 253.

8.9.1 SMOKING

The U.S. Surgeon General has published two extensive reports on the risk of smoking (HEW, 1964, 1979). However, these reports do not include a formal

Table 8.7

Leading Causes of Death in the United States[a]

Cause	No. of Deaths per 100,000	Deaths per Year
Heart failure	372	820,000
Malignancies	160	352,000
Cerebrovascular failure	106	233,000
Accidents	58	128,000
Pneumonia	34	75,000
Infant deaths	22	48,000
Diabetes	19	42,000
Arteriosclerosis	17	37,000
Bronchitis	17	37,000
Liver cirrhosis	14	31,000
Influenza	3	7,000
Automobile accidents	28	62,000
Suicide	11	24,000
Accidental falls	10	22,000
Homicide (New York City)	9	20,000
Fire	4	9,000

[a] From the U.S. National Center for Health Statistics (1971).

numerical evaluation of the risk of experiencing various adverse effects that members of different population groups encounter. Smoking combines acute and chronic risks to smokers as well as to those who live or interact with smokers. Furthermore, smoking combines almost all the other adverse risk factors discussed above (Shepard, 1981).

Active Smoking. A very strong case has been made by the U.S. Surgeon General (1979) for linking cigarette smoking with a high risk of experiencing dozens of adverse health effects. Smokers are exposed to everything from radioactive ^{222}Rn, which accumulates on the tobacco leaf in the field, to cadmium metal, alkaloids, polynuclear aromatic carcinogens, and particulates. Furthermore, smokers inhale significant carbon monoxide concentrations (Section 7.2.2). Inhaled smoke contains 200–1500 ppm of CO. Carbon monoxide is not only readily measured in the gas phase, but its presence in the body is reflected in the carboxyhemoglobin (COHb) blood burden; it competes with O_2 for binding to the hemoglobin. The partial pressure of oxygen must be 25 times larger than that of CO for a 50% hemoglobin load. Normal healthy people have blood concentrations of 0.4–0.7 COHb. Woebkenberg et al. (1981) found levels of 7.43 ± 0.37 and 6.27 ± 0.30 in smokers in Cleveland and Elyria. In non-smokers the levels in these industrial cities were 3.21 ± 0.17 and 2.00 ± 0.15. In industrial areas the level is increased up to 1.2% by pollution. A typical one-pack-a-day smoker has 5–6% COHb and a heavy smoker 7–9% COHb. The presence of so much COHb causes a measurable increase on the visual light threshold. At a level of 16% serious headache sets in, and at 30%, nausea. In people with severe heart disease all clinical responses are shifted and patients with angina pectoris feel chest pain at 5–9%. The observation of a COHb body burden establishes a direct link between exposure levels and clinical symptoms.

Nitric oxide is another pollutant present in smoke in large concentrations, sometimes up to 1200 ppm (Section 7.2.3). On the moist mucus of the lung, nitric oxide can form a mixture of nitrous acid (HNO_2) and nitric acid (HNO_3). Both are strong and corrosive acids that attack the alveoli.

The corrosive effect of smoke is so strong that smokers living in unpolluted areas still have a four times greater incidence of respiratory and circulatory diseases than nonsmokers living in the most polluted areas of the world. In fact, tobacco smoke is such a strong pollutant that the passive smoker (anyone living around the smoker) has a significantly increased rate of disease.

Extensive studies on over 1 million men and women who were enrolled between October 1, 1959, and March 1960 and traced through September 1965 (see Surgeon General for: Hammond, 1966) clearly established the risk for several types of cancer. In the 45–64 age group the ratio of smokers to nonsmokers among men was 7.84 for lung cancer, 9.9 for cancer of the pharynx, 6.09 for cancer of the larynx, and 4.17 for cancer of the esophagus. No relationship was found for cancer of the colon, prostate, or kidney, or for leukemia; thus, smoking does not seem to increase the risk for these diseases. Yet more telling is the dose–response pattern of the ratio of smokers to nonsmokers. For lung cancer the ratio of the mortality increases from 4.62 for those who smoke 1–9 cigarettes to 8.6 for 1-pack-a-day smokers to 14.7 for 1.5-pack-a-day smokers to 18.8 for those who smoke 2 or more packs a day. Furthermore, the rate decreases measurably within 1–5 years among those who stop smoking. Lung cancer among smokers most often manifests itself at the age of 56–58. It is believed that all of these cases of lung cancer, which would correspond to 70,000 people a year in the United States, could be prevented. Smoking is held responsible for 85% of all lung cancer. Of 100 men at age 50, 17 will die during the next 10 years; of these, 1 out of 15 will die from lung cancer and 1 out of 3 from arteriosclerosis. Both diseases are influenced by smoking. Finally, cigarette smoking increases the mortality from other diseases. The frequency of lung cancer among asbestos workers who smoke is up to 30 times greater than among nonsmoking asbestos workers; among uranium miners it is 50 times higher; and among women who take birth control pills it is 10 times higher.

The importance of environmental factors has been discussed in preceding sections. Alcohol, too, is strongly implicated; it greatly increases the predisposition due to smoking. Cancer of the mouth or pharynx is 2.5 times more frequent among nonsmokers who drink 1.5 or more ounces of alcohol a day than among nonsmoking people who abstain. The synergism between drinking alcoholic beverages and smoking increases this factor as a function of both habits. For those who smoke two packs of cigarettes a day and drink 1.5 or more ounces of alcohol, the risk is increased 15.5-fold. This corresponds to a quadratic synergism (Rothman, 1972). It is not clear what causes the cancer, but it is feasible that it is due to impurities in the alcohol. This is suggested by the fact that the evidence of esophogeal cancer is highest in Puerto Rico, where local rum is popular, and in certain parts of France, where wines have a special bouquet.

Epidermoid carcinoma in the bronchus of smokers who died from other symptoms has been established beyond reasonable doubt (see Surgeon General for: Koss, 1975). However, these lesions are difficult to recognize clinically and almost impossible to remove. In contrast, laryngeal lesions can be readily identified through biopsy or cytological studies, and they can be removed. Unfortunately, squamous and epidermoid cancers of the oral cavity are difficult to recognize, even for well-trained dentists, because they look like either an ordinary inflammation or a benign leukoplakia.

Passive Smoking. According to the U.S. Surgeon General (1979), 78% of nonsmokers and 35% of smokers feel that it is annoying to be near a person who is smoking cigarettes. Similar results were obtained in polls abroad (Schmeltz, et al., 1975; Fischer, 1978; Weber-Tschopp, 1978): 70% reported eye irritation, 32% headaches, 30% nasal symptoms, and 25% cough. Smoking is specially annoying in aircrafts because of the dry air (Section 7.1.1; FAA, 1971; Barad, 1979). Contact lens wearers are especially inconvenienced (Stick, 1980). Table 8.3 lists some components of smoke; all are irritants. Levels of CO, BaP, and other indicators are listed in the preceding section and section 7.1.5.3. Recently, N-nitrosamines have been identified in smoke (Hoffmann et al., 1979). Field exposure levels have been recorded by Repace and Lowrey (1980) and other authors (NRC, 1981, Table 7.5) and reviewed by Sterling (1982).

While the smoker himself inhales for only 4 sec/min, his surroundings receive the full dose. Recent observations indicate that this exposure takes its toll. The incidence of lung cancer among nonsmoking women was 3.4 times higher for those who were married to smokers than that for nonsmoking couples (Trichopoulos, et al. 1981); and in a study of 91,000 Japanese women the incidence of lung cancer in nonsmokers was 2.08 times higher for those married to smokers than for nonsmoking couples. The rate for emphysema was 1.49 times higher (Hirayama, 1981).

Parental smoking has been invoked as cause for increased respiratory illness among children (Colley, et al. 1974; Fergusson, 1981; Melia, 1981). There are several similar connections. In one case, two wives of the same chain smoker contracted Raynard's syndrome (Bocanegra and Espinoza, 1980). However, the Tobacco Institute rejects all data on adverse effects from passive smoking as insufficiently established (Bock, 1982).

8.9.2 FORMALDEHYDE

As explained in Section 8.7.8, formaldehyde vapor is an irritant. Occupational work indicates that a level above 1 ppm is noticeable and irritating to all people. At this level 100% of the occupants would be at risk. Indoor air levels are restricted by OSHA standards to 3 ppm, and the recommended NIOSH standard is 1.2 ppm. The dose-equivalent indoor level would be 0.4 ppm because people spend four times more time at home than at work.

Following 15 years of complaints from consumers and state attorney's offices, the U.S. Interagency Regulatory Liaison Group (IRLG), comprising the U.S. Consumer Product Safety Commission, U.S. Department of Energy, HUD, the EPA, and NIOSH, has compiled a risk assessment for formaldehyde-caused cancer. Cohn (1981) prepared a separate risk assessment for foam insulation.

On the basis of the CIIT rat study of 1980 (Section 8.7) the risk at 14 ppm exceeds 90%. Depending on the dose–response curve chosen (Figure 8.1), one can derive a risk curve (Figure 8.16). The dose can be calculated from the exposure levels (Section 7.3) or read directly from Table 7.13 (Section 7.4). A more detailed breakdown for residents of homes with indoor concentrations between 0.05 and 0.7 ppm is given in Table 8.8. Table 8.9 shows urban air levels and other components. The cumulative exposure is obtained by adding occupational, home, and ambient air levels (Table 8.10). The same table also

8.9 Risk Assessment

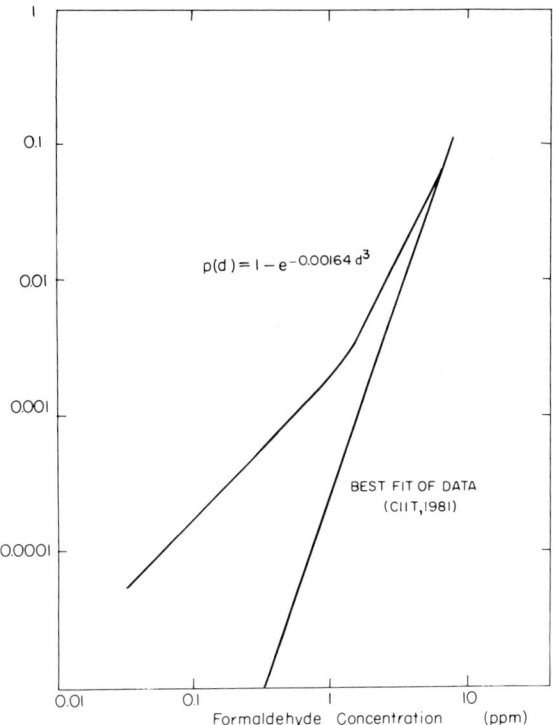

Figure 8.16. Cancer risk from formaldehyde derived from rat exposure (Cohn, 1981). The curves constitute the upper 95% confidence limit from EPA data (Dailey, 1981, unpublished, and Cohn, 1981).

shows the 95% upper confidence limit on carcinogenic risk. As can be seen, the risk for occupants of homes with 1 ppm is about 0.15%; for a heavy smoker who is exposed for 6 hours daily to his own cigarette fumes the risk is 4.5%. For occupants in office trailers or schoolchildren in temporary quarters with 0.1 ppm the risk is 0.02%, and for embalmers about 0.3%. Obviously, these risks are highly speculative, since no human cancer has yet been correlated to formaldehyde despite a total population exposure of 140,000 ppm hr^{-1} year^{-1} in the United States alone; but the values reliably reflect relative exposure and upper relative risk limits.

8.9.3 ASBESTOS

A risk assessment compiled by the EPA (1979) showed a significant risk for all those who are exposed to more than five fibres per milliliter air for any

Table 8.8

Total Formaldehyde Exposure for Seven Population Groups at Five Indoor Concentrations[a]

Exposure Concentration (ppm)	Age (years)						Total
	5	5–17	18–64[b]	18–64[c]	18–64[d]	64	
0.05	149	480	596	441	316	300	2,282
0.05–0.099	510	1,643	2,040	1,506	1,079	1,024	7,802
0.10–0.29	1,771	5,707	7,077	5,229	3,747	3,555	27,086
0.30–0.49	901	2,905	3,601	2,661	1,907	1,810	13,785
0.50–0.69	368	1,185	1,468	1,085	778	738	5,622
Total exposure time (10^6 person-hr/year)	3,699	11,920	14,782	10,922	7,827	7,426	56,577
Total exposures (10^6 ppm per person-hr/year)	976	3,146	3,900	2,882	2,066	1,960	14,930
People exposed (10^3)	541	2,268	2,909	1,598	1,441	1,087	9,844

[a] An exponential reduction of formaldehyde emission with home age is assumed; exposures are in units of 10^6 person-hr/year; from JRB-report and Dailey (EPA, 1981, unpublished).
[b] All subjects were men.
[c] Subjects were female homemakers.
[d] Subjects were women employed outside the home.

Table 8.9

Summary of Total Formaldehyde Exposure for Four Groups, and Cumulative Dose for Seven Population Groups[a]

Exposure	1+3+4	1+3+5	2+4+5	2+3+5
Formaldehyde producers	2,773– 3,771	3,088– 4,086	4,540– 5,538	4,855– 5,853[a]
UF resin manufacturing	2,689– 3,979	3,004– 4,294	4,456– 5,746	4,771– 6,061
UFFI manufacturers	568–11,675	883–11,990	2,335–13,442	2,650–13,757
UF distributors	537– 8,867	852– 9,182	2,304–10,634	2,619–10,949
UFFI dealers	911– 5,279	1,226– 5,594	2,678– 7,046	2,993– 7,361
UFFI installers	1,067– 6,891	1,382– 7,206	2,834– 8,658	3,149– 8,973
Wood product manufacture[b]	2,523– 5,643	2,838– 5,958	4,290– 7,410	4,605– 7,725
Funeral services	505– 8,659	820– 8,974	2,272–10,426	2,587–10,741
Pathologists	3,875–11,051	4,190–11,366	5,642–12,818	5,957–13,133
Fertilizer manufacture	755– 3,407	1,070– 3,722	2,522– 5,174	2,837– 5,489
Textile cotton finishing, all resins[c]	651– 3,355	966– 3,670	2,428– 5,122	2,733– 5,437
Biology instructor (higher education)	3,719–23,531	4,034–23,846	5,486–25,298	5,801–25,613
Majors in biological sciences, health and medical profession	2,627–15,835	2,042–16,150	4,394–17,602	4,709–17,917
Iron and steel foundries	755–28,991	1,070–29,306	2,522–30,758	2,837–31,073
Average for all use categories[d]	2,173	2,488	3,940	4,255
Not occupationally exposed	443	758	2,210	2,525

[a] From Dailey (EPA, 1981, unpublished); worst assumptions (exposure in units of ppm hr^{-1} year^{-1}).
Key:
1, conventional residence minus UFFI; 2, mobile home/conventional residence plus UFFI; 3, occupational; 4, urban air; 5, rural air.
[b] Includes sum of plywood and particleboard categories.
[c] Includes sum of all cotton apparel resin user categories.
[d] Mean levels of exposure obtained for each category and used to calculate average exposure level.

length of time. Asbestos is still used in over 5,000 products including construction and consumer products, and even in some temporary dental fillings. Thus, the entire population is exposed to some extent. The risk is highest for people who live or work in buildings with friable asbestos tiles or insulation. Figure 8.17 shows the risk as a function of integrated concentration, that is exposure level times time.

8.9.4 RADON

The risk assessment for radon is not yet complete. It is currently estimated that 1 rem delivered to 1 million people will cause 100 cancers over 25 years. The calculation for the radon risk in U.S. residences diverges widely because the indoor levels are poorly known. The prediction range from zero to about ten thousand excess cancer death per annum (Hartley, 1981).

Table 8.10

Risk Assessment for Cancer from Formaldehyde[a]

Group	Level (ppm)	Duration (hr/week)	Period (years)	Risk[b] (%)
Embalmers, high	4.0	20	10	0.304
low	0.52	20	10	0.012
Textile workers, high	1.4	40	10	0.081
low	0.1	40	10	0.004
Pathologists	4.8	30	10	0.71
Biology instructor (high school)	8.3	5	10	0.0035
Office trailer	0.02	40	15	0.0013
	0.1	40	15	0.0066
School trailer	0.02	30	12	0.0008
	0.1	30	12	0.004
Home residents	0.01	125	5	0.0008
	0.10	125	5	0.010
	1.00	125	5	0.142
	2.54	125	5	0.36
Smoking adds	40	6	10	4.554
Ambient air	0.03	168	70	0.04
Rural ambient air	0.005	168	70	0.006

[a] From EPA (unpublished, 1981), and Cohn, 1981.
[b] Worst case (CPSC, 1982).

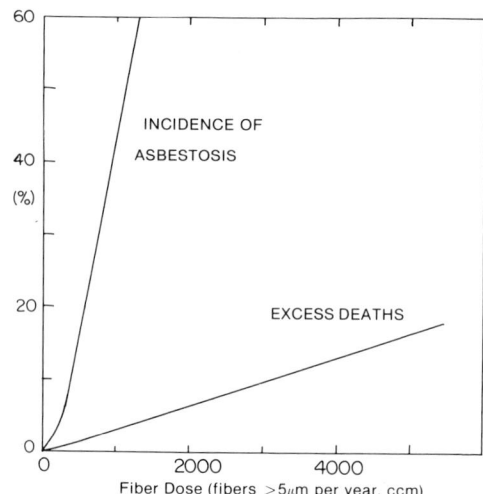

Figure 8.17. Asbestos risk. Incidence of asbestosis and excess deaths as a function of fiber dose (after EPA, 1979).

9. Control

The indoor habitat is by definition artificial. Thus, the question is not whether indoor air should be natural, but how the indoor climate may best be controlled to provide a comfortable habitat for a building's occupants. As Section 3.1 shows, this requires clean air, an operative temperature between 68 and 79°F, a room humidity of 40–60%, and an air movement between 0.15 and 0.3 m/sec. Temperature, humidity, and air movement have not only a large influence on our perception of pollutants, but also on their chemical, physical, and physiological actions. Since the outdoor climate constantly undergoes diurnal and seasonal cycles, and since a building's occupants and their activities change during the day, the chore of the comfort control system is comparable to that of a ship's captain who has to maintain a steady course against a heavy crosscurrent and strong gusty winds while carrying a shifting cargo. The success of indoor air quality control depends on the design of the building structure, the skill of the building manager, and the activities of the building's occupants. There are two types of tools: passive tools, such as the heat content of the building; and active tools, such as the heater and the ventilation system.

9.1 Active versus Passive Control

Passive control includes all design factors: the location in relation to air currents, sun, and neighboring buildings, landscaping, architectural style, and choice of building materials to provide the most desirable sun exposure, temperature, humidity, and air movement at all times of the day and during all seasons (American Institute of Architects, 1974; Manning, 1965). The U.S. Department of Housing and Urban Development and the American Institute of Architects' Research Institute have thoroughly reviewed the potential for passive climate control by improved building design (HUD, 1980). They established 13 climate regions (Figure 9.1) and identified the potential for indoor climate improvement by better solar design. Table 9.1 lists 16 typical cities, the

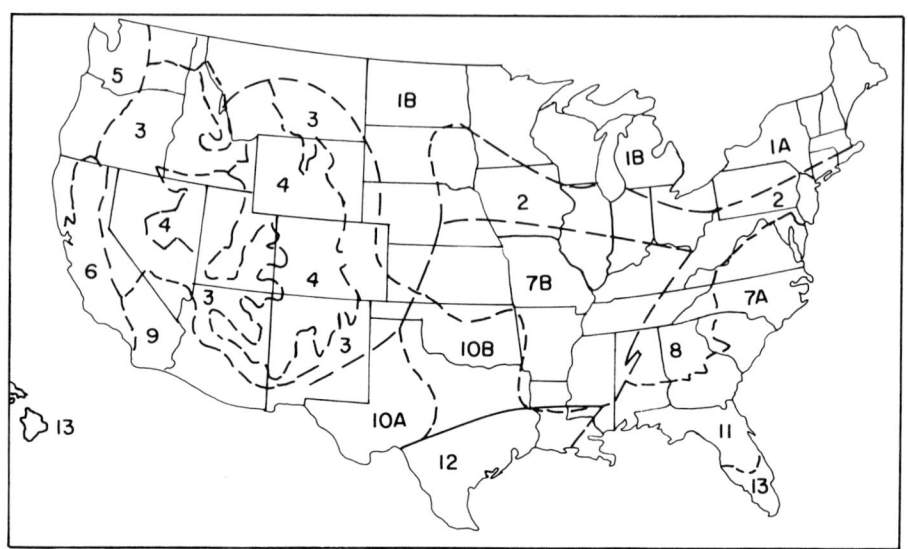

Figure 9.1. Thirteen U.S. climate regions for passive solar design (HUD/American Institute of Architects' Research Institute, 1978).

basic climatic problems, the best reasonable potential climate modification by passive solar design, and the percentage of improvement. The potential for improvement is about 50% in Miami, FL, Midland, TX, Phoenix, AZ, and Fresno, CA; about 40% in Fort Worth, TX, Salt Lake City, UT, Hartford, CT, Knoxville, TN, and New Orleans, LA; about 35% in Little Rock, AR, Medford, OR, Charleston, SC, Houston, TX, and Indianapolis, IN. Passive design helped least, but still 28%, in Madison, WI, and Ely, NV.

In each city the degree of improvement depends on a skillful combination of use of temperature, wind, moisture, and sun. In the case of Madison, WI, for example, the situation is illustrated by Figure 9.2. The seasonal temperature range is shown in Figure 9.2a. The periods of natural comfort are June and August. July is too hot, and the rest of the year is too cold. The humidity (Figure 9.2b) is comfortable at noon all year round, but is always very high during the night. Figure 9.2c shows the energy balance, including solar heat gain, on vertical and horizontal building parts. In winter the temperature, humidity, and wind are liabilities and the sun is a desirable heat source. In summer the temperature and sun are liabilities, and the wind is an asset for night cooling. The wind velocity all year round is 8–12 mph and sufficient for natural ventilation.

In a significant part of the current housing stock the building does not aid indoor temperature control; on the contrary, it excludes air when the latter

9.1 Active versus Passive Control

Table 9.1
Basic Climate Conditions in 13 U.S. Cities; Potential Improvement via Passive Solar Design[a]

Climate Region	City	Basic Climate Condition[b]		Potential Climate Condition[c]		Improved Comfort (%)
		Too Cool	Too Hot	Too Cool	Too Hot	
1A	Hartford, CT	75	13	63	0	37
1B	Madison, WI	76	12	65	7	28
2	Indianapolis, IN	66	20	59	9	32
3	Salt Lake City, UT	77	11	61	1	38
4	Ely, NV	92	0	76	0	24
5	Medford, OR	79	8	66	0	34
6	Fresno, CA	62	17	43	5	52
7A	Charleston, SC	46	42	33	33	34
7B	Little Rock, AR	52	35	42	24	34
8	Knoxville, TN	56	28	45	19	36
9	Phoenix, AZ	48	37	30	18	52
10A	Midland, TX	55	26	37	10	53
10B	Fort Worth, TX	47	39	33	24	43
11	New Orleans, LA	36	52	25	39	36
12	Houston, TX	35	54	25	43	32
13	Miami, FL	35	69	7	39	54

[a] From American Institute of Architects (1978).
[b] Percentage of conditioning per year based on temperature and humidity.
[c] Percentage of conditioning per year based on simple design modifications of the building in response to the climate.

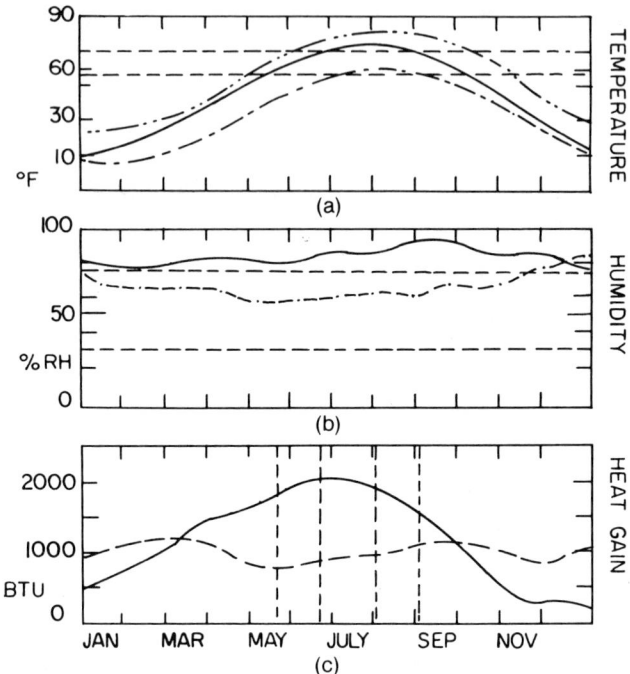

Figure 9.2. Range of comfortable (a) temperature, (b) humidity, and (c) peak heat gain in Madison, WI (American Institute of Architects, 1978). Dashed lines indicate comfortable range; dot–dash lines the daily range. In part (c) the solid line represents horizontal gain per square foot, dashed line is gain on vertical surface. The indoor climate is comfortable without air conditioning only in June and August.

could equilibrate temperature and humidity. Therefore, the atmosphere inside many houses is actually hotter and more humid in summer than the ambient air. For example, mobile homes or ramblers in new suburban areas without trees can build up high summer indoor temperatures and the temperature in an unventilated attic can reach 160°F, thus contributing an unnecessary heating load to the rooms below.

Very important factors in passive building design are the provision for natural ventilation and the willingness of occupants to take the trouble to adjust that ventilation as necessary. Until the advent of air conditioning, everybody was well aware of the benefit of natural ventilation (Carrington, 1914). During the last 30 years, much of this skill has been lost in North America, because abundant energy and reasonably priced air-conditioning equipment have caused occupants of commercial and residential buildings to become overly reliant on ventilation appliances, which provide instant heating or cooling without dependence on the vagaries of the outdoor climate. In commercial buildings, the management of natural ventilation requires instrumental comparison of indoor and outdoor temperature and humidity, and equipment for switching from natural ventilation to forced ventilation or recirculation of air. With modern electronics the design of such equipment is trivial and the cost of operating it can be more than recovered (Turiel et al., 1979), but the use of natural ventilation has not yet been widely rediscovered. The Commercial and Residential Conservation program of the U.S. Department of Energy does not promote the design of improved instrumentation or equipment to be used for natural ventilation.

9.2 Temperature Control

The building temperature is determined by the energy flow. In most buildings temperature control is considered the first priority. The temperature is measured at some chosen location and heating or cooling are adjusted as necessary. In poorly designed or insulated buildings the temperature variation between the occupant and the thermostat can be very large, even if adequate air movement and air mixing are provided. A special problem exists in glass curtain wall buildings. In severe climates the radiative heat component can easily reach 10°C, even if occupants remain 5 feet from the wall. In this type of situation it is impossible to provide adequate comfort without modification of the wall. This can be achieved by installing curtains, or louvres.

Comfort depends on vertical as well as horizontal temperature gradients. In the past three decades work places were expected to be kept within a very narrow temperature range, close to ideal comfort. And people were not accustomed to be tolerant of temperature drifts. As a result of the energy price increases, thermal comfort has been redefined and the comfort range has been expanded. This is reflected in the revised ASHRAE standard 55-1981 which extends the tolerable range considerably and allows temperature drifts and ramps between 67.5°F and 84°F, as long as the changes do not exceed $1°F\ hr^{-1}$. Two situations are common:

1. In old buildings the temperature and ventilation are separately controlled. Heating is provided by the central heating system, fireplaces, or an electric wallboard heating system; ventilation is achieved in each room individually as necessary to prevent stale air by opening or closing windows. This system is still in use in most European buildings regardless of size, even in hospitals.

2. In almost all large commercial buildings and in North American residences, especially suburban ramblers built since the 1950s, heat is circulated from a central location in the form of conditioned air. In this type of building air purity, airflow and heat flow are rigidly linked together.

It would take a treatise to compare the relative merits of the two systems. The advantage of the first is that the occupants can ventilate their rooms without throwing the heat supply in the entire building off balance. Since air is not recirculated, there is less danger of extended exposure to recirculated pollutants. For example, smoke or stale air can be rapidly purged after a meeting or during breaks without continuous heating. The advantage of the second is that air is periodically mixed as long as the ventilation is in operation. The main problem with the second is that any modification of airflow in any room will affect the entire building. Furthermore, air quality in these type of buildings is *de facto* linked to the need for cooling or heating. Thus, in winter, air quality is coupled to the outside temperature, and the quality of wall insulation. If the building is poorly insulated, a brisk air flow is guaranteed, because it is necessary to maintain an adequate temperature; if the building is well insulated, or retrofitted with insulation, ventilation is no longer necessary, and, since air quality is not yet widely recognized as a necessity, this type of structure is pre-destined to accumulate stale air and become an "ill building."

During the last 20 years it was fashionable either to design buildings without doors between rooms or to remove doors during remodeling. This open design converts entire floors into one giant room in which air is mixed. The advantage of this floor plan is that all spaces are equally comfortable; however, field studies (Spengler et al., 1978; Moschandreas, 1978; Hollowell et al., 1980) have shown that in such buildings odors and pollutants spread rapidly over the entire building. Thus, nitric oxide is inhaled not only by the cook during the hours of meal preparation, but by all occupants, regardless of activity. Obviously, while this spreading decreases local concentrations, it can increase the exposure to the pollutant by a large factor. Control of this problem requires removal of contaminated air to the outside with the help of local fans.

In large buildings the indoor air climate is an important part of the engineering design and constitutes such a large fraction of the cost that highly sophisticated sensors and feedback systems can be employed to provide a very comfortable climate for all occupants at reasonable costs.

9.3 Humidity and Moisture

In outdoor air the humidity content varies on a daily basis (Figure 9.2b) and water vaporizes and condenses in regular cycles. This cycle is an important driving force for vegetation. In the indoor environment this cycle is undesirable, since moisture cannot sink into the ground and run off as it would on a lawn or in a field. High humidity causes swelling of wood, rot, and many other problems.

Condensation occurs in buildings when air is loaded with water vapor beyond saturation and the building envelope is below the dew point of room air. The effect occurs in poorly vented or densely populated rooms, and in poorly insulated homes when hot humid air comes in contact with cold surfaces. Both effects may occur simultaneously in the bathroom or shower stall and above the cooking stove in the kitchen; if a clothes dryer is vented into indoor spaces; and when air is conditioned.

If condensation is to be avoided, water must be removed at the same rate as that at which it is generated by the occupants' bodies and their activities. This physical law of mass conservation can only be fulfilled if buildings are ventilated at an adequate rate. This problem is particularly severe in poorly insulated homes in which forced air heating causes excessively dry air that must be humidified (Klein and Kunze, 1971) because the moisture will continuously condense on the building envelope all winter, while the occupants feel dry.

Many older homes depend on leaks and cracks for moisture transport and on small eddy currents of air to prevent moisture accumulation in wall cavities. In homes with brick walls moisture transport involves transport through the porous and water-vapor-permeable membranes (Figure 4.12). Modern energy-efficient and superinsulated houses transmit moisture on only a small part of the surface. Thus, accurate moisture flow control is mandatory. If homes are retrofitted with insulation and moisture barriers, the water transport must be readjusted. This fact has been widely neglected in current retrofitting programs and will inevitably cause substantial structural damage to the buildings in coming years.

A different problem exists in aircraft, where humidity is 9% or even lower, and moisture must be contained by recirculating the perspiration of the passengers and crew as the only source of humidity in cabin air.

9.4 Ventilation

The term ventilation has two different meanings: (1) a system providing fresh outdoor air, and (2) circulation of air within a building. This double meaning causes basic misunderstandings that imperil indoor air quality. Not all ventilation is fresh air ventilation; in fact, most air-conditioning systems recirculate 80–100% of the air. Indoor air quality depends not only on how much fresh air is admitted or on how effective filters are, but also on how air is mixed. The chemical laws of dilution prove that it is better to purge the entire air volume periodically as a batch rather than to slowly admix fresh air. The laws of thermodynamics show that doing this also saves energy, especially if the temperature gradient between indoors and outdoors is great. Slow admixing of outdoor air is ineffective and constitutes a constant thermal leak. Regular purging keeps the exposure to indoor pollution sources down because the total dose is smaller and because purging allows periodic recovery periods during which the body burden can be reduced. Many of the current problems in air-conditioned buildings would be eliminated if the air were periodically purged rather than continuously mixed.

Engineers distinguish between natural ventilation, forced ventilation, and infiltration. Infiltration is caused by inadvertent leaks through the building envelope. Infiltration depends on wind velocity and direction, but rarely mixes or purges air and thus leaves indoor air stale. Current energy conservation efforts correctly concentrate on reducing infiltration (DOE, 1974, 1978–1980; Dubin and Long, 1978; Blomsterberg and Harrje, 1979; Harrje, Dutt, and Beyea, 1979), but do not always sufficiently stress the need for compensatory ventilation (Turiel et al., 1979).

The effects of the different forms of air mixing are described in Section 7.3. If the outdoor air is clean, it is desirable to exchange air at a high rate because indoor air recirculation only mixes pollutants; it does not remove them unless the system contains special chemical filters, which in any case must be carefully and regularly maintained.

From Lavoisier's pioneering work in 1770 until the 19th century it was accepted that the purpose of ventilation was to remove metabolic water and carbon dioxide. Von Pettenkofer (1873) recognized that moisture and heat, rather than carbon dioxide, were responsible for stale air. Accordingly, the estimates of the recommended minimum air exchange rates increased from 4 ft^3/min (Tredgold, 1836) to 30 ft^3/min (Billings, 1893), Figure 1.8 and stayed at that value (which was accepted formally by 25 states) until 1946, when the American Society of Heating, Refrigerating and Air-Conditioning Engineers (ASHRAE) reduced it to 10 ft^3/min on the basis of the research of Yaglou (1936), which related stale air to body odor and heat. In 1973 ASHRAE reduced the recommended value to 5 ft^3/min in order to conserve energy. At these low levels it was rapidly rediscovered that air quality can be adversely affected by accumulation of chemical and physical contaminants. A revised standard, ANSI/ASHRAE 62-1981 (Table 9.2 and 9.3), now specifies a "ventilation rate procedure" to determine the need for ventilation rates above 5 ft^3/min to handle smoke and other special problems. This can be done only by variable ventilation (Turiel et al., 1979). The design of an air ventilation system is a mature field (Haines, 1961; Emerick, 1969; Bond and Straub, 1972; Eads, 1974; Croome-Gale and Roberts, 1975; McDermott, 1976; ASHRAE, 1974–1977; Air Monitoring and Conditioning Association, 1977; Goldschmidt and Didion, 1974; Walberg and Sallvik, 1977; Dossat, 1978; Hollowell et al., 1979; Roose, 1978). The use of summer attic fans was described by Reppert (1979).

Energy-saving devices, including heat exchangers, have been described by Kays and London (1964), SMACCNA (1978), Shinskey (1978), Roseme et al. (1979), Hollowell et al. (1980), and Fisk et al. (1981). Heat exchangers have been used for many decades in industry. In recent years heat exchangers for commercial and residential buildings have become available. They operate by bringing exhaust air and fresh makeup air close to a surface through which heat can be transmitted. Since a large amount of heat is necessary to adjust the moisture content of air when it is heated or cooled (Section 3.3), heat

9.4 Ventilation

Table 9.2

Recommended Commercial Ventilation Rates

Commercial Area	Estimated Persons per 1000 ft² Floor Area	ASHRAE 62-1973 Outdoor Required Ventilation Air per Human Occupant				ASHRAE 62-1981	
		Minimum		Recommended		Smoking	Non-Smoking
		cfm	l/s	cfm	l/s	cfm	cfm
Shoe repair shops (combined workrooms/trade areas)	10	10	5	15–20	7.5–10	15	10
Garages, auto repair shops, service stations							
Parking garages (enclosed)	—	1.5	7.5	2–3	10–15	1.5	1.5
Auto repair workrooms (general)	—	1.5	7.5	2–3	10–15	1.5	1.5
Theaters							
Auditoriums (no smoking)	150	5	2.5	5–10	2.5–5	35	7
Auditoriums (smoking permitted)	150	10	5	10–20	5–10	35	7
Ballrooms (public)	100	15	7.5	20–25	10–12.5	35	7
Bowling alleys (seating area)	70	15	7.5	20–25	10–12.5	35	7
Gymnasiums and arenas							
Playing floors—minimal or no seating	70	20	10	25–30	12.5–15	—	20
Locker rooms	20	30	15	40–50	20–25	35	15
Spectator areas	150	20	10	25–30	12.5–15	35	7
Amusement parlors and pool rooms	25	20	10	25–30	12.5–15	35	7
Swimming pools	25	15	7.5	20–25	10–12.5	—	0.5
Iceskating, curling, and roller rinks	70	10	5	15–20	7.5–10	—	20
Transportation							
Waiting rooms	50	15	7.5	20–25	10–12.5	35	7
Ticket and baggage areas, corridors, and gate areas	50	15	7.5	20–25	10–12.5	35	7
Concourses	150	10	5	15–20	7.5–10	35	7
Offices							
General office space	10	15	7.5	15–25	7.5–12.5	20	5
Conference rooms	60	25	12.5	30–40	15–20	35	7
Doctors' consultation rooms	—	10	5	10–15	7.5–10	35	7
Waiting rooms	30	10	5	15–20	7.5–10	35	7
Educational Facilities							
Classrooms	—	—	—	—	—	25	5
Laboratories	—	—	—	—	—	—	10
Libraries	—	—	—	—	—	—	5

exchangers are porous to moisture as well as to heat (i.e., the exchange surfaces are paper membranes; see Figure 9.3).

In residences built during the last 30 years 100% of the air is recirculated; fresh makeup air is all infiltration. Air filters are used to reduce dust resuspension. In new energy-efficient buildings, the position of the conduit for fresh makeup air is often not carefully considered. The design and location of the air inlet should be as thoughtfully determined as that of chimneys.

In commercial buildings, recirculated air accounts for anywhere from 80 to 100% of ventilation, depending on design (DeRoos et al., 1978; NIOSH, 1978) and on operation (Holcombe and Kalika, 1972; Kusuda, 1976; Szczepanski, 1980; Young et al., 1981). It is very difficult to achieve energy savings in such systems without major changes in the airflow patterns. Thus, a lower setting of

Table 9.3

Residential Ventilation Standards

Residential Area	ASHRAE Standard 62-73: Single-Unit Dwellings ($ft^3 \, min^{-1} \, person^{-1}$)		ANSI/ASHRAE Standard 62-1981: Single or Multiple Units ($ft^3 \, min^{-1} \, room^{-1}$)
	Minimum	Recommended	Minimum
General living areas	5	7–10	10
Bedrooms	5	7–10	10
Kitchens	20	30–50	100
Bathrooms	20	30–50	50
All other rooms	NA	NA	10
Basements, utility rooms	5	5	NA
Garage			100

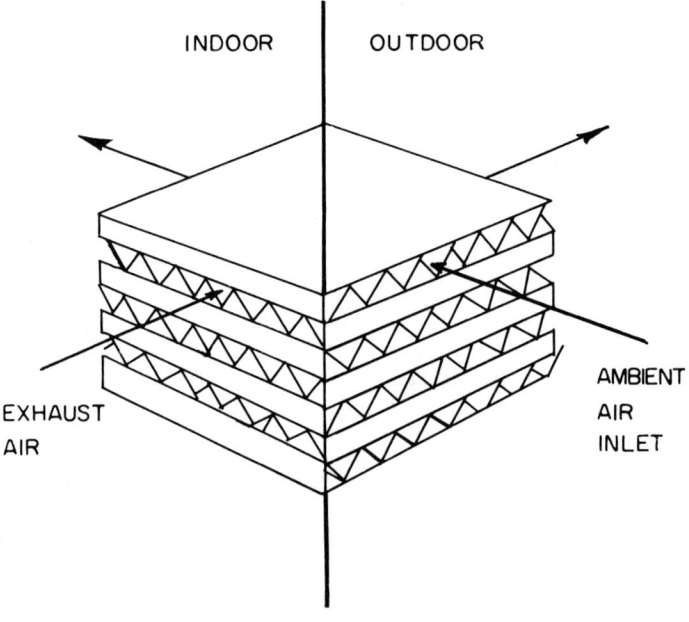

Figure 9.3. Residential air heat exchanger; size of space is 0.06 in. (after Mitsubishi, 1980).

the thermostat or "economizer cycle" adjustments inevitably cause air quality problems. An example of such a situation has been carefully analyzed by Sterling (1980).

9.5 Air Pollutant Control

There are three sources of indoor air pollutants: outdoor air, building materials, and indoor activities that generate pollutants. The main source is usually indoor activities.

9.5.1 CONTROL OF OUTDOOR AIR POLLUTION

The ambient air quality has tremendously improved during the last 10 years. For example, as a result of cooperation between industry and federal agencies, sulfur dioxide (SO_2) has been reduced at its source by a factor of 10 throughout most of the industrial nations (Section 7.2). The remaining problems are economic rather than technical. A description of this field goes beyond this book. Methods for controlling SO_2, CO, nitric oxides, particulates, and other agents listed in the Clean Air Act are described in many dozen books (American Industrial Hygiene Association, 1972; Strauss, 1971, 1975; Danielson, 1973, EPA, 1970, 1971, 1973–1981; Vesilind, 1975; Lear, 1976; Stern, 1977, 1978); asbestos control is described by Gerber and Rossi (1977), Mirick (1980) and Rajhans and Bragg, 1978); radon control is described by Culot (1978), Hess et al. (1979), the Atomic Energy Control Board of Canada (1978), Eaton (1980), Silberstein (1980), Richardson (1980), Ingersoll (1981), Lepman et al. (1981), and Moschandreas and Rector (1981). Trace contaminant removal is discussed by Deitz (1975). Particulates are discussed by Higgins et al. (1979); dust by Dorman (1974). The EPA provided manuals for hydrocarbons and organic solvents (1974), mercury (1973), and many other pollutants.

Chapter 7 explains that indoor concentrations of ambient pollutants such as ozone, sulfur dioxide, and particulates are almost always reduced because of absorption, deposition or natural decay. If they must be actively eliminated, as in hospitals or in heavily contaminated areas, indoor air must be filtered. Such filters can be combined with air-conditioners or heat-exchangers. Filters must be carefully maintained, or they will become secondary air pollution sources.

Earth gases. In some locations the soil itself has the potential to release undesirable gases. Among the most common are radon gas from uranium-containing soil, such as is found in mining areas; methane from natural gas lines or biodegradation in swamps and above abandoned cemeteries; and organic fumes from solid chemical waste disposal sites. These gases might not be noticed in the ambient air, but if the source is trapped below a house, they can accumulate in the basement and can be spread through the house by the heating system or air conditioner. They also can enter the groundwater system and then be introduced into homes with the water supply. In several instances, radon, methane, pesticides, PCBs, and other known toxicants have been measured in the residential environment at a level above safe industrial limits.

The best control method is to seal the building by installing a vapor-tight apron below the floor and providing separate ventilation for the crawl space.

Extensive experience with radon abatement has been gained in Canada, where sealants, ventilation, and filtration have been tested for several years (Atomic Energy Control Board of Canada, 1978). Louvers and fans have been used in Uranium City, Saskatchewan. This approach requires additional heat insulation in crawl spaces; two coats of epoxy sealant applied to the floor and walls of the basement can achieve the same result. An electrostatic air filter is effective even if it is used only intermittently, during occupancy. However, its plates must be regularly cleaned. Room humidity in excess of 40% reduces the efficiency of such devices by a factor of 2-3. In extreme cases soil excavation around the home might be necessary to reduce the source. In many cases, the best removal method remains increased ventilation (Gesell and Prichard, 1980).

Radon in well water can be effectively purged with aerators. Spraying water against a drip wall reduces Rn by about a factor of 4. Thus, well water with 1300 pCi/liter reduces to 315, 80, 40, 20, 10, and below background in six subsequent spray stages, respectively. The air-water distribution factor is about 1:110. Thus, water with 10^5 pCi/liter is in equilibrium with air of 900 pCi/liter. Radon also is removed effectively by boiling water (Gesell and Prichard, 1980).

9.5.2 METABOLIC POLLUTANTS

Reliable procedures and guidelines for the removal of metabolic products have been established by ASHRAE (1977, 1981). Body odor is reliably removed at a ventilation rate of 5 ft^3/min (Cain et al., 1981). Human wastes are well characterized (Sections 3.2 and 5.2) and efficient methods have been designed for their continuous control in submarines (Lin, 1975; Arnest, 1961; Decorpo et al., 1980), spacecraft (Mitchell et al., 1976), and commercial aircraft (Bulloch, 1979; Crabtree et al., 1980; Giles, 1980; Newman et al., 1980; Timby, 1980), as well as commercial supersonic aircraft (Lemaire, 1975; Turner, 1967). Air cleaning is important in hospital clean rooms in order to control microbes (DeRoos et al., 1978; Wanner, 1974). However, humidity and particulate control are very important in order to check microbial growth on air conditioners and other moist surfaces (Sections 5.1, 5.3, 7.1, and 8.5); and regular maintenance of these devices is essential to prevent them from becoming potent secondary sources.

9.5.3 INDOOR AIR CONTAMINANTS

Smoke. By far the most difficult pollutant to control or eliminate is tobacco smoke. Smoke subsides only slowly, and it adsorbs on all indoor surfaces and thus lingers for days (Narasaki, 1976; Brunnemann, 1978). In hotel rooms, bars, and other entertainment places, the heavier components of the smoke penetrate the furniture, bedding, and carpets (Crawshaw, 1976), from which they cannot be fully removed. This effect is enhanced if air-conditioning filters are not regularly maintained, because once they are saturated they become an active source and spread the odor throughout the room.

The persistence of cigarette odor, originally studied by Yaglou (1937), has recently attracted new attention (Duffee et al., 1980; Cain et al., 1981; Ingersoll et al., 1981), and a panel of observers has been used to correlate odor to particulate (Figure 9.4a), and carbon monoxide (Figure 9.4b) concentrations. Usually, butanol is used as a reference. Figure 9.4c shows the acceptance of tobacco smoke. For comparison, Figure 9.4d shows body odor acceptance.

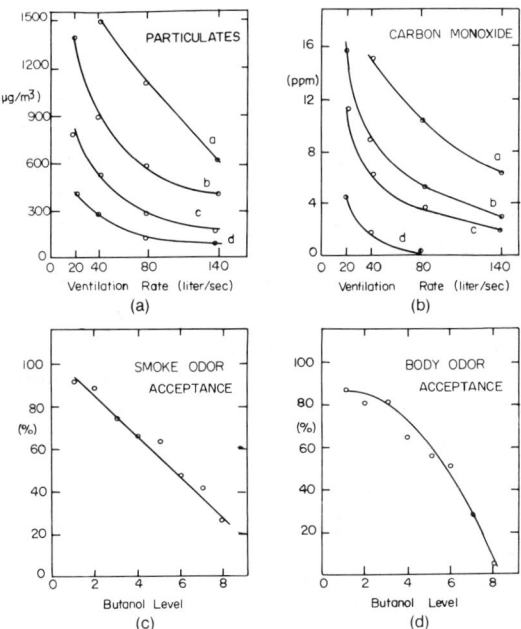

Figure 9.4. Particulate mass, carbon monoxide, and the odor of tobacco smoke as a function of ventilation rate; for comparison the odor curve for body odor is given (after Cain et al., 1981). In Figures (a) and (b) curve a represents a load of 24 cigarettes/hr; b, 16 cigarettes/hr; c, 8 cigarettes/hr; d, 4 cigarettes/hr.

Dust. For sensitive people, especially those with allergies, the control of viable and inorganic dust is most important (Creip and Green, 1936; F. Silver, 1979; Pfeiffer and Nikel, 1980). This can be achieved with filters (Holcombe and Kalika, 1972), but their efficiency must be constantly checked. Other control measures involve the cleaning of floors and other places where dust accumulates. However, not all cleaning methods are equally good. Normal vacuum cleaners retain mainly large debris and dust particles, resuspending in the air respirable particles with sizes of from 0.1 to 5 μm, thus creating a worse situation than before cleaning and rug shampoos may leave irritating vapors (Kreiss, 1981). Special equipment is available with water filters that eliminate this problem. Indoor air cleaning equipment is described by Yano and Oikawa (1974), Klotz and Wanner (1974), Kobayashi et al. (1974), and Balocco (1979).

Dust may contain any combination of outdoor and indoor debris, from mites, animal feces, human scales, and plants to asbestos and pesticides (Yeh et al., 1977). A booklet by the Center for Science in the Public Interest (1976) details dozens of other undesirable components. In residences pets, plants, and hobby crafts are heavy contributors of dust. In a centrally ventilated building dust must be filtered carefully to prevent endless resuspension. In recent years filters with an efficiency of 80–95% have become available. The best method for controlling dust is the prevention of its deposition.

Microbes. A very special indoor problem, microbes thrive in dark and humid places. Air conditioners, which inevitably condense water, are ideal breeding grounds for mold and bacteria unless they are carefully cleaned at regular intervals. Today an array of sophisticated dust control devices is available. Mechanical filters are still the most popular, but cyclone separators, scrubbers, and electrostatic precipitators are becoming increasingly more efficient and popular, even for residential use. However, periodic outbreaks of infectious diseases from air conditioners in hospitals, hotels, and homes can be prevented only by regular filter maintenance scheduling (Luciano, 1977; DeRoos et al., 1978; Sections 5.3, 7.2, and 8.5). To some extent microbes can be controlled with uv light (McLean, 1961; McNall, 1975).

Gaseous Pollutants. The most obnoxious gases are cooking odors; cigarette smoke; combustion gases from gas ranges, kerosene heaters, wood stoves, or fireplaces; and spray-can propellants.

Disposable spray-can aerosols are a very substantial source of gaseous pollution (Sections 5.6 and 7.2). One spray of oven cleaner, for example, will release 10 g of propellant and about 1 g of corrosive cleaner, of which a large fraction never finds its target, but lingers in the air at a concentration of 200 ppm or higher. The same holds for pesticide cans, hair sprays, and dusting agents (Table 7.9). These products often contain chemicals at a multiple of the level permitted in the workplace. Furthermore, the unused product mixes with

9.5 Air Pollutant Control

dust and can be resuspended by the building ventilation. In many cases potent pest bombs sold by veterinarians to fumigate homes in order to exterminate lice or fleas have caused acute poisoning to home owners who ignored instructions.

Combustion gases include carbon monoxide, nitric oxide, and smoke (Section 7.2). Nitric oxide is a strong irritant and is readily noticed. Although obnoxious and undesirable, its presence is a good indicator of carbon monoxide, which is not noticeable, but viciously poisonous. Natural gas is intentionally scented to indicate its presence. Combustion gases from kitchen ranges must be vented by directional ventilation.

Cooking vapors are currently suspected of including potentially carcinogenic nitrosamines and sootlike polyaromatics. Because of their odors, the latter are well recognized and controlled (McNall, 1975).

The best way to manage gaseous pollutants is by adequate ventilation with fresh air. Especially in residences, it is almost always cheaper to purge indoor air than to remove gaseous components. If removal is necessary, carbon scrubbers are the best and most reliable device. Carbon filters are commercially available. They consist of 2-cm-thick beds made of from 1- to 5-mm-diameter granules. Their efficiency is up to 95% with an air pressure drop of only 50 Torr/cm of a 1-m^2 bed at a flow rate of 25 cm/sec. Such filters must be regularly replaced to prevent breakthrough and possibly even desorption. Activated carbon can be treated with chemicals to increase the absorption of certain gases. Thus, in industrial respirators carbon is impregnated with lead acetate to absorb hydrogen sulfide, with phosphoric acid to absorb mercury, with sulfite to absorb formaldehyde, and with metal oxides or noble metals to oxidize phosgene, carbonyl sulfide, ozone, and mercaptans.

Activated alumina can be used as a replacement for carbon in some cases. If it is packed with permanganate, it oxidizes many organic pollutants, such as formaldehyde, and renders them harmless.

Personal Comfort Products. Commercial buildings and residences are crammed with latent sources of air pollutants. Myerson (1977) and Fritsch (1977) list entire volumes of household pollutants. The most reliable control of pollution is to remove unnecessary sources, such as spent aerosol spray cans. Clinical ecologists who specialize in controlling indoor air pollution start their work with an inspection of closets and shelves and usually find piles of forgotten, half-spent containers of products that should have been removed long ago. Typical examples are cosmetic sprays, paint solvents, pesticides, and cleaning agents (Pfeiffer and Nikel, 1980; F. Silver, 1979).

In commercial buildings the most common sources are multilith copy machines which use ammonia or methanol, or electrostatic copiers, which may produce ozone. Control should be achieved by direct ventilation to the outdoors.

9.6 Building Materials

During the last decades many new materials with excellent performance characteristics have become widely available for construction (Dietz, 1969; Malloy, 1969; Seymour, 1975; ASTM, 1978; Hornbostel, 1978). Problems arise from exposure to these materials if their properties are not sufficiently known, or if they are not used properly. These problems can be prevented by design and material standards. Only a few materials are discussed here.

Asbestos. A fireproof inorganic insulation material that is superior for the insulation of hot appliances, asbestos is also easy to apply. Thus, it was extensively used to spray-coat structural steel in high-rise buildings, to coat air ducts in schools and large buildings, to manufacture brake linings, and even as electrical insulation in hair dryers. It has since been discovered, however, that asbestos fibers lodge permanently in the lung and increase the risk of cancer 50- to 100-fold. Thus, asbestos is no longer being used for any of these applications. As a temporary measure asbestos can be sealed by coating or impregnating it with polymers.

Fiberglass. A fireproof inorganic insulation, fiberglass is used in paper-backed bats that are placed in wall voids or on attic floors. Fiberglass dust is irritating to many people but is probably far less carcinogenic than asbestos.

Wall-to-Wall Carpeting. Some carpets are glued to the floor with organic solvents that subsequently persist in the rag and slowly fumigate the room. Furthermore, carpeting acts as an adsorbant for smoke, fumes, and many gases, and must be periodically cleaned (Yano, 1978) or, eventually, replaced. Carpets absorb many pollutants (Walsh, Black, Morgan, and Crawshaw, 1977; Crawshaw, 1978). Carpet shampoos may leave residual vapors (Kreiss, 1981).

Plywood and Particleboard. Exterior grade plywood and particleboard bonded with phenolic formaldehyde resin, (which has a deep red color) can be used in any application without producing odors. In contrast, indoor grade products bonded with colorless urea-formaldehyde resins have a potential for formaldehyde release when they are new or when they are exposed to condensation from stale, moist air. Particleboard and indoor plywood are used for flooring, wall paneling, shelves and other cabinet work, doors, and almost all furniture, whether it is upholstered or has an artificial wood finish. The potential problem with these products has been described in Section 7.2.8.2 and 7.4.3. Malodorous products can be professionally fumigated with ammonia or sulfur dioxide, but the best remedy is adequate circulation of air at 50% RH.

Wall Insulation. All of the organic wall insulation materials approved by the U.S. Department of Energy can cause problems if they are not properly installed (Malloy, 1969; DOE, 1978). Polyurethane and styrofoam can burn, releasing deadly fumes; shredded cellulose can settle and attract moisture; and

9.6 Building Materials

urea-formaldehyde foam insulation (UFFI), can release irritating formaldehyde vapors if not properly manufactured and installed. Unfortunately, the installation of UFFI requires careful planning, and a skilled technician rather than the hasty installation that it is often accorded. Once wall insulation is installed, the installation can be checked only with expensive infrared equipment, and improperly installed material is difficult to remove.

Whenever wall insulation is properly installed infiltration is reduced. In buildings where fresh air is supplied only by infiltration, this reduction inevitably causes air quality problems. Thus, whenever insulation is installed, infiltration should be eliminated entirely throughout the building and replaced by intentional ventilation to provide an adequate energy-efficient source of fresh air, because the best cure for indoor air impurity problems remains circulation of air at about 50% relative humidity and 15–20°C at a speed of 0.1 to 0.3 m/sec, with enough fresh make-up air for steady state removal of contaminants at the rate at which they are produced.

10. Legislative Control, Regulation, Codes, and Guidelines

This chapter deals with regulatory provisions that directly or indirectly affect indoor air quality. Emphasis is on the current U.S. situation which represents a case study of similar situations in other countries. The quoted laws and regulations are published in the U.S. Federal Register (FR), the U.S. Code of Federal Regulations (CFR), and the United States Code (USC). Since pollution is man made, the need for air quality regulation is directly related to population density (Figure 7.5a). In Australia, with a population density of 2 people/km^2, there is no need yet for a nationwide ambient air law. However, any city in which people crowd into subways and theaters will quickly recognize the need to regulate even such activities as smoking. England, with 220 people/km^2, and other European countries with similar population densities, have controlled air quality for almost 200 years. In the United States, with 22 residents/km^2, air quality control was the prerogative of states and local authorities until 1967, when the Federal Air Quality Act became law.

The regulatory situation is strongly influenced by the current economic, political, and societal uncertainties, which cause not only fluctuations in priorities, but the transfer of responsibilities among government agencies. In some cases these discontinuities threaten to make laws and regulations counterproductive or obsolete before they are fully implemented. This situation is undesirable for industry, government and consumer, because it leads to overreaction, followed by piecemeal retraction, and regulatory insecurity. The history of ambient air regulations has been reviewed by Stern (1982) for the U.S., and Powell (1982) for Canada.

10.1 Indoor Air

The concept of residential and nonoccupational indoor air quality regulation is not yet widely accepted. Only densely populated and industrialized countries such as Japan (Japanese Law #20, 1970) and Sweden (Wahren, 1980) have acquired any experience with such laws (McFadden et al., 1978; Yanagisawa, 1980). In the United States only two federal regulations currently deal explicitly with indoor air—one concerns radiation, the other ozone; both are conditional. The most stringent and comprehensive voluntary guidelines are those implemented in the ASHRAE standard 62-1981 which sets ventilation standards for acceptable indoor air quality.

Ozone Standards. The Food and Drug Administration (FDA) of the U.S. Health and Human Services Department (HHS), formerly the Department of Health, Education and Welfare, established in 1972 a limit of 0.05 ppm by volume for ozone in houses, apartments, hospitals, and offices (21 CFR 801-415). It applies only to ozone as a waste product. In 1980 the Federal Aviation Administration implemented a time weighted average ozone concentration limit of 0.1 ppm in commercial aircraft for flights longer than four hours (14 FR 25.832 and 14 CFR 121.220).

Radiation Standards. This standard, implemented by the EPA on April 27, 1980 (40 CFR 192.10), states:

> Remedial action shall be conducted so as to provide reasonable assurance that the levels of radioactivity in any occupied or occupiable building shall not exceed either an average annual indoor radon decay product concentration — including background — of 0.015 working levels, or indoor gamma radiation — above background — of 0.2 milliroentgen per hour.

The regulation has the title "Environmental Standards for Cleanup of Open Lands and Buildings Contaminated with Residual Radioactive Materials from Inactive Uranium Processing Sites." The intent was to assure that people who spend up to 20 hours daily in the home will not be exposed to a higher integrated radiation dose than that allowed for 8-hour daily occupation exposure. However, it was recently discovered that radon levels in the basement of homes in certain other geological areas are exposed to natural radon levels exceeding the regulation limits. The EPA, the U.S. Department of Energy, and HUD currently work jointly with the states of Maine, Montana and Florida to reduce these levels with methods developed in Canada (Findlay, 1980), where similar problems were recognized and abated in the Provinces of Saskatchewan and Ontario.

All other federal air quality standards deal with ambient outdoor air, with hospitals and public buildings, or with occupational levels. Federal, state, and local regulations restricting asbestos and smoking in public places are discussed later.

Several state legislatures have developed or contemplated bills similar to those in effect in Sweden, Germany, Denmark and Holland to regulate formaldehyde levels in indoor air. Minnesota was the first state to pass legislation (Minnesota 4 S.R., 6-23, 1980) establishing a "Rule Governing the Level of Formaldehyde in Residential Units," (MCAR S.1.448, Formaldehyde in Housing Units) which forces home builders to maintain levels at or below 0.4 ppm at 75°F and 40–60% relative humidity in new homes, as measured by the NIOSH method 77-157-A. In April 1981 Wisconsin passed a similar indoor ambient air quality standard (WC 14.03) which provides for a time weighted average limit of 0.4 ppm. Both have been challenged in court. Formaldehyde levels apply only to homes that are at least six months old. Other states, including New York, New Jersey, West Virginia, Arizona, and California (Greene, 1980), have introduced legislation for control of air quality via other means. The Commonwealth of Massachusetts banned the use of urea-formaldehyde foam insulation in November 1979 and enacted a retroactive strict repurchase law. The ban was invalidated in December 1981.

The World Health Organization and federal, state, and local agencies in the United States have carefully monitored other indoor air quality problems. In 1980 six federal agencies formed the Interagency Regulatory Liaison Group in order to develop the national research plan for indoor air quality (45 FR 72781, and Hartley, 1981). The program comprises an inventory of current research on (1) identifying pollutants, (2) monitoring health effects and control, and (3) recommendations for future work (EPA, 1981). The State of California Consumer Affairs office has commissioned a directory of experts in indoor air (Gerszhov, 1982). The EPA has received from the National Research Council a report on indoor air quality (NRC, 1981), and the Comptroller of the U.S. General Accounting Office presented a report to Congress titled "Indoor Air Pollution: An Emerging Health Problem" (GAO, 1980). However, the federal government has been hesitant to set indoor air standards for office buildings and residences because of practical problems connected with the enforcement of such standards in private buildings. The latter became evident in July 1979 when President Carter used his executive power to implement the Energy Building Temperature Regulations (44 FR 41205) discussed below.

10.2 Ambient Air

For purposes of the Clean Air Act, ambient air is defined as "that portion of the atmosphere external to buildings, to which the general public has access" (40 CFR 50.1(e)). Federal activities in this field are based on the Clean Air Act of 1967, Public Law 90-148 (42 CFR 50). The Amendment of 1970 (PL 91-604) mandated the establishment of air quality standards. The first set of National Ambient Air Quality Standards were published on April 30, 1971, after extensive preparation, review, and public comment. They currently cover six pollutants (42 CFR 412): sulfur dioxide, particulates, carbon monoxide, ozone, hydrocarbons, lead, and nitric oxides (see Table 10.1). The standards also prescribe the analytical methods that are to be used (40 CFR 51). A set of secondary standards limits sulfur dioxide emission from new sources to 13 mg (i.e., 0.5 ppm for a 3-hour period). Particulate guidelines are 50 μg, but may not exceed 150 μg/day once per year. The EPA is also responsible under the Clean Air Act for establishing a policy and procedures for identifying, assessing, and regulating airborne substances posing a risk of cancer (40 CFR 61). So far asbestos, vinyl chloride, and benzene have been listed.

The standards are based on detailed Air Quality Documents which are prepared and reviewed according to procedures, which likewise were open to

Table 10.1
National Ambient Air Quality Standards[a]

Contaminant	Long Term Concentration ($\mu g/m^3$)	Long Term Averaging Time	Short Term Concentration ($\mu g/m^3$)	Short Term Averaging Time
Sulfur oxides, measured as sulfur dioxide	80	1 year	365[b]	24 hours
Particulate matter	75[c]	1 year	260[b]	24 hours
Carbon monoxide	—	—	10,000[b]	8 hours
			40,000[b]	1 hour
Ozone	—	—	235[d]	1 hour
Hydrocarbons	—	—	160	3 hours[e]
Nitrogen dioxide	100	1 year	—	—
Lead	1.5	3 month[f]	—	—

[a] From the U.S. Environmental Protection Agency (40 CFR 50). These values are continually revised. Changes are published in the U.S. Federal Register.
[b] May be exceeded only once per year.
[c] Geometric mean.
[d] Standard is attained when the expected number of days per calendar year with maximal hourly average concentrations above 0.12 ppm (235 $\mu g/m^3$) is equal to or less than 1, as determined by Appendix H to subchapter C, 40 CFR 50.
[e] The 3-hour period is 6 A.M. to 9 A.M.
[f] The 3-month period is a calendar quarter.

public scrutiny by all affected parties. The authors of the criteria documents rely on peer review of the best available data (46 FR 61502) on the health effects of air pollutants. The ambient standards are derived from four types of data: (1) epidemiological studies; (2) industrial standards for 8-hour exposure by active healthy adults; (3) human laboratory experiments; and (4) animal laboratory experiments. The enforcement of the standards is largely left to the individual states.

The correlation between animal studies and human health is based on procedures developed by the Food and Drug Administration (FDA) (21 CFR 121.5, 1969) for the testing of pharmaceuticals. This agency decided to apply a safety factor of 10 for translating animal data to man, because it is well known that species differ substantially in their responses to individual insults. Another factor of 10 is used to protect sensitive individuals. Thus, a total translational factor of 100 is used. An exception is carbon monoxide, for which the EPA decided on a safety factor of 55.

The EPA also regulates radiation exposure. It has set a limit on the use of uranium at a level that prevents exposure of the population in the vicinity of plants to 25 mrem over the whole body and 75 mrem in the thyroid gland. Nuclear power plants are limited to the emission of 50,000 Ci of radioactive krypton per gigakillowatt a year, with the radioactive iodine isotope 129 not to exceed 85 mCi and plutonium 239 not to exceed 0.5 mCi. The total radiation from all other transuranium elements is also limited.

In December 1978 the FDA banned the use of fluorocarbon propellants in pressurized containers (16 CFR 1401).

10.3 Occupational Air Quality Standards

Control of air pollutants at the workplace is well established. Coal mine ventilation has been regulated for almost 200 years. The ventilation of cotton mills was pioneered in England by Dr. Percival, who was the driving force behind the English Health and Morals of Apprentices Act of 1802.

Occupational air standards in the United States are based on the Williams-Steiger Occupation Safety Act (PL 91-596, 1970; 29 CFR-651). Its intent is "to assure for working men and women in the Nation safe and healthful working conditions and preserve our human resources." The regulation builds on the "Threshold Limit Values (TLV) of Airborne Contaminants" by the American Conference of Government and Industrial Hygienists (ACGIH).

These standards are all based on the several decades of industrial and government experience that went into the formulation of the ACGIH's "Documentation of the Threshold Limit Values for Substances in Working Air." The values are valid for 8-hour exposure periods followed by 16-hour recovery periods; they apply to "nearly all workers." Thus, the ACGIH clearly exempts sensitive individuals. The ACGIH work builds on the earlier experience of the American National Standards Institute (ANSI), formulated in the 2-37 standard series which is periodically updated.

NIOSH has supplemented these standards with detailed criteria documents for over 200 chemicals. The data are regularly updated and published as part Z of 29 CFR 1910.1000. Excerpts of this data are reproduced in Table 10.2. The values are reviewed by NIOSH and enforced by OSHA (37 CFR 202). NIOSH also provides directives for the handling of dangerous or carcinogenic substances. Among those are asbestos (29 CFR 1900.1001), coal tar pitch volatiles (29 CFR 1900.1002), benzidine (29 CFR 1900.1010), vinyl chloride (29 CFR 1900.1017), lead (29 CFR 1900.1025) and others. The air in coal mines and other mines is covered by the Federal Mine Safety and Health Amendment Act of 1977 (PL 95-164; 30 CFR 75.300); this law describes minimum ventilation rates and maximum levels for methane and other toxic or explosive gases (30 CFR 75.301.4).

10.3 Occupational Air Quality Standards

Table 10.2

Select Workplace Safety and Health Standards[a]

Contaminant	Concentration[b] (ppm)	(mg/m³)
Ammonia	50	35
Carbon dioxide	5000	9000
Carbon monoxide	50	55
Cresol	5	22
Formaldehyde	2	3
Furfuryl alcohol	50	200
Nitric oxide	25	30
Nitrogen dioxide	5	9
Octane	500	2350
Ozone	0.1	0.2
Propane	1000	1800
Sulfur dioxide	5	13
Trichloro methane	50	240
Inert or nuisance dust, respirable fraction	—	5
Asbestos	[c]	[c]
Coal dust	—	2.4

[a] From OSHA and NIOSH (29 CFR 1910.1000, Appendix Z).
[b] Values are 8-hour time-weighted averages, except values for nitrogen dioxide, which are ceiling values.
[c] Fewer than two fibers longer than 5 μm in each cubic centimenter.

Submarine Standards. The document "Limits and Measuring Methods for Atmospheric Constituents in Nuclear Submarines," developed by the U.S. Naval Research Lab (NAVSEA 0938-011-4010, 1970, Revision B), covers regulations necessary to protect both the submarine and the crew against corrosion, explosions and health hazards.

Aerospace Standards. When long-term space flight became possible, the U.S. Air Force commissioned the National Academy of Sciences/National Research Council (NAS/NRC) to establish a Committee on Air Quality Standards in Space Flight. The first report was published in 1967. NASA followed up with a three-volume criteria document, "Compendium on Human Responses to the Aerospace Environment" (NASA Report CR-1205), prepared by the Lovelace Foundation's Inhalation Toxicity Research Institute in 1968. This data is reflected in the regulation of ventilation and control of harmful concentrations of gases and vapors in the cockpit of airforce and commercial aircraft (14 CFR 25.83).

Other laws and the implementation of laws are still left largely to the states, which have developed their own programs with the approval of the corresponding federal agency. A detailed study of these state health programs was undertaken by El-Ahraf (1980), who analyzed programs in 50 states. He found four administrative trends:

> Trend I. States in which the main thrust has remained with agencies derived from the state public health services, where responsibility has rested since Massachusetts took the lead with its Sanitary Commission in 1850; this model is represented by California.
>
> Trend II. States in which responsibility is split between EPA-type organizations and public health departments; the latter are usually responsible for food and sanitation. This model is represented by Oregon.
>
> Trend III. States in which almost all programs are the responsibility of a single department; this model is represented by New Mexico.
>
> Trend IV. States in which environmental programs are part of natural resource organizations; this model is represented by Pennsylvania.

The management of these programs is so much in flux that the organization charts for some states were obsolete before the paper appeared in print. An example of state action is the New York State School Asbestos Safety Act of 1979, Bill 4724, of March 6, 1979, in which the state made "plans to identify and eliminate asbestos and set minimum safe levels for air." Almost all states have adopted similar programs.

10.4 Laws Regulating Occupancy

A wide variety of laws influence indoor air quality indirectly. Among them are laws covering occupancy, fire hazards, smoking, or ventilation. Occupancy laws are local laws, based on fire safety. The National Fire Protection Association (NFPA) national fire code has been accepted by almost all local governments. The NFPA regulates the floor area per person necessary for various activities in order to assure safe evacuation in case of fire (Table 7.2). The values correlate to ASHRAE ventilation standards, which are based on the human need for fresh air.

10.5 Smoking Regulations

The effect of smoking on human health has been studied by the U.S. Surgeon General (1979) for the last 20 years (HEW, 1964 and 1979). The public attitude toward smoking, which is defined as "the carrying of a lighted pipe, cigar, or cigarette of any kind, or any other lighted smoking equipment of any plant in a state of combustion" (Feldman, Huang, Lyman, and Sobeck 1978), has undergone periodic changes. In the late 19th century, smoking was prohibited in 14 states. In Washington it was considered a hazard to sawmills, and Massachusetts "intended to guard the city against the damage of fire" (53 Mass. 231, 1847). In the lawsuit Austin vs. Tennessee (101 Tenn. 563, 1898), the state court upheld a statute banning cigarette sales, identifying the "character of cigarettes as being . . . wholly noxious and deleterious to health. Their use is harmful, never beneficial. They possess no virtue, but are inherently bad, and bad only." The U.S. Supreme Court upheld this ban, but by 1914 antismoking ordinances were invalidated in state courts in Kentucky and Illinois, and subsequently in other state courts, as an unreasonable curtailment of personal freedom.

For the next half century smokers could smoke wherever and whenever they wished. The attitude was that any hazards would affect only the smokers themselves or unduly sensitive people. Few people were aware at the time that research on industrial hygiene had long established that smoking increased the incidence of lung cancer and cancer of the tongue among those exposed to dust from rock grinding and anthracitic coal. Even when, in 1964, the U.S. Surgeon General found a compelling correlation between chronic smoking and lung cancer and declared that smoking might be harmful to health, few people considered smoking a threat to the health of those who did not smoke. Thus, it came as a surprise when in 1974 the Surgeon General determined that epidemiological evidence suggested that passive smoking (i.e., inhalation of smoke by nonsmokers) enhances chronic disease, including cancer. This report reflects a general change in public attitude. By 1980 six states limited smoking in public places: California, Connecticut, Maryland, Montana, Nebraska, Oregon and Rhode Island. Currently, the strictest law is that of Minnesota: it regulates smoking not only in restaurants, sports arenas and commercial buildings, but also at the workplace.

In 1973 the Civil Aeronautics Board (CAB) promulgated rules (14 CFR 252, revised January 1, 1980) that required air carriers to provide separate sections for nonsmokers. During the rule-making process the CAB received over 31,000 comments. Originally airlines resisted the rules, and flight attendants were often reluctant to enforce them. The situation improved after April 28, 1976 (CAB Order 76-4-160), when Allegheny Airlines agreed during enforcement proceedings to offer passengers a choice of smoking or no-smoking accommodations, to make preboarding and on-board announcements concerning the location of the respective sections, to disembark at the next scheduled stop any passenger who persisted in smoking in no-smoking areas, to discipline crew members who failed to enforce the rule, and to assist the CAB in monitoring airline compliance. All other airlines immediately adopted the same procedure, but the CAB has not yet dropped enforcement proceedings against six other U.S. air carriers. In October 1981 the rule was modified (46

FR 45934). Passengers must now arrive at the airport before a deadline to be set individually by each airline, to be seated in the no-smoking section.

The Interstate Commerce Commission (ICC) had a harder time regulating smoking on interstate buses. In 1971 the ICC enacted regulations that required all noncharter passenger motor carriers to restrict smoking to 20% of the seats located at the rear of each bus; but it took five years of defense against litigation by the National Association of Motor Bus Owners (NAMBO), a trade organization, before the amended rule, 49 CFR 1061, was completed on October 19, 1976. It was amended again on October 1, 1980.

In the meantime the ICC also restricted smoking in intercity passenger trains (49 CFR 1124.21, 1976). Amtrak had already banned cigar and pipe smoking in 1972, just after Chief Justice of the Supreme Court Warren E. Burger complained to the U.S. Secretary of Transportation about the pollution from cigar smoke that he had to endure on a Metroliner from Washington DC to New York (Feldman et al., 1978). Almost all local governments have now banned all smoking in local public commuter buses and rapid transit systems.

On July 1, 1980, the U.S. Department of Defense (DOD) initiated strict rules "recognizing the right of individuals working or visiting in DOD occupied buildings to an environment reasonably free from contaminants," barred smoking in auditoriums, shuttle buses, elevators, and the like, and provided for no-smoking sections in eating facilities (32 CFR 203). Parallel to federal action, several states and municipalities took local action. Connecticut and Minnesota, for example, both enacted a Clean Indoor Air Act. The Connecticut Public Act 79-410 of January 10, 1979, prohibits smoking in elevators, in health care institutes, in public schools, during government meetings, in retail food stores, and in part of any restaurant with 25 or more seats. A fine of $5 is mandated. Chapter 211 of the Minnesota Clean Air Act of June 2, 1975, specifically includes offices and commercial establishments, as well as public conveyances, auditoriums, arenas, and the workplace. Other federal and local agencies have been more reluctant to establish smoking restrictions at the workplace, except when work-related hazards are involved. Thus, smoking restrictions are mainly voluntary.

It has been claimed that the OSHA 35-ppm ceiling for CO in effect restricts smoking at the workplace because other work activities frequently preempt permissible CO concentrations, but the enforcement of limits on contaminants from simultaneous separate sources is not easy. Likewise, invocation of the Workmen's Compensation Benefits and Unemployment Compensation Benefits requires that those suffering from smoke engage in litigation. The legal aspects of smoking regulations have been described by Feldman and her co-workers (1978).

10.6 Ventilation

The need for adequate ventilation was recognized long before 1771, when Priestley and Lavoisier identified oxygen and determined the nature of human respiration, because accidents on English ships had periodically caused death when the hatch of the crowded, unventilated steerage cabin was nailed shut to prevent water leaks during heavy storms. The 19th-century U.S. Code specifically required "adequate light, air and accommodations for steerage passengers" (46 CFR 153) and provided for regular inspections and reports on the ventilation of vessels by the customs collectors (46 CFR 160). The regulations for ventilation of vessels are now contained in 46 CFR 72.15.

In the 19th century it was increasingly realized that humid, cold houses fostered various diseases, such as tuberculosis. Studies by the Barracks and Hospital Commission by Roscoe in 1857 proposed an air volume of 20 ft^3/min per person. De Chaumont proposed 50 ft^3/min per person to eliminate body odors. Max von Pettenkofer measured CO_2 production and computed a need for removing 0.6 ft^3 hour^{-1} person^{-1} to maintain CO_2 levels below 0.3%. From this he computed a need of 30 ft^3/min for each person. In 1894 the American Society of Heating and Ventilation Engineers was founded and established itself promptly as a leader by promoting a ventilation rate of 30 ft^3/min (Figure 1.5). This value has proven workable since then. In 1931 the New York City Commission on ventilation requested at least 15 ft^3/min per person. In 1936 Professor Yaglou engaged in a series of benchmark studies that led him to propose ventilation values of 15 ft^3/min per person for sedentary workers, 33 ft^3/min for laborers, and 18-38 ft^3/min for schoolchildren. His recommendations were based on minimum needs necessary to eliminte body odors. Yaglou continued his studies until 1957, when he published a study on smoking in which he proposed a need for an air volume of 40 ft^3/min for smokers.

The first widely accepted ventilation standards were formulated in 1946 by the American Standards Association)ASA Standard A53.1, "Light and Ventilation"). In 1965 the American Society of Heating, Refrigeration and Air-Conditioning Engineers (ASHRAE) undertook the monumental task of revising the standard for "National and Mechanical Ventilation" with the cooperation of experts in industry, academia, and government. It took eight years before Standard 62-73 was promulgated in 1973 (Tables 9.2 and 9.3). This standard was immediately accepted by almost all parties and all political entities in the United States. However, as a consequence of the energy crisis of the last 10 years, ASHRAE immediately set out to revise Standard 62-73 to reduce the recommended values to the minimum values necessary to maintain indoor air quality. In some cases the proposed reduction is up to 50% of the airflow. In order to "provide safe, healthful and comfortable indoor environments by using materials and methods that optimize efficiency of energy utilization," ASHRAE appointed a committee of engineers, physicians, chemists, and psychologists to "include a ventilation rate procedure that indirectly controls indoor air quality, and a new procedure to permit direct control of the indoor air quality." This goal was achieved by using EPA National Ambient Air Quality Standards, specially selected standards for 27 other chemicals, or 10% of OSHA standards to define acceptable outdoor air quality. A minimum value of 5 ft^3/min per person was chosen to eliminate CO_2, and special provisions were made for removing tobacco smoke, "because tobacco smoke is one of the most difficult contaminants to control." The document also lists guidelines for limits for the indoor air concentration of many chemicals in order to ensure acceptable air quality. Some of these are listed in Table 10.3. A formal appeal of the formaldehyde limit of 0.1 ppm was unanimously rejected.

According to many experts, the new ASHRAE 62-1981 standard (Tables 9.2 and 9.3) is reasonable (Cain et al., 1981; Ingersoll, Stitt, and Zapalac, 1981), even though it returns to the lowest ventilation level since 1825 (Figure 1.3). However, the standard makes no provision for the safety margin necessary to maintain the minimum ventilation when buildings are remodeled or building maintenance is substandard. In fact, the new standard may necessitate a regular building and monitoring program that could be costly and exceed the capability of current building managers (McNall, 1981).

10.7 Energy Conservation Standards

Abundant domestic fuel supplies in the United States led to an exponential increase in urban energy consumption after World War II. This caused strains on electricity-providing utility companies, resulting in several blackouts and brownouts in New York City and several northeastern metropolitan areas; it also necessitated increased importation of oil and helped precipitate the energy crisis of the 1970s.

The 1973 oil embargo in the Middle East led to a public reassessment of National goals and a sequence of emergency government actions. The Office of Economic Opportunity funded energy fuel supplies and home weatherization. The latter program continued through 1978. In 1976 the Energy Conservation and Production Act, Public Law 94-385, established a national residential weatherization assistance program for low-income people. Parallel with this development, Congress initiated a fundamental reorganization of several departments dealing with energy. This reorganization clearly will continue for several more years.

The first step was the Federal Non-Nuclear Energy Act of 1974, Public Law 93-438, and the Federal Energy Act of 1974, Public Law 93-275, which established the Energy Research and Development Agency, which three years later

Table 10.3
Guidelines for Acceptable Indoor Air Quality[a]

Contaminant	Concentration	Exposure Time
Acetone[b]	—	—
Ammonia[b]	—	—
Asbestos	—	—
Benzene[b]	—	—
Carbon dioxide	4.5 g/m^3	Continuous
Chlordane[b]	5 µg/m^3	Continuous
Chlorine	—	—
Cresol[b]	—	—
Dichloromethane[b]	—	—
Formaldehyde[b]	120 µg/m^3	Continuous
Hydrocarbons, aliphatic[b]	—	—
Hydrocarbons, aromatic[b]	—	—
Mercury	—	—
Ozone[b]	100 µg/m^3	Continuous
Phenol[b]	—	—
Radon	0.01 WL	Annual average
Tetrachloroethylene[b]	—	—
Trichloroethane[b]	—	—
Turpentine[b]	—	—
Vinyl chloride[b]	—	—

[a] Derived from ANSI/ASHRAE, and ACGIH which state: "If the air is thought to contain any contaminant not listed guidance on acceptable exposure...should be obtained by reference to the standards of the Occupational Safety and Health Administration. For application to the general population the concentration of these contaminants should not exceed 1/10 of the limits which are used in industry. ...In some cases, this procedure may result in unreasonable limits. Expert consultation may then be required."

[b] This contaminant has an odor at concentrations sometimes found in indoor air. The tabulated concentrations do not necessarily result in odorless conditions.

(on October 1, 1977) was transformed into the U.S. Department of Energy (PL 95-91), which comprises the Bureau of Mines and the Atomic Energy Commission, including the seven national laboratories. In a quick sequence of energy-related legislation, Congress next passed six laws, Public Laws 95-617 to 95-621, which together formed the National Energy Act. The Energy Tax Act (PL 95-618) and the National Energy Conservation Policy Act (PL 95-619) of November 9, 1978 (92 Stat. 3206 et seq.), which in Part I, Title II, charged the U.S. Department of Energy with implementing a Residential Conservation Service (RCS) Program to encourage energy conservation measures in existing houses. The goal of residential energy conservation was to save 1.4 quads (i.e., 242 million barrels) of oil by the year 2000. The Energy Security Act of June 30, 1980 (PL 96-294) included an amendment in Title V, Subtitle D, providing for a similar program for commercial buildings and multifamily dwellings. The residential conservation program was unusually explicit, and provided for detailed steps of implementation. It required utility companies with sales exceeding 750 million kilowatt-hours and other heating fuel suppliers to provide eligible customers with (1) information about estimated savings on energy costs for selected energy conservation measures; (2) voluntary energy audits; (3) voluntary arrangements for the purchase, installation, and financing of select energy conservation measures; (4) lists of suppliers, contractors, and lenders who agreed to comply with the program standards; and (5) inspections after installation. Many energy advisory services were set up, and advice was available on how to set up such a service. Amazingly, this program did not address the subject of indoor air quality.

A wealth of information has been produced by the Department of Energy in cooperation with many private organizations. The results are being published in books, journal articles and government reports, and in the *Federal Register*. Many of these are available in public libraries or are distributed free of charge. Much information has been exchanged with the International Energy Agency in Vienna, which represents many countries where energy costs have always been many times higher than in the United States. Many of these countries have long experience in energy conservation; many suffered jointly with the United States from the 1973 oil embargo and were therefore eager to engage in joint projects. Among the most active countries currently are Great Britain, Sweden, and Denmark, whose governments have all sponsored energy conservation programs.

The energy audit provisions have been tested in a series of pilot programs. An energy audit is an on-site inspection by a trained auditor whose function is to evaluate and provide useful information on the energy-related aspects of a building. The audit may deal with improved energy efficiency and may include corrective actions to improve efficiency or to make other energy-related improvements in the building. The Tennessee Valley Authority and several national laboratories, together with the Princeton University house doctor (Dutt and Harrje, 1981), Carnegie-Mellon, and other universities, have assisted utilities in developing programs. The Association of State Attorneys and Metropolitan Attorneys have established a task force to prevent consumer fraud by overeager home remodeling firms, and publish a newsletter that is sponsored by the U.S. Department of Energy (Villano, 1980–1981). Amazingly, the audits do not provide for checking the adequacy of the fresh air supply, or the indoor air quality, but audit strategies have been proposed (Traynor, 1981).

10.7 Energy Conservation Standards

The costs of this program are to be borne largely by the utilities, who in turn are entitled to recover these expenses by various means. For example, they may sell the residential fuel savings to the highest industrial bidder. Energy audits are performed at no cost or nominal cost, and U.S. consumers can recover the cost of energy devices by direct tax deductions. Warranties, inspections, and complaint resolution procedures are available to consumers. So far, however, only 1.5 million of the 70 million home owners in the United States have taken advantage of this option.

The residential conservation steps are spelled out in the law. They include (a) caulking and weather stripping to reduce the infiltration of outside air; (b) vent dampers and modified ignition systems to increase furnace efficiency; (c) insulation of ceilings, walls, floors, pipes, ducts, and water heaters; (d) storm and thermal windows and doors; (e) clock thermostats; and (f) more efficient central air-conditioners. Other provisions include active and passive solar devices. The program includes provisions for the U.S. Department of Energy to promulgate standards. The material standards are to be part of the regulations, 10 CFR 456.800 et seq.; they cover cellulose fibers, mineral fibers, mineral blankets and bats, vermiculite, perlite, cellular polystyrene, cellular polyurethane and isocyanate, urea-formaldehyde foam, aluminum foil, caulked sealing materials, storm and thermal windows, furnace dampers, and load control devices. A second series of regulations, 10 CFR 456.900 et seq., covers installation standards. This program was designed to have an immediate and lasting influence.

Careful environmental impact statements were prepared by the Department of Energy's Assistant Secretary for Environment (DOE/EIS-0150-DS). These documents recognize that such a drastic program, involving 70 million houses, would thoroughly modify the indoor environment. The reduced air infiltration combined with reduced heat flow would cause a lower ventilation rate and thus higher concentrations of indoor pollutants. Furthermore, it was recognized that some insulation materials might contribute to indoor pollution. Thus, the Department sponsored several research projects at national laboratories in Berkeley, Oak Ridge, and elsewhere to study indoor air quality. This program has now operated for five years and has produced outstanding results, published in the form of over 50 research reports. The pollutants singled out for their high impact were radon, carbon monoxide, formaldehyde, nitric oxides, and airborne bacteria. The environmental impact reports also review the health danger to production workers in industries that supply the necessary materials. These chemicals include asbestos, phenols, and benzene. On the other hand, the energy savings would reduce the need for oil and coal combustion and would reduce outdoor nitric oxide pollution around power plants by some 80,000 tons, sulfur dioxide by 18,000 tons, carbon monoxide by 6000 tons, and particulates by 4000 tons for every 10^{12} Btu saved.

In the second part of the program, commercial buildings with monthly energy needs of 4000 kWh electricity, 1000 therms of natural gas, or 100 million Btu of oil were included. In 1981 the emphasis of both the residential and commercial conservation programs shifted from mandated to voluntary actions, and the federal funding was reduced by eighty percent for 1982. If the Department of Energy is reorganized, the conservation program will likely be returned to HUD.

10.8 Building Codes

The air quality in buildings is strongly affected by location, the design of the structure, building materials, and the choice of structural parameters (Chapter 4). Therefore, indoor air quality is influenced by building codes and standards for building materials. Such guidelines and restrictions are promulgated in three forms: as governmentally mandated regulations, as codes, and as privately developed standards. The last are usually developed by a procedure involving the consensus of organizations and specialists who are interested in and knowledgeable about the field.

The role of the federal government in this field has been limited because buildings have traditionally been designed within the framework of local communities and with the goal of fitting local climate and custom. However, when the federal government started to sponsor housing projects for the poor, it instituted minimum standards for federally funded housing. The Minimum Property Standards (MPS) are incorporated into 24 CFR 200. They apply to all housing aided by federal financing, such as Farmers Home Administration Guarantee and Veteran's Administration loans. They include provisions for ventilation and lighting. HUD also published a *Manual of Acceptable Practice*. In regard to ventilation, it has followed the ASHRAE consensus standards in this field. It is currently reviewing these standards to eliminate possible conflicts between recent energy conservation standards and the earlier standards, which were concerned only with air quality.

The U.S. Department of Housing and Urban Development has been concerned about indoor air quality for many years. In 1977 it published, jointly with the EPA, a review of the status of indoor air pollution research (Shearer and Bromberg, 1977), and issued a guide for rapid assessment of air quality at housing sites. At that time the main concern was still to keep outdoor pollution out of homes.

10.8.1 MOBILE HOME AND MANUFACTURED HOUSING REGULATIONS

The situation changed when prefabricated housing and manufactured housing became available, and when such shelters were moved and sold over

10.8 Building Codes

state and local borders. In the early 1970s one of four new homes was a mobile home. These homes are sold within a radius of many hundred miles. Thus, mobile home manufacturers found it increasingly difficult to offer a product that met the requirements of sometimes conflicting local building codes. The consumer, on the other hand, found it difficult to handle questions and complaints about builders who were not under local jurisdiction. In 1974 Congress promulgated the National Mobile Home Construction and Safety Standards Act, Public Law 93-383. The correlating regulations are published in 42 CFR 5401 et seq. The goal of the law was to clarify overlapping jurisdictions and to "reduce the number of personal injuries and death and the amount of insurance costs and property damage resulting from mobile home accidents and to improve the quality and durability of mobile homes." In reaction to this legislation, HUD established an Office of Mobile Home Standards within the Office of Neighborhoods, and Consumer Protection. Since August 1981 it is called Office of Manufactured Housing Standards. Section 626(a) of the law requires HUD to prepare a biennial report to the President that is available to the public and includes much valuable information. These reports contain seven sections dealing with statistical data, a list of current construction and safety standards, a summary of compliance, a summary of all research activities, policy evaluations, the status of current enforcement actions,and a listing of technical information that has been made available. HUD also regulates mortgage rates for manufactured housing.

The federal standards cover the entire design and manufacturing process, including planning, construction, durability, and safety considerations. They apply to all mobile homes fabricated after June 1976. These federal standards explicitly preempt all state and local requirements. However, according to HUD's 1980 report to Congress, six years after the law was enacted some 30 states were still enforcing local requirements that conflicted with federal standards. The states and local agencies participate in the enforcement, and all activities are reviewed by an advisory council that includes representatives of the public and of consumer and manufacturing groups. Mobile home manufacturers are inspected by a Design Approval Primary Inspection Agency and a Production Inspection Agency. The National Conference of States on Building Codes and Standards is charged with monitoring the plant reviews of the two agencies. Finished mobile homes carry certification labels and consumer manuals that reflect the status of compliance.

So far, HUD has published 37 sets of standards for 42 CFR 280 et seq. It also has issued some 50 Interpretative Bulletins, which cover everything from smoke detectors to carpet applications. Noteworthy are bulletins on condensation control, air infiltration, heat loss certificates.

There is no question that the federal mobile home program greatly improved the quality of mobile homes, and thus has helped both the industry and the consumer, and is responsible for the rapidly increasing popularity of this type of housing. In 1979 HUD started an evaluation of the problem of formaldehyde release from soft-wood plywood and particleboard and of the vapor condensation problem caused by human activities in tightly sealed mobile homes. HUD participates in the interagency regulatory liaison group task force on formaldehyde on the basis of its contractor report (Moschandreas and Rector, 1981). It has recently published an advance notice of proposed rulemaking to collect advice whether regulatory actions might be advisable (46 FR 43466).

HUD has been under pressure to act slowly in enforcing the recent Department of Energy's energy performance standards. Industry representatives have argued that HUD's mobile home enforcement program is far more efficient than the Department of Energy's implementation of energy standards via local building codes. This forces mobile home manufacturers to act earlier than builders of traditional homes and thus unfairly places mobile home manufacturers at a disadvantage. Other problems have arisen because the Department of Energy's Energy Conservation Act requires implementation of the "best available technology," whereas the Mobile Home Act, administered by HUD, prescribes only "reasonable standards."

Since HUD is charged with encouraging adequate housing for all people, the mobile home office is currently promoting mobile homes as a source of affordable housing for low- and middle-income families. In most states mobile homes are still taxed as personal property rather than as real estate, and until recently mobile homes depreciated rapidly.

The division between HUD and the state and local agencies in enforcing mobile home regulations has left some gaps and uncertainties that prompted the Federal Trade Commission to propose a labeling regulation that after five years of development is currently nearing completion. One problem that the FTC noticed was that manufacturers and movers still work as separate groups. Thus, it may be difficult for a consumer to determine which of the two should be responsible in cases of warranty problems. The FTC estimates that as many as 70–80% of new mobile homes need warranty service, and that about half of the complainants have had trouble obtaining adequate service. Among the problems are leaking plumbing and building envelopes, both of which can cause indoor air problems.

10.8.2 FEDERAL BUILDING ENERGY PERFORMANCE STANDARDS (BEPS)

Title III of the Federal Energy Conservation and Production Act of 1978, the Energy Conservation Standards for New Buildings Act, 42 USC 6801 and 42 CFR 6831-6840, provides for building energy efficiency standards to "minimize the design energy budget" and "maximize the practical improvement in energy efficiency." The goal of this "performance standard" was to set "an energy consumption goal or goals to be met without specification of the method, materials and processes to be employed in achieving that goal or goals, but including statements of the requirements, criteria and evaluation methods to be used and any necessary comment" (PL 94-385, para 303.(9)).

In 1977 HUD and the American Institute of Architects (AIA) signed a memorandum of understanding to evaluate building design. The project was later transferred from HUD to the Department of Energy. The AIA has accumulated a substantial computer inventory and derived the energy needs per square foot and year for each building use type in the climate zone in each of the official federal regions. Unfortunately, these climate zones do not take into account sunlight, humidity, or wind. Opponents also claim that the BEPS use an energy evaluation scheme that discourages the use of electricity. Since electricity can be produced with coal, this action disfavors the use of the most abundant energy resource. The energy performance principle constitutes a theoretically and philosophically ideal solution, but the translation of such standards into a finished building requires a team of energy specialists and architects, and requires access to substantial computer capacity. The problem with BEPS is that such standards are vague and do not provide consumers or builders with specific information about the methods or material to select.

While the BEPS have not yet been implemented, the ASHRAE consensus standard 90-75 provides a link between BEPS and the HUD minimum standards. It incorporates performance requirements for the exterior envelope and for the heating, ventilation, and heat distribution systems. It is not easy for a small contractor to interpret this standard. Undoubtedly, much work is still necessary before the concept of performance standard is widely accepted and implemented.

10.8.3 CODES DEVELOPED BY PRIVATE INITIATIVE

Building construction is regulated by literally hundreds of local jurisdictions. It has been estimated that almost 90% of all new buildings are constructed under the authority of some kind of code. So far only 20 states have statewide building codes. About half of the nation has adopted one of the three leading model building codes developed by private initiative.

The Southern Building Code Congress International (SBCCI) has been adopted by over 2500 governments in 25 states. It is composed of southern state and local government representatives who formulate a set of standard codes that cover buildings, housing, fire prevention, and other matters.

The International Conference of Building Officials (ICBO) is legislated by governmental representatives of Western states and those in the North Central region. Its Uniform Building Code is published every three years.

The Building Officials and Code Administrators International (BOCA) covers most of the northeastern United States. Its basic code is complemented by many reports and periodicals.

All three codes provide detailed prescriptions for materials and other standards. Thus, these codes are most useful for the small contractor and for governmental officials who have to approve and inspect buildings.

Many efforts are currently under way to coordinate government and industry activities on all levels. In 1970 the National Bureau of Standards took the lead in organizing a conference on research and innovation in the building regulating process (NBS Special Publication 552, 1979). Other initiatives came from the American Insurance Association (originator of the well-established National Building Code), ASHRAE, and the AIA. On a national level, the National Conference of State Building Codes and Standards, called 'NixBix," and the Council of American Building Officials (CABO) are other progressive and important forces.

The problem with consensus standards is that their development takes many years, during which building technology and people's life-styles change so strongly that code provisions may be obsolete before the standards are implemented.

10.8.4 BUILDING TEMPERATURE REGULATION

The purpose of heating buildings is to improve the comfort of users and to prevent condensation of moisture, which might damage the structure were it to remain unchecked. As explained in Chapters 1, 3, and 8, the definition of comfortable indoor temperature changes with time and place. During the sixties, many East Coast office buildings were maintained at 75°F (24°C) 24 hours a day, 365 days a year, so that workers would feel comfortable in short-sleeved shirts or blouses. Maintaining such a temperature during a cold winter requires vast energy supplies. After the oil shortages during the cold winter of 1976, the DOE drafted the Emergency Building Temperature Regulation (10 CFR 1490), which gave the President authority to introduce a set of measures whenever an energy shortage was anticipated. The provisions limited building thermostat settings to 68°F in winter (to keep the temperature above 65°F) and to 77°F in summer (to keep the temperature around 78°F). One provision is that buildings that remain unoccupied for more than eight hours may not be heated if the outdoor temperature is 50°F. Another requires that water heaters be set at 105°F, or the lowest possible setting. Predictably, a large segment of industry and the population were surprised by, and complained about, the sudden and drastic threat to a life-style that had evolved over the last 50 years. As it happens, many modern appliances, such as commercial dishwashers, cannot operate safely at 105°F (40.5°C), but the revised regulation was made final on May 7, 1979 (44 FR 39354), and on July 17, 1979, President Carter invoked his emergency power (44 FR 41205) and kept the regulation in effect by three declarations (4667, 4750, 4813), each valid for nine months. Most public agencies and many private citizens and organizations followed the instructions. Many people were surprised at how easily they could function at the reduced winter temperature, but few realized how big the energy savings really were, because fuel costs increased so rapidly that fuel bills continued to increase. Furthermore, some building managers abused the regulations, for example, by mounting thermostats on the inside wall of poorly insulated buildings, which left the occupants with excessively cold living or working areas. The building temperature regulations were canceled on February 18, 1981, by proclamation 4820 of President Reagan (46 FR 12941), who branded them an "excessive regulatory burden" and found that "voluntary restraint and market incentives will achieve the same."

Parallel with federal action, several private organizations also reconsidered the definition of indoor comfort, and ASHRAE issue a revised standard for thermal comfort (ASHRAE 55-1981) which allows building operators to extend the range of indoor air temperatures to 68–79°F, depending upon season and air flow, section 3.1.

10.9 Material and Installation Standards

10.9.1 BUILDING MATERIALS

As explained in Section 4.2, indoor air quality depends as much on building materials as on the building design. There are currently no standards, voluntary or otherwise for release of odors or chemicals from building materials in the United States. The unanticipated sixfold reduction in ventilation rates has revealed problems with several materials, especially radon from masonry work (Section 7.3) and formaldehyde from urea-formaldehyde foam insulation, particleboard, and plywood. Currently, an intensive effort is under way to develop standards such as already exist in Sweden and Japan, and in the United States for textiles. As a stop-gap measure, several U.S. manufacturers have introduced an internal, voluntary grading of selected forest products (Morschauser, 1979; Groah, 1981; NPA–HPMA, 1981).

In the past it was usually taken for granted that the inventor and manufacturer of products were best able to determine how a material should be made and used, and material standards were developed by the American National Standards Institute (ANSI), the American Society of Mechanical Engineers, the American Society of Testing and Materials (ASTM), the American Society of Heating, Refrigerating, and Air-Conditioning Engineers (ASHRAE), and the National Fire Protection Association. All used a consensus approach in developing standards (i.e., they consulted individuals, industries, associations, and governments that have long-range experience and know-how in the field).

The Energy Emergency Conservation Act, as implied by its title, created an unexpected and urgent demand on building performance that strained normal procedures for the development of standards. Therefore, the National Energy

Conservation Policy Act (PL 950619, subpart H) explicitly charged the U.S. Department of Energy to develop materials and installation standards for all materials necessary for the Residential and Commercial Conservation Program. Fortunately, ASTM, ASHRAE, ANSI, and the other organizations already had detailed standards available for many materials. Plywood and particleboard standards had been developed by manufacturers' associations. The Department of Energy developed others, as necessary, in cooperation with these bodies. However, energy conservation required that materials be used under conditions that had never been contemplated. For example, nobody had used large quantities of urea-formaldehyde resin-bonded particleboard and plywood paneling in small, poorly ventilated, and tightly sealed rooms, that is, under conditions that the inventors and manufacturers of the materials thought were long part of the past. The problem of quality assurance to prevent potential slow formaldehyde release from these products is described in Section 4.2. Other problems were encountered with dust, moisture absorption, fungi resistance, and the settling of cellulose and fiberglass insulation. Cellulose material had already led to many complaints, which were resolved by the U.S. Consumer Product Safety Commission standards 16 CFR 1209 and 16 CFR 1404. The use of polyurethane foam and polystyrene poses serious fire prevention problems, but the most serious immediate problems are associated with the use of urea-formaldehyde insulation foam, a material which is manufactured on site, and which, despite excellent physical, chemical, and fire resistance properties, requires considerable skill on the part of its installers. If the installation fails, not only does the house remain poorly insulated, but the penetrating and lasting odor of formaldehyde may permeate the structure and indoor air spaces. In some cases, indoor air levels of formaldehyde have exceeded occupational safety levels. At such levels formaldehyde can cause acute respiratory and eye discomfort. Furthermore, at very high levels it has caused cancer in mice and rats (Section 8.9). Canada and the Commonwealth of Massachusetts banned this product in November 1979 (CoM, 1980). When an ASTM committee, after seven years, could not agree on a proposed voluntary material standard, the U.S. Consumer Products Safety Commission published a thorough analysis of the situation and in March 1982 also passed a ban (46 FR 11188 and 46 FR 56762), whereas the Department of Energy remained hopeful that the material could eventually be regulated and safely used (45 FR 63786 and 46 FR 8996).

The use of asbestos-containing products has had to be severely restricted (16 CFR 1145) and friable asbestos to be banned (16 CFR 1305) because it produces airborne respirable fibers that can cause cancer of the lung. The regulatory trend is shown in Table 10.4.

Voluntary Standards. While the ASHRAE standard 62-1981 on ventilation rates for acceptable indoor air quality provides guidelines for threshold concentrations for many chemicals, these air quality norms must still be translated

10.9 Material and Installation Standards

Table 10.4

Trend of Exposure Limits for Asbestos

Limit	Time-Weighted (8 hr/day) Average (fibers/ml)	Ceiling (fibers/ml)
OSHA		
Original, 1972	5.0	10.0
Present, 1976	2.0	10.0
Proposed, 1975	0.5	5.0
NIOSH revised, 1977	0.1	0.5
ACGIH adopted, 1980		
Amosite	0.5	—
Chrysotile	2.0	—
Crocidolite	0.2	—
Other forms	2.0	—

into appropriate material standards. The problem with standards for many chemicals is that their release mechanism is not yet fully understood. The urea-formaldehyde foam insulation industry has striven for seven years to develop a material standard for the ASTM (Rossiter, 1980), but so far without agreeing on a proposal. The particleboard and plywood industries are currently working with HUD to devise a voluntary formaldehyde standard that would achieve an indoor level of less than 0.4 ppm. Such a level would prevent the few extreme cases that caused the majority of the publicized incidents of the last 10 years, but it would still leave an estimated 30% of mobile home residents exposed to an integrated 24 hr doses equivalent to the 1.2 ppm limit by NIOSH which is based on an annual dose of 40 work weeks times 40 hours. This would translate to a level of about 0.3 ppm for occupants who use their home the typical 5000 hr/year. On the basis of the current technical capability of UF resin (Sundin, 1980; Meyer, 1979), a level of 0.03 ppm should now be possible.

The radon problem differs from the formaldehyde situation in three ways. First, radon is clearly a human carcinogen (Section 7.2); second, radon can be readily measured with current technology; and third, anyone with reasonable skills can monitor radon in building materials. Thus, the radon problem is simply one of enforcement. No steps have been taken to translate these uncontested facts into a standard.

Many other chemicals, such as carbon monoxide and nitric oxide from natural gas cannot be regulated by establishing material standards, since this pollutant results from combustion conditions which cannot be factory controlled. This problem must be solved by ventilation standards for ranges.

10.9.2 PESTICIDES, TOXIC SUBSTANCES, AND SOLID WASTES

The EPA has programs in the areas of pesticides, toxic substances, and solid wastes to prevent the marketing or use of materials that could endanger the general population. Sections 3.4 and 7.3 list several dozen of the many thousand chemicals that can enter the home. The testing of a chemical follows a strict protocol, developed by independent panels of experts and reviewed by all affected parties (Calabrese, 1978).

In 1976 the Toxic Substances Control Act (TSCA) was enacted. It charged the EPA with coordinating the action of all federal agencies that screen potentially toxic material. Originally, eight agencies were involved. Their interaction is fluid, but their functions have developed as follows.

EPA. The U.S. Environmental Protection Agency was established on December 2, 1970, by combining 15 components from five departments— Interior, Agriculture, Health, Education and Welfare (since May 4, 1980, Health and Human Services), the National Air Pollution Control Administration, the Bureau of Solid Waste Management, and the Food and Drug Administration. The regulatory activities of the EPA follow four steps: (1) schedule of a development plan, formation of work group, and notification; (2) development plan for classification, purpose, participation, regulatory analysis, publication schedule, and notification of interested parties; (3) preparation of decision package for involvement of all concerned, impact analysis, rule preamble, recommended action; and (4) review by steering committee, a so-called red border review, and final review by administration. The EPA handles pesticides separately from other toxic substances. Airborne contaminants are reviewed by the Office of Air Quality Planning and Standards or the Office of Mobile Air Source Pollution. Both belong to the Office of Air, Noise and Radiation. EPA is also in charge of a $1.6 billion clean-up program, set up by Congress in December 1980, for some 400 toxic waste sites where chemicals dumped during the last forty years have started to leak into soil and air, and, in such places as Love Canal, New York, penetrated homes.

HHS. The U.S. Department of Health and Human Services was established on May 4, 1980, from the former Department of Health, Education and Welfare. The Public Health Service, one of its four components, is responsible for conducting research, correlating illness and disease, and ensuring the safety of food and drugs. It includes the Food and Drug Administration, founded in 1930 as an extension of the Food, Drug and Cosmetics Act of 1906; the Center for Disease Control, which is charged with the epidemiology and case studies of diseases, and the National Institutes of Health. The National Center for Toxicological Research is part of the FDA. The National Institutes of Health include the National Cancer Institute, the National Institute of Allergy and Infectious Disease, the National Institute of Environmental Health Sciences, and the National Toxicology Program (NTP, 1982), as well as eight other institutes.

10.9 Material and Installation Standards

The National Institute for Occupational Safety and Health, NIOSH, is part of the Center of Disease Control; it consists of 11 programs and divisions located in Rockville, MD, Cincinnati, OH, and Morgantown, WV.

The U.S. Department of Labor protects worker rights and supervises workplace safety via the Occupational Safety and Health Administration (OSHA), which was founded in 1970 to supervise its title act. The latter consists of several agencies. It has taken special interest in controlling metals and carcinogens. Since 1975 the U.S. Department of Transportation has been regulating the safe transportation of hazardous materials. The U.S. Department of Agriculture, formed in 1862, is responsible for food safety, but has relinquished control of pesticides to the EPA.

The Council of Environmental Quality (CEQ) was created in 1969 to advise the President. The current administration is reconsidering its function.

CPSC. The Consumer Product Safety Commission was established on May 14, 1973, to help industry develop voluntary safety standards, to issue and enforce mandatory safety standards, or to ban unsafe products in order to reduce unreasonable risk of injury associated with consumer products. This small commission has been very effective in identifying hazards and working with consumers and industry toward engineering and scientific solutions. In 1978 industry and the commission agreed to a voluntary exchange of certain asbestos-containing products. In 1979, jointly with the EPA and FDA, the CPSC was active in the ban of chlorofluorocarbon sprays. It is currently considering rules for controlling benzene, formaldehyde, benzidine-based dyes, deodorizers, hydrocarbon propellants, hexane, lead, and other pigments, and has investigated carbon monoxide production in unvented fossil fuel heaters and appliances. It also has prepared a list of strong sensitizers (16 CFR 1500.12) which lists formaldehyde (concentration above 1%), oil of bergamot (2%+) powdered orris root (used in some European fragrances), p-phenylene diamine, and ethylene diamine-epoxy.

The Department of Energy (DOE) was formed in October 1977 by merger of several agencies with the Energy Research and Development Agency (ERDA) which itself had been formed in 1974 by merger of the Atomic Energy Commission (AEC), the Bureau of Mines and several agencies. DOE is currently sharing responsibility for almost anything remotely related to energy research, development, application and regulation. Since the first impact of public energy awareness has subsided, this department will surely relinquish many of its functions relating to buildings and air quality to other agencies. In fact, this department will likely be thoroughly restructured.

The work of all the above agencies is coordinated by the toxic substances advisory committee, a clearinghouse on environmental carcinogens, a committee to coordinate environmental programs, a genetic toxicology program and steering group, an interagency collaborative group on environmental carcinogenesis, and the Interagency Regulatory Liaison Group (IRLG). The IRLG maintains several levels of committees, including 28 chemical-specific work groups that form a regulatory development work group. The specialty groups deal with acrylonitrile, arsenic, asbestos, benzene, benzidine dyes, beryllium, cadmium, chlorofluorocarbons, chlorinated solvents, chromates, coke oven emissions, dibromochloropropane (DBCP), diethylstilbestrol (DES), dioxins, ethylene dibromide, ethylene oxide, formaldehyde, lead, nitrosamines, mercury, ozone, polychlorinated biphenyls (PCBs), radiation, sulfur dioxide, vinyl chloride, and waste utilization. Billick (1981) has provided a case study of the functioning of the IRLG using the case of lead.

There are several toxic substance subcommittees collecting data and coordinating research for pesticides and every aspect of every active concern. Details are reviewed in a regularly updated directory (J. Collé, 1980).

10.9.3 TEXTILES AND CLOTHING

Formaldehyde release from textiles and clothing is measured by the American Association of the Textile and Cotton Council test method AATCC-112-1978. The Japanese Case 112 of October 1973 revised 1979 limits formaldehyde release from two categories of products: (1) diapers, sleepwear, gloves, bedding and other articles in contact with infants less than 24 months must yield a color absorbance of less than 0.05 with the JIS-test, using acetylacetone, and (2) household products, such as underwear, sleepwear, socks, gloves and hosiery, as well as adhesives used on skin to mount wigs, artificial eyelashes or beards must release less than 75 micrograms per gram of the product in the test.

Appendices

APPENDIX I Units and Conversion Factors

This book encompasses fields that employ different units and notations. All scientific data are in metric units; engineering data are in either metric or U.S. units, depending on the custom among practitioners in a given subspecialty.

As regards numeration, the American system is used for large numbers.

Numbers
    1 billion (United States) = 10^9 = 1 milliard (Great Britain, France, Germany)
    1 trillion (United States) = 10^{12} = 1 billion (Great Britain, France, Germany)

Concentrations
    1 ppm = 1 part per million = 10^{-6}
    1 ppb = 1 part per billion = 10^{-9}
    1 ppt = 1 part per trillion = 10^{-12}

Length
    1 ft = 0.3048 m

Volume
    1 cf = 1 ft^3 = 28.32 liters

Weight
    1 gal (United States) = 3.785 liters
    1 lb = 0.4536 kg

Pressure
    1 psi = 6893 pascal = 0.068 atm

Ventilation rate
    1 cfm = 1 ft^3/min = 0.472 liter/sec
    1 ach = 1 air change per room per hour

Heat
    1 Btu = 1055 J = 252 cal
    1 quad = 172.9 barrels of oil

Moisture permeability
    1 perm = 1 grain/hr ft^2 inch(H$_2$O) = 5.75 10^{-7} g/Pa m^2 sec = 275 g/m^2 hr atm

Metabolic heat dissipation
   1 met = 58.2 W m^{-2} = 50 kcal hr^{-1} m^{-2} = 18.4 Btu hr^{-1} ft^{-2}

Clothing insulation
   1 clo = 0.155 m^2 °C W^{-1}

Temperature conversion

Celsius	Fahrenheit	Celsius	Fahrenheit	Celsius	Fahrenheit	Celsius	Fahrenheit
−6.7	20	7.2	45	21.1	70	35.0	95
−6.1	21	7.8	46	21.7	71	35.6	96
−5.5	22	8.3	47	22.2	72	36.1	97
−5.0	23	8.9	48	22.8	73	36.7	98
−4.4	24	9.4	49	23.3	74	37.2	99
−3.9	25	10.0	50	23.9	75	37.8	100
−3.3	26	10.6	51	24.4	76	38.3	101
−2.8	27	11.1	52	25.0	77	38.9	102
−2.2	28	11.7	53	25.6	78	39.4	103
−1.7	29	12.2	54	26.1	79	40.0	104
−1.1	30	12.8	55	26.7	80	40.6	105
−0.6	31	13.3	56	27.2	81	41.1	106
.0	32	13.9	57	27.8	82	41.7	107
0.6	33	14.4	58	28.3	83	42.2	108
1.1	34	15.0	59	28.9	84	42.8	109
1.7	35	15.6	60	29.4	85	43.3	110
2.2	36	16.1	61	30.0	86	43.9	111
2.8	37	16.7	62	30.6	87	44.4	112
3.3	38	17.2	63	31.1	88	45.0	113
3.9	39	17.8	64	31.7	89	45.6	114
4.4	40	18.3	65	32.2	90	46.1	115
5.0	41	18.9	66	32.8	91	46.7	116
5.5	42	19.4	67	33.3	92	47.2	117
6.1	43	20.0	68	33.9	93	47.8	118
6.7	44	20.6	69	34.4	94	48.3	119

APPENDIX II Acronyms and Abbreviations

AATCC	American Association of the Textile and Cotton Council
ACGIH	American Conference of Governmental Industrial Hygienists
ACS	American Chemical Society
AECB	Atomic Energy Control Board of Canada
AGA	American Gas Association
AIA	American Institute of Architects
AIHA	American Industrial Hygiene Association
AMCA	Air Monitoring and Conditioning Association
ANSI	American National Standards Institute
APCA	Air Pollution Control Association
APHA	American Public Health Association
ASHRAE	American Society of Heating, Refrigerating and Air-Conditioning Engineers; formerly ASVHE
ASTM	American Society for Testing and Materials
ASVHE	American Society of Ventilation and Heating Engineers
BaP	Benzo-alpha-pyrene, a component of combustion fumes
BEPS	Building Energy Performance Standards (currently administered by the DOE)
BOCA	Building Officials and Code Administrators International
CAB	Civil Aeronautics Board (a division of the U.S. Department of Transportation)
CABO	Council of American Building Officials
CAMP	Continuous Air Monitoring Program (operated by the EPA)
CDC	Center for Disease Control (a division of NIOSH)
CFR	U.S. Code of Federal Regulations (updated annually)
CHESS	Community Health Surveillance System (operated by the EPA)
CIIT	Chemical Industry Institute of Toxicology (Research Triangle, NC)
CPSC	U.S. Consumer Product Safety Commission
DOE	U.S. Department of Energy
ECETOC	European Chemical Industry Ecology and Toxicology Center, Brussels
EPA	U.S. Environmental Protection Agency

EPRI	Electric Power Research Institute (Palo Alto, CA)
ESCA	Electron spectroscopy for chemical analysis
ET*	Effective temperature scale
FAA	Federal Aviation Agency (a division of the U.S. Department of Transportation)
FDA	U.S. Food and Drug Administration
FESYP	European Federation of Particleboard Manufacturers
FR	U.S. Federal Register
FSP	Fine suspended particulates
FTC	U.S. Federal Trade Commission
GC–MS	Gas chromatography followed by mass spectroscopy
HEW	U.S. Department of Health, Education and Welfare; now HHS
HHS	U.S. Department of Health and Human Services
HPLC	High-pressure liquid chromatography
HPMA	Hardwood Plywood Manufacturers Association (McLean, VA)
HUD	U.S. Department of Housing and Urban Development
ICBO	International Conference of Building Officials
IRLG	Interagency Regulatory Liaison Group (coordinates regulatory agencies of the U.S. government)
LBL	Lawrence Berkeley National Laboratory (operated by the University of California for the DOE)
NAPCA	National Air Pollution Control Administration, now EPA
NAS	National Academy of Sciences (United States)
NASA	National Aeronautics and Space Administration
NASN	National Air Surveillance Network (operated by the EPA)
NCRP	National Council on Radiation Protection and Measurements
NFPA	National Fire Protection Association (a member of ANSI)
NIOSH	U.S. National Institute for Occupational Safety and Health (part of HHS)
NPA	National Particleboard Association
NRC	National Research Council (research branch of NAS)
NRCC	National Research Council of Canada (Ottawa)
NSF	National Science Foundation
NTIS	U.S. National Technical Information Service (Springfield, VA; distributes research reports of all federal agencies)
ORNL	Oak Ridge National Laboratory (operated by Union Carbide for the DOE.)
OSHA	Occupational Safety and Health Administration (a division of the U.S. Department of Labor)

PAH	Polynuclear aromatic hydrocarbons
PCBs	Polychlorinated biphenyls
PCPs	Polychlorinated phenols
PIXE	Proton-induced x-ray emission
RSP	Respirable particles
SBCCI	Southern Building Code Congress International
SMACCNA	Sheet Metal and Air Conditioning Contractors' National Association
SRI	Stanford Research Institute (Menlo Park, CA)
TCDD	2,3,7,8-Tetrachlorodibenzo-p-dioxin (an impurity in some herbicides)
TSCA	Toxic Substances Control Act (administered by the EPA)
UNSCAR	United Nations Scientific Committee on the Effect of Atomic Radiation
USAF	U.S. (Department of the) Air Force
USN	U.S. (Department of the) Navy
WBGT	Wet-bulb globe temperature
WHO	World Health Organization (part of UNESCO, Geneva)
WMO	World Metereological Organization (part of UNESCO, Geneva)

References and Bibliography

References and Bibliography

Introduction

This section comprises the works cited in the text and a bibliography of significant material relating to air quality. The period covered is from 1960 to March 1982, although some exceptionally important material published before 1960 is also included.

The works cited in the text and the bibliography constitute a single list that is arranged alphabetically according to the first author's surname. Arrangement of works by a specific author, whether working alone or with colleagues, is chronological. Moreover, references by the same author that were published in the same year are not otherwise distinguished: the reader seeking the entry intended by the text citation "Sterling, 1981" will find in the list works by Sterling, Weinkam, and Sterling and by Sterling and Kobayashi that appeared in 1981. In such instances, the specific work meant may be determined from the title of the work.

Government documents are listed by agency name or by the name of the contractor who conducted the work for the agency. Text citations use the agency acronym (see Appendix II), whereas entries in the list use the full formal name. Thus, text citations of work published by the EPA appear in the list under U.S. Environmental Protection Agency, that by the CAB is under U.S. Civil Aeronautics Board, and so on.

Government regulations or laws do not appear in the list because the text citations are sufficient to enable readers to locate these items in any law library or in the government or business section of most public libraries throughout the United States.

U.S. government reports can be ordered from the National Technical Information Service (NTIS), Department of Commerce, Springfield, VA 22161.

However, each agency has its own rules for publishing, distributing, and referencing internal reports, contract reports, and proceedings of conferences that it co-sponsors. Since these rules are currently in flux, uniform referencing is not possible, and in some cases the most reliable way to locate a reference is through its author.

Abeles, F. J. 1970. Assessing the life support systems of the Ben Franklin submarine. Paper presented at Am. Soc. Mech. Eng. Design Eng. Conf., Chicago, May 11-14.

Abelles, F. B., Craker, L. E., Forrence, L. E., and Leather, G. R. 1971. Fate of air pollutants: Removal of ethylene, sulfur dioxide and nitrogen dioxide by soil, *Science* **173**, 914.

Air Infiltration Centre. Quarterly. *Air Infiltration Review,* Berkshire, England.

Air Monitoring and Conditioning Association. 1977. *AMCA Fan Application Manual.* Arlington Heights, IL.

Air Pollution Control Association. 1975. Continuous Monitoring of Stationary Pollution Sources. (Proc. APCA Conf., St. Louis, MO, March 20-21.) Pittsburgh, PA.

Albrechtsen, O. 1979. The effect of electric fields on mental work. In *Indoor Climate, Effects on Human Comfort, Performance, and Health in Residential, Commercial, and Light-Industry Buildings* (P. O. Fanger and O. Valbjørn, eds.), p. 345. Copenhagen: Danish Building Research Institute.

Albrechtsen, O., Østerballe, O., and Weeke, B. 1979. Influence of small atmospheric ions on the air-ways in patients with bronchial asthma. In *Indoor Climate, Effects on Human Comfort, Performance, and Health in Residential, Commercial, and Light-Industry Buildings* (P. O. Fanger and O. Valbjørn, eds.), p. 377. Copenhagen: Danish Building Research Institute.

Allan, G., Dutkiewicz, J., and Gilmartin, E. J. 1980. Long-term stability of urea-formaldehyde foam insulation, *Env. Sci. Technol.* **14**, 1235.

Allen, R. J., Wadden, R. A., and Ross, E. D. 1978. Characterization of potential indoor sources of ozone, *Am. Ind. Hyg. Assoc. J.* **39**, 466.

Allen, R. J., and Wadden, R. A. 1982. Analysis of indoor concentrations of carbon monoxide and ozone in an urban hospital, *Env. Res.* **27**, 136.

Amdur, A. O. 1959. The physiological response of guinea pigs to atmospheric pollutants, *J. Air Pollution* **1**, 170.

American Association of the Textile and Cotton Council. 1978. Formaldehyde Odor in Resin Treated Fabric, Determination of: Sealed Jar Method. AATCC Test Method 112-1978.

American Chemical Society. 1974. *Reprints 167th National Meeting,* Vol. 14, No. 1. (Symposium on Environmental Quality Monitoring and Symposium on Sources and Evolution of the Atmospheric Aerosol, Los Angeles, CA, April 1-5.) Pittsburgh, PA.

American Conference of Governmental Industrial Hygienists–American Industrial Hygiene Association, Aerosol Hazards Evaluation Committee. 1975. Background documentation on evaluation of occupation exposure to airborne asbestos, *Am. Ind. Hyg. Assoc. J.* **35**, 91.

American Conference of Governmental Industrial Hygienists–American Industrial Hygiene Association, Aerosol Hazards Evaluation Committee. 1975. Recommended procedures for sampling and counting asbestos fibers: Procedures for the evaluation of occupational exposures to airborne asbestos, *Am. Ind. Hyg. Assoc. J.* **35**, 83.

American Conference of Governmental Industrial Hygienists. 1976. *Documentation of the Threshold Limit Values for Substances in Workroom Air,* 3d ed. Cincinnati, OH: ACGIH.

American Conference of Governmental Industrial Hygienists. 1978. *Air Sampling Instruments for Evaluation of Atmospheric Contaminants,* 5th ed. Cincinnati, OH: ACGIH.

American Conference of Governmental Industrial Hygienists. 1978. *Industrial Ventilation: A Manual of Recommended Practice,* 15th ed. Lansing, MI: ACGIH.

American Conference of Governmental Industrial Hygienists. 1979. *Air Sampling Instruments for Evaluation of Atmospheric Contaminants,* 5th ed. Cincinnati, OH: ACGIH.

American Conference of Governmental Industrial Hygienists. 1980. *Threshold Limit Values for Chemical Substances and Physical Agents in the Workroom Environment with Intended Changes for 1980.* Cincinnati, OH: ACGIH.

American Gas Association–British Gas Corporation–Gaz de France. 1970. *Oxides of Nitrogen: A Critical Survey.*

American Industrial Hygiene Association. 1972. *Air Pollution Manual,* 2nd ed. Westmont, NJ: AIHA.

American Institute of Architects. 1974. *Energy Conservation in Building Design.* Washington, DC: AIA.

American Public Health Association. 1977. *Methods of Air Sampling and Analysis* (M. Katz, ed.), 2nd ed. Washington, DC: APHA.

American Society for Testing and Materials. 1974. Standard Test Method for Concentration of Formaldehyde Solutions. ANSI/ASTM D 2194-65 (reapproved 1974).

American Society for Testing and Materials. 1976. *Calibration in Air Monitoring.* (Symposium, Univ. of Colorado at Boulder, August 5-7, 1975.) ASTM Special Tech. Publ. 598. Philadelphia, PA.

American Society for Testing and Materials. 1978. Test Method for Measurement of Odor. Standard ASTM D 1391. Philadelphia, PA.

American Society for Testing and Materials. 1978. Significance of Tests and Properties of Concrete and Concrete-Making Materials. ASTM Special Tech. Publ. 169B. Philadelphia, PA.

American Society of Heating, Refrigerating and Air-Conditioning Engineers. 1963. ASHRAE Psychrometric Chart No. 1: Normal Temperature, Barometric Pressure 29.921 Inches of Mercury. New York.

American Society of Heating, Refrigerating and Air-Conditioning Engineers. 1963. ASHRAE Psychrometric Chart No. 2: Low Temperature, Barometric Pressure 29.921 Inches of Mercury. New York.

American Society of Heating, Refrigerating and Air-Conditioning Engineers. 1963. ASHRAE Psychrometric Chart No. 3: High Temperature, Barometric Pressure 29.921 Inches of Mercury. New York.

American Society of Heating, Refrigerating and Air-Conditioning Engineers. 1963. ASHRAE Psychrometric Chart No. 4: High Altitude Normal Temperature, 5,000 ft. (24.89 in Hg), 32 F to 120 F Dry Bulb. New York.

American Society of Heating, Refrigerating and Air-Conditioning Engineers. 1963. ASHRAE Psychrometric Chart No. 5: High Altitude Normal Temperature, 7,500 ft. (24.89 in Hg), 32 F to 120 F Dry Bulb. New York.

American Society of Heating, Refrigerating and Air-Conditioning Engineers. 1974. *ASHRAE Handbook and Product Directory,* Vol. 3: *Applications.* New York.

American Society of Heating, Refrigerating and Air-Conditioning Engineers. 1975. *ASHRAE Handbook and Product Directory,* Vol. 1: *Equipment.* New York.

American Society of Heating, Refrigerating and Air-Conditioning Engineers. 1976. *ASHRAE Handbook and Product Directory,* Vol. 2: *Systems.* New York.

American Society of Heating, Refrigerating and Air-Conditioning Engineers. 1977. *ASHRAE Handbook and Product Directory,* Vol. 4: *Fundamentals.* New York.

American Society of Heating, Refrigerating and Air-Conditioning Engineers. 1980. Energy Conservation in New Building Design. ASHRAE Standard ANSI/ASHRAE/IES 90A-1980; ASHRAE/IES 90B-1975; ASHRAE 90C-1977. New York.

American Society of Heating, Refrigerating and Air-Conditioning Engineers. 1981. Thermal Environmental Conditions for Human Occupancy. ASHRAE Standard ANSI-ASHRAE 55-1981. Atlanta, GA.

American Society of Heating, Refrigerating and Air-Conditioning Engineers. 1981. Standard for Ventilation Required for Minimum Acceptable Indoor Air Quality. ASHRAE Standard ANSI-ASHRAE 62-1981. Atlanta, GA.

Amoore, J. E. 1970. *Molecular Basis of Odor.* Springfield, IL: Charles C Thomas.

Anders, J. M. 1887. *House-Plants as Sanitary Agents.* Philadelphia, PA: J. B. Lippincott.

Andersen, I. 1972. Relationships between outdoor and indoor air pollution, *Atmos. Env.* **6,** 275.

Andersen, I., Lundqvist, G. R., and Proctor, D. F. 1972. Human nasal mucosal function under four controlled humidities, *Am. Rev. Respir. Dis.* **106,** 438.

Andersen, I., Lundqvist, G. R., and Jensen, P. L. 1974. Human response to 78-hour exposure to dry air, *Arch. Env. Health* **29,** 319.

Andersen, I., Lundqvist, G. R., and Molhave, L. 1974. Formaldehyde in the atmosphere in Danish homes, *Ugeskr. Laeg.* **136**(38), 2133.

Andersen, I., Lundqvist, G. R., and Molhave, L. 1975. Indoor air pollution due to chipboard used as a construction material, *Atmos. Env.* **9,** 1121.

Andersen, I., Lundqvist, G. R., and Molhave, L. 1976. Effect of air humidity and sulfur dioxide on formaldehyde emission from construction materials, *Holzforsch. Holzverwertung* **28**(5), 120.

Andersen, I. 1977. The ambient air. In *Respiratory Defense Mechanisms* (J. Brain, D. F. Proctor, and L. M. Reid, eds.), Part 1, pp. 24–62. New York: Marcel Dekker.

Andersen, I. 1979. Formaldehyde in the indoor environment—health implications and the setting of standards. In *Indoor Climate, Effects on Human Comfort, Performance, and Health in Residential, Commercial, and Light-Industry Buildings* (P. O. Fanger and O. Valbjørn, eds.), p. 65. Copenhagen: Danish Building Research Institute.

Anderson, H. A., Lilis, R., Daum, S. M., and Selikoff, I. J. 1979. Asbestosis among household contacts of asbestos factory workers, *Ann. N.Y. Acad. Sci.* **300,** 387.

Anderson, H. A. 1980. Personal communication.

Anderson, H. A., Hanrahan, L. P., Dally, K. A., and Rankin, J. 1981. Irritant symptomatology, clinical observations and formaldehyde exposure among Wisconsin mobile home residents. Preprint, *Proc. Int. Symp. Indoor Air Pollution, Health and Energy Conserv.,* Univ. of Massachusetts, October 13–16: Amherst, MA.

Andrewes, F. W. 1902. Examination of the atmosphere of the central London railway. In London County Council Report to the Parliamentary Committee, No. 615.

Andrews, B. A. K., and Harper, R. J., Jr. 1980. Formaldehyde release—Are current test methods realistic?, *Textile Res. J.* **50,** 177.

Andrews, B. A. K., Harper, R. J., Jr., Smith, R. D., and Reed, J. W. 1980. Lowering formaldehyde release with polyols, *Textile Chemist and Colorist* **12**, 287.

Andrews, B. A. K., Harper, R. J., Jr.,and Vail, S. L. 1980. Variables that influence formaldehyde release from cottons finished for durable press, *Textile Res. J.* **50**, 315.

Andrews, B. A. K. 1982. Formaldehyde Release from Textiles. Report for CPSC.

Ange, C. R., and McIntire, M. S. 1964. Lead teratogenicity, *Am. J. Dis. Child* **108**, 436.

Angus, T. 1968. *The Control of the Indoor Climate.* New York: Pergamon Press.

Annis, J. C., and Annis, P. J. 1973. Vacuum cleaning—A source of residential airborne dust, *ASHRAE Symposium Bull.* **72-1**.

Anthon, D. W., Fanning, L. Z., Hollowell, C. D., and Lin, C. 1979. Air Sampling Using Pneumatic Flow Controllers. LBL Report 9403; EEB-Vent 79-13.

Appel, B. R., Kothny, E. L., Hoffer, E. M., and Wesolowski, J. J. 1977. Comparison of Wet Chemical and Instrumental Methods for Measuring Airborne Sulfate. Final report. EPA-600/7-77-128.

Archer, V. E. 1980. Effects of low levels of radon on man, *Proc. Meeting Assessment of Radon Exposure to Man.* Rome, Italy.

Arcos, J. C., Argus, M. F., and Wolf, G. 1968. *Chemical Induction of Cancer; Structural Bases and Biological Mechanisms,* Vol. 1. New York: Academic Press.

Arnest, R. T. 1961. Atmosphere Control in Closed Space Environment (Submarine). Bureau of Medicine and Surgery Report 367. Washington, DC: U.S. Govt. Printing Office.

Aronow, W. 1978. Effects of passive smoking on angina pectoris, *New England J. Med.* **299**, 21.

Arthur, R. D., Cain, J. D., and Barrentine, B. F. 1976. Atmospheric levels of pesticides, *Bull. Env. Contamination Toxicol.* **15**, 129.

Atherton, M. J. 1981. Estimation of personal exposure to carbon monoxide from fixed and portable monitoring. Preprint, *Int. Symposium Indoor Air Pollution, Health and Energy Conserv.,* Univ. of Massachusetts, October 13–16: Amherst, MA.

Atomic Energy Control Board of Canada. 1976. Report on the Preliminary Investigation of the Technical and Economic Factors for the First Stage Remedial Measures at Port Hope, Ontario. Canadian Govt. Report AECB-1212.

Atomic Energy Control Board of Canada. 1978. Report on Investigative and Remedial Measures, Radiation Reduction and Radioactive Decontamination, in Uranium City, Saskatchewan (Keith Consulting, Regina, Prince Albert, Lethbridge, Edmonton, March 1978). Canadian Govt. Report AECB-1198.

Atomic Energy Control Board of Canada. 1980. Report on Investigation and Implementation of Remedial Measures for the Radiation Reduction and Radioactive Decontamination of Elliot Lake, Ontario (DSMA Atcon Ltd., February 1980). Canadian Govt. Report AECB-1211.

Attia, A. 1979. Room ventilation in hospitals. In *Indoor Climate, Effects on Human Comfort, Performance, and Health in Residential, Commercial and Light-Industry Buildings* (P.O. Fanger and O. Valbjørn, eds.), p. 237. Copenhagen: Danish Building Institute.

Atwater, W. O., and Benedict, F. G. 1905. *A Respiration Calorimeter with Appliances for the Direct Determination of Oxygen.* Washington, DC: Carnegie Institution.

Ayers, S. M., Evans, R., Licht, D., Griesback, V., Reimold, F., Ferrand, E. F., and Criscitiello, A. 1973. Health effects of exposure to high concentrations of automotive emissions: Studies in bridge and tunnel workers in New York City, *Arch. Env. Health* **27**(3), 167.

Bach, W., and Daniels, A. 1975. *Handbook of Air Quality in the United States.* Honolulu, HI: Oriental Publishing.

Balocco, L. 1979. Suction of dust from pneumatic and electric manual equipment for preventing silicosis in indoor personnel, *Inquinamento* **21**(6), 81.

Banaszak, E. F., et al. 1970. Hypersensitivity pneumonitis due to contamination of an air conditioner, *New England J. Med.* **283**, 271.

Barad, C. B. 1979. Smoking on the job: The controversy heats up, *Occupational Health and Safety,* (1), p. 21.

Bardana, E. J., Jr. 1980. Formaldehyde; Hypersensitivity and irritant reactions at work and in the home, *Immun. Allergy Prac.* **2**(3), 60.

Barrett, R. E., Miller, S. E., and Locklin, D. W. 1973. *Field Investigation of Emissions from Combustion Equipment for Space Heating.* EPA-R2-73-084a; API Publ. 4180. Columbus, OH: Battelle Columbus Laboratories.

Barrie, L. 1981. Aerosols at Mould Bay, *Atmos. Env.* **15**(9).

Bartels, T. T., and Crump, N. L. 1977. Toxic gas contaminants in enclosed environments, *Contamination Control* **9**, 20.

Barthel, W. F., et al. 1969. Determination of PCP in blood, urine, tissue and clothing, *J. Assoc. Anal. Chem.* **52**, 294.

Bartlett, D., Jr. 1968. Pathophysiology of exposure to low concentrations of carbon monoxide, *Arch. Env. Health* **16**, 719.

Bates, M., and Christie, P. 1971. *Respiratory Function in Disease.* Philadelphia, PA: W. B. Saunders.

Baumann, H., and Schmidt, G. 1958. Überprüfung der Reizlosigkeit eines Puders und einer neuen Wundauflage aus Kunstharzschaumstoff ('OBC-Puder'), *Fortschr. Med.* **76**(2), 59.

Beall, J. R., and Ulsamer, A. G. 1981. Toxicity of volatile organic compounds present indoors, *Bull. N.Y. Acad. Med.* **57**, 978.

Beasley, R. K., Hoffmann, C. E., Rueppel, M. L., and Worley, J. W. 1980. Sampling of formaldehyde in air with coated solid sorbent and determination by high performance liquid chromatography, *Anal. Chem.* **52**, 1110.

Beck, G. J., and Bouhuys, A. 1980. Community studies of obstructive lung disease, *Toxicol. Res. Projects Directory* **5**(5).

Bedford, T. 1937. *Modern Principles of Ventilation and Heating.* London: H. K. Lewis.

Belanger, B. C. (ed.). 1978. *Calibration and Related Measurement Services of the National Bureau of Standards.* NBS Special Publ. 250. Washington, DC.

Bellin, P., and Spengler, J. D. 1980. Indoor and outdoor carbon monoxide measurements at an airport, *J. Air Pollution Control Assoc.* **30**, 392.

Benarie, M., Chuong, B. T., and Nonat, A. 1977. Indoor and outdoor pollution by suspended particles, *Sci. Total Env.* **7**(3), 283.

Benes, M., and Zahomek, J. 1981. Nitric oxides level from natural gas and city gas fueled kitchen stoves. *Pryn.* **61**(7), 198.

Benson, F. B., Henderson, J. J., and Caldwell, D. E. 1972. *Indoor–Outdoor Air Pollution Relationships: A Literature Review.* EPA Publ. AP-112.

Benzinger, T. H. 1979. The physiological basis for thermal comfort. In *Indoor Climate Effects on Human Comfort, Performance, and Health in Residential, Commercial, and Light-Industry Buildings* (P. O. Fanger and O. Valbjørn, eds.), p. 441. Copenhagen: Danish Building Research Institute.

Berge, A., Mellegaard, B., Hanetho, P., and Ormstad, E. B. 1980. Formaldehyde release from particleboard—Evaluation of a Mathematical Model, *Holz Roh-Werkstoff* **38**, 251.

Berger, J. M., and Lamm, S. H. 1981. A Medical and Scientific Assessment of Regulatory Positions on Formaldehyde, Health and Cancer. Report. Silver Spring, MD: Professional Consultants.

Berglund, B., and Lindvall, T. 1979. Olfactory evaluation of indoor air quality. In *Indoor Climate, Effects on Human Comfort, Performance, and Health in Residential, Commercial, and Light-Industry Buildings* (P.O. Fanger and O. Valbjørn, eds.), Copenhagen: Danish Building Research Institute.

Berglund, B., Johansson, I., and Lindvall, T. 1981. A longitudinal study of air contaminants in a newly built preschool. Preprint, *Int. Symposium Indoor Air Pollution, Health and Energy Conserv.,* Univ. of Massachusetts, October 13–16: Amherst, MA.

Berglund, B., Johansson, I., and Lindvall, T. 1981. The influence of ventilation on indoor/outdoor air contaminants in an office building. Preprint, *Int. Symposium Indoor Air Pollution, Health and Energy Conserv.,* Univ. of Massachusetts, October 13–16: Amherst, MA.

Berg-Munch, B., and Fanger, P. O. 1981. The influence of air temperature on the perception of body odor. Preprint, *Int. Symposium Indoor Air Pollution, Health and Energy Conserv.,* Univ. of Massachusetts, October 13–16: Amherst, MA.

Berlad, A. L., Tutu, N., Jaung, R., and Yeh, Y-J. 1979. Energy Transport in Porous Insulator Systems. ORNL/SUB-7551/1.

Biersteker, K., deGaff, H., and Nass, C. A. G. 1965. Indoor air pollution in Rotterdam homes, *Int. J. Air and Water Pollution* **9**, 343.

Billick, H. I. 1981. Lead, a case study in interagency policy-making. *Env. Health Perspectives* **42**(12), 730.

Billings, J. S. 1893. Ventilation and heating, *Engineering Record,* pp. 120–130.

Billings, J. S., Mitchell, S. W., and Bergey, D. H. 1895. The composition of expired air and its effect upon animal life, *Smithsonian Contribution to Life* **29**, 1.

Binder, R. E., Mitchell, C. A., Hosein, H. R., and Bouhuys, A. 1976. Importance of the indoor environment in air pollution exposure, *Arch. Env. Health* **31**(6), 277.

Birky, M. M., and Clarke, F. B. 1981. Inhalation of toxic products from fires, *Bull. N.Y. Acad. Med.* **57**, 997.

Birmingham, D. 1900. Death rate among ganisters, *J. Sanit. Inst. England.*

Blanchard, R. L. 1969. Rn-222 decay concentrations in uranium mine atmospheres, *Nature* **223**, 287.

Blomsterberg, A. K., and Harrje, D. T. 1979. Approaches to evaluation of air infiltration energy losses in buildings, *ASHRAE Trans.* **85**(I).

Bocanegra, T. S., and Espinoza, L. R. 1980. Raynaud's phenomenon in passive smokers, *New England J. Med.* **303**, 1419.

Bock, F. G. 1982. Non-smoker and cigarette smoking: A modified perception of risk, *Science* **215**, 197.

Boegel, M. L., Nazaroff, W. W., and Ingersoll, J. G. 1979. Instructions for operating LBL passive environmental radon monitor (PERM), LBL Report 10246; EEB-Vent 79-14.

Boggess, W. R., and Wixson, B. G. (eds.) 1977. *Lead in the Environment* (a report and analysis of research at Colorado State Univ., Univ. of Illinois at Urbana–Champaign, and Univ. of Missouri at Rolla). NSF/RA-770214. Washington, DC: National Science Foundation.

Bond, R. G., and Straub, C. D. (eds.) 1972. *Handbook of Environmental Control,* Vol. I: *Air Pollution.* Cleveland, OH: CRC Press.

Booz, Allen, and Hamilton, 1979. Alternative Metering Practices: Implications for Conservation in Multifamily Residences. HCPIM 1693-03. Washington, DC: U.S. Department of Energy.

Boren, H. G. 1970. Pulmonary cell kinetics after exposure to cigarette smoke. In *Inhalation Carcinogenesis* (M. G. Hanna, Jr., P. Nettesheim, and J. R. Gilbert, eds.), p. 229. Washington, DC: National Cancer Institute/AEC Technical Information Division.

Bouhuys, A., Mitchell, C. A., Hosein, H. R., Schoenberg, J. B., Binder, R. E., and Schilling, R. S. 1975. Epidemiological studies of lung disease in urban and rural communities, *Proc. Int. Symposium Recent Adv. Assessment of Health Effects of Env. Pollution* **2,** 669.

Bowen, R. P., Shirtliffe, C. J., and Chown, G. A. 1981. Urea-Formaldehyde Insulation: Problem Identification and Remedial Measures for Wood-Frame Construction. Building Practice Note 23, ISSN 0701-5216, National Research Council of Canada.

Boyer, R. S., Klock, L. E., Schmidt, C. D., Hyland, L., Maxwell, K., Gardner, R. M., and Renzetti, A. D., Jr. 1974. Hypersensitivity lung disease in the turkey raising industry, *Am. Rev. Respir. Dis.* **109,** 630.

Boyle, R. 1662. *A Defense of the Doctrine Touching the Spring and Vegetables of the Air.* London: Robin.

Boyle, R. 1675. *Experiments and Observations about the Mechanical Production of Odors.* London: S. Flesher.

Boyle, R. 1692. *General History of the Air.* London: Churchill.

Brain, J. D., Proctor, D. F., and Reid, L. M. (eds.) 1977. *Respiratory Defense Mechanisms,* Part 1 (Vol. 5 in the series *Lung Biology in Health and Disease*). New York: Marcel Dekker.

Brandom, W. F., Saccomanno, G., Archer, V. E., Archer, P. G., and Bloom, A. D. 1978. Chromosome aberrations as a biological dose–response indicator of radiation exposure in uranium miners, *Rad. Res.* **76,** 159.

Brenchley, D. L., Turley, C. D., and Yarmac, R. F. 1973. *Industrial Source Sampling.* Ann Arbor, MI: Ann Arbor Science.

Breslow, N. E., and Whittemore, A. S. (eds.) 1979. *Energy and Health.* (Proc. SIAM-SIMS Conf., Alta, UT, June 26-30, 1978; SIAM-SIMS Conf. Ser., Vol. 6.) Philadelphia, PA: Society for Industrial and Applied Mathematics.

Breysse, P. A. 1977. Formaldehyde in mobile and conventional homes, *Env. Health Safety News* **26,** 1.

Bridbord, K., Brubaker, P. E., Bay, B., Jr., and French, J. G. 1975. Exposure to halogenated hydrocarbons in the indoor environment, *Env. Health Perspectives* **11,** 215.

Bridbord, K. 1977. *Human Exposure to Lead from Motor Vehicle Emissions.* HEW (NIOSH) Publ. 77-145. Cincinnati, OH.

Briggs, T. M., Overstreet, M., Kolhari, A., and Devitt, T. W. 1974. *Air Pollution Considerations in Residential Planning,* Vols. I and II. EPA-450/3-74-046-a.

Briscoe, W. A., and King, T. K. C. 1974. The Bohr integral, *Am. J. Med.* **57,** 349.

Brookman, E. T., and Birenzvige, A. 1980. Domestic Combustion Sources. EPA-600/7-80-084.

Broome, C. V., and Fraser, D. W. 1979. Epidemiologic aspects of Legionellosis, *Epidemiologic Revs.* **1,** 1.

Brown, S. S., Mitchell, F. L., and Young, D. S. (eds.) 1979. *Chemical Diagnosis of Disease.* Amsterdam: Elsevier/North-Holland.

Brun, M. J. 1974. Influence of Air Pollution in Indoor Air Quality. Centre Interprofessional Technique & Etudes de la Pollution Atmospherique, Paris E.I.T.E.P.A. Etudes Documentaires 43, CI 812:22.

Brundrett, G. W. 1979. Window ventilation and human behaviour. In *Indoor Climate, Effects on Human Comfort, Performance, and Health in Residential, Commercial, and Light-Industry Buildings* (P. O. Fanger and O. Valbjørn, eds.), p. 317. Copenhagen: Danish Building Research Institute.

Brunnemann, K. D., Adams, J. D., Ho, H. P. S., and Hoffmann, D. 1978. The influence of tobacco smoke on indoor atmospheres, II: Volatile and tobacco specific nitrosamines in main- and sidestream smoke and their contribution to indoor pollution, *Proc. 4th Int. Conf. Sens. Env. Pollution,* pp. 876–878. New Orleans, Nov. 6–11, 1977.

Brunnemann, K. D., and Hoffmann, D. 1978. Chemical Studies on Tobacco Smoke, LIX: Analysis of Volatile Nitrosamines in Tobacco Smoke and Polluted Indoor Environments. IARC Sci. Publ. 19; *ISS Env. Aspects N-Nitroso Compounds,* pp. 343–356.

Brunner, A., Russenberger, H. J., and Wanner, H. U. 1977. Keim und Partikelausstreuung in Abhangigkeit der Tatigkeit und Bekleidung, *Chem. Rundschau* **30**(37), 7.

Brunner, K. 1980. Studies on the manufacture of particleboard with low formaldehyde emission. In *Technical Workshop on Formaldehyde* (B. Meyer and J. Fandey, eds.), Vol. 2, p. 125. Washington, DC: Baker, Hames and Burkes.

Bryan, W. R., and Shimkin, M. B. 1940. Probit evaluation, *J. Natl. Cancer Inst.* **1,** 807.

Buchbinder, L., Solowey, M., and Solotorovsky, M. 1945. Comparative quantitative studies of bacteria in air of enclosed places (Part I of air pollution survey report), *J. Am. Soc. Heat. Ventil. Eng.* **7,** 389.

Buchland, F. E., and Tyrrell, D. A. 1964–1965. Experiments on the spread of colds, *J. Hyg. Cond.* **62,** 365; **63,** 327.

Budde, W. L., and Eichelberger, J. W. 1979. *Organics Analysis Using Gas Chromatography Mass Spectroscopy: A Techniques and Procedures Manual.* Ann Arbor, MI: Ann Arbor Science.

Bulloch, C. 1979. Survivability in aircraft fires: New standards are needed, *Interavia* **34,** 557.

Burch, D. M., and Hunt, C. M. 1978. *Retrofitting an Existing Wood-Frame Residence for Energy Conservation: An Experimental Study.* NBS Building Sci. Ser. 105. Washington, DC.

Burch, G. E. 1974. Toxic agents, cardiovascular disease and the polluted home, *Am. Heart J.* **87,** 679.

Burge, H. P., Boise, J. R., Solomon, W. R., and Bandera, E. 1978. Fungi in libraries: An aerometric survey, *Mycopathologia* **64**(2), 67.

Burris, B. L. (ed.) 1978. *Publications of the National Bureau of Standards, 1977 Catalog: A Compilation of Abstracts and Key Word and Author Indexes.* NBS Special Publ. 305, Suppl. 9. Washington, DC: National Bureau of Standards.

Burton, W. E., and Miller, R. L. 1963. The role of aerobiology in dentistry, *Proc. 1st Int. Symposium Aerobiol.* Oakland, CA: U.S. Naval Supply Center.

Butcher, S. S., and Charlson, R. J. 1972. *An Introduction to Air Chemistry.* New York: Academic Press.

Cadle, R. D. 1975. *The Measurement of Airborne Particles.* New York: Wiley.

Cahill, T. A., et al. 1977. Analysis of respirable fractions in atmospheric particulates via sequential filtration, *J. Air Pollution Control Assoc.* **27,** 675.

Cain, W. S. 1980. Chemoreceptive modalities of odor and irritation, *Nature* **284,** 255.

Cain, W. S., Isseroff, B., Leaderer, P., Lipsitt, E. D., Huey, H. J., Perlman, D., Bergland, L. G., and Dunn, J. D. 1981. Ventilation Requirements for Control of Occupancy Odor and Tobacco Smoke Odor: Laboratory studies final report. LBL Report 12589; UC-41.

Cairns, T., and Siegmund, E. G. 1981. PCBs: Regulatory history and analytical problems, *Anal. Chem.* **53,** 1183A.

Calabrese, E. J. 1978. *Methodological Approaches to Deriving Environmental and Occupation Health Standards.* New York: Wiley-Interscience.

Campbell, I. R., and Mergard, E. G. 1972. *Biological Aspects of Lead: An Annotated Bibliography. Literature from 1950 through 1964,* Parts I and II. EPA Publ. AP-104.

Candas, V., Libert, J. P., Vogt, J. J., Ehrhart, J., and Muzet, A. 1979. Body temperature during sleep under different thermal conditions. In *Indoor Climate, Effects on Human Comfort, Performance, and Health in Residential, Commercial, and Light-Industry Buildings* (P. O. Fanger and O. Valbjørn, eds.), p. 763. Copenhagen: Danish Building Research Institute.

Caplan, K. T., Doemeny, L. J., and Sorenson, S. D. 1977. Performance characteristics of the 10 mm cyclone respirable mass sampler, Part I: Monodisperse studies, *Am. Ind. Hyg. Assoc. J.* **38,** 83.

Caplan, K. T., Doemeny, L. J., and Sorenson, S. D. 1977. Performance characteristics of the 10 mm cyclone respirable mass sampler, Part II: Coal dust studies, *Am. Ind. Hyg. Assoc. J.* **38,** 162.

Carbone, R. D. 1978. Formaldehyde Exposure in Mobile Homes. Master's thesis. Univ. of Washington, Seattle.

Carhart, H. W., and Johnson, J. E. 1980. A history of the Naval Research Laboratory contributions to submarine life support systems. Paper presented at Am. Soc. Mech. Eng. Intersoc. Conf. Env. Systems, San Diego, CA, July 14–17.

Carlin, A. P. 1977. Hazardous Wastes: A Risk–Benefit Framework Applied to Cadmium and Asbestos. EPA-600/5-77-002.

Carrington, T. S. 1914. *Fresh Air and How to Use it,* 2nd ed. New York: National Association for the Study and Prevention of Tuberculosis.

Cartwright, J., and Nagelschmidt, G. 1961. The size and shape of dust from human lungs and its relation to selective sampling. In *Inhaled Particles and Vapors*. Oxford: Pergamon Press.

Center for Disease Control. 1981. Heat stroke during heatwave of 1980, *Morbidity and Morality Weekly Reports* **29**(54), 109 and **30**(6), 20.

Chaddock, J. B. 1976. Energy consumption in buildings in the USA and possibilities for energy conservation. In *Energy Conservation in the Built Environment* (Courtney, R. G. ed.), p. 167. Lancaster: The Construction Press, CIB.

Chamberlain, A. C. 1970. Deposition and uptake by cattle of airborne particles, *Nature* **225**, 99.

Chambre Syndicale de l'Amiante. 1964. Rapports et discussions, Congrès International sur l'Asbestore. Paris: Chambre Syndicale de l'Amiante.

Chapin, F. S., Jr. 1974. *Human Activity Patterns in the City*. New York: Wiley.

Chemical Industry Institute of Toxicology. 1980. Symposium on Health Effects of Formaldehyde, November 3-7, Research Triangle Park, NC: CIIT.

Chemical Industry Institute of Toxicology. 1980. *Formaldehyde Toxicity*. (Proc. 3d CIIT Conf. Toxicol., Raleigh, NC, November 20-21.

Cheremisinoff, P. N., and Morresi, A. C. (eds.) 1978. *Air Pollution Sampling and Analysis Deskbook*. Ann Arbor, MI: Ann Arbor Science.

Christensen, H. E., and Luginbyhl, T. T. (eds.) 1974. *The Toxic Substances List*. HEW (NIOSH) Publ. 74-134. Rockville, MD.

Christensen, H. E., and Luginbyhl, T. T. (eds.) 1975. *Suspected Carcinogens: A Subfile of the NIOSH Toxic Substances List*. HEW (NIOSH) Publ. 75-188. Rockville, MD.

Christensen, H. E., and Fairchild, E. J. 1976. *Registry of Toxic Effects of Chemical Substances*. HEW (NIOSH) Publ. 76-191.

Christensen, H. E., and Fairchild, E. J. (eds.) 1976. *Suspected Carcinogens: A Subfile of the NIOSH Registry of Toxic Effects of Chemical Substances,* 2nd ed. HEW (NIOSH) Publ. 77-149. Cincinnati, OH.

Coblentz, C. W., and Achenback, P. R. 1963. Field measurements of air infiltration in ten electrically heated houses, *ASHRAE J.* **5**(7), 69.

Coburn, R. F. (ed.) 1970. Biological effects of carbon monoxide. *Ann. N. Y. Acad. Sci.* **174**, Article 1.

Cohn, M. S. 1981. Revised carcinogenic risk assessment of urea-formaldehyde foam insulation: Estimation of cancer risk due to inhalation of formaldehyde released by UFFI. Report for CPSC.

Collé, J., and Slike, K. A. (eds.) 1980. *Directory of Federal Coordinating Groups for Toxic Substances,* 2nd ed. EPA-560/13-80-008.

Collé, J., Schosman, E., and Slike, K. (eds.) 1980. *Federal Activities in Toxic Substances*. EPA-560/13-80-015.

Collé, R. 1980. The physics and interaction properties of radon and its progeny. In *Radon in Buildings* (R. Colle and P. E. McNall, Jr., eds.). NBS Special Publ. 581. Washington, DC.

Collé, R., and McNall, P. E., Jr. (eds.) 1980. *Radon in Buildings*. NBS Special Publ. 581. Washington, DC.

Colley, J. R. T., Holland, W. W., and Corkhill, R. T. 1974. Influence of passive smoking and parental phlegm on pneumonia and bronchitis in early childhood, *Lancet* **2**(7888), 1031.

Colligan, M. J. 1981. The psychological effects of indoor air pollution, *Bull. N.Y. Acad. Med.* **57**, 1014.

Collins, K. R. 1979. Hypothermia and thermal responsiveness in the elderly. In *Indoor Climate, Effects on Human Comfort, Performance, and Health in Residential, Commercial, and Light-Industry Buildings* (P. O. Fanger and O. Valbjørn, eds.), p. 819. Copenhagen: Danish Building Research Institute.

Collins, R. D., and Decries, D. M. 1973. Air concentrations and residues from use of Shell's No-Pest Insecticide Strip, *Bull. Env. Contamination Toxicol.* **9**, 227.

Colome, S. D., Spengler, J. D., and McCarthy, S. 1981. Indoor-outdoor air quality comparisons in ten residential environments, Preprint, *Int. Symposium Indoor Air Pollution, Health and Energy Conserv.*, Univ. of Massachusetts, October 13–16; Amherst, MA. *Env. International* **8**, in press.

Compton, B., Bazydlo, D. P., and Zweig, G. 1972. *Field Evaluation of Methods of Collection and Analysis of Airborne Pesticides,* Vol. 1: *Field Evaluation and Analysis.* Syracuse, NY: University Research Corp.

Comstock, G. W., Meyer, M. B., Helsing, K. J., and Tockman, M. S. 1972. Respiratory effects of household exposures to tobacco smoke and gas cooking.

Consumer Reports. 1972. Household Insecticides, *Consumer Reports.* August, 522.

Conway, R. A. 1982. *Environmental Risk Analysis for Chemicals.* New York: Van Nostrand.

Cooper, H. B. H., Jr., and Rossano, A. T., Jr. 1971. *Source Testing for Air Pollution Control.* Wilton, CT: Environmental Science Services Division.

Cornet, G. 1889. Die Verbreitung der Tuberkelbacillen ausserhalb des Körpers, *Z. Hyg. Infekstionskrankh.* **5**, 191.

Cortese, A. D., and Spengler, J. D. 1975. Determination of environmental carbon monoxide exposures through personal monitoring—Fixed vs. personal and transportation mode relationships. Paper presented at 68th Ann. Meeting, Air Pollution Control Assoc., Boston, June 15–20.

Cote, W. A., Wade, W. A., III, and Yocom, J. E. 1974. A Study of Indoor Air Quality. EPA-650/4-74-042.

Couch, R. B. 1981. Viruses and indoor air pollution, *Bull. N.Y. Acad. Med.* **57**, 907.

Courtney, R. G. (ed.) 1976. *Energy Conservation in the Built Environment.* Lancaster, England: Construction Press/CIB.

Covelli, H. D., Kleeman, J., Martin, J. E., Landau, W. L., and Hughes, R. L. 1973. Bacterial emission from both vapor and aerosol humidifiers, *Am. Rev. Respir. Dis.* **108**, 698.

Crabtree, R. E., Saba, M. P., and Strang, J. E. 1980. The cabin air conditioning and temperature control system for the Boeing 767 and 757 airplanes. Paper presented at Am. Soc. Mech. Eng. Intersoc. Conf. Env. Systems, San Diego, CA, July 14–17.

Crawshaw, G. H. 1976. The role of wool carpets in the control of indoor air pollution, *Textilveredlung* **11**(6), 267.

Crawshaw, G. H. 1978. The role of wool carpets in controlling indoor air pollution, *Textile Inst. Ind.* **16**(1), 12.

Creech, C., and Mack, M. 1975. Cancerogenic properties of vinyl chloride, *Ann. N.Y. Acad. Sci.* **246**, 88.

Creip, L. H., and Green, M. A. 1936. Air cleaning as an aid in the treatment of hay fever and bronchial asthma, *J. Allergy* **7**, 120.

Croome-Gale, D. J., and Roberts, B. M. 1975. *Air Conditioning and Ventilation of Buildings* (Int. Ser. Monographs in Heating, Ventilation and Refrigeration, Vol. 10). New York: Pergamon Press.

Crowley, T. P. 1978. Contaminated humidifiers, *J. Am. Med. Assoc.* **240**, 348.

Culot, R. 1978. Field applications of a radon barrier to reduce indoor airborne radon progeny, *Health Phys.* **34**, 498.

Dadaian, J. H., Yin, S., and Laurenzi, G. A. 1971. Studies of mucus flow in the mammalian respiratory tract, II: The effects of serotonin and related compounds on respiratory tract mucus flow, *Am. Rev. Respir. Dis.* **103**, 808.

Dally, K. A., Hanrahan, L. P., Woodbury, M. A., and Kanarek, M. S. 1981. Formaldehyde exposure in non-occupational environments. *Arch. Env. Health* **36**(6), 277.

Dally, K. A., Hanrahan, L. P., Anderson, H., Eckmann, A., and Kanarek, M. S. 1981. A follow-up study of indoor air quality in Wisconsin homes. Preprint, *Int. Symposium Indoor Air Pollution, Health and Energy Conserv.*, Univ. of Massachusetts, October 13–16; Amherst, MA.

Danielson, T. A. (ed.) 1973. *Air Pollution Engineering Manual.* EPA Publ. AP-40.

Darlow, H. M., and Bale, W. R. 1959. Infective hazards of water-closets, *Lancet* **1**(7084), 1196.

Dasgupta, P. K., DeCesare, K., and Ullrey, J. C. 1980. Determination of atmospheric sulfur dioxide without tetrachloromercurate(II) and the mechanism of the Schiff reaction, *Anal. Chem.* **52**, 1912.

Davies, C. N. (ed.) 1961. *Inhaled Particles and Vapors,* Vol. I. New York: Pergamon Press.

Davies, J. E., Edmundson, W. F., and Raffonelli, A. 1975. The role of house dust in human DDT pollution, *Am. J. Public Health* **65**(1), 53.

Davies, P. O. A. L., and Moore, D. J. 1964. Experiments on the behavior of effluent emitted from stacks at or near the roof level of tall reactor buildings, *Int. J. Air and Water Pollution* **8**, 515.

Davis, J. H., Davies, J. E., Raffonelli, A., and Reich, G. 1973. The investigation of fatal acrylonitrate intoxications. In *Pesticides and the Environment: A Continuing Controversy.* (Proc. Inter-American Conf. Toxicol. and Occupational Med.)

Davison, R. L., Natusch, D. F. S., Wallace, J. R., and Evans, C. A. 1974. Trace elements in fly ash, *Env. Sci. Technol.* **8**, 1107.

Daws, L. F. 1967. Movement of air streams indoors. In *Airborne Microbes* (P. H. Gregory and J. L. Monteith, eds.), Cambridge: Cambridge Univ. Press.

Dean, J. A. (ed.) 1973. *Lange's Handbook of Chemistry,* 11th ed. New York: McGraw-Hill.

Debeir, M., and Judd, M. D. 1979. Atmospheric trace contaminants: Their measurement and identification for Spacelab, *ESA Bull.* **17**, 66.

DeBruin, A. 1976. *Biochemical Toxicology of Environmental Agents.* Amsterdam: Elsevier.

Decorpo, J. J., Wyatt, J. R., and Saalfeld, F. E. 1980. Atmospheric monitoring in submersibles. Paper presented at Am. Soc. Mech. Eng. Intersoc. Conf. Env. Systems, San Diego, CA, July 14–17.

Decoufle, P., Stanislawczyk, K., Houten, L., Bross, I. D. J., and Viadana, E. 1977. A Retrospective Survey of Cancer in Relation to Occupation. HEW (NIOSH) Publ. 77-178. Cincinnati, OH.

Deimel, M. 1976. Erfahrungen über Formaldehyd-Raumluft-Konzentrationen in Schulneubauten. Vortragstagung "Organische Verunreinigungen in der Innen- und Aussenluft." Berlin.

Deitz, V. R. (ed.) 1975. *Removal of Trace Contaminants from the Air.* (Proc. a symposium co-sponsored by Division of Colloid and Surface Chemistry and Division of Environmental Chemistry at 168th Ann. Meeting, American Chemical Society, Atlantic City, NJ, September 10–11, 1974.) ACS Symposium Ser. 17. Washington, DC.

Derham, R. L., Petersen, G., Sabersky, R. H. and Shair, F. H. 1974. On the relation between indoor and outdoor concentrations of nitrogen oxides, *J. Air Pollution Control Assoc.* **24**, 158.

DeRoos, R. L., Banks, R. S., Rainer, D., Anderson, J. L., and Michaelsen, G. S. 1978. Hospital Ventilation Standards and Energy Conservation: A Summary of the Literature with Conclusions and Recommendations. FY 78 Final report. LBL Report 8316.

Desaedeleer, G. G., and Winchester, J. W. 1975. Trace metal analysis of atmospheric aerosol particle size fractions in exhaled human breath, *Env. Sci. Technol.* **9**, 971.

Desaedeleer, G. G. 1976. Traffic restrictions and the abatement of ambient urban lead and bromine concentrations, unpublished.

Desrosiers, A. E., and Farber, S. A. 1980. The radiological health aspects of solar home heating. In *Radon in Buildings* (R. Colle and P. E. McNall, Jr., eds.). NBS Special Publ. 581. Washington, DC.

Dickey, L. D. (ed.) 1976. *Clinical Ecology.* Springfield, IL: Charles C Thomas.

Dickgiesser, N. 1978. Examinations about the behavior of grampositive and gram-negative bacteria in dry and moist atmosphere, *Zentralbl. Bakteriol.* **B167**, 48.

Dietz, A. G. H. 1969. *Plastics for Architects and Builders.* Cambridge, MA: MIT Press.

Dmitriev, M. T., Gubernskii, Y. D., Thiem, T. H., Zakharchenko, M. P., and Preobrazhenskii, L. I. 1979. Ozone index as an integral indicator of air pollution, *Gig. Sanit.* **11**, 38.

Dockery, D. W., and Spengler, J. D. 1981. Indoor–outdoor relationships of respirable sulfates and particles, *Atmos. Env.* **15**, 335.

Dockery, D. W., and Spengler, J. D. 1981. Personal exposure to respirable particulates and sulfates, *J. Air Pollution Control Assoc.* **31**, in press.

Doll, R. 1979. *Long Term Hazards from Environmental Chemicals.* Royal Society Discussions, 1977. London: Royal Society.

Dorman, R. G. 1974. *Dust Control and Air Cleaning.* (Int. Ser. Monographs on Heating, Ventilation and Refrigeration, Vol. 9.) New York: Pergamon Press.

Dossat, R. J. 1978. *Principles of Refrigeration,* 2nd ed. New York: Wiley.

Dotto, L., and Schiff, H. I. 1978. *The Ozone War.* New York: Doubleday.

Douglas, R. L. 1977. Radiological survey at the inactive uranium mill site near Riverton, Wyoming. NTIS PB 271 332.

Dravnieks, A. 1977. Organic Contaminants in Indoor Air and Their Relation to Outdoor Contaminants: Final Report on Phase I. Development of Methodology and Initial Field Trials. ASHRAE RP-183; IITRI Project C8276. Chicago, IL: Illinois Institute of Technology, Research Institute.

Druckrey, H., Schmahl, D., Beuthner, H., and Muth, F. 1960. Vergleichende Prüfung von Tabakrauchkondensaten, Benzpyren und Tabakextrakt auf carcinogene Wirkung an Ratten, *Naturwiss.* **47**, 605.

Duan, N. 1981. Micro-environment types: A model for human exposure to air pollution. Preprint, *Int. Symposium Indoor Air Pollution, Health and Energy Conserv.*, Univ. of Massachusetts, October 13-16; Amherst, MA.

Dubin, F. S., and Long, C. G., Jr. 1978. *Energy Conservation Standards for Building Design, Construction, and Operation.* New York: McGraw-Hill.

Duffee, R. A., Jann, P. R., Flesh, R. D., and Cain, W. S. 1980. Odor/Ventilation Relationships in Public Buildings. APCA Report 80-61.3.

Duffee, R. A., and Jann, P. 1981. Ventilation/Odor Study. Field study. LBL Report 12632; EEB-Vent 81-10.

Duncan, D. L., Gesell, T. F., and Johnson, R. H., Jr. 1977. Radon-222 in potable water, Proc. 10th Midyear Topical Symposium Health Phys. Soc., Northeastern New York Chapter, Rensselaer Polytechnic Institute, Troy, NY, October 11-13, 1976.

Duncan, J. R., and Morkin, K. M. 1980. Air quality impact of residential wood burning stove, *Proc. APCA Conf.* **80,** 7.

DuPont (E. I. du Pont de Nemours & Co.) 1981. PRO-TEK Colorimetric air monitoring badge systems operating and analytical procedures, laboratory validation report, formaldehyde badge, ser. 11, type C-60. Wilmington, DE.

Dutt, G., Harrje, D. T., Lavine, M., Linteris, G., and Socolow, R. 1981. Results of the modular retrofit experiment: A test of the house doctor concept. Preprint, *Int. Symposium Indoor Air Pollution, Health and Energy Conserv.*, Univ. of Massachusetts, October 13-16; Amherst, MA.

Duwe, M. P. 1979. *The Diurnal Variation in Radon Flux from the Soil Due to Atmospheric Pressure Change and Turbulence.* Ph.D. thesis. Univ. of Wisconsin, Madison, 1976.

Dzubay, T. G. (ed.) 1977. *X-ray Fluorescence Analysis of Environmental Samples.* Ann Arbor, MI: Ann Arbor Science.

Eads, W. G. 1974. *Testing, Balancing and Adjusting of Environmental Systems.* Vienna, VA: Sheet Metal and Air Conditioning Contractors National Association.

Eaton, R. S. 1980. Radon control in housing in Canada. In *Radon in Buildings* (R. Collé and P. E. McNall, Jr., eds.). NBS Special Publ. 581. Washington, DC.

Eaton, W. C., Howard, J. N., Jr., Burton, R. M., Benson, F., and Ward, G. H. 1972. A Preliminary Study of Indoor Air Pollution in a Home Using a Gas Stove. Part I, Oxides of Nitrogen; Part II, Condensation Nuclei. EPA-

Eckert, E. R. J., and Goldstein, R. J. 1976. *Measurements in Heat Transfer,* 2nd ed. (Ser. in Thermal and Fluids Engineering.) Washington, DC: Hemisphere Publishing; New York: McGraw-Hill.

Eckmann, A. D., Dally, K. A., Hanrahan, L. P., and Anderson, H. A. 1980. Comparison of the chromotropic acid and pararosaniline methods for HCHO determination using various collection techniques. Preprint, *Int. Symposium Indoor Air Pollution, Health and Energy Conserv.*, Univ. of Massachusetts, October 13-16; Amherst, MA.

Eckmann, B. H., Schaefer, G. L., and Huppert, M. 1964. Bedside interhuman transmission of coccidioidomycosis via growth on fomites: An epidemic involving six persons, *Am. Rev. Respir. Dis.* **89,** 175.

Edmonds, R. L. (ed.) 1979. *Aerobiology: The Ecological Systems Approach.* US/IBP Synthesis Ser. 10. Stroudsburg, PA: Dowden, Hutchinson & Ross.

Eichholz, G. G., and Poston, J. W. 1979. *Principles of Nuclear Radiation Detection.* Ann Arbor, MI: Ann Arbor Science.

Eickhoff, T. C. 1979. Epidemiology of Legionnaires' disease, *Ann. Intern. Med.* **90,** 499.

Eisenbud, M. 1963. *Environmental Radioactivity.* New York: McGraw-Hill.

Eisenbud, M. 1978. *Environment, Technology, and Health: Human Ecology in Historical Perspective.* New York: New York Univ. Press.

Eisenreich, S. J., Looney, B. B., and Thornton, J. D. 1981. Airborne organic contaminants in the Great Lakes ecosystem, *Env. Sci. Technol.* **15,** 30.

El-Ahraf, P. H., and Baca, T. E. 1980. The administration of state and local environmental health programs: Who is responsible?, *J. Env. Health* **43,** 86.

Elfers, L., and Richter, H. 1975. *Monitoring Vinyl Chloride around Polyvinyl Chloride Fabrication Plants.* Cincinnati, OH: PEDCo Environmental Specialists.

Elkins, R. H., Zawacki, T. S., and Macriss, R. A. 1973. Detailed energy usage profile in the Canton test homes. Report on Project HC-4-14, Chicago, IL: Institute of Gas Technology.

Elkins, R. H., Ng, D. Y. C., Zimmer, J., and Macriss, R. A. 1974. A survey of carbon monoxide and nitrogen dioxide levels in the indoor environment. Paper presented at 67th Ann. APCA Meeting, Denver, CO, June 13.

Elliott, L. J. 1981. Formaldehyde field measurement. Personal communication.

Elliott, L. P., and Rowe, D. R. 1975. Air quality during public gatherings, *J. Air Pollution Control Assoc.* **25,** 635.

Ellis, E. C., and Margeson, J. H. 1975. Evaluation of Gas Phase Titration Technique as Used for Calibration of Nitrogen Dioxide Chemiluminescence Analyzers. EPA-650/4-75-021.

Ellis, E. C. 1976. Technical Assistance Document for the Chemiluminescence Measurement of Nitrogen Dioxide. EPA-600/4-75-003; NTIS PB 268 456.

Ellis, F. P. 1972. Mortality from heat illness and heat-aggravated illness, *Env. Res.* **5,** 1.

Emerick, R. H. 1966. *Handbook of Mechanical Specifications for Buildings and Plants: A Checklist for Engineers and Architects.* New York: McGraw-Hill.

Emerick, R. H. 1969. *Troubleshooters' Handbook for Mechanical Systems.* New York: McGraw-Hill.

Eng, W. G. 1979. Survey on eye comfort in aircraft, I: Flight attendants, *Aviation Space Env. Med.* p. 401.

Enterline, P. E. 1976. Estimating the health risks in studies of the health effects of asbestos, *Am. Rev. Respir. Dis.* **113,** 175.

Erickson, M. D., Kelner, L., Bersey, J. T., Rosenthal, D., Zweidinger, R. A., and Pellizzari, E. D. 1980. Methods for the analysis of polybromated biphenyls by GC-MS. *Biomed Mass Spectros.* **7**(3), 99.

Etheridge, D. W., and Nevrala, D. J. 1979. Air infiltration in the U.K. and its impact on the thermal environment. In *Indoor Climate, Effects on Human Comfort, Performance, and Health in Residential, Commerical, and Light-Industry Buildings* (P. O. Fanger and O. Valbjørn, eds.), p. 293. Copenhagen: Danish Building Research Institute.

European Chemical Industry. 1981. Assessment of Data on the Effects of Formaldehyde on Humans. Tech. Report No. 1. Brussels, Belgium: Ecology and Toxicology Center (ECETOC).

European Chemical Industry. 1981. The Mutagenic and Carcinogenic Potential of Formaldehyde. Tech. Report No. 2. Brussels, Belgium: Ecology and Toxicology Center (ECETOC).

Everett, J. J., and Mathur, J. 1981. Indoor air quality audits in energy efficient residences. Preprint, *Int. Symposium Indoor Air Pollution, Health and Energy Conserv.*, Univ. of Massachusetts, October 13–16; Amherst, MA.

Fairchild, E. J. (ed.) 1977. *Registry of Toxic Effects of Chemical Substances,* Vol. I. HEW (NIOSH) Publ. 78-104-A. Cincinnati, OH.

Falk, H. 1980. Pentachlorophenol in log homes—Kentucky, *Morbidity and Mortality Weekly Reports* **29**(36), 431.

Fanger, P. O. 1970. *Thermal Comfort: Analysis and Applications in Environmental Engineering.* New York: McGraw-Hill.

Fanger, P. O., and Valbjørn, O. (eds.) 1979. *Indoor Climate, Effects on Human Comfort, Performance, and Health in Residential, Commercial, and Light-Industry Buildings.* Copenhagen: Danish Building Research Institute.

Fanning, L. Z. 1979. Formaldehyde in Office Trailers. LBL Report LBID-084; EEB-Vent 79-17.

Farber, S. M., and Wilson, R. H. L. (eds.) 1961. *The Air We Breathe: A Study of Man and His Environment.* Springfield, IL: Charles C Thomas.

Feigley, E. C., and Chastein, J. B. 1982. Experimental comparison of three different samplers exposed to concentration profiles of organic vapors. *Am. Ind. Hyg. Ass. J.* **43**(4), 227.

Feldman, J., Huang, S., Lyman, J., and Sobeck, E. 1978. *The Legal Aspects of Smoking Regulation.* Stanford, CA: Stanford Environmental Law Society.

Fergusson, D. M., Horwood, L. J., Shannon, F. T., and Taylor, B. 1981. Parental Smoking and lower respiratory illness in the first three years of life. *J. Epidem. and Community Health* **35**, 180.

Ferrand, E. F., and Moriates, S. 1981. Health aspects of indoor air pollution: Social, legislative, and economic considerations, *Bull. N.Y. Acad. Med.* **57**, 1061.

Ferris, B. G., Jr., Speizer, F. E., Spengler, J. D., Dockery, D., Bishop, Y. M. M., Wolfson, M., and Humble, C. 1979. Effects of sulfur oxides and respirable particles on human health, *Am. Rev. Respir. Dis.* **120**, 767.

Ferris, B. G., Jr., Speizer, F. E., Bishop, Y. M. M., Spengler, J. D., and Ware, J. H. 1980. The six city study: A progress report. In Atmospheric Sulfur Deposition: Environmental Impact and Health Effects (Shriner, Richmond & Lindberg, eds.), Ann Arbor, MI: Ann Arbor Science Publishers.

Fincher, E. L. 1969. Aerobiology and hospital sepsis. Chapter 17 in *An Introduction to Experimental Aerobiology* (R. L. Dimmick and A. B. Akers, eds.). New York: Wiley-Interscience.

Findlay, W. O. 1980. Application of radon standards to new and existing housing in Elliot Lake, Ontario. In *Radon in Buildings* (R. Colle and P. E. McNall, Jr., eds.). NBS Special Publ. 581. Washington, DC.

Fink, J. N., et al. 1971. Interstitial pneumonitis due to hypersensitivity to an organism contaminating a heating system, *Ann. Int. Med.* **74**, 80.

Fischbein, A., Rice, C., Sarkozi, L., Kon, S. H., Petrocci, M., and Selikoff, I. J. 1979. Exposure to lead in firing ranges, *J. Am. Med. Assoc.* **241**, 1141.

Fischer, T., Weber, A., and Grandjean, E. 1978. Air pollution due to tobacco smoke in restaurants, *Int. Arch. Occupational Env. Health* **41,** 267.

Fisher, A. A., Kanof, N. B., and Iondi, E. M. 1962. Free formaldehyde in textiles and paper, *Arch. Dermatol.* **86,** 753.

Fisk, W. J., Roseme, G. D., and Hollowell, C. D. 1981. Test Results and Methods: Residential Air-to-Air Heat Exchangers for Maintaining Indoor Air Quality and Saving Energy. LBL Report 12280. *Env. International* **8,** in press.

Fisk, W. J., and Turiel, I. 1982. Residential Air-to-Air Heat Exchangers: Peformance, Energy Savings, and Economics. LBL Report 13843.

Fitzgerald, G. R., Barniville, G., Black, J., Silke, B., Carmody, M., and O'Dwyer, W. F. 1978. Paraquat poisoning in agricultural workers, *J. Irish Med. Assoc.* **71,** 336.

Flachsbart, P. G., and Phillips, S. 1980. An index and model of human response to air quality, *J. Air Pollution Control Assoc.* **30,** 759.

Flachsbart, P. G., and Ott, W. 1981. Field surveys of carbon monoxide in commerical settings using personal exposure monitors. Chapter 6 in *A "Hot" Office Building.* EPA report.

Flamm, W. G., 1978. *Mutagenesis (Adv. Toxicol.* **5)**, New York: Wiley.

Flindt, M. L. H. 1969. Pulmonary disease due to inhalation of derivatives of *Bacillus subtilis* containing proteolytic enzyme, *Lancet* **1**(7607), 1177.

Flowers, E. S. 1974. Relationship between exposure to asbestos, collagen formation, ferruginous bodies, and carcinoma, *Am. Ind. Hyg. Assoc. J.* **35,** 724.

Flugge, K. 1896. *Die Microorganismus,* Vol. I. Leipzig: Vogel.

Flugge, K. 1897. Ueber Luftinfektion, *Z. Hyg. Infektionskrankh.* **25,** 179.

Flugge, K., and Heymann, B. 1905. Ueber Luftverunreinigungen, Waermestauung und Lüftung in geschlossenen Raeumen, *Z. Hyg. Infektionskrankh.* **49,** 363, 388, 405, 433.

Flügge, K. 1916. *Grossstadtwohnung und Kleinhaussiedlung in ihrer Einwirkung auf die Volksgesundheit.* Jena; G. Fischer.

Fomin, A. G., Genin, A. M., and Voronin, G. I. 1965. Physiological-hygienic evaluation of the life protection systems of the "Vostok" and "Voskhod" spacecraft. In *Problems of Space Biology* (V. N. Chernigovski, ed.), Vol. 7. (Proc. 2nd Int. Symposium Basic Problems of Man in Space, Paris, June 14–18.)

Foote, R. S. 1972. Mercury vapor concentrations inside buildings, *Science* **177,** 513.

Foster, J. F., and Beatty, G. H. 1974. *Interlaboratory Cooperative Study of the Precision and Accuracy of the Measurement of Nitrogen Dioxide Content in the Atmosphere Using ASTM Method D1607.* ASTM Data Ser. Publ. DS 55. Philadelphia, PA.

Foster, J. F., Beatty, G. H., and Howes, J. E., Jr. 1974. *Interlaboratory Cooperative Study of the Precision and Accuracy of the Measurement of Total Sulfation in the Atmosphere using ASTM Method D2010.* ASTM Data Ser. Publ. DS 55-S2. Philadelphia, PA.

Foster, J. F., Beatty, G. H., and Howes, J. E., Jr. 1974. *Interlaboratory Cooperative Study of the Precision and Accuracy of the Measurement of Dustfall Using ASTM Method D1739.* ASTM Data Ser. Publ. DS 55-S4. Philadelphia, PA.

Foster, M. 1891. *Physiology.* New York: D. Appleton.

Fox, R. W., and McDonald, A. T. 1978. *Introduction to Fluid Mechanics,* 2nd ed. New York: Wiley.

Frank, N. H., Hunt, F., Jr., and Cox, M. 1977. Population exposure: An indicator of air quality improvement. Paper 77-44.2 presented at 70th Ann. Meeting, Air Pollution Control Assoc., Toronto, Canada.

Frazier, J. A. (ed.) 1981. *Indoor Pollution.* NRC report for EPA. Washington, DC: NAS Press.

Fritsch, A. J. (ed.) 1977. *The Household Pollutants Guide.* Garden City, NY: Anchor Books/Doubleday.

Fuerst, R. G., Scaringelli, F. P., and Margeson, J. H. 1976. Effect of Temperature on Stability of Sulfur Dioxide Samples Collected by the Federal Reference Method. EPA-600/4-76-024.

Fugas, M., 1975. Assessment of total exposure to an air pollutant, *Proc. Int. Conf. Env. Sensing and Assessment,* 2, 109; IEEE 75-CH 1004-1 ICESA. New York: IEEE.

Fugas, M. 1981. Results of personal exposure versus monitoring station data for respirable particles. Preprint, *Int. Symposium Indoor Air Pollution, Health and Energy Conserv.,* Univ. of Massachusetts, October 13–16; Amherst, MA. *Env. International* **8.**

Fulwiler, R. D., Abbott, J. C., and Darcy, F. J. 1972. The evaluation of detergent enzymes in air, *Am. Ind. Hyg. Assoc. J.* **33,** 231.

Fumarola, D. 1981. Toxic activities of Legionellaceae: New aspects, *Am. J. Clin. Pathol.* **76,** 245.

Gage, S. 1980. Personal communication.

Gagge, A. P., Nishi, Y., and Nevins, R. G. 1976. Role of clothing in meeting FEA energy conservation guidelines, *ASHRAE Trans.* **82**(2).

Gagge, A. P. 1979. The role of humidity during thermal comfort. In *Indoor Climate, Effects on Human Comfort, Performance, and Health in Residential, Commercial, and Light-Industry Buildings* (P. O. Fanger and O. Valbjørn, eds.), p. 527. Copenhagen: Danish Building Research Institute.

Gardner, D. E., Hu, E. P. C., and Graham, J. A. (eds.) 1980. *Experimental Models for Pulmonary Research,* Vol. 35.

Gary, V. F., Oatman, L., Pleus, R., and Gray, D. 1980. Formaldehyde in the home: Some environmental disease perspectives, *Minnesota Med.* **63,** 107.

Geiger, R. 1965. *The Climate near the Ground.* Cambridge, MA: Harvard Univ. Press.

Geisling, K. L., Miksch, R. R., and Rappaport, S. M. 1982. The Generation of Dry Formaldehyde at ppb–ppm Levels by the Vapor-Phase Depolymerization of Trioxane. *Anal. Chem.* **54,** 140.

Geisling, K. L., Miksch, R. R., Rappaport, S. M., and Tashima, M. K. 1981. A New Passive Monitor for Determining Formaldehyde in Ambient Air. LBL Report 12560.

Gelperin, A. 1973. Humidification and upper respiratory infection incidence, *Heating, Piping and Air Conditioning* **45**(3).

General Electric Co. 1972. *Indoor-Outdoor Carbon Monoxide Pollution Study.* EPA R4 73-020; NTIS PB 220 428.

Gentilizza, M. 1977. Rate of concentration of sulfur dioxide in the indoor and outdoor atmospheres, *Arh. Hig. Rada. Toksikol.* **28,** 279.

George, A. C., and Breslin, A. J. 1980. The distribution of ambient radon and radon daughters in residential buildings in New Jersey–New York area, *Natural Radiation Environment,* Vol. 3. DOE Special Symposium Ser. 51, DOE CONF-780422.

Gerber, R. M., and Rossi, R. C. 1977. *Evaluation of Electron Microscopy for Process Control in the Asbestos Industry.* Environmental Protection Technol. Ser. EPA-600/2-77-059; NTIS PB 266 091.

Gerzoff, S. 1982. *Indoor Environmental Quality: A Directory of Experts.* Report for the California Department of Consumer Affairs, Sacramento, CA.

Gesell, T. F., and Prichard, H. M. 1975. The technologically enhanced natural radiation environment, *Health Phys.* **28**, 361.

Gesell, T. F., and Prichard, H. M. 1980. The contribution of radon in tap water to indoor radon concentration, *Natural Radiation Environment,* Vol. 3, p. 1347. DOE Special Symposium Ser. 51, DOE CONF-780422.

Gibson, J. E. 1982. Nasal Cancer in the Rat Induced by Gaseous Formaldehyde, paper presented at the Chem. Ind. Inst. Toxicology Conf., Raleigh, NC: CIIT.

Giles, G. 1980. Improved MPG for the BAe 146 feeder-jet, *Aircraft Eng.* **52**, 28.

Giles, W. H., Lee, R. E., and Dworetzky, L. H. 1974. The influence of traffic-generated carbon monoxide on indoor air quality of an air rights lane and a canyon structure, Part I, *Proc. Inst. Env. Sci.* **20**, 318.

Gip, L. 1966. Investigation of the occurrence of dermatophytes on the floor and in the air of indoor environments, *Acta Dermatol. Venereol. (Stockholm)* **46** (Suppl. 58), 1-54.

Girman, J. R., Apte, M. G., Traynor, G. W., and Hollowell, C. D. 1981. Pollutant Emission Rates from Indoor Combustion Appliances and Sidestream Cigarette Smoke. LBL Report 12562.

Givoni, B. 1969. *Man, Climate, and Architecture.* New York: Elsevier.

Gjessing, D. T. 1978. *Remote Surveillance by Electromagnetic Waves for Air-Water-Land.* Ann Arbor, MI: Ann Arbor Science.

Godin, G., Wright, G. R., and Shepherd, R. J. 1972. Urban exposure to carbon monoxide, *Arch. Env. Health* **25**, 305.

Godish, T. 1981. Survey of indoor formaldehyde levels and apparent building-related illness in conventional housing in Delaware County, Indiana. Preprints, *Int. Symposium Indoor Air Pollution, Health and Energy Conserv.* Univ. of Massachusetts, October 13-16; Amherst, MA.

Goldschmidt, V. W., and Didion, D. (eds.) 1974. *Proceedings of the Conference on Improving Efficiency in HVAC Equipment and Components for Residential and Small Commercial Buildings.* Ray W. Herrick Laboratories, Purdue Univ. October 7-8. Lafayette, IN: Purdue Research Foundation.

Goldstein, B. D., Melia, R. J. W., and Florey, C. V. 1981. Indoor nitrogen oxides, *Bull. N.Y. Acad. Med.* **57**, 873.

Goldwater, L. J., Manoharan, A., and Jacobs, M. B. 1961. Suspended particulate matter: Dust in domestic atmospheres, *Arch. Env. Health* **2**, 511.

Gonzales, R. R. 1979. Role of natural acclimatization (cold and heat) and temperature: Effect on health and acceptability in a built environment. In *Indoor Climate, Effects on Human Comfort, Performance, and Health in Residential, Commercial, and Light-Industry Buildings* (P. O. Fanger and O. Valbjørn, eds.), p. 737. Copenhagen: Danish Building Research Institute.

Granier, M., et al. 1980. Humidifier lung: An outbreak in office workers, *Chest* **77**, 183.

Gray, J. M. L. 1974. *An Evaluation of an Aldehyde-Specific Detector for Gas Chromatography and Some Reactions of Sulfur Dioxide-Nitrogen Dioxide-Aldehyde Mix-*

tures in Air. Ph.D. thesis. Univ. Colorado, Boulder, CO. *Diss. Abstr. Int.* **35B**, 1529.

Green, A. R. 1979. An updated assessment of the critical environmental factors involved in the prevention of allergic disease, *Ann. Allergy* **42**, 372.

Green, B. G. 1978. Referred thermal sensations: Warmth versus cold, *Sensory Processes* **2**, 220.

Green, G. H. 1979. Field studies of the effect of air humidity on respiratory disease. In *Indoor Climate, Effects on Human Comfort, Performance, and Health in Residential, Commercial, and Light-Industry Buildings* (P. O. Fanger and O. Valbjørn, eds.), p. 207. Copenhagen: Danish Building Research Institute.

Greene, T. 1980. Problems with formaldehyde at State level. In *Technical Workshop on Formaldehyde* (B. Meyer and J. Fandey, eds.), Vol. 1, p. 142. Washington, DC: CPSC.

Greenspon, M. E. 1977. Environmental control of Washington metro—Subway station air conditioning, *ASHRAE Trans.* **83**,(1), 299.

Gregory, J. 1981. Office air quality, tight buildings, and job stress—Impact on women office workers' occupational health. Paper presented at AIHA–ACGH Conf., Portland, OR, May 25.

Gregory, P. H., and Monteith, J. L. (eds.) 1967. *Airborne Microbes. (Proc. Symposium Soc. Gen. Microbiol.* **17**.) New York and London: Cambridge Univ. Press.

Gregory, P. H. 1973. *The Microbiology of the Atmosphere,* 2nd ed. New York: Wiley.

Grieble, H. G., Colton, F. R., Bird, T. J., Toigo, A., and Griffith, L. G. 1970. Fine-particle humidifiers: Source of *Pseudomonas aeruginosa* infections in a respiratory-disease unit, *N. England J. Med.* **282**, 531.

Griscom, J. H. 1850. *The Uses and Abuses of Air: Showing its Influence in Sustaining Life, and Producing Disease; with Remarks on the Ventilation of Houses,* 2nd ed. Clinton Hall, NY: J. S. Redfield.

Grob, R. L. (ed.) 1977. *Modern Practice of Gas Chromatography.* New York: Wiley-Interscience.

Grot, R. A., and Clark, R. E. 1979. Air Leakage Characteristics and Weatherization Techniques for Low-Income Housing. DOE/ASHRAE Conf. on Thermal Performance of Exterior Envelopes of Buildings, Orlando, FL.

Gubernskii, Y. D., Dmitriev, M. T., Orlova, N. S., and Kaliniva, N. V. 1981. Atmospheric hygiene in high capacity public buildings. *Vest. Akad. Med. Nauk. SSSR,* (1), 71.

Gudzinowicz, B. J., Gudzinowicz, M. J., and Martin, H. F. 1976. *Fundamentals of Integrated GC–MS,* Part I: *Gas Chromatography.* New York: Marcel Dekker.

Gullino, P. 1977. Doubling time of cancer cells, *Cancer* **39**, 2697.

Gunter, B. J., Lawry, L. K., and Philbin, E. J. 1977. Hazard evaluation and technical assistance. NTIS PB 278 835.

Hachenberg, H. 1973. *Industrial Gas Chromatographic Trace Analysis.* London: Heyden & Son.

Hackney, J. D., Linn, W. S., Buckley, R. D., Collier, C. R., and Mohler, J. G. 1973. Respiratory and biological adaptations in men repeatedly exposed to ozone. In *Environmental Stress: Individual Human Adaptations* (L. J. Folinsbee, J. A. Wagner, et al., eds.). New York: Academic Press.

Haines, J. E. 1961. *Automatic Control of Heating and Air Conditioning*, 2nd ed. New York: McGraw-Hill.

Halpern, M. 1978. Indoor-outdoor air pollution exposure continuity relationships, *J. Air Pollution Control Assoc.* **28**, 689.

Hamilton, A., and Hardy, H. L. 1974. *Industrial Toxicology*, 3d ed. Acton, MA: Publishing Sciences Group.

Hammad, Y. Y., and Corn, M. 1971. Hygienic assessment of airborne cotton dust in a textile manufacturing facility, *Am. Ind. Hyg. Assoc. J.* **32**, 662.

Hanna, M. G., Jr., Nettesheim, P., and Gilbert, J. R. (eds.) 1970. *Inhalation Carcinogenesis*. Washington, DC: National Cancer Institute/AEC Technical Information Division.

Hanrahan, L. P., Anderson, H. A., Dally, K. A., Eckmann, A. D., and Kanarek, M. S. 1981. An investigation of the offgassing decay function in mobile homes with climate corrected formaldehyde readings. Preprint, *Int. Symposium Indoor Air Pollution, Health and Energy Conserv.*, Univ. of Massachusetts, October 13–16; Amherst, MA.

Hanrahan, L., Anderson, H., Dally, K., Eckmann, A., and Kanarek, M. 1981. A multivariate analysis of health effects in a cohort of mobile home residents exposed to formaldehyde. Preprint, *Int. Symposium Indoor Air Pollution, Health and Energy Conserv.*, Univ. of Massachusetts, October 13–16; Amherst, MA.

Harke, H. P., Baars, A., Frahm, B., Peters, H., and Schultz, C. 1972. Passive smoking concentration of smoke constituents in the air of large and small rooms as a function of number of cigarettes smoked and time, *Int. Arch. Arbeitsmed.* **29**, 323.

Harley, J. H. 1981. Radioactive emissions and radon, *Bull. N.Y. Acad. Med.* **57**, 883.

Harrington, J. B. 1965. Atmospheric diffusion of ragweed pollen in urban areas. In *Atmospheric Pollution by Aeroallergens: Meteorological Phase* (E. W. Hewson, ed.). Ann Arbor: Univ. of Michigan Press.

Harris, J. C., Rumack, B. H., and Aldrich, F. D. 1981. Toxicology of urea formaldehyde and polyurethane foam insulation, *J. Am. Med. Assoc.* **245**, 243.

Harrison, J. W., Timmons, M. L., Denyszyn, R. B., Decker, C. E. 1977. *Evaluation of the EPA Reference Method for Measurement of Non-Methane Hydrocarbons*. EPA-600/4-77-033.

Harrje, D. T., Dutt, G. S., and Beyea, J. 1979. Locating and eliminating obscure but major energy losses in residential housing, *ASHRAE Trans.* **85**, 800.

Harrje, D. T., Dutt, G. S., and Gadsby, K. J. 1980. Isolating the building thermal envelope. Paper presented at Thermosense III Conf., Minneapolis, MN.

Hart, G. W., et al. 1981. *To Breathe Clean Air*. Report of the National Commission on Air Quality, Washington, DC.

Harting, F. H., and Hess, W. 1879. Der Lungenkrebs, die Bergkrankheit in den Schneeberger Gruben, *Vierteljahrsschr. gerichtl. Med. öffentl. Gesundheitwesen* **30**, 296; **31**, 102 and 313.

Hartley, R. 1981. *Indoor Air Quality Research* (Proc. Interagency Workshop, Leesburg, VA), December 2–5, 1980. EPA Report, EPA-600/7-81-119.

Harvey, M. C. 1979. *Carbon Fiber Behavior in an Enclosed Volume*. NASA-CR-159076.

Hasselblad, V. 1977. *Lung Function in Schoolchildren: 1971–1972 Chattanooga Study*. EPA-600/1-77-002; NTIS PB 262 378.

Hatch, T. F., and Gross, P. 1964. *Pulmonary Deposition and Retention of Inhaled Aerosols*. New York: Academic Press.

Hawthorne, A. R., Dougherty, J. M., and Metcalfe, C. E. 1980. *DUVA (Derivative UV-Absorption Spectrometer): Instrument Description and Operating Manual.* ORNL-5688.

Hawthorne, A. R., Gammage, R. B., Matthews, T. G., Blackman, G. S., Howell, T. C., and Allen, R. J. 1981. *An Evaluation of Formaldehyde Emission Potential from Urea-Formaldehyde Foam Insulation: Panel Measurements and Modeling.* ORNL/TM-7959.

Hazleton Laboratories, Inc. 1972. *Chronic Exposure of Cynomolqus Monkeys to Sulfur Dioxide Singly and in Binary Combination with Certain Airborne Pollutants.* APRAC Project CAPM-6-68; NTIS PB 212 792. Vienna, VA: Hazleton Laboratories.

Heck, W. W., Fox, F. L., Brandt, C. S., and Dunning, J. A. 1969. *Tobacco: A Sensitive Monitor for Photochemical Air Pollution.* NAPCA Publ. AP-55. Cincinnati, OH: HEW.

Heller, S. R., and Milne, G. W. A. 1978. *Indexes to EPA/NIH Mass Spectral Data Base.* NSRDS-NBS 63; NTIS PB 290 661.

Henderson, D. A. 1976. The eradication of smallpox, *Sci. Am.* **235**(4), 25.

Henderson, J. J., Benson, F. B., and Caldwell, D. E. 1973. *Indoor-Outdoor Air Pollution Relationships,* Vol. II: *An Annotated Bibliography.* EPA Publ. AP-112b.

Hendricks, S. L., Borts, I. H., Heeren, R. H., Hausler, W. J., and Held, H. R. 1962. Brucellosis outbreak in an Iowa packing house, *Am. J. Public Health* **52**, 1166.

Henning, H. 1915. Der Geruch, *Z. Psychol.* **73**, 161.

Hers, J. F. P., Masurel, N., and Gans, J. C. 1969. Acute respiratory disease associated with pulmonary involvement in military servicemen in the Netherlands: A serologic and bacteriologic survey, January 1967 to January 1968, *Am. Rev. Respir. Dis.* **100**, 499.

Hershaft, A., Morton, J., and Shea, G. 1976. *Critical Review of Air Pollution Dose-Effect Functions.* Environmental Control, Inc., report for the Council on Environmental Quality, Washington, DC.

Hesketh, H. E. 1977. *Fine Particles in Gaseous Media.* Ann Arbor, MI: Ann Arbor Science.

Hess, C. T., Norton, S. A., Brutseart, W. F., Casparius, R. E., Coombs, E. G., and Hess, A. L. 1979. Radon-222 in potable water supplies in Maine: The geology, hydrology, physics and health effects. Land and Water Resources Center, Univ. of Maine at Orono.

Hess, C. T., Norton, S. A., Brutsaert, W. F., Casparius, R. E., Coombs, E. G., and Hess, A. L. 1980. Radon-222 in potable water supplies of New England, *New England Water Works Assoc.* **94**(2), 113.

Hewitt, W. 1977. *Microbiological Assay: An Introduction to Quantitative Principles and Evaluation.* New York: Academic Press.

Heyder, J., and Davies, C. N. 1971. The breathing of half-micron aerosols, II: Dispersion of particles in the respiratory tract, *J. Aerosol Sci.* **2**, 437.

Higgins, E. A., Lategola, M. T., McKenzie, J. M., Melton, C. E., and Vaughan, J. A. 1979. *Effects of Ozone on Exercising and Sedentary Adult Men and Women Representative of the Flight Attendant Population.* FAA-AM-79-20, Civil Aeromedical Institute, FAA, Oklahoma City, OK.

Higgins, E. A., Lategola, M. T., Melton, C. E., and Vaughan, J. A. 1980. *Effects of Ozone (0.30 parts per million) on Sedentary Men Representative of Airline Passengers and Cockpit Crewmembers.* Civil Aeromedical Institute, FAA, Oklahoma City, OK.

Higgins, I. T. T. (ed.) 1979. *Airborne Particles.* Baltimore, MD: University Park Press.

Hill, A. C. 1971. Vegetation: A sink for atmospheric pollutants, *J. Air Pollution Control Assoc.* **21**(6), 341

Hinds, W. C., and First, M. W. 1975. Concentration of nicotine and smoke in public places, *New England J. Med.* **292,** 844.

Hinkle, L. E., Jr., and Loring, W. C. (eds.) 1977. *The Effect of the Man-Made Environment on Health and Behavior.* HEW Publ. (CDC) 77-8318. Atlanta, GA: CDC.

Hinkle, L. E. 1981. Regulation and legislation of the indoor environment. A panel discussion. *Bull. N.Y. Acad. Med.* **57,** 1067.

Hirayama, T. 1981. Non-smoking wives of heavy smokers have a higher risk of lung cancer: A study from Japan, *Brit. Med. J.* **282,** 183.

Hittmann Associates Inc. 1974. Residential Energy Consumption. HUD-PDR-29.2.

Hodges, G. R., Fink, J. N., and Schlueter, D. P. 1974. Hypersensitivity pneumonitis caused by a contaminated cool-mist vaporizer, *Ann. Intern. Med.* **80,** 501.

Hoetjer, J. J., and Koerts, F. 1981. Verfahren zur Bestimmung der Formaldehydabgabe aus Spanplatten, *Holz Roh-Werkstoff* **39,** 391.

Hoffmann, D., Adams, J. D., Brunnemann, K. D., and Hecht, S. S. 1979. Assessment of tobacco-specific *N*-nitrosamines in tobacco products, *Cancer Res.* **39,** 2505.

Hofmann, W., Steinhausler, F., and Pohl, E. 1978. Age-, sex- and weight-dependent dose distribution patterns for human organs and tissues due to inhalation of natural radioactive nuclides, *Natural Radiation Environment,* Vol. 3, p. 215. DOE Special Symposium Ser. 51, DOE CONF-780422.

Holcombe, J. K., and Kalika, P. 1972. Controlling interior air pollution, *Plant Engineering* (March 23 and April 20).

Holleman, J. W., Ryon, M. G., and Hammons, A. S. 1980. Chemical Contaminants in Nonoccupationally Exposed U.S. Residents. EPA-600/1-80-001.

Hollowell, C. D., Budnitz, R. J., and Traynor, G. W. 1976; 1977. Combustion-generated indoor air pollution, *Proc. 4th Int. Clean Air Congr.* 684 (1977); LBL Report 5918 (1976).

Hollowell, C. D., Berk, J. V., and Traynor, G. W. 1978. Indoor Air Quality Measurements in Energy-Efficient Buildings. LBL Report 7831; CONF-780636-12.

Hollowell, C. D., Berk, J. V., Lin, C. I., and Turiel, I. 1979. Indoor Air Quality in Energy-Efficient Buildings. LBL Report 8892; CONF-790523-2.

Hollowell, C. D., Berk, J. V., and Traynor, G. W. 1979. Impact of reduced infiltration and ventilation on indoor air quality, *ASHRAE J.* **21**(7), 49.

Hollowell, C. D., Berk, J. V., Boegel, M. L., Ingersoll, J. G., Krinkel, D. L., and Nazaroff, W. W. 1980. Radon in Energy-Efficient Residences. LBL Report 9560; EEB-Vent 79-16.

Hollowell, C. D. 1981. Indoor Air Quality. LBL Report 12887.

Hollowell, C. D., and Miksch, R. E. 1981. Sources and Concentrations of Organic Compounds in Indoor Environments. *Bull. N.Y. Acad. Med.* **57,** 962.

Holman, C. W., and Muschenheim, C. (eds.) 1972. *Bronchopulmonary Diseases and Related Disorders.* New York: Harper & Row.

Holman, J. P. 1976. *Heat Transfer,* 4th ed. New York: McGraw-Hill.

Holt, P. F. 1980. Dust elimination from pulmonary alveoli, *Env. Res.* **23,** 224.

Honigman, B., Cromer, R., and Kurt, T. L. 1982. Carbon monoxide levels in athletes during exercise in an urban environment, *J. Air Pollution Control Assoc.* **32,** 77.

Honma, M., and Crosby, H. J. (eds.) 1963. *Toxicity in the Closed Ecological System.* (Compilation of papers presented at symposium, July 29–31.) Palo Alto, CA: U.S. Navy Special Projects Office and the Lockheed Missiles and Space Company.

Horie, Y., Chaplin, S., and Helfenbein, D. 1977. *Population Exposure to Oxidants and Nitrogen Dioxide in Los Angeles.* Vol. II: *Weekday/Weekend and Population Mobility Effects.* EPA-450/3-77-004b.

Hornbostel, C. 1978. *Construction Materials: Types, Uses, and Applications.* New York: Wiley-Interscience.

Horvath, C. 1980. *High-Performance Liquid Chromatography: Advances and Perspectives,* Vol. 1. New York: Academic Press.

Hove, G. M. 1977. *A World Geography of Human Diseases.* New York: Academic Press.

Huber, G., and Wanner, H. 1981. Indoor air quality and minimum ventilation rate. Preprint, *Int. Symposium Indoor Air Pollution, Health and Energy Conserv.,* Univ. of Massachusetts, October 13–16; Amherst, MA.

Huff, J. E., Moore, J. A., Saracci, R., and Tomatis, L. 1980. Long-term hazards of polychlorinated dibenzodioxins and polychlorinated dibenzofurans, *Env. Health Perspectives* **36,** 221.

Hultqvist, B. 1956. Studies of naturally occurring ionizing radiation, *Vetenskaps Handl.* Stockholm: Kgl. Svenska Akad.

Humphreys, M. A. 1979. The influence of season and ambient temperature on human clothing behaviour. In *Indoor Climate, Effects on Human Comfort, Performance, and Health in Residential, Commercial, and Light-Industry Buildings* (P. O. Fanger and O. Valbjørn, eds.), p. 699. Copenhagen: Danish Building Research Institute.

Huntington, E. 1919. *World Power and Evolution.* New Haven, CN: Yale Univ. Press.

Imperato, P. J. 1981. Legionellosis and the indoor environment, *Bull. N.Y. Acad. Med.* **57,** 922.

Ingersoll, J. G., Stitt, B. D., and Zapalac, G. H. 1981. A Fast and Accurate Method for Measuring Radon Exhalation Rates from Building Materials. LBL Report 12526.

Ingersoll, R. B. 1971. *Soil as a Sink for Atmospheric Carbon Monoxide.* SRI Project SCU-8799; APRAC Project CAPA-4-68; NTIS PB 205 890. Menlo Park, CA: Stanford Research Institute.

International Council of Scientific Unions. 1975. Scientific Committee on Problems of the Environment (SCOPE). *Environmental Pollutants: Selected Analytical Methods.* Ann Arbor, MI: Ann Arbor Science.

Irvine, R. W., and Carter, N. (eds.) 1971. *Instrumentation in the Pulp and Paper Industry,* Vol. 12 (Proc. 12th Int. ISA Pulp and Paper Instrumentation Symposium, Lancaster, PA, May 3–7). Pittsburgh, PA: Instrument Society of America.

Ishido, S., Kamada, K., and Nakagawa, T. 1956. Microbe indoor air concentration, *Bull. Dept. Home Econ., Osaka City Univ.* **4,** 31.

Jaeger, R. J. 1981. Carbon monoxide in houses and vehicles, *Bull. N.Y. Acad. Med.* **57,** 860.

Janssen, J. E., Woods, J. E., Hill, T. J., and Maldinado, E. 1981. Ventilation for control of indoor air quality. Preprint, *Int. Symposium Indoor Air Pollution, Health and Energy Conserv.,* Univ. of Massachusetts, October 13–16; Amherst, MA.

Japanese Law #20. April 14, 1970, Sanitary Environment for Buildings. Japanese Cabinet Order #304, October 1970.

Jarke, F. H. 1979. Organic Contaminants in Indoor Air and Their Relation to Outdoor Contaminants. ASHRAE. Final report 183; IITRI Project C8276.

Jennings, W. 1978. *Gas Chromatography with Glass Capillary Columns.* New York: Academic Press.

Jennison, M. W. 1942. Atomizing of mouth and nose secretions into the air as visualized by high-speed photography. In *Aerobiology* (Am. Assoc. Advancement of Sci. Publ. 17), p. 106. Washington, DC.

Jewell, R. A. 1981. Reduction of formaldehyde levels in mobile homes. In *Symposium on Wood Adhesives—Research, Applications, and Needs.* Madison, WI: U.S. Forest Products Laboratory. Unpublished.

Johns, W. E. 1980. Wood interaction with moisture and formaldehyde. In *Technical Workshop on Formaldehyde* (B. Meyer and J. Fandey, eds.), p. 256. Washington, DC: CPSC.

Johnson, C. J., Moran, J. C., Paine, S. C., Anderson, H. W., and Breysse, P. A. 1975. Abatement of toxic levels of carbon monoxide in Seattle ice skating rinks, *Am. J. Public Health* **65**(10), 1087.

Johnson, C. J., Moran, J., and Pekich, R. 1975. Carbon monoxide in school buses, *Am. J. Public Health* **65**(12), 1327.

Johnson, R. E., and Sargent, F. 1968. The Physical and Chemical Properties of Human Sweat and Factors Affecting the Water Balance in Confined Spaces. Urbana, IL: Illinois Univ. (available through NTIS).

Jonassen, N. 1979. Static electricity in indoor environments. In *Indoor Climate, Effects on Human Comfort, Performance, and Health in Residential, Commercial, and Light-Industry Buildings* (P. O. Fanger and O. Valbjørn, eds.), p. 363. Copenhagen: Danish Building Research Institute.

Jopke, W. H., and Hass, D. R. 1970. Contamination of dishwashing facilities, *Hospitals* **44**(6), 124.

Joshi, S., and Wanner, H. U. 1975. Relation between outdoor and indoor air pollution, *Chem. Rundschau* **28**(22), 11, 13.

Joshi, S., and Wanner, H. U. 1975. Emissions from combustion of indoor dust, *Zentralbl. Bakteriol.* **160**, 499.

Ju, C., and Spengler, J. D. 1981. Respirable particle variation in residents, *Env. Res.* **15**, 593.

Junge, C. E. 1963. *Air Chemistry and Radioactivity.* New York: Academic Press.

Kahn, A., et al. 1975. A Study of Carbon Monoxide Sources in the St. Louis Metro-region. Cuers Report 4. Edwardsville, IL: Southern Illinois Univ.

Kaluzny, R. L. 1980. A Survey of Homeowner Experience with New Residential Housing Construction. HUD-PDR-622.

Kanarek, D. J., Wainer, R. A., Chamberlin, R. I., Weber, A. L., and Kazemi, H. 1973. Respiratory illness in a population exposed to beryllium, *Am. Rev. Respir. Dis.* **108**, 1295.

Kaneko, F., Nakadoi, T., Yamaoka, S., et al. 1977. Indoor air pollution levels in winter, *Seikatsu Eisei* **21**(5), 153.

Kaneko, F., Oka, M., Nakadoi, T., et al. 1979. Indoor air pollution: Nitrogen oxide levels from cooking, heating, *Seikatsu Eisei* **23**(1), 13

Katz, M. 1969. *Measurement of Air Pollutants: Guide to the Selection of Methods.* Geneva: WHO.

Katz, M. 1980. Advances in the analysis of air contaminants: A critical review, *J. Air Pollution Control Assoc.* **30,** 528.

Kays, W. M., and London, A. L. 1964. *Compact Heat Exchangers,* 2nd ed. New York: McGraw-Hill.

Kelley, J. S., and Sophocleus, G. J. 1978. Retinal hemorrhages in subacute carbon monoxide poisoning: Exposures in homes with blocked furnace flues, *J. Am. Med. Assoc.* **239,** 1515.

Kerr, H. D., Kulle, T. J., McIlhaney, M. L., and Swidersky, P. 1978. *Effects of Nitrogen Dioxide on Pulmonary Function in Human Subjects: An Environmental Chamber Study.* EPA-600/1-78-025; NTIS PB 281 186.

Keyes, D. L. 1980. Population Redistribution. In *Population Redistribution and Public Policy* (J. L. Berry and L. P. Silverman, eds.). Washington, DC: NAS.

Kim, M. K., and Schodek, D. L. 1981. The technique of urban mapping of potential indoor air pollution intensity for public health management. Preprint, *Int. Symposium Indoor Air Pollution, Health and Energy Conserv.,* Univ. of Massachusetts, October 13-16; Amherst, MA.

Kim, W. S., Geraci, C. L., and Kupel, R. E. 1980. Solid sorbent tube sampling and ion chromatographic analysis of formaldehyde, *Am. Ind. Hyg. Assoc. J.* **41,** 334.

Kimura, K. 1973. On the spectra analytical filter paper dust meter, *J. Sci. Labour* **49,** 493.

Kinne, H., and Cooley, A. M. 1913. *Shelter and Clothing.* New York: Macmillan.

Klein, H. -J., and Kunze, M. 1971. Experimental investigations on the spread of *Pseudomonas aeruginosa* by a cold aerosol apparatus for moistening of the room atmosphere, *Zentralbl. Bakteriol.* **A216**(2), 199.

Klotz, F., and Wanner, H. U. 1974. Air cleaning equipment: Its effect upon the quality of indoor air. *Zentralbl. Bakteriol.* **A227**(1-4), 559.

Knight, V. 1973. Airborne transmission and pulmonary deposition of respiratory viruses. In *Viral and Mycoplasmal Infections of the Respiratory Tract* (V. Knight, ed.), pp. 1-9. Philadelphia, PA: Lea & Febiger.

Knoll, G. F. 1979. *Radiation Detection and Measurement.* New York: Wiley.

Kobayashi, M., Miyaji, E., and Suzuki, Y. 1974. Air Cleaning Plan of Office Building, Part 3: Indoor Environment after Cleaning. Preprint, Japan Society of Architecture No. 155-60.

Kober, G. M., and Hayhurst, E. R. (eds.) 1924. *Industrial Health.* Philadelphia, PA: Blakiston.

Koch, E. I. 1976. Carbon monoxide pollution in schoolbuses: HR 12954 is legislation to correct this problem, *Congressional Record House* (18 April), p. 2949.

Koepsel, W. C. 1977. Sources and sinks of carbon monoxide and ozone in the indoor atmosphere. Marquette Univ. Milwaukee, WI. *Diss. Abstr. Int.* **B37**(11), 5539.

König, H. L., Kirmaier, N., Schauerte, W., Beierlein, H. R., and Breidenbach, H. 1979. Subjects in motocar praxistest: Influence of air-electrical pulse fields. In *Indoor Climate, Effects on Human Comfort, Performance, and Health in Residential, Commercial, and Light-Industry Buildings* (P. O. Fanger and O. Valbjørn, eds.), p. 333. Copenhagen: Danish Building Research Institute.

Korsgaard, J. 1979. The effect of the indoor environment on the house dust mite. In *Indoor Climate, Effects on Human Comfort, Performance, and Health in Residen-*

tial, Commercial, and Light-Industry Buildings (P. O. Fanger and O. Valbjørn, eds.), p. 187. Copenhagen: Danish Building Research Institute.

Korsgaard, J. 1981. Changes in indoor climate after tightening of apartments. Preprint, *Int. Symposium Indoor Air Pollution, Health and Energy Conserv.,* Univ. of Massachusetts, October 13-16; Amherst, MA.

Kramer, F. 1980. Handling of consumer formaldehyde problems. In *Technical Workshop on Formaldehyde* (B. Meyer and J. Fandey, eds.), Vol. 1, p. 88. Washington, DC: CPSC.

Kraybill, H. F., and Mehlman, M. A. 1977. *Environmental Cancer (Adv. Mod. Toxicol.* 3). New York: Wiley.

Kreiss, K., Gonzalez, M. G., Conright, K. L., and Scheere, A. R. 1981. Respiratory irritation due to carpet shampoos: Two outbreaks. Preprint, *Int. Symposium Indoor Air Pollution, Health and Energy Conserv.,* Univ. of Massachusetts, October 13-16; Amherst, MA.

Kreith, F. 1973. *Principles of Heat Transfer,* 3d ed. (Ser. in Mech. Eng.) New York: IEP-Dunn-Donnelly.

Kreith, F., and West, R. E. (eds.) 1980. *Economics of Solar Energy and Conservation Systems,* Vol. I: *General Principles.* Boca Raton, FL: CRC Press.

Krost, K. J., Pellizzari, E. D., Walburn, S. G., and Hulbrand, S. A. 1982. Collection and analysis of hazardous organic emissions. *Anal. Chem.* **54,** 810.

Kurzel, R. B., and Cetrulo, C. L. 1981. The effect of environmental pollutants on human reproduction, including birth defects, *Env. Sci. Technol.* **15,** 626.

Kusuda, T. 1976. Control of ventilation to conserve energy while maintaining acceptable indoor air quality, *ASHRAE Trans.* **82,** 1169.

Kusuda, T., Hund, C. M., and McNall, P. E. 1976. Radioactivity (radon and daughter products) as a potential factor in building ventilation, *ASHRAE J.* **21**(7), 30.

Kusuda, T., Bean, J. W., and McNall, P. E. 1979. Potential energy savings using comfort-index controls for building heating and cooling systems. In *Indoor Climate, Effects on Human Comfort, Performance, and Health in Residential, Commercial, and Light-Industry Buildings* (P. O. Fanger and O. Valbjørn, eds.), p. 797. Copenhagen: Danish Building Research Institute.

Kusuda, T., Silberstein, S., and McNall, P. E., Jr. 1982. Modeling of Radon and Daughter Concentrations in Ventilated Spaces. Washington, DC: NBS.

LaBastille, A. 1981. Acid rain, how great a menace?, *Natl. Geographic* **160,** 652.

LaBelle, C. W., and Brieger, H. 1955. Synergistic effects of aerosols, *Arch. Ind. Health* **12,** 623.

LaFleur, P. D. (ed.) 1976. *Accuracy in Trace Analysis: Sampling, Sample Handling, Analysis,* Vol. I. (Proc. 7th Materials Research Symposium, Gaithersburg, MD, October 7-11, 1974.) NBS Special Publ. 422.

Lammers, J. T. H., and Hoen, P. J. J. 1979. Simultaneous establishment of thermal comfort for different categories of men in a swimming pool. In *Indoor Climate, Effects on Human Comfort, Performance, and Health in Residential, Commercial, and Light-Industry Buildings* (P. O. Fanger and O. Valbjørn, eds.), p. 779. Copenhagen: Danish Building Research Institute.

Landrigan, P. J., McKinney, A. S., Hopkins, L. C., et al. 1975. Chronic lead absorption: Result of poor ventilation in an indoor pistol range, *J. Am. Med. Assoc.* **234,** 394.

Lane, D. J. (ed.) 1976. *Respiratory Disease.* London: Heinemann.

Langkilde, G. 1979. The influence of the thermal environment on office work. In *Indoor Climate, Effects on Human Comfort, Performance, and Health in Residential, Commercial, and Light-Industry Buildings* (P. O. Fanger and O. Valbjørn, eds.), p. 835. Copenhagen: Danish Building Research Institute.

Langton, N. H. (ed.) 1969. *Space Research and Technology,* Vol. 1: *The Space Environment.* New York: American Elsevier.

Lao, Y. J., Smith, R. W., Rich, T. L., and Davis, T. G. 1982. CO levels in homes with fuel-burning space heaters. *J. Envir. Health.* **44,** 180.

Las, Y. L., and Blackwell, F. O. 1979. Lead in newsprint, *J. Env. Health* **42,** 197.

Laskin, S., Kuschner, M., and Drew, R. T. 1970. Studies in pulmonary carcinogenesis. In *Inhalation Carcinogenesis* (M. G. Hanna, Jr., P. Nettesheim, and J. R. Gilbert, eds.), p. 321. Washington, DC: National Cancer Institute/AEC Technical Information Division.

Lave, L. B., and Seskin, E. P. 1972. Air pollution, climate, and home heating: Their effects on U.S. mortality rates, *Am. J. Public Health* **62**(7), 909.

Lave, L. B., and Seskin, E. P. 1977. *Air Pollution and Human Health.* Baltimore, MD: Johns Hopkins Univ. Press.

Lavoisier, A. 1801. Second mémoire sur la transpiration des animaux, *Traité de Chimie,* 2nd ed. Paris.

Lawrence Berkeley Laboratory. 1973–1976. Environmental Instrumentation Group. *Instrumentation for Environmental Monitoring,* Vols. 1–4. LBL-1. Berkeley, CA.

Leadbetter, M. R., and Corn, M. 1972. Particle size distribution of rat lung residues after exposure to fiberglass dust clouds, *Am. Ind. Hyg. Assoc. J.* **33,** 511.

Lear, C. W. 1976. *Charged Droplet Scrubber for Fine Particle Control: Laboratory Study.* EPA-600/2-76-249a; NTIS PB 258 823.

Leary, J. S. 1974. Safety evaluation of PVC resin strip containing dichlorvos, *Arch. Env. Health* **29,** 308.

Lebowitz, M. D. 1980. Effects of Cosmetic Aerosols on Respiratory Physiology. FDA report.

Lebowitz, M. D. 1981. The adverse health effects of biological aerosols, other aerosols, and micro-climate indoors on asthmatics and non-asthmatics. Preprint, *Int. Symposium Indoor Air Pollution, Health and Energy Conserv.,* Univ. of Massachusetts, October 13–16; Amherst, MA.

Lebowitz, M. D., Armet, D., and Knudson, R. J. 1981. The effect of passive smoking on pulmonary function in children. Preprint, *Int. Symposium Indoor Air Pollution, Health and Energy Conserv.,* Univ. of Massachusetts, October 13–16; Amherst, MA.

Lebret, E., Brunekreef, B., and Boleij, J. S. M. 1981. Indoor carbon monoxide pollution in the Netherlands. Preprint, *Int. Symposium Indoor Air Pollution, Health and Energy Conserv.,* Univ. of Massachusetts, October 13–16; Amherst, MA.

Lebrun, J., and Marret, D. 1979. Differences in comfort sensations in spaces heated in different ways: Belgian experiments. in *Indoor Climate, Effects on Human Comfort, Performance, and Health in Residential, Commercial, and Light-Industry Buildings* (P. O. Fanger and O. Valbjørn, eds.), p. 627. Copenhagen: Danish Building Research Institute.

Ledbetter, J. O. 1972. *Air Pollution* (in two parts), Part A: *Analysis.* New York: Marcel Dekker.

Lee, A. M., and Fraumeni, J. F., Jr. 1969. Arsenic and respiratory cancer in man: An occupational study, *J. Natl. Cancer Inst.* **42,** 1045.

Lee, D. H. K., and Kotin, P. (eds.) 1972. *Multiple Factors in the Causation of Environmentally Induced Disease.* New York and London: Academic Press.

Lee, R. E., Jr., and Mage, D. T. 1979. Personal exposure monitors—A survey. Paper 79-14.1 presented at 72nd Ann. Meeting, Air Pollution Control Assoc., Cincinnati, OH.

Lefcoe, N. M., and Inculet, I. I. 1975. Particulates in domestic premises, II: Ambient levels and indoor–outdoor relationships, *Arch. Env. Health* **30,** 565.

Lehmann, G., Müller, E. A., and Spitzer, H. 1950. Der Calorienbedarf bei gewerblicher Arbeit, *Arbeitsphysiol.* **14,** 166.

Lehmann, W. F. 1982. Correlations between various formaldehyde tests. *Proc. 16th Int. Particleboard Symposium* (T. Maloney, ed.), p. 333. Pullman, WA: Wash. State Univ.

Leithe, W. 1971. *The Analysis of Air Pollutants* (R. Kondor, trans.). Ann Arbor, MI: Ann Arbor Science.

Lemaire, R. 1975. Air conditioning in the SST, *Rev. Méd. Aéronaut. Spatiale* **9** (2nd quarter), 65.

Lepman, S. R., Boegel, M. L., and Hollowell, C. D. 1981. *Radon: A Bibliography.* LBL Report 12200; EEB-Vent 81-5.

Levetin, E., and Hurewitz, D. 1978. A one-year survey of the airborne molds of Tulsa, Oklahoma, II: Indoor survey, *Ann. Allergy* **41**(1), 25.

Levine, M. S. 1981. Investigation of health effects and environmental measures in a large office building. Preprint, *Int. Symposium Indoor Air Pollution, Health and Energy Conserv.,* Univ. of Massachusetts, October 13–16; Amherst, MA.

Lewis, H. E., Foster, A. R., Mullan, B. J., Cox, R. N., and Clark, R. P. 1969. Aerodynamics of the human microenvironment, *Lancet* **1**(7609), 1273.

Lieberman, A., and Schipma, P. 1969. *Air-Pollution-Monitoring Instrumentation: A Survey.* NASA SP-5072. Washington, DC.

Liebich, H. M., Bertsch, W., Zlatkis, A., and Schneider, H. J. 1975. Volatile organic components in the Skylab 4 spacecraft atmosphere, *Aviation, Space and Env. Med.* **46,** 1002.

Limaye, D. R., Sharko, J. R., Price, J. P., Orlando, J. A., Oelschlager, F., Wander, J., Hunt, G. G., Wright, T., Schossler, L., and Hyder, A. 1975. Comprehensive Evaluation of Energy Conservation Measures. EPA-230/1-75-003; NTIS PB 250 824.

Lin, C. H. 1975. Analysis of a regenerable polyethyleneimine carbon dioxide and humidity control system, *Proc. Intersoc. Conf. Env. Systems.*

Lin, C.-I., Anaclerio, R. N., Anthon, D. W., Fanning, L. Z., and Hollowell, C. D. 1979. Indoor/Outdoor Measurements of Formaldehyde and Total Aldehydes. LBL Report 9397; EEB-Vent 79-7.

Linko, Y-Y., Johnson, J. A., and Miller, B. S. 1962. The origin and fate of certain carbonyl compounds in white bread, *Cereal Chem.* **39**(6), 468.

Linnaeus, C. 1756. Odores medicamentorus, *Amoenitates Acad.* **22,** 325.

Lipschutz, R. D., Girman, J. R., Dickinson, J. B., Allen, J. R., and Traynor G. W. 1981. Infiltration and Indoor Air Quality in Energy Efficient Houses in Eugene, Oregon. LBL Report 12924; UC-95d.

Lister, J. 1868. An address on antiseptic system treatment in surgery, *Brit. Med. J.* **11**, 53.

Liu, B. Y. (ed.) 1976. *Fine Particles: Aerosol Generation Measurement, Sampling, and Analysis.* New York: Academic Press.

Long, K. R., Pierson, D. A., Brennan, S. T., Frank, C. W., and Hahne, R. A. 1979. *Problems Associated with the Use of Urea-Formaldehyde Foam for Residential Insulation,* Part I: *The Effects of Temperature and Humidity on Formaldehyde Release from Urea-Formaldehyde Foam Insulation.* ORNL/SUB-7559/1.

Løwenstein, H., Gravesen, S., and Schwartz, B. 1979. Airborne allergens—identification problems and the influence of temperature, humidity and ventilation. In *Indoor Climate, Effects on Human Comfort, Performance, and Health in Residential, Commercial, and Light-Industry Buildings* (P. O. Fanger and O. Valbjørn, eds.), p. 111. Copenhagen: Danish Building Research Institute.

Luciano, J. R. 1977. *Air Contamination Control in Hospitals.* New York: Plenum Press.

Lundgren, D. A., Harris, F. S., Jr., Marlow, W. H., Lippmann, M., Clark, W. E., and Durham, M. D. (eds.) 1979. *Aerosol Measurement.* Gainesville, FL: Univ. Presses of Florida.

Lundin, F. E., Jr., Lloyd, J. W., Smith, E. M., Archer, V. E., and Holaday, D. A. 1969. Mortality of uranium miners in relation to radiation exposure, hard-rock mining, and cigarette smoking—1950 through September 1967, *Health Phys.* **16**, 571.

Lundqvist, G. R. 1979. The effect of smoking on ventilation requirement. In *Indoor Climate, Effects on Human Comfort, Performance, and Health in Residential, Commercial, and Light-Industry Buildings* (P. O. Fanger and O. Valbjørn, eds.), p. 275. Copenhagen: Danish Building Research Institute.

MacLeod, K. E. 1981. Sources of Emissions of Polychlorinated Biphenyls into the Ambient Atmosphere and Indoor Air. *Env. Sci. Technol.* **15**, 926.

Macriss, R. A., and Elkins, R. H. 1977. Control of the level of nitric oxides (NO_x) in the indoor environment. DOE Report CONF-770557-2.

Maddalon, R. F., and Quinlivan, S. C. 1977. *Technical Manual for Inorganic Sampling and Analysis.* EPA-600/2-77-024; NTIS PB 266 842.

Mage, D. T., and Wallace, L. 1979. Personal Monitors for Exposure and Health Effect Studies. Reports of a conference. EPA-600/9-79-032.

Malloy, J. F. 1969. *Thermal Insulation.* New York: Van Nostrand.

Malmstroem, T. 1981. Aspects of efficient ventilation in office rooms. Preprint, *Int. Symposium Indoor Air Pollution, Health and Energy Conserv.,* Univ. of Massachusetts, October 13-16; Amherst, MA.

Manning, P. 1965. *Office Design: A Study of Environment.* Liverpool, England: The Pilkington Research Unit.

Mantel, N., and Schneidermann, M. A. 1975. Estimating 'safe' levels, a hazardous undertaking, *Cancer Res.* **35**, 1379.

Mantovani, A. 1978. The role of animals in the epidemiology of the mycoses, *Mycopathologia* **65**, 61.

Marier, G. 1974. Blood fluorocarbon level following exposure to household aerosols, *Canad. Med. Assoc. J.* **111**, 39.

Marsch, A. 1947. *Smoke.* London: Faber & Faber.

Marshall, F. J. 1980. Health Investigations of Urea-Formaldehyde Foam, December 1977 to January 1980. Trenton, NJ: New Jersey State Health Department.

Martell, E. A. (ed.) 1977. *Production of nuclei in the atmosphere by alpha interactions, Proc. 9th Int. Conf. Atmospheric Aerosols, Condensation and Ice Nuclei.* Galway (Ireland): Univ. College.

Martell, E. A., and Poet, W. E. 1981. Radon progeny on biological surfaces and their effects. Paper presented at Special Symposium on Natural Radiation Environment, Bombay, India, January.

Martin, A. E., Kaloyanova, F., and Maziarka, S. 1976. *Housing, the Housing Environment and Health—An Annotated Bibliography.* World Health Organization Offset Publ. Geneva.

Marx, P., and Schluter, G. 1980. An electronic analyzer for indoor climate, *Env. International* **3**, 265.

Commonwealth of Massachusetts. 1980. Regulations Concerning Hazardous Substances, 105 CMR 650.220(3) and 650.222, and Repurchase of Banned Hazardous Substances, 650.220. Boston, MA.

Matthews, T. G., Hawthorne, A. R., Metcalfe, C. E., Howell, T. C., Schneider, M. T., and Gammage, R. B. 1980. The Formaldehyde Monitoring Program, Development of Sampling and Analysis Procedures. Oak Ridge National Laboratory Report to CPSC.

Maxwell, C. M., Bohn, R., Caiazza, R., and Cowherd, C., Jr. 1977. Development of HATREMS Data Base and Emission Inventory Evaluation. EPA-450/3-77-011; NTIS PB 267 634.

May, K. R., and Pomeroy, N. P. 1973. Bacterial dispersion from the body surface. In *Airborne Transmission and Airborne Infection* (J. F. Hers and K. C. Winkler, eds.), p. 426. New York: Wiley.

Mayrsohn, H., and Crabtree, J. H. 1976. Source Reconciliation of Atmospheric Hydrocarbons in the South Coast Air Basin, 1975. RD-77-001. El Monte, CA: California Air Resources Board.

Mayrsohn, H., Kuramoto, M., Crabtree, J. H., Sothern, R. D., and Mano, S. H. 1976. Atmospheric Hydrocarbon Concentrations, June–September, 1976. El Monte, CA: California Air Resources Board.

McCarroll, J. R., Goldman, R. F., and Denniston, J. C. 1979. Food intake and energy expenditure in cold weather military training, *Military Med.* **144**(9), 606.

McClenny, W. A., and Stevens, R. K. 1981. New instrumentation adaptable to indoor air pollution studies. Preprint, *Int. Symposium Indoor Air Pollution, Health and Energy Conserv.,* Univ. of Massachusetts, October 13–16; Amherst, MA.

McDermott, H. J. 1976. *Handbook of Ventilation for Contaminant Control (Including OSHA requirements).* Ann Arbor, MI: Ann Arbor Science.

McEwen, F. L., and Stephenson, G. R. 1979. *The Use and Significance of Pesticides in the Environment.* New York: Wiley.

McFadden, J. E., Beard, J. H., and Moschandreas, D. J. 1978. Survey of Indoor Air Quality Health Criteria and Standards. EPA-600/7-7-027.

McFadden, W. 1973. *Techniques of Combined Gas Chromatography/Mass Spectrometry: Applications in Organic Analysis.* New York: Wiley.

McGinnis, J. P., Mincer, H. H., and Hembree, J. H., Jr. 1974. Mercury vapor exposure in a dental school environment, *J. Am. Dental Assoc.* **88**, 785.

McIntyre, D. A. 1979. The effect of air movement on thermal comfort and sensation. 1979. In *Indoor Climate, Effects on Human Comfort, Performance, and Health in Residential, Commercial, and Light-Industry Buildings* (P. O. Fanger and O. Valbjørn, eds.), p. 541. Copenhagen: Danish Building Research Institute.

McIntyre, D. A. 1980. *Indoor Climate*. London: Applied Science Publishers, Ltd.

McKee, H. C., Childers, R. E., and Parr, V. B. 1975. Collaborative Study of Reference Method for Measurement of Photochemical Oxidants in the Atmosphere (Ozone–Ethylene Chemiluminescent Method). EPA-650/4-75-016.

McLean, R. L. 1961. The effect of ultraviolet radiation upon the transmission of epidemic influenza in long-term hospital patients, *Am. Rev. Respir. Dis.* **83,** 36.

McNall, P. E., Jr. 1975. Practical methods of reducing airborne contaminants in interior spaces, *Arch. Env. Health* **30,** 552.

McNall, P. E. 1981. Building ventilation measurements, predictions, and standards, *Bull. N.Y. Acad. Med.* **57,** 1027.

McQuiston, F. C., and Parker, J. D. 1977. *Heating, Ventilating, and Air Conditioning: Analysis and Design*. New York: Wiley.

Meinke, W. W., and Scribner, B. F. (eds.) 1967. Trace characterization: Chemical and physical. In *Lectures and Discussions, 1st Materials Research Symposium*. (Gaithersburg, MD, October 3–7, 1966). NBS Monograph 100. Washington, DC.

Melia, F., and Swan, A. V. 1981. Respiratory Illness in British School Children. *J. Epidem. and Community Health* **35,** 161.

Melia, R. J. W., et al. 1972. Association between gas cooking and respiratory disease in children, *Brit. Med. J.* **2,** 149.

Menzies, R. 1791. Essay physiologique sur la respiration, *Ann. Chim.* **8,** 211.

Mercer, T. T., Bates, D. V., Fish, B. R., Hatch, T. F., and Morrow, P. E. 1966. Deposition and retention models for internal dosimetry of the human respiratory tract, *Health Phys.* **12,** 173.

Mes, J., Davies, D. J., and Turton, D. 1982. PCPs in the adipose tissues of Canadians. *Bull. Envir. Control Toxicol.* **28**(1), 97.

Meyer, B. 1977. *Sulfur, Energy, and Environment*. Amsterdam: Elsevier.

Meyer, B. 1979. *Urea-Formaldehyde Resins*. Reading, MA: Addison-Wesley.

Meyer, B. 1979. Formaldehyde release from urea-formaldehyde systems, *Proc. 13th Int. Particleboard Symposium* (T. Maloney, ed.), p. 343. Pullman, WA: Wash. State Univ.

Meyer, B., Johns, W. S., and Woo, J. K. 1980. Formaldehyde release from urea-formaldehyde resins, *Forest Prod. J.* **30**(3), 24.

Meyer, B., and Fandey, J. (eds.) 1980. *Technical Workshop on Formaldehyde*. Washington, DC: CPSC.

Meyer, B., and Hartley, R. P. 1981. *Inventory of Current Indoor Air Quality Related Research; Interagency Energy and Environment Research and Development Program*. EPA-600/7-81-119.

Meyer, B., and Nunlist, R. 1981. C-13 NMR Identification of Urea-Formaldehyde Resins. LBL Report 11762; *Polymer Preprints* **22,** 130.

Meyer, B. 1981. Formaldehyde Release from Phenolic Bonded Wood Panels. Report, American Plywood Association, Tacoma, WA.

Meyer, B., Koshlap, K., Geisling, K. L., and Miksch, R. R. 1982. Comparison of wet and dry test methods for formaldehyde emission. LBL-14259.

Meyers, B. F. 1981. Soldier of orange: The administrative, diplomatic, legislative and litigatory impact of herbicide Agent Orange in South Vietnam. In *New York State Temporary Commission on Dioxin Exposure, Preliminary Report*. Albany, NY: State of New York Temporary Commission on Dioxin Exposure.

Meyers, R. E., et al. 1978. Constraints on Coal Utilization with Respect to Air Pollution Production and Transport over Long Distances: Summary. Upton, NY: Brookhaven National Laboratory.

Michaelson, G. S. 1979. Ventilation as a means of contamination control in hospitals and laboratories. In *Indoor Climate, Effects on Human Comfort, Performance, and Health in Residential, Commercial, and Light-Industry Buildings* (P. O. Fanger and O. Valbjørn, eds.), p. 225. Copenhagen: Danish Building Research Institute.

Miksch, R. R., Anthon, D. W., and Fanning, L. Z. 1980. A modified pararosaniline procedure for the determination of formaldehyde in air. *Anal. Chem.* **53**(13), 2118.

Miksch, R. R., Hollowell, C. D., Fanning, L. Z., Newton, A., and Schmidt, H. 1980. Trace organic contaminants in indoor air environments. Paper presented at ACS Meeting, San Francisco, CA, August. LBL Report 10777.

Miksch, R. R., and Anthon D. W., 1981. A Recommendation for Combining the Standard Analytical Methods for the Determinations of Formaldehyde and Total Aldehydes in Air. LBL Report 13194.

Miller, R. L., Buron, W. E., and Spore, R. W. 1963. Aerosols produced by dental instrumentation, *Proc. 1st Int. Symposium Aerobiol.*, p. 97.

Mills, C. A. 1942. *Climate Makes the Man*. New York: Harper.

Milne, G. W. A., and Heller, S. R. 1980. NIH-EPA Chemical Information System, *J. Chem. Inf. Computer Sci.* **20**, 204.

Mincu, P., and Diaconescu, M. L. 1978. Indoor pollution, *Rev. Ig., Bacteriol., Virusol., Parazitol., Epidemiol., Pneumofiziol., Ig.* **27**(3), 201.

Mintz, S., Hosein, R., Batten, B., and Silverman, F. 1981. Design of a personal sampler of three respiratory irritants. Preprints, *Int. Symposium Indoor Air Pollution, Health and Energy Conserv.*, Univ. of Massachusetts, October 13-16; Amherst, MA.

Miquel, P. 1878-1899. Annual reports in *Annu. Obs. Montsouris*, Paris.

Miquel, P. 1883. *Les Organismes Vivants de l'Atmosphère*. Paris: Gauthier-Villars.

Miquel, P., and Benoist, L. 1890. De l'enregistrement des poussieres atmosphériques brute et organisées, *Ann. Micrographie* **1**, 572.

Mirick, W. 1980. Asbestos sealant evaluation, *Toxicol. Res. Projects Directory* **5**(2).

Mitchell, K. L., Sessions, B. W., and Turner, L. D. 1976. Spacelab environmental control system. Paper presented at Intersoc. Conf. Env. Systems, July 12-15. San Diego, CA.

Mitchell, W. J., and Midgett, M. R. 1975. Method for Obtaining Replicate Particulate Samples from Stationary Sources. EPA-650/4-75-025.

Mitre Corporation. 1973. *Compendium of Analytical Methods*, Vol. II: *Method Summaries*. EPA-R4-73-027b; NTIS PB 228 425.

Mitsubishi Electric Corporation. 1980. *Lossnay Heat Exchanger Manual*. Tokyo.

Miyazaki, T., Kaneko, F., Yamaoka, S., and Fukuda, M. 1977. Indoor air pollution by cigarettes, mosquito-sticks and incense-sticks, *Seikatsu Eisei* **21**(5), 167.

Mogabgab, W. J. 1968. Acute respiratory illnesses in university (1962-1966), military and industrial (1962-1963) populations, *Am. Rev. Respir. Dis.* **98**, 359.

Møhave, L., and Møller, J. 1979. The atmospheric environment in modern Danish dwellings—measurement in 39 flats. In *Indoor Climate, Effects on Human Comfort, Performance, and Health in Residential, Commercial, and Light-Industry Buildings* (P.O. Fanger and O. Valbjørn, eds.), p. 171. Copenhagen: Danish Building Research Institute.

Mollier, R. 1923, Ein neues Diagram fur Dampfluftgemische, *Zentralbl. Verb. Deut. Ind.* **67,** 869.

Morgan, M. S., and Morris, S. C. 1977. Individual air pollution monitors, 2: Examination of some nonoccupational research and regulatory uses and needs. Report: ISS BNL-50637.

Morken, D. A. 1975. The biological effect of radon on the lung. In *Noble Gases,* p. 501. ERDA TIC-CONF-730915.

Morken, D. A. 1980. The biological and health effects of radon: A review, In *Radon in Buildings* (R. Colle and P. E. McNall, Jr., eds.), p. 300. NBS Special Publ. 581. Washington, DC.

Morris, A. L., and Barras, R. C. (eds.) 1978. Air quality meteorology and atmospheric ozone. In *Methods of Sampling and Analysis of Atmospheres.* ASTM Special Publ. 653. Philadelphia, PA.

Morris, J. E. W. 1972. Microbiology of the submarine environment, *Proc. Roy. Soc. Med.* **65,** 799.

Morris, J. E. W., and Fallon, R. J. 1973. Studies on the microbial flora in the air of submarines and the nasopharyngeal flora of the crew, *J. Hyg.* **71,** 761.

Morrow, P. E. 1970. Models for the study of particle retention and elimination in the lung. In *Inhalation Carcinogenesis* (M. G. Hanna, Jr., P. Nettesheim, and J. R. Gilbert, eds.). Washington, DC: National Cancer Institute/AEC Technical Information Division.

Morschauser, C. R. 1979. Status of formaldehyde regulation in the particleboard industry. *Proc. 12th Int. Particleboard Symposium* (T. Maloney, ed.), p. 227. Pullman, WA: Wash. State Univ.

Morveau, Messrs., Maset and Duhamel. 1786. *Encyclopedie de Chymie, Pharmacie et Metallurgie.* Paris: Panchoucke.

Moschandreas, D. J., Courtney, W. J., Pilotte, J. O., et al. 1978. Indoor and outdoor sources of particulate air pollution in a residential environment, *Proc. 4th Int. Conf. Sens. Env. Pollution,* p. 370.

Moschandreas, D. J., and Stark, J. W. C. 1978. The GEOMET Indoor–Outdoor Air Pollution Model. EPA-600/7-78-106.

Moschandreas, D. J., Stark, J. W. C., McFadden, J. E., and Morse, S. S. 1978. *Indoor Air Pollution in the Residential Environment,* Vols. I and II. EPA-600/7-78-229a and b.

Moschandreas, D. J., and Morse, S. 1979. Exposure estimation and mobility patterns. Paper 79-14.4 presented at 72nd Ann. Meeting, Air Pollution Control Assoc., Cincinnati, OH.

Moschandreas, D. J., Winchester, J. W., Nelson, J. W., and Burton, R. M. 1979. Fine particle residential indoor air pollution, *Atmos. Env.* **13,** 1413. See also **15,** 205 (1981).

Moschandreas, D. J., Zabransky, J., and Pelton, D. J. 1980. Indoor air quality characteristics of the office environment. Paper presented at Air Pollution Control Assoc. Meeting, Montreal, Quebec.

Moschandreas, D. J., Moyer, R. H., Ward, J. R., Rector, H. E., Schakenbach, J. T., Heavner, R., Stone, R., Tucker, B., Samuels, R., and DePaso, D. 1980. An Evaluation of Formaldehyde Problems in Residential Mobile Homes. GEOMET Report ESF-797 prepared for HUD.

Moschandreas, D. J. 1981. Exposure to pollutants and daily time budgets of people, *Bull. N.Y. Acad. Med.* **57**, 845.

Moschandreas, D. J., Pelton, D. J., and Cade, R. C. 1981. Carbon monoxide and aerosol concentrations in public access buildings. *Proc. Int. Symposium Indoor Air Pollution, Health, and Energy Conservation,* October 13–16, University of Massachusetts, Amherst, MA.

Moschandreas, D. J., and Rector, H. E. 1981. Radon and Aldehyde Concentrations in the Indoor Environment. LBL Report 12590; UC-41.

Most, R. S. 1981. Persistence of systems from UFFI home insulation, *J. Env. Health* **43**, 251.

Muchtarova, M., and Dimov, N. 1978. Gas chromatographic identification of some indoor air pollutants using correlation equations, *J. Chromatog.* **148**(1), 269.

Muir, C. S., and Wagner, G. (eds.) 1978. *Directory of on-going research in cancer epidemiology 1978.* IARC Scientific Publ. 26. Lyon, France: International Agency for Research on Cancer, WHO.

Muir, D. C. F. 1972. *Clinical Aspects of Inhaled Particles.* London: Heinemann.

Muir, D. C. F. 1976. The response of the lung to inhaled particles. Chapter 6 in *Respiratory Disease* (D. J. Lane, ed.). London: Heinemann.

Mulik, J. D., and Sawicki, E. (eds.) 1979. *Ion Chromatographic Analysis of Environmental Pollutants,* Vol. 2. Ann Arbor, MI: Ann Arbor Science.

Muysers, K., and Smidt, U. 1958. *Respirations-Massenspektrometrie.* Stuttgart: Schattauer.

Myers, G. E., and Nagaoka, M. 1980. Formaldehyde from UF-bonded panels—Its measurement and its relation to air contamination. Paper presented at symposium, Wood Adhesives, Research, Applications, and Needs, U.S. Forest Products Laboratory, Madison, WI.

Myers, G. E. 1981. Formaldehyde emission: Methods of measurement and effects of several particleboard variables, *Wood Sci.* **13**(3), 140.

Myers, G. E., and Nagaoka, M. 1981. Emission of formaldehyde by particleboard: Effect of ventilation rate and loading on air contamination, *Forest Prod. J.* **31**(7), 39.

Myers, G. E. 1982. Investigation of urea formaldehyde polymer cure by infrared, *J. Appl. Polymer Sci.* **26**, 747.

Myers, G. E. 1982. Hydrolytic stability of cured urea-formaldehyde resins, *Wood Sci.* **14**.

Myers, G. E., and Nagaoka, M. 1982. Formaldehyde emission: Methods of measurement and effects of some particleboard variables, *Forest Prod. J.* **32**.

Myers, G. E., Seymour, J. W., and Khan, T. 1982. Formaldehyde air contamination in mobile homes: Variation with interior location and time of day, *Env. Sci. Technol.*

Myerson, B. (ed.) 1977. *Consumer Guide to Product Information.* New York: Bristol-Myers.

Nakanishi, E., Pereira, N. C., Fan, L. T., and Hwang, C. L. 1973. Simultaneous control of temperature and humidity in a confined space, I: Mathematical modeling of the dynamic behavior of temperature and humidity in a confined space, *Building Sci.* **8**, 39.

Narasaki, H. 1976. Control of indoor air pollution due to smoking, *Kuki Seijo* **14**(4), 12.

Narat, J. K. 1925. Experimental production of malignant growth by simple chemicals, *J. Cancer Res.* **9**, 135.

National Council on Radiation Protection and Measurements. 1975. *Natural Background Radiation in the United States: Recommendations of the National Council on Radiation Protection and Measurements.* NCRP Report 45. Washington, DC.

National Fire Protection Association. 1976. *Safety to Life from Fire in Buildings.* NFPA/ANSI Code 101-1976, Chicago, IL.

National Fire Protection Association. 1977. *Manual on Mobile Home Heating and Cooling Load Calculations* (*National Fire Code* **16**). NFPA 501 BM-1. Boston, MA.

National Particleboard Association. 1981. Equilibrium Jar Method Colorimetric Determination of Formaldehyde Chromotropic Acid Reagent for Product Testing. Silver Springs, MD: NPA.

National Research Council. 1969. *Effects of Chronic Exposure to Low Levels of Carbon Monoxide on Human Health, Behavior and Performance.* Washington, DC: NAS.

National Research Council. 1971. *Biological Effects of Atmospheric Pollutants—Fluorides.* Washington, DC: NAS.

National Research Council. 1971. *Asbestos.* Washington, DC: NAS.

National Research Council. 1972. *Lead: Airborne Lead in Perspective.* Washington, DC: NAS.

National Research Council. 1972. *Particulate Polycyclic Organic Matter.* Washington, DC: NAS.

National Research Council. 1974. *Chromium.* Washington, DC: NAS.

National Research Council. 1975. *Nickel.* Washington, DC: NAS.

National Research Council. 1976. *Vapor-Phase Organic Pollutants.* Washington, DC: NAS.

National Research Council. 1976. *Medical and Biological Effects of Environmental Pollutants.* Washington, DC: NAS.

National Research Council. 1977. *Airborne Particles.* EPA-600/1-77-053; NTIS PB 276 723.

National Research Council. 1977. *Ammonia.* EPA-600/1-77-054; NTIS PB 278 182.

National Research Council. 1977. *Principles and Procedures for Evaluating the Toxicity of Household Substances.* Revison of NAS Publ. 1138. Washington, DC: NAS.

National Research Council. 1977. *Carbon Monoxide.* Washington, DC: NAS.

National Research Council. 1978. *Chloroform, Carbon Tetrachloride, and Other Halomethanes: An Environmental Assessment.*, Washington, DC: NAS.

National Research Council. 1978. *An Assessment of Mercury in the Environment.* Washington, DC: NAS.

National Research Council. 1979. *Protection against Depletion of Stratospheric Ozone by Chlorofluorocarbons.* Washington, DC: NAS.

National Research Council. 1980. *Formaldehyde and Other Aldehydes.* Washington, DC: NAS.

National Research Council. 1980. *Formaldehyde—An Assessment of Its Health Effects.* Washington, DC: NAS.

National Research Council. 1980. The Effects on Populations of Exposure to Low Levels of Ionizing Radiation. Washington, DC: NAS Press.

National Research Council. 1981. *Indoor Pollutants*. Report for EPA. Washington, DC: NAS Press.

National Research Council. 1982. *Causes and Effect of Stratospheric Ozone Reduction by Fluorocarbons*. Washington, DC: NAS.

Nazaroff, W. W. 1981. An Improved Technique for Measuring Working Levels of Radon Daughters in Residences. LBL Report 9986; EEB-Vent 79-15. Also *Health Phys.* **39**, 683.

Neal, A. D., Wadden, R. A., and Rosenberg, S. H. 1978. Evaluation of indoor particulate concentrations for an urban hospital, *Am. Ind. Hyg. Assoc. J.* **39**, 578.

Nefedov, I. U., Ryzhkova, V. E., Savina, V., and Sokolov, N. L. 1969. Investigation of human expired air for contaminants, *Env. Space Sci.* **3**, 387.

Nefedov, I. U., Savina, V. P., and Sokolov, N. L. 1972. Expired air as a source of spacecraft environment carbon monoxide contamination, *Proc. 23d Int. Astron. Congr.* Vienna, Oct. 8-15, 1972.

Nefedov, I. U., Zaloguev, S. N., and Savina, V. P. 1976. The problem of habitability in spaceships. In *Space Activity Impact on Science and Technology*. Oxford: Pergamon Press.

Nelson, C. L. 1979. Environmental bacteriology in the unidirectional (horizontal) operating room, *Arch. Surg.* **114**, 778.

Nelson, G. O. 1971. *Controlled Test Atmospheres: Principles and Techniques*. Ann Arbor, MI: Ann Arbor Science.

Nero, A. V. 1981. Indoor Radon Sources, Concentrations, and Standards, LBL Report 12932.

Newman, W. H., Viele, M. R., and Hrach, F. J. 1980. Reduced bleed air extraction for DC-10 cabin air conditioning, *Proc. 16th Int. Propulsion Conf.* Hartford, CT, June 30, 1980.

Newmark, F. M. 1968. Pollen aerobiology, *Ann. Allergy* **26**, 358.

New York State Commission on Ventilation. 1923. *Report on Ventilation*. New York: Dutton.

Niemelä, R., and Vainio, H. 1981. Formaldehyde exposure in the work and general environment, *Scand. J. Env. Health* **33**, 444.

Nikolov, S. K., Kambulin, N. A., and Kolupaeva, M. V. 1976. Problems of labor hygiene in modern stock breeding, *Gig. Sanit.* **12**, 80.

Nishi, Y. 1979. Moisture and heat transfer characteristics of the indoor clothing. In *Indoor Climate, Effects on Human Comfort, Performance, and Health in Residential, Commercial, and Light-Industry Buildings* (P. O. Fanger and O. Valbjørn, eds.), p. 615. Copenhagen: Danish Building Research Institute.

Noll, K. E., and Davis, W. T. 1976. *Power Generation: Air Pollution Monitoring and Control*. Ann Arbor, MI: Ann Arbor Science.

Noll, K. E., and Miller, T. L. 1977. *Air Monitoring Survey Design*. Ann Arbor, MI: Ann Arbor Science.

Norris, K. P., and Harper, G. J. 1970. Windborne dispersal of foot and mouth virus, *Nature* **225**, 98.

Noyes Data Corporation. 1972. *Pollution Analyzing and Monitoring Instruments*. Park Ridge, NJ: Noyes Data.

O'Callaghan, P. W. 1978. *Building for Energy Conservation.* New York: Pergamon Press.
O'Hare, D. 1942. *Airborne Infection.* New York: Commonwealth Fund.
Oikawa, K. 1977. *Trace Analysis of Atmospheric Samples.* New York: Wiley.
Olesen, B. W., and Thorshauge, J. 1979. Differences in comfort sensations in spaces heated by different methods: Danish experiments. In *Indoor Climate, Effects on Human Comfort, Performance, and Health in Residential, Commercial, and Light-Industry Buildings* (P. O. Fanger and O. Valbjørn, eds.), p. 645. Copenhagen: Danish Building Research Institute.
Olesen, B. W., Schøler, M., and Fanger, P. O. 1979. Discomfort caused by vertical air temperature differences. In *Indoor Climate, Effects on Human Comfort, Performance, and Health in Residential, Commercial, and Light-Industry Buildings* (P. O. Fanger and O. Valbjørn, eds.), p. 561. Copenhagen: Danish Building Research Institute.
Olgyay, V., and Olgyay, A. 1963. *Design with Climate.* Princeton, NJ: Princeton Univ. Press.
Oliver, H. R., and Mayhead, G. J. 1974. Wind measurements in a pine forest during a destructive gale, *Agr. Meteorol.* **14,** 347.
Olsen, D., and Haynes, J. L. 1969. *Preliminary Air Pollution Survey of Organic Carcinogens: A Literature Review.* NAPCA Publ. APTD 69-43. Raleigh, NC: HEW.
O'Quinn, S. E., and Kennedy, B. 1965. Contact dermatitis due to formaldehyde in clothing textiles, *J. Am. Med. Assoc.* **194,** 503.
Osborne, S. 1980. Formaldehyde Release from Eight Commercial Urea-Formaldehyde Foam Insulation Materials. Report for CPSC. Philadelphia, PA: Franklin Institute Research Laboratory.
Ott, W. R. 1971. An Urban Survey Technique for Measuring the Spatial Variation of Carbon Monoxide Concentrations in Cities. Ph.D. dissertation, Department of Civil Engineering, Stanford University.
Ott, W. R. 1980. Concepts of Human Exposure to Environmental Pollution. SIMS Tech. Report 32, Department of Statistics. Menlo Park, CA: Stanford University. *Env. International* **7,** in press.
Ott, W. R. 1981. Exposure Estimates Based on Computer Generated Activity Patterns. Paper presented at 74th Ann. Meeting, Air Pollution Control Assoc., June 21-26. Philadelphia, PA.
Ott, W. R., and Willits, N. 1981. Modeling the dynamic response of an automobile for air pollution exposure studies, *Proc. Environmetrics-81.,* p. 104. Philadelphia, PA: Soc. Ind. Appl. Math.
Ott, W. R., and Flachsbart, P. G. 1982. Measurements of indoor and outdoor carbon monoxide concentrations using personal monitors. *Env. International,* **8.**
Ower, E., and Parkhurst, R. C. 1977. *The Measurement of Air Flow,* 5th ed. New York: Pergamon Press.
Ozkaynak, H., Ryan, B., Turner, I. O., and Allan, G. 1982. Indoor air pollution simulations using dynamic multicompartment models. *Env. International,* **8.**

Pack, D. H. (ed.) 1977. Air Pollution Measurement Techniques. Special Env. Report 10, WMO 460. Geneva: World Meteorological Organization.
Palmes, E. D., and Gunnison, A. F. 1973. Personal monitoring device for gaseous contaminants, *Am. Ind. Hyg. Assoc. J.* **34,** 78.

Palmes, E. D., Gunnison, A. F., DiMattio, J., and Tomczyk, C. 1976. Personal sampler for nitrogen dioxide, *Am. Ind. Hyg. Assoc. J.* **37,** 570.

Palmes, E. D., Tomczyk, C., and March, A. W. 1979. Relationship of indoor NO_2 concentrations to use of unvented gas appliances, *J. Air Pollution Control Assoc.* **29,** 392.

Palmes, E. D., and Tomczyk, C. 1982. Personal samplers for NO_x, CO and CO_2. BUMINES-OFR-137; NTIS PB82-123597.

Parker, A. (ed.) 1978. *Industrial Air Pollution Handbook.* London: McGraw-Hill.

Parker, J. F., Jr., and West, V. R. (eds.) 1973. *Bioastronautics Data Book,* 2nd ed. NASA Publ. SP-3006.

Partridge, L. J., Nayak, P. R., Stricoff, R. S., and Hagopian, J. H. 1978. *A Recommended Approach to Recirculation of Exhaust Air.* HEW (NIOSH) Publ. 78-124.

Peatman, J. B. 1977. *Microcomputer-Based Design.* New York: McGraw-Hill.

Pellizzari, E. D. 1978. Measurement of Carcinogenic Vapors in Ambient Atmospheres. Final report. EPA-600/7-78-062; NTIS PB 283 023.

Pellizzari, E. D., and Bunch, J. E. 1979. Ambient Air Carcinogenic Vapors—Improved Sampling and Analytical Techniques and Field Studies. EPA-600/2-79-081.

Pellizzari, E. D., Erickson, M. D., Hartwell, T. D., Williams, S. R., Sparaciro, C. M., and Waddell, R. D. 1981. Toxic Chemicals in Environmental and Human Samples. Research Triangle Park, NC: EPA.

Pennington, J. H., Lumley, J., and O'Grady, F. 1966. The growth of *Pseudomonas pyocyanea* in Garthur condenser humidifiers: An experimental study, *Anaesthesia* **21,** 211.

Pepys, J. 1981. The role of inhaled chemical dusts, vapours and fumes in the production of asthma. Preprint, *Int. Symposium Indoor Air Pollution, Health and Energy Conserv.,* Univ. of Massachusetts, October 13-16; Amherst, MA.

Perera, F. P. 1979. *Respirable Particles.* Cambridge, MA: Ballinger Publ. Co.

Perry, R., and Young, R. J. (eds.) 1977. *Handbook of Air Pollution Analysis.* New York: Wiley.

Pesez, M., and Bartos, J. 1974. *Colorimetric and Fluorimetric Analysis of Organic Compounds and Drugs,* Vol. I. New York: Marcel Dekker.

Petersen, G. A., and Sabersky, R. H. 1975. Measurements of pollutants inside an automobile, *J. Air Pollution Control Assoc.* **25,** 1028.

Pfeiffer, G. O., and Nikel, C. M. (eds.) 1980. *The Household Environment and Chronic Illness; Guidelines for Constructing and Maintaining a Less Polluted Residence.* Springfield, IL: Charles C Thomas.

Phibbs, B. P., Sundin, R. E., and Mitchell, R. S. 1971. Silicosis in Wyoming bentonite workers, *Am. Rev. Respir. Dis.* **103,** 1.

Pickrell, J. A., Griffis, L. C., and Hobbs, C. H. 1981. Release of formaldehyde from various consumer products. Report by the Lovelace Inhalation Toxicology Research Institute for the CPSC. Albuquerque, NM.

Pocchiari, F., Silano, V., and Zampieri, A. 1979. Human health effects from accidental release of tetrachlorodibenzo-*p*-dioxin (TCDD) at Seveso, Italy, *Proc. N.Y. Acad. Sci.* **1979,** 311.

Pohl-Rüling, J., and Fischer, P. 1979. The dose-effect relationship of chromosome aberrations due to an increased burden of natural radioactivity, *Rad. Res.* **80,** 61.

Port, R., Schmähl, D., and Wahrendorf, J. 1976. Some examples of dose-response studies in chemical carcinogenesis, *Oncology* **33,** 66.

Powell, R. J., and Wharton, L. M. 1982. Development of the Canadian Clean Air Act, *J. Air Pollution Control Assoc.* **32**, 62.

Pressler, C. L. 1981. Urea-Formaldehyde Foam Insulation: Static Measurement of Gaseous Formaldehyde at Ambient and Elevated Temperature Conditions. Washington, DC: CPSC.

Prichard, H. M., and Gesell, T. F. 1977. Rapid measurements of ^{222}Rn concentrations in water with a commercial liquid scintillation counter, *Health Phys.* **33**, 577.

Priestley, J. 1790. Experiments and observations on different kinds of air, etc., *Ann. Chim.* **7**, 133.

Priestley, J. 1790. *Experiments and Observations on Different Kinds of Air, and other Branches of Natural Philosophy Connected with the Subject.* Birmingham: Pearson. New York: Kraus Reprint, 1970.

Pruett, J. G., and Winslow, S. G. 1980. *Asbestos in air: A bibliography with abstracts, 1965–1980.* Report NLM/TIRC-80/2. Toxicology Information Response Center, Oak Ridge National Laboratory.

Rabinowitz, M., Wetherill, G., and Kopple, J. 1975. Absorption, storage and secretion of lead by normal humans, *Proc. Univ. Mo. Ann. Conf. Trace Subst. Env. Health* **9**, 361.

Radford, E. P., and Martell, E. A. 1977. Polonium-210 : Lead-210 ratios as an index of residence times of insoluble particles from cigarette smoke in bronchial epithelium. In *Inhaled Particles* (W. H. Walton, ed.), Vol. 4, Part 2. Oxford: Pergamon Press.

Rajhans, G. S., and Bragg, G. M. 1978. *Engineering Aspects of Asbestos Dust Control.* Ann Arbor, MI: Ann Arbor Science.

Ramazzini, B. 1713. *Diatribes de morbis Antificum* (W. C. Right, trans.). Chicago: Univ. of Chicago Press, 1940.

Randolph, T. H., and Moss, R. W. 1980. *An Alternative Approach to Allergies.* New York: Lippincott and Crowell.

Rati, E., and Ramalingam, A. 1979. Toxic strains among air-borne isolates of *Aspergillus flavus, Indian J. Exp. Biol.* **17**(1), 97.

Rea, W. J., Smiley, R. E., Sprague, D. E., Edgar, R. T., Fenyves, E. J., Greenberg, M., and Johnson, A. R. 1981. Environmental control of indoor air pollution (challenge testing in humans with ambient chemicals). Preprint, *Int. Symposium Indoor Air Pollution, Health and Energy Conserv.,* Univ. of Massachusetts, October 13–16; Amherst, MA.

Real Estate Research Corp. (New York). 1974. *The Cost of Sprawl.* Washington, DC: U.S. Govt. Printing Office.

Reed, C. E. 1981. Allergic agents, *Bull. N.Y. Acad. Med.* **57**, 897.

Reed, G. H. 1974. *Refrigeration: A Practical Manual for Apprentices,* 3d ed. Barking, England: Applied Science Publishers.

Repace, J. L., and Lowry, A. H. 1980. Indoor air pollution, tobacco smoke and public health, *Science* **208**, 464, see also *Science* **215**, 197.

Repace, J. L., Ott, W. R., and Wallace, L. A. 1980. Total Human Exposure to Air Pollution. Paper presented at 73d Ann. Meeting, Air Pollution Control Assoc., June 22–27, Montreal, Canada.

Repace, J. L. 1982. The problem of passive smoking, *Bull. N. Y. Acad. Med.* **57**, 936.

Repace, J. L., and Lowrey, A. H. 1982. Tobacco Smoke, Ventilation, and Indoor Air Quality. *ASHRAE Trans.* **88**(1).

Reppert, M. H. (ed.) 1979. *Summer Attic and Whole-House Ventilation* (Proc. Workshop, Gaithersburg, MD, July 13, 1978. Washington, DC: NBS.

Reznikov, M., Leggo, J. H., and Dawson, D. J. 1971. Investigation by seroagglutination of strains of the *Mycobacterium intracellulare–M. scrofulaceum* group from house dusts and sputum in southeastern Queensland, *Am. Rev. Respir. Dis.* **104**, 951.

Richardson, A. C. B. 1980. Control of radioactive hazardous wastes pursuant to the Resource Conservation and Recovery Act (RCRA); In *Radon in Buildings* (R. Collé and P. E. McNall, Jr., eds.), p. 68. NBS Special Publ. 581, Washington, DC.

Riley, D. J., and Saldana, M. 1973. Pigeon breeder's lung: Subacute course and the importance of indirect exposure, *Am. Rev. Respir. Dis.* **107**, 456.

Riley, E. C., Murphy, G., and Riley, R. L. 1978. Airborne spred of measles in a suburban elementary school, *Am. J. Epidermiol.* **107**, 421.

Riley, R. L., Mills, C. C., O'Grady, F., Sultan, L. U., Wittstadt, F., and Shivpuri, D. N. 1962. Infectiousness of air from a tuberculosis ward: Ultra-violet irradiation of infected air—Comparative infectiousness of different patients, *Am. Rev. Respir. Dis.* **85**, 511.

Riley, R. L. 1981. Airborne infection. Preprint, *Int. Symposium Indoor Air Pollution Health and Energy Conserv.*, Univ. of Massachusetts, October 13–16; Amherst, MA.

Rimington, C., Stillwell, D. E., and Maunsell, K. 1947. The allergens of house-dust, *Brit. J. Exp. Pathol.* **28**, 309.

Robinson, J. P. 1977. *How Americans Use Time: A Social-Psychological Analysis of Everyday Behavior.* New York: Praeger.

Rockette, H. 1977. Mortality among Coal Miners Covered by the UMWA Health and Retirement Funds. HEW (NIOSH) Publ. 0. 77-155.

Roddis, R. J. 1960. *The Law Relating to Caravans.* London: Shaw & Son.

Roffael, E. 1980. Measurement and mechanism of formaldehyde release from chipboard. In *Technical Workshop on Formaldehyde* (B. Meyer and J. Fandey, eds.), Vol. 2, p. 145. Washington, DC: CPSC.

Rogozen, M. B. 1980. Dynamic Simulation of Radon Daughter Exposure in Apartments Using Solar Rockbed Heat Storage. Tech. Note 922-41-08, SAI-1-068-81-545. Report for Solar Energy Research Institute by Science Applications Inc., Los Angeles, CA.

Rohl, A. N. 1979. Asbestos-containing material in school buildings. EPA report.

Rohles, F. H., Woods, J. E., and Nevins, R. N. 1973. The influence of clothing and temperature on sedentary comfort, *ASHRAE Trans.* **79**(2), 71.

Rohles, F. H., Woods, J. E., and Nevins, R. N. 1973. The effect of air movement and temperature on the thermal senses of sedentary man, *ASHRAE Trans.* **80**(1), 100.

Roose, R. W. (ed.) 1978. *Handbook of Energy Conservation for Mechanical Systems in Buildings.* (Compilation of articles and information published previously in *Heating, Piping and Air Conditioning.*) New York: Van Nostrand Reinhold.

Rosebury, T. 1962. *Microorganisms Indigenous to Man.* New York: McGraw-Hill.

Roseme, G. D., Hollowell, C. D., Meier, A., Rosenfeld, A., and Turiel, I. 1979. Air-to-Air Heat Exchangers: Saving Energy and Improving Indoor Air Quality. LBL Report-9381; EEB-Vent 79-11.

Rosenzweig, A. L. 1970. Contaminated humidifiers, *New England J. Med.* **283**, 1056.

Ross, H., and Berg, D. 1981. Workshop on Indoor Air Quality Research Needs. Interagency Energy–Environment Research and Development Program report. EPA-600/7-81-118.

Ross, J. H. 1852. *Hints and Helps to Health and Happiness; or Long Life and Little Physic,* 3d ed. Auburn, NY: Derby and Miller.

Rossiter, W. J. 1980. ASTM task force on urea formaldehyde foam insulation. In *Technical Workshop on Formaldehyde* (B. Meyer and J. Fandey, eds.), Vol. 3, p. 77. Washington, DC: CPSC.

Roth, E. M. (ed.) 1968. *Compendium of Human Responses to the Aerospace Environment,* Vol. 3, Sections 10–16. NASA CR-1205 (III). Washington, DC.

Rothman, L. S., Goldman, A., Gillis, J. R., Tipping, R. H., Brown, L. R., Margolis, J. S., Maki, A. G., and Young, L. D. G. 1981. AFGL trace gas compilation: 1980 version, *Appl. Optics* **20**, 1323.

Rothman, P. 1972. The effect of joint exposure to alcohol and tobacco on the risk of cancer of mouth and larynx, *J. Chronic Dis.* **25**, 711.

Ruecker, M. R., and Shaver, B. 1974. Mass spectrometers for atmospheric monitoring and control in the Shuttle era. *Proc. Intersoc. Conf. Env. Systems* (Seattle, WA, July 29–Aug. 1), New York: AICHE.

Rueden, H., and Langer, H. 1976. Indoor and outdoor measurements of carbon monoxide, *Gesund.-Ing.* **97**(9), 205.

Rundo, J., Markun, F., and Plondke, N. J. 1979. Observation of high concentrations of radon in certain houses, *Health Phys.* **36**, 729.

Russenberger, H. J., Scholz, E., and Wanner, H. U. 1974. Bakterielle Kontamination der Raumluft in Intensivpflegestationen, *Soz. Praventivmed.* **19**, 337.

Russenberger, H. J., and Wanner, H. U. 1974. Keimgehalt der Raumluft in Abhangigkeit des Luftwechsels, *Zentralbl. Bakteriol.* **A227**, 564.

Rutkowska, I., and Sazynska, M. 1981. Formaldehyde levels in new appartments. *Rocz. Panstw. Zahl. Hig.* **31**(6), 623.

Rye, W. A. 1973. Human response to isocyanate exposure, *J. Occupational Med.* **15**, 306.

Rylander, R., Myrback, K.-E., Verner-Carlson, B., and Öhrström, M. 1974. Bacteriological investigation of wall-to-wall carpeting, *Am. J. Public Health* **62**(2), 163.

Rylander, R. 1976. The response of the lung to inhaled particles. Chapter 7 in *Respiratory Disease* (D. J. Lane, ed.). London: Heinemann.

Rylander, R., et al. 1978. Humidifier fever and endotoxin exposure, *Clin. Allergy* **8**, 511.

Sadler, W. S. 1925. *The Essentials of Healthful Living.* New York: Macmillan.

Saeltzer, A. 1872. *A Treatise on Acoustics in Connection with Ventilation; and an Account of the Modern and Ancient Methods of Heating and Ventilation.* New York: Van Nostrand.

Salvaggio, J. E., and Karr, R. M. 1979. Hypersensitivity pneumonitis: State of the art, *Chest* **75**, 270.

Samet, J. M., Speizer, F. E., Bishop, Y. M. M., and Spengler, J. E. 1981. The relationship between air pollution and emergency room visits in an industrial community, *J. Air Pollution Control Assoc.* **31**, 236.

Sanctorius. 1728. *Medicina Statica* (John Quincy, trans.), 4th ed. London.

Sansone, E. B., and Slain, M. W. 1973. Redispersion of indoor surface contamination: A review, *J. Hazard. Mater.* **2**(4), 347.

Satish, J., and Wanner, H. U. 1975. Emissionen bei der Versengung von Raumluftstaub, *Zentralbl. Bakteriol.* **B160,** 499.

Satish, J., and Wanner, H. U. 1976. Source et importance de la pollution de l'air à l'intérieur des bâtiments, *Soz. Präventivmed.* **21,** 124.

Savage, E. P. 1975. Pesticide Residues in Houses Using Forced Air Plenum Distribution Systems. Report, Colorado State Univ., Ft. Collins, CO.

Savage, E. P., Malberg, J. W., Wheeler, H. W., and Tessari, J. D. 1976. Accidental carbon monoxide poisoning in Colorado and Wyoming, 1971–1973, *Public Health Reports* **91,** 560.

Savage, E. P., Keefe, T. J., Wheeler, H. W., Mounce, L., Helnic, L., Applehans, F., Goes, E., Goes, T., Mihlan, G., Rench, J., and Taylor, D. K. 1982. Household pesticide usage in the U.S. *Arch. Env. Health* **36,** 304.

Savina, V. P., and Kuznetsova, T. I. 1980. Sources of micro-impurity contamination of cabin atmospheres and their toxicological evaluation—in spacecraft. In *Sanitary, Hygienic and Physiological Aspects of Manned Spacecraft,* p. 11. Moscow: Izdatel/stvo Nauka.

Sawicki, E., and Sawicki, C. R. 1978. *Aldehydes: Photometric Analysis,* Vol. 5: *Formaldehyde Precursors.* New York: Academic Press.

Sawyer, R. N. 1979. Indoor asbestos pollution: Application of hazard criteria, *Ann. N.Y. Acad. Sci.* **330,** 579.

Sax, N. I. 1979. *Dangerous Properties of Industrial Materials,* 5th ed. New York: Van Nostrand Reinhold.

Sayer, W. J., MacKnight, N. M., and Wilson, H. W. 1972. Hospital airborne bacteria as estimated by the Andersen sampler *versus* the gravity settling culture plate, *Am. J. Clin. Pathol.* **58,** 558.

Scales, J. W. (ed.) 1972. *Air Quality Instrumentation,* Vol. 1. Pittsburgh, PA: Instrument Society of America.

Schaefer, O., Eaton, R. D. P., Timmermans, F. J. W., and Hildes, J. A. 1980. Respiratory function impairment and cardiopulmonary consequences in long time residents of the Canadian Arctic, *Canad. Med. Assoc. J.* **123**(10), 997.

Schaplowsky, A. F., Polk, L. G., Oglesbay, F. B., Morrison, J. H., Gallagher, R. E., and Bergman, W., Jr. 1973. *Carbon Monoxide Contamination of the Living Environment: A National Survey of Home Air Specimens and Children's Blood Samples.* HEW (CDC), Bureau of State Services Community Environmental Management. Preventive Health Systems Section, Atlanta, GA.

Schenker, M. B., and Weiss, S. T. 1981. Health effects of indoor formaldehyde exposure. Preprint, *Int. Symposium Indoor Air Pollution, Health and Energy Conserv.,* Univ. of Massachusetts, October 13–16; Amherst, MA.

Schmähl, D. 1976. Threshold doses in chemical carcinogenesis, *Oncology* **33**(2), 51.

Schmähl, D. 1980. Activity of the research council on smoking and health, *Munch. Med. Wochschr.* **122**(II), 3.

Schmahl, D. 1980. Combination effects in chemical carcinogenesis, *Arch. Toxicol.* Suppl. 4, p. 29.

Schmeltz, I., Hoffmann, D., and Wynder, E. L. 1975. The influence of tobacco smoke on indoor atmospheres, *Preventive Med.* **4,** 66.

Schmidt, H. E., Hollowell, C. D., Miksch, R. R., and Newton, A. S. 1980. Trace Organics in Offices. LBL Report 11378; EEB-Vent 80-7.

Schoenberg, J. B., and Mitchell, C. A. 1975. Airway disease caused by phenolic (phenol-formaldehyde) resin exposure, *Arch. Env. Health* **30**, 574.

Schönborn, C., and Winden, F. 1973. Occurrence of fungi in the air and dust of clinic rooms, *Mykosen* **16**, 385.

Schrag, M. P., and Rao, A. K. 1976. Fine Particle Emissions Information System: Summary Report (Summer 1976). EPA-600/2-76-147; NTIS PB 258 825.

Schuette, F. J. 1967. Plastic bags for collection of gas samples, *Atmos. Env.* **1**(4), 515.

Schuetzle, D., Lee, F. S.-C, Prater, T. J., and Tejada, S. B. 1981. The identification of polynuclear aromatic hydrocarbon (PAH) derivatives in mutagenic fractions of diesel particulate extracts, *Int. J. Env. Anal. Chem.* **9**, 99.

Schulze, H.-D. 1975. Measuring formaldehyde concentration in closed space, *Z. Ges. Hyg. Grenzgeb.* **21**(4), 311.

Schutte, W. C., Cole, R. S., Frank, C. W., and Long, K. R. 1981. Problems Associated with the Use of Urea-Formaldehyde Foam for Residential Insulation. ORNL/Sub-7559/3.

Scott, B. W., and Stuart, J. H. 1980. *Atmospheric Contaminant Sensor.* Book 1: *Design and Fabrication of Mass Spectrometer for Submarine Atmosphere Monitoring.* Pomona, CA: Perkin-Elmer Corp. NTIS.

Scott, C. C., and Jacobson, I. 1970. Pseudomonas in ventilators, *Lancet* **1**,(7640), 239.

Sears, D. R. 1977. Preliminary Environmental Assessment of Solar Energy Systems. EPA-600/7-77-086.

Sebben, J., Pimm, P., and Sheppard, R. J. 1977. Cigarette smoke in enclosed public facilities, *Arch. Env. Health* **32**(2), 53.

Seguin, M. 1790. Extrait de l'article air (rédigé par Morveau), *Ann. Chim.* **7**, 46.

Sehmel, G. A. 1980. Particle resuspension: A review, *Env. International* **4**, 107.

Seisaburo, S., Kiyoko, K., and Tatsuko, N. 1959. Free dust particles and airborne microflora, *Bull. Dept. Home Econ., Osaka City Univ.* **4**, 31.

Selikoff, I. J. 1981. Household risks with inorganic fibers, *Bull. N.Y. Acad. Med.* **57**, 947.

Selway, M. D., Allen, R. J., and Wadden, R. A. 1980. Ozone production from photocopying machines, *Am. Ind. Hyg. Assoc. J.* **41**, 455.

Seppanen, O., and Punttila, A. 1981. Air quality control of ventilation in an office building. Preprint, *Int. Symposium Indoor Air Pollution, Health and Energy Conserv.*, Univ. of Massachusetts, October 13-16; Amherst, MA.

Sevcova, H., Sevcova, J., and Thomas, J. 1978. Alpha radiation of the skin and the possibility of late effects, *Health Phys.* **35**, 803.

Sewall, H. 1897. Hearing, cutaneous and muscular sensibility, equilibrium, smell and taste. In *An American Textbook of Physiology* (W. H. Howell, ed.). Philadelphia, PA: W. B. Saunders.

Sexton, K., and Repetto, R. 1981. Indoor air pollution and public policy. Preprint, *Int. Symposium Indoor Air Pollution, Health and Energy Conserv.*, Univ. of Massachusetts, October 13-16; Amherst, MA. *Env. International* **8**, in press.

Seymour, J. W. 1979. Formaldehyde problems in manufactured housing, *Proc. 33d Ann. Meeting, Forest Products Research Society* (San Francisco, CA, July 1979). Madison, WI: Mead and Hunt.

Seymour, J. W. 1980. Formaldehyde testing in mobile homes: Source materials, test methods, and conditions, *Proc. Symposium Wood Adhesives—Research, Applications, and Needs.* Madison, WI: U.S. Forest Products Laboratory, unpublished.

Shair, F. H., and Heitner, K. L. 1974. Theoretical model for relating indoor pollutant concentrations to those outside, *Env. Sci. Technol.* **8,** 444.

Shapira, Y., Pandolf, K. B., Avellini, B. A., Pimental, N. A., and Goldman, R. A. 1980. Physiological responses of men and women to humid and dry heat, *J. Appl. Physiol.* **49**(1), 1.

Shearer, S. D., Jr., and Bromberg, S. M. 1977. The Status of Indoor Air Pollution Research 1976. EPA-600/4-77-029.

Sheehy, J. W. 1980. Evaluation of Exhaust Emissions Data for Diesel Engines Used in Underground Mines. NIOSH Tech. Report. DHHS (NIOSH) Publ. 80-146.

Sheet Metal and Air Conditioning Contractors' National Association. 1978. *Energy Recovery Equipment and Systems: Air to Air.* Vienna, VA: SMACCNA.

Shepard, R. J. 1981. *The risks of passive smoking.* London: Croom-Heller.

Shepelev, E., Meleshko, G., Fofanov, V. I., and Tsitovich, S. I. 1974. Some results of studying a simple bioregenerative life support system, *Proc. Int. Astronaut. Fed. 25th Congr.* Amsterdam, Sept. 30, 1974.

Sheppard, D., Wong, W. S., Uehara, C. F., Nadel, J., and Boushey, H. 1980. Lower threshold and greater bronchomotor responsiveness of asthmatic subjects to SO_2, *Am. Rev. Respir. Dis.* **122,** 873.

Sheppard, D., Saisho, A., Nadel, J., and Boushey, H. 1981. Exercise increases SO_2-induced bronchoconstriction in asthmatic subjects, *Am. Rev. Respir. Dis.* **123,** 486.

Shettigara, P. T., and Morgan, R. W. 1975. Asbestos, smoking and laryngeal carcinoma, *Arch. Env. Health* **30,** 517.

Shinskey, F. G. 1978. *Energy Conservation through Control.* New York: Academic Press.

Shurcliff, W. A. 1980. *Superinsulated Houses and Double-Envelope Houses: A Preliminary Survey of Principles and Practice,* 2nd ed. Cambridge, MA: Shurcliff.

Sidorenko, G. I. 1967. Data on the distribution of *Clostridium perfringens* in the environment of man, Communication 1, *J. Hyg. Epidemiol. Microbiol. Immunol. (Prague)* **11,** 171.

Silberstein, S. 1980. The effects of ventilation on radon and daughter activities, *Radon in Buildings* (R. Collé and P. E. McNall, Jr., eds.), p. 61. NBS Special Publ. 581. Washington, DC.

Silver, F. 1979. Buildings affect your health, *Human Ecologist* **1,** 22. Chicago, IL: Human Ecology Action League (HEAL).

Silver, I. H. (ed.) 1970. *Aerobiology.* London: Academic Press.

Simon, C. G., and Bidleman, T. F. 1979. Sampling airborne polychlorinated biphenyls with polyurethane foam—Chromatographic approach to determining retention efficiencies, *Anal. Chem.* **51,** 1110.

Sinsheimer, R. L. 1980. Why Smoking on Airliners Should be Totally Banned. *New York Times* Letters to the Editor, September 13. New York.

Sittig, M. 1974. *Pollution Detection and Monitoring Handbook.* Park Ridge, NJ: Noyes Data.

Sittig, M. 1977. *Particulates and Fine Dust Removal.* Park Ridge, NJ: Noyes Data.

Sizemore, M. M., Clark, H. O., and Ostrander, W. S. 1979. *Energy Planning for Buildings.* Washington, DC: American Institute of Architects.

Skaret, E. 1981. Building, ventilation and indoor contaminants, *Int. Symposium Indoor Air Pollution, Health and Energy Conserv.*, Univ. of Massachusetts, October 13-16; Amherst, MA.

Smith, M. B. 1976. *Handbook of Ocular Toxicity*. Acton, MA: Publishing Sciences Group.

Smith, N. J. 1976. Evaluation of both outdoor and indoor exposure to carbon monoxide and their health significance in Calgary, Alberta, Canada, *Diss. Abstr. Int.* **B39**(6), 2756.

Smith, P. W. 1977. Room humidifiers as the source of *Acinetobacter* infections, *J. Am. Med. Assoc.* **237**, 795.

Smyth, W. F. 1980. *Electroanalysis in Hygiene, Environmental, Clinical and Pharmaceutical Chemistry* (Analytical Chemistry Symposia Ser., Vol. 2), Amsterdam: Elsevier.

Solomon, P., et al. (eds.) 1971. *Sensory Deprivation*. Cambridge, MA: Harvard Univ. Press.

Solomon, R. L., and Hartford, J. H. 1976. Lead and cadmium in dusts and soils in a small urban community, *Env. Sci. Technol.* **10**, 773.

Solomon, W. R. 1975. Assessing fungus prevalence in domestic interiors, *J. Allergy Clin. Immunol.* **56**, 235.

Spedding, D. J., and Rowlands, R. P. 1970. Sorption of sulphur dioxide by indoor surfaces, I: Wallpapers, *J. Appl. Chem.* **20**, 143.

Spedding, D. J. 1974. Air pollution (Oxford Chemistry Ser., Vol. 20). New York: Oxford Univ. Press.

Spedding, D. J., and Edmondson, D. L. 1980. Formaldehyde emission from building materials. *Clean Air (Melbourne)*, **14**(4), 35.

Speers, R., Jr., O'Grady, F. W., Shooter, R. A., Bernard, H. R., and Cole, W. R. 1966. Increased dispersal of skin bacteria into the air after shower baths: The effect of hexachlorophene, *Lancet* **1**(7450), 1298.

Speizer, F. E., Ferris, B., Jr., Bishop, Y. M. M., and Spengler, J. 1980. Health effects of indoor NO_2 exposure: Preliminary results. Chapter 22 in *Nitrogen Oxides and Their Effects on Health* (S. D. Lee, ed.). Ann Arbor, MI: Ann Arbor Science.

Speizer, F. E., Ferris, B., Jr., Bishop, Y. M. M., and Spengler, J. 1980. Respiratory disease rates and pulmonary function in children associated with NO_2 exposure, *Am. Rev. Respir. Dis.* **121**, 3.

Spendlove, J. C. 1975. Penetration of structures by microbial aerosols, *Dev. Ind. Microbiol.* **16**, 427.

Spengler, J. D., Stone, K. R., and Lilley, F. W. 1978. High carbon monoxide levels measured in enclosed skating rinks, *J. Air Pollution Control Assoc.* **28**(8), 776.

Spengler, J. D., Ferris, B. G., Jr., Dockery, D. W., and Speizer, F. E. 1979. Sulfur dioxide and nitrogen dioxide levels inside and outside homes and the implications on health effects research, *Env. Sci. Technol.* **13**, 1276.

Spengler, J. E., Dockery, D. W., Turner, W. A., Wolfson, J. M., and Ferris, B. G., Jr. 1981. Long-term measurements of respirable sulfates and particles inside and outside homes, *Atmos. Env.* **15**, 23.

Spicer, C. W., and Schumacher, P. M. 1977. Phase I: Studies of the Effect of Environmental Variables on the Collection of Atmospheric Nitrate and the Development of a Sampling and Analytical Nitrate Method. Report for U.S. Environmental Protection Agency. Columbus, OH: Battelle Columbus Laboratories.

Spiegelmann, J., Friedman, H., and Blumstein, G. I. 1963. The effects of central air-conditioning on pollen, mold and bacteria concentrations, *J. Allergy Clin. Immunol.* **34**, 426.

Spirtas, R., and Levin, H. J. 1970. Characteristics of Particulate Patterns, 1957–1966. NAPCA Publ. AP-61. Raleigh, NC: HEW.

Spivey, G. H., and Radford, E. P. 1979. Inner-city housing and respiratory disease in children: A pilot study, *Arch. Env. Health* **34**(1), 23.

Stahl, W. H. (ed.) 1973. *Compilation of Odor and Taste Threshold Values Data.* Philadelphia, PA: ASTM.

Stanton, M. F., Blackwell, R., and Miller, E. 1969. Experimental pulmonary carcinogenesis with asbestos, *Am. Ind. Hyg. Assoc. J.* **30**, 236. See also *J. Natl. Cancer Inst.* **58**, 587 (1977).

Stedman, R. L. 1968. The chemical composition of tobacco and tobacco smoke, *Chem. Rev.* **68**(2), 153.

Sterling, D. A., Clark, C., and Bjornson, S. 1981. The effect of air control systems on the indoor distribution of viable particulates. Preprint, *Int. Symposium Indoor Air Pollution, Health and Energy Conserv.,* Univ. of Massachusetts, October 13–16; Amherst, MA.

Sterling, E., and Sterling, D. 1980. Office Environment Analysis. Report for the Legal Services Society of British Columbia, Vancouver, B.C.

Sterling, T. D., and Kobayashi, D. M. 1977. *Env. Res.* **13**, 1.

Sterling, T. D., and Sterling, E. 1979. Carbon monoxide in kitchen and homes with gas cookers, *J. Air Pollution Control Assoc.* **29**, 238.

Sterling, T. D., Weinkam, J., and Sterling, E. 1981. The case for entirely removing the gas range from indoors. Preprint, *Int. Symposium Indoor Air Pollution, Health and Energy Conserv.,* Univ. of Massachusetts, October 13–16; Amherst, MA.

Sterling, T. D., and Kobayashi, D. 1981. Use of gas ranges for cooking and heating in urban dwellings, *J. Air Pollution Control Assoc.* **31**, 162.

Sterling, T. D., Dimich, H., and Kobayashi, D. 1982. Indoor byproduct levels of tobacco smoke: A critical review of the literature, *J. Air Pollution Control Assoc.* **32**, 250.

Stern, A. C. (ed.) 1977. *Air Pollution,* 3d ed. New York: Academic Press.

Stern, A. C. 1982. History of air pollution legislation in the United States, *J. Air Pollution Control Assoc.* **32**, 44.

Stevens, J. C., and Green, B. G. 1978. Temperature-touch interaction: Weber's phenomenon revisited, *Sensory Processes* **2**, 206.

Stevens, R. K., and Herget, W. F. 1974. *Analytical Methods Applied to Air Pollution Measurements.* Ann Arbor, MI: Ann Arbor Science.

Stewart, R. D., and Hake, C. L. 1976. Paint-removal hazard, *J. Am. Med. Assoc.* **235**(4), 398.

Stewart, R. D., Peterson, J. E., Baretta, E. D., Bachand, R. T., Hosko, M. J., and Herrman, A. A. 1969. *Experimental Human Exposure to Carbon Monoxide: Environmental Medicine.* APRAC Project CAPM-3-68; NTIS PB 195 432. Milwaukee, WI: Marquette School of Medicine.

Stick, S. 1980. The perils of second-hand smoking, *New Scientist* **88**(12), 10.

Stockman, J. D., and Fochtman, E. G. (eds.) 1978. *Particle Size Analysis.* Ann Arbor, MI: Ann Arbor Science.

Stolwijk, J. A. J. 1979. Health Aspects Related to Indoor Air Quality. EURO Reports and Studies 21. Geneva: WHO.

Stolwijk, J. A. J. 1979. Physiological responses and thermal comfort in changing environmental temperature and humidity. In *Indoor Climate, Effects on Human Comfort, Performance, and Health in Residential, Commercial, and Light-Industry Buildings* (P. O. Fanger and O. Valbjørn, eds.), p. 491. Copenhagen: Danish Building Research Institute.

Stone, R., Tucker, B., DePaso, D., Shepard, E., Villarreal, E., and To, T. 1981. An Evaluation of Formaldehyde Problems in Residential Mobile Homes. HUD.

Stranden, E. 1977. Population doses from environmental gamma radiation in Norway, *Health Phys.* **33**(4), 319.

Strauss, W. (ed.) 1971. *Air Pollution Control,* Part I. New York: Wiley-Interscience.

Strauss, W. 1975. *Industrial Gas Cleaning: The Principles and Practice of the Control of Gaseous and Particulate Emissions,* 2nd ed. (Int. Ser. in Chem. Eng., Vol. 8). New York: Pegamon Press.

Sugden, T. M., and West, T. F. (eds.) 1980. *Chlorofluorocarbons in the Environment: The Aerosol Controversy.* London: Ellis Horwood Ltd.

Sullivan, R. J. 1979. Indoor–Outdoor Air Pollution Exposure Study. EPA report, Contract 68-02-2493

Summer, W. 1971. *Odor Pollution of Air: Causes and Control.* London: Leonard Hill.

Sundell, J., and Spengler, J. D. 1981. Nordic guidelines for building regulation regarding indoor air quality. Preprint, *Int. Symposium Indoor Air Pollution, Health and Energy Conserv.,* Univ. of Massachusetts, October 13–16; Amherst, MA.

Sundin, B. 1978. Formaldehyde emission from particleboard and other building materials, *Proc. 12th Int. Symposium on Particleboard* (T. Maloney, ed.), p. 251. Pullman, WA: Wash. State Univ.

Sundin, B. 1980. Formaldehyde testing methods and standards and experiences from Sweden. In *Technical Workshop on Formaldehyde* (B. Meyer and J. Fandey, eds.), Vol. 2, p. 158. Washington, DC: CPSC.

Sundin, B. 1982. Status of formaldehyde problem, *Proc. 16th Int. Symposium on Particleboard* (T. Maloney, ed.), p. 300. Pullman, WA: Wash. State Univ.

Surgeon General (U.S.) 1979. *Smoking and Health.* HEW Publ. (PHS) 79-50066.

Suta, B. E. 1980. Population Exposures to Atmospheric Formaldehyde inside Residences. SRI project; EPA Publ. EGU-5794.

Swift, D. L. 1979. The role of the upper respiratory airways in the modification and removal of airborne particles and vapors. In *Indoor Climate, Effects on Human Comfort, Performance, and Health in Residential, Commercial, and Light-Industry Buildings* (P. O. Fanger and O. Valbjørn, eds.), p. 159. Copenhagen: Danish Building Research Institute.

Swinne-Desgain, D. 1975. *Cryptococcus neoformans* of saprophytic origin, *Sabouraudia* **13,** 303.

Switzer, P. 1981. Statistical descriptions of carbon monoxide exposures and their relation to traffic volume, *Environmetrics-81 Proc.* p. 106. Philadelphia, PA: Am. Soc. Ind. Appl. Math.

Sykora, J. L., Karol, M. H., Keleti, G., and Novak, D. 1981. Amoebae as sources of hypersensitivity pneumonitis. Preprint, *Int. Symposium Indoor Air Pollution, Health and Energy Conserv.,* Univ. of Massachusetts, October 13–16; Amherst, MA.

Szalai, A. (ed.) 1972. *The Use of Time: Daily Activities of Urban and Suburban Populations in Twelve Countries.* Paris: Mouton.

Szczepanski, C. Z. 1980. Rapid Evaluation of Air Pollution Impact on Residential Sites and Estimation of Project Emissions. HUD Format AP 1.

Tabershaw, I. R., Doyle, H. N., Gaudette, L., Lamm, S. H., and Wong, O. 1979. Review of the formaldehyde problem in mobile homes. Report to the National Particleboard Association. Rockville, MD: Tabershaw Occupational Medicine Associates.

Tager, I. B., et al. 1979. Association between parental smoking and pulmonary disease in children, *Am. J. Epidemiol.* **110,** 15.

Takeushi, K., Okamoto, S., and Kuhn, R. R. 1974. Seasonal variation of metallic content in suspended particulate matter of indoor air, *Kuki Seijo* **12**(2), 33.

Taylor, P. R. 1980. Illness in workers in a building with sealed windows. Paper presented at *Epidemic Intelligence Service 29th Ann. Conf.*, April 25, 1980. Atlanta, GA: Center for Disease Control.

Thacker, S. B., Bennett, J. V., Tasai, T. F., Fraser, D. W., McDade, J. E., Shepard, C. C., Williams, K. H., Jr., Stuart, W. H., Dull, H. B., and Eickhoff, T. C. 1978. An outbreak in 1965 of severe respiratory illness caused by the Legionnaires' disease bacterium, *J. Infectious Dis.* **138,** 512.

Thaxton, P., Yonushonis, W. P., and Baughman, G. R. 1975. Synergistic relationship of temperature, air velocity, and mercury in the chicken, *J. Appl. Physiol.* **38**(6), 969.

Thomas, A. A. (ed.) 1968. *Proc. 4th Ann. Conf. on Atmospheric Contaminants in Confined Spaces.* Aerospace Med. Res. Lab. Report AMRL-TR-68-175. Wright-Patterson Air Force Base, OH.

Thomas, A. A. 1971. Man's tolerance to trace contaminants NASA Langley Res. Center Biotechnology, N71-28526 16-05, p. 89.

Thompson, B. 1977. *Fundamentals of Gas Analysis by Gas Chromatography.* Palo Alto, CA: Varian.

Thomson, W. A. R. 1979. *A Change of Air: Climate and Health.* New York: Scribner.

Thun, M. J., Lakat, M. F., and Altman, R. 1980. New Jersey Urea Formaldehyde Foam Insulation Study. Trenton: New Jersey State Health Department.

Timby, E. A. 1980. Airflow Rate Requirements in Passenger Aircraft. NTIS 40P. London: Aeron. Res. Council.

Timm, W., and Smith, P. M. 1980. Test for formaldehyde off-gassing rates from insulation and other building materials, *J. Therm. Insul.* **4**(10), 137.

Tokyo Air Filtering System Designing Committee. 1967. Studies concerning the effects of atmospheric pollution on the indoor environment and measures to prevent pollution; the method to evaluate the indoor dust concentration in the building ventilated by the equipment with air filters, *Air Cleaning (Tokyo)* **4,** 1.

Tomita, B. 1980. Chemical analysis of urea formaldehyde resins by C-13 NMR, and methods for the reduction of formaldehyde liberation. In *Technical Workshop on Formaldehyde* (B. Meyer and J. Fandey, eds.), Vol. 2, p. 22. Washington, DC: CPSC.

Touchstone, J. C., and Rogers, D. 1980. *Thin Layer Chromatography: Quantitative Environmental and Clinical Applications.* New York: Wiley.

Tovey, E. R., and Vandenberg, R. A. 1978. Effect of reagins and allergen extracts on radioallergosorbent assays for mite allergen, *Clin. Allergy* **8**, 329.

Trattner, R. B., Kimmel, H. S., and Dias, D. D. 1976. Air quality on an indoor firing range, *J. Env. Sci. Health* **A11**(4–5), 321.

Traynor, G. W. 1981. Indoor Air Quality: Potential Audit Strategies. LBL Report 12387.

Traynor, G. W., Apte, M. G., Girman, J. R., and Hollowell, C. D. 1981. Indoor Air Pollution from Domestic Combustion Appliances. LBL Report 12886.

Treado, S. J., Burch, D. M., and Hunt, C. M. 1979. An Investigation of Air-Infiltration Characteristics and Mechanisms for a Townhouse. NBS Tech. Note 992.

Tredgold, T. 1836. *The Principles of Warming and Ventilation—Public Buildings*. London: Taylor.

Trichopoulos, D., Kalandidi, A., Sparros, L., and MacMahon, B. 1981. Lung cancer and passive smoking, *Int. J. Cancer* **27**, 1.

Truhaut, R. 1977. *Science for Better Environment* (Proc. Int. Conf. Human Env., Kyoto, 1975). New York: Pergamon Press.

Turiel, I., Hollowell, C. D., and Thurston, B. E. 1979. Variable Ventilation Control Systems: Saving Energy and Maintaining Indoor Air Quality. LBL Report 9380; *Changing Energy Use Futures* **3**, 1258.

Turk, A., Johnston, J. W., Jr., and Moulton, D. G. (eds.) 1974. *Human Responses to Environmental Odors*. New York: Academic Press.

Turner, B. B. 1967. Airline considerations for the SST environmental control system. Paper presented at SAE-Aeronaut. Space Eng. and Mfg. Meeting, Los Angeles, CA, October 2–6.

Turner, W. A., Spengler, J. D., Dockery, D. W., and Colome, S. D. 1979. Design and performance of a reliable personal monitoring system for respirable particulates, *J. Air Pollution Control Assoc.* **29**, 747.

Turner-Warwick, H. 1976. The response of the lung to inhaled particles. Chapter 8 in *Respiratory Disease* (D. J. Lane, ed.). London: Heinemann.

Tyrrell, D. A. J. 1967. The spread of viruses of the respiratory tract by the airborne route. In *Airborne Microbes* (P. H. Gregory and J. L. Monteith, eds.). New York and London: Cambridge Univ. Press.

United Nations Environment Programme. 1976. Air Quality in Selected Urban Areas, 1973–1974. WHO Offset Publ. 30. Geneva.

United Nations Environment Programme. 1977. *Environmental Health Criteria 5: Nitrates, Nitrites and N-Nitroso Compounds*. Geneva: WHO.

United Nations Environment Programme. 1977. *Air Monitoring Programme Design for Urban and Industrial Areas*. WHO Offset Publ. 33. Geneva.

United Nations Scientific Committee on the Effects of Atomic Radiation. 1977. Sources and Effects of Ionizing Radiation. UNSCAR report to the General Assembly. New York: United Nations.

U.S. Center for Disease Control. 1980. *Morbidity and Mortality Weekly Reports* **29**, 517.

U.S. Center for Disease Control. 1981. Heat stroke during the heat wave of 1980, *Morbidity and Mortality Weekly Reports* **30**, 37.

U.S. Civil Aeronautics Board. 1981. Smoking aboard Aircraft. Final rule, 46 FR 45934, September 16.

U.S. Consumer Product Safety Commission. 1981. Urea-Formaldehyde Foam Insulation (UFF1); Proposed ban; Denial of petition. 45 FR 34032, May 21, 1980; 45 FR 39434, June 10, 1980; and 46 FR 11188, February 5, 1981.

U.S. Consumer Product Safety Commission. 1982. Ban of urea formaldehyde foam insulation. 47*FR* 14366, April 2.

U.S. Department of Agriculture. 1974. *Wood Handbook: Wood as an Engineering Material.* Agriculture Handbook No. 72.

U.S. Department of the Air Force. 1978. *Engineering Weather Data.* Air Force Manual AFM 88-29; Army Tech. Manual TM 50785; Navy Manual NAVFAC P-89.

U.S. Department of Commerce. 1973. Monthly Normals of Temperature, Percipitation, and Heating and Cooling Degree Days 1941–1970. Climatography of the United States No. 81. Asheville, NC: National Climatic Center.

U.S. Department of Energy. 1978. *Energy Information Data Base: Subject Thesaurus.* TID-7000-R3. Oak Ridge, TN.

U.S. Department of Energy. 1978. *Project Retro-Tech: Home Weatherization.* DOE/CS-0040, Vols. 1–4.

U.S. Department of Energy. 1978. *Energy Audit Workbook for Apartment Buildings.* DOE/CS-0041/1.

U.S. Department of Energy. 1978. *Energy Audit Workbook for Schools.* DOE/CS-0041/2.

U.S. Department of Energy. 1978. *Energy Audit Workbook for Hospitals.* DOE/CS-0041/3.

U.S. Department of Energy. 1978. *Energy Audit Workbook for Hotels and Motels.* DOE/CS-0041/4.

U.S. Department of Energy. 1978. Material Criteria and Installation Practices for the Retrofit Application of Insulation and Other Weatherization Materials: A Technical Report. DOE/CS-0051.

U.S. Department of Energy. 1979. Energy Conservation in Existing Office Buildings in New York City. Contract EY-76-C-02-2799.000. New York: Tishman Assoc.

U.S. Department of Energy. 1979. Air Infiltration in Buildings. DOE/CS-0099-D.

U.S. Department of Energy. 1979. *Project Retro-Tech: Home Weatherization Manual.* DOE/CS-0106. Conservation Paper 28C. Revised.

U.S. Department of Energy. 1980. Urea-Formaldehyde Foam Insulation (UFFI) Material Standard. Interim final rule, 45 FR 63786, September 25.

U.S. Department of Energy. 1980. Residential Conservation Service Program: Expansion to Multifamily and Commercial Buildings. Environmental Impact Statement Draft Supplement. DOE/EIS-0050-DS.

U.S. Department of Energy. 1981. UFFI Material Standard. Proposed amendments, 46 FR 8996, January 27.

U.S. Department of Health, Education and Welfare. 1969. *Air Quality Criteria for Sulfur Oxides.* NAPCA Publ. AP-50. Washington, DC: U.S. Govt. Printing Office.

U.S. Department of Health, Education and Welfare. 1970. *Air Quality Criteria for Carbon Monoxide.* NAPCA Publ. AP-62. Washington, DC: U.S. Govt. Printing Office.

U.S. Department of Health, Education and Welfare. 1970. *Air Quality Criteria for Photochemical Oxidants.* NAPCA Publ. AP-63. Washington, DC: U.S. Govt. Printing Office.

U.S. Department of Health, Education and Welfare. 1970. *Air Quality Criteria for Hydrocarbons.* NAPCA Publ. AP-64. Washington, DC: U.S. Govt. Printing Office.

U.S. Department of Health, Education and Welfare. 1971. *Air Quality Criteria for Nitrogen Oxides.* NAPCA Publ. AP-84. Washington, DC: U.S. Govt. Printing Office.

U.S. Department of Health, Education and Welfare. 1979. *Smoking and Health.* Report of the Surgeon General. HEW Publ. (PHS) 79-50066.

U.S. Department of Health and Human Services. 1965–1982. *Smoking and Health Bulletin.* Office of Smoking and Health.

U.S. Department of Housing and Urban Development. 1973. *Manual of Acceptable Practices* (Vol. 4 of the HUD Minimum Property Standards).

U.S. Department of Housing and Urban Development. 1974. Residential Energy Conservation. Summary report. HUD-HAI-8.

U.S. Department of Housing and Urban Development. 1975. Mobile Home Standards. 40 FR 58752, December 18.

U.S. Department of Housing and Urban Development 1976. *Air Quality Considerations in Residential Planning,* Vol. 1: *Guide for Rapid Assessment of Air Quality at Housing Sites.* HUD-Report, unpublished.

U.S. Department of Housing and Urban Development. 1980. *Fourth Biannual Report to Congress on Mobile Homes.* Title VI of the Housing and Community Development Act of 1974. HUD-NVACP-564.

U.S. Department of Housing and Urban Development. 1980. Fire Performance Evaluation of the Federal Mobile Home Construction and Safety Standard. HUD-NVACP-627.

U.S. Department of Housing and Urban Development. 1980. Regional Guidelines for Building Passive Energy Conserving Homes. HUD-PDR-355(2).

U.S. Department of Housing and Urban Development. 1980. Mobile Home Research: Economic Cost–Benefit and Risk Analysis of Results of Mobile Home Safety Research. HUD-PDR-656.

U.S. Department of Housing and Urban Development. 1981. Mobile Home Research: Thermal Envelope Systems. HUD-PDR-636.

U.S. Department of Housing and Urban Development. 1981. Mobile Home Research: Mobile Home Heating, Cooling and Fuel Burning Systems. HUD-PDR-634.

U.S. Department of Housing and Urban Development. 1981. Manufactured Home Construction and Safety Standards. Advanced notice of proposed rule making, 46 FR 43466, August 28.

U.S. Department of the Interior. 1978. Federal Mine Safety and Health Act of 1977, Public Law 91-173, as amended by Public Law 95-164.

U.S. Department of Labor. 1980. *Regulations and Standards Applicable to Metal and Nonmetal Mining and Milling Operations.* 30 CFR Parts 40, 41, 43, 44, 48, 50, 55, 56, and 57. Mine Safety and Health Administration. Metal and Nonmetal Mine Safety and Health.

U.S. Department of Labor. 1979. *General Industry: OSHA Safety and Health Standards (29 CFR 1910)*. OSHA 2206. Revised November 7, 1978.

U.S. Department of the Navy. 1962. *Submarine Atmosphere Habitability Data Book*. Washington, DC: Bureau of Ships.

U.S. Department of Transportation; FAA; HEW; NIOSH. 1971. *Health Aspects of Smoking in Transport Aircraft*. Publ. AD-736097. Washington, DC: U.S. Govt. Printing Office.

U.S. Environmental Protection Agency. 1971. *Air Quality Criteria for Nitrogen Oxides*. Air Pollution Control Office Publ. AP-84.

U.S. Environmental Protection Agency. 1971. *Photochemical Oxidants and Air Pollution: An Annotated Bibliography,* Part 1: *Categories A through F*. Air Pollution Control Office Publ. AP-88.

U.S. Environmental Protection Agency. 1971. *Photochemical Oxidants and Air Pollution: An Annotated Bibliography,* Part 2: *Categories G through N and Indexes*. Air Pollution Control Office Publ. AP-88.

U.S. Environmental Protection Agency. 1972. *Directory of Air Quality Monitoring Sites 1972*. EPA-450/2-73-006. Research Triangle Park, NC.

U.S. Environmental Protection Agency. 1973. *Control Techniques for Mercury Emissions from Extraction and Chlor-Alkali Plants*. EPA Publ. AP-118.

U.S. Environmental Protection Agency. 1973–1982. *Bibliography of R&D Research Reports*. EPA-600/5-73-002.

U.S. Environmental Protection Agency. 1973. *Nitrogenous Compounds in the Environment*. EPA-SAB-73-001.

U.S. Environmental Protection Agency. 1974. *Lead and Air Pollution: A Bibliography with Abstracts*. EPA-450/1-74-001.

U.S. Environmental Protection Agency. 1974. *Atmospheric Sampling*. Manual, Training Course 435.

U.S. Environmental Protection Agency. 1974. Health Consequences of Sulfur Oxides: A Report from CHESS, 1970–1971. EPA-650/1-74-004.

U.S. Environmental Protection Agency. 1976. *Quality Assurance Handbook for Air Pollution Measurement Systems,* Vol. I: *Principles*. EPA-600/9-76-005.

U.S. Environmental Protection Agency. 1977. *Air Quality Criteria for Lead*. EPA-600/8-77-017.

U.S. Environmental Protection Agency. 1977. *Industrial Process Profiles for Environmental Use,* Chapter 10 (Plastics and Resins Industry). EPA 600/2-77-023j.

U.S. Environmental Protection Agency. 1977–1982. *EPA Publications Bibliography: Quarterly Abstract Bulletin*. NTIS UB/C/042-004.

U.S. Environmental Protection Agency. 1978. Guidance issued to Governor of Florida: Maximum Recommended Radon Concentrations in Homes. Washington, DC: Office of Radiation Programs.

U.S. Environmental Protection Agency. 1979. Pollution Control in the Forest Products Industry. Report on technology transfer. EPA 625/3-79-010.

U.S. Environmental Protection Agency. 1980. *Air Quality Criteria for Particulate Matter and Sulfur Oxides,* Vol. I: *Summary and Conclusions*. External Review Draft No. 1.

U.S. Environmental Protection Agency. 1980. *Air Quality Criteria for Particulate Matter and Sulfur Oxides,* Vol. II: *Air Quality*. External Review Draft No. 1.

U.S. Environmental Protection Agency. 1980. *Air Quality Criteria for Particulate Matter and Sulfur Oxides,* Vol. III: *Welfare Effects.* External Review Draft No. 1.

U.S. Environmental Protection Agency. 1980. *Air Quality Criteria for Particulate Matter and Sulfur Oxides,* Vol. IV: *Health Effects.* External Review Draft No. 1.

U.S. Environmental Protection Agency. 1980. Addendum to "The Health Consequences of Sulfur Oxides: A Report from CHESS, 1970–1971." (May 1974). EPA-600/1-80-021.

U.S. Environmental Protection Agency. 1980. Monitoring Methods Development in the Beaumont–Lake Charles Area. Interim report. EPA 600/4-80-046.

U.S. Environmental Protection Agency. 1981. *Proc. Conf. Wood Combustion Emiss. Assessment.* EPA-600/9-81-029. NTIS PB81-248155.

U.S. General Accounting Office. 1980. *Report to Congress: Indoor Air Pollution: An Emerging Health Problem.* CED-80-111.

U.S. Interagency Regulatory Liaison Group. 1980. Scientific bases for the identification of potentially carcinogenic substances, and estimation of risks, *Ann. Rev. Public Health* **1,** 345.

U.S. National Air Pollution Control Administration. 1969. *Guidelines for the Development of Air Quality Standards and Implementation Plans.* Washington, DC: HEW.

U.S. National Air Pollution Control Administration. 1969. *Air Quality Criteria for Particulate Matter.* NAPCA Publ. AP-49. Washington, DC: HEW.

U.S. National Air Pollution Control Administration. 1970. *Control Techniques for Carbon Monoxide, Nitrogen Oxide, and Hydrocarbon Emissions from Mobile Sources.* NAPCA Publ. AP-66. Washington, DC: HEW.

U.S. National Air Pollution Control Administration. 1970. *Control Techniques for Nitrogen Oxide Emissions from Stationary Sources.* NAPCA Publ. AP-67. Washington, DC: HEW.

U.S. National Air Pollution Control Administration. 1970. *Control Techniques for Hydrocarbon and Organic Solvent Emissions from Stationary Sources.* NAPCA Publ. AP-68. Washington, DC: HEW.

U.S. National Air Pollution Control Administration. 1970. *Nitrogen Oxides: An Annotated Bibliography.* NAPCA Publ. AP-72. Raleigh, NC: HEW.

U.S. National Air Pollution Control Administration. 1970. *Nationwide Inventory of Air Pollutant Emissions 1968.* NAPCA Publ. AP-73. Raleigh, NC: HEW.

U.S. National Institute for Occupational Safety and Health. 1972. *Criteria for a Recommended Standard: Occupational Exposure to Carbon Monoxide.* Washington, DC: HEW.

U.S. National Institute for Occupational Safety and Health. 1974. *Criteria for a Recommended Standard: Occupational Exposure to Ammonia.* HEW (NIOSH) Publ. 74-136. Washington, DC.

U.S. National Institute for Occupational Safety and Health. 1974. *Criteria for a Recommended Standard... Occupational Exposure to Sulfur Dioxide.* Washington, DC: U.S. Govt. Printing Office.

U.S. National Institute of Occupational Safety and Health. 1974. *Criteria for a Recommended Standard—Occupational Exposure to Sulfuric Acid.* HEW (NIOSH) Publ. 74-128. Washington, DC: U.S. Govt. Printing Office.

U.S. National Institute for Occupational Safety and Health. 1975. *Criteria for a Recommended Standard: Occupational Exposure to Xylene.* HEW (NIOSH) Publ. 75-168.

U.S. National Institute for Occupational Safety and Health. 1976. *Criteria for a Recommended Standard: Occupational Exposure to Carbon Dioxide.* HEW (NIOSH) Publ. 76-194.

U.S. National Institute for Occupational Safety and Health. 1976. *Registry of Toxic Effects of Chemical Substances.* HEW (NIOSH) Publ. 76-191.

U.S. National Institute for Occupational Safety and Health. 1976. *Criteria for a Recommended Standard—Occuptional Exposure to Formaldehyde.* HEW (NIOSH) Publ. 77-126.

U.S. National Institute for Occupational Safety and Health. 1977. *NIOSH Manual of Analytical Methods,* Part I: NIOSH Monitoring Methods, Vol. 1, 2nd ed. HEW (NIOSH) Publ. 77-157-A.

U.S. National Institute for Occupational Safety and Health. 1977. *NIOSH Manual of Analytical Methods,* Part II: *Standards Completion Program Validated Methods,* Vol. 2, 2nd ed. HEW (NIOSH) Publ. 77-157-B.

U.S. National Institute for Occupational Safety and Health. 1977. *NIOSH Manual of Analytical Methods,* Part II: *Standards Completion Program Validated Methods,* Vol. 3, 2nd ed. HEW (NIOSH) Publ. 77-157-C.

U.S. National Institute for Occupational Safety and Health. 1977. *NIOSH Manual of Sampling Data Sheets.* HEW (NIOSH) Publ. 77-159.

U.S. National Institute for Occupational Safety and Health. 1978. *The Recirculation of Industrial Exhaust Air.* HEW (NIOSH) Publ. 78-141.

U.S. National Institute for Occupational Safety and Health. 1978-1981. *NIOSH Manual of Analytical Methods.* Vol. 4-7. HEW (NIOSH) Publ. 78-175.

U.S. National Institute for Occupational Safety and Health. 1978. *Summary of NIOSH Recommendations for Occupational Health Standards.* Cincinnati, OH.

U.S. National Institute for Occupational Safety and Health. 1979. *NIOSH Manual of Analytical Methods,* Vol. 5, 2nd ed. Cincinnati, OH.

U.S. National Institute for Occupational Safety and Health. 1980. *Workplace Exposure to Asbestos.* DHHS (NIOSH) Publ. 81-103.

U.S. National Institute for Occupational Safety and Health. 1980. New Developments in Occupational Stress. DHHS (NIOSH) Publ. 81-102.

U.S. National Toxicology Program. 1982. *Second Annual Report on Carcinogens* (R. E. Shapiro, ed.). DHHS (NIEHS) Publ. 81-43. Research Triangle, NC.

U.S. Veterans Administration. 1981. Agent Orange: A Critical Review of Health Effects. Report by JRB-Associates, McLean, VA.

Valbjørn, O., Nielsen, P. A., and Kjaer, J. 1981. Indoor climate problems in Danish dwellings: Complaints and diseases referred to the type and materials of dwellings and the living habits. Preprint, *Int. Symposium Indoor Air Pollution, Health and Energy Conserv.,* Univ. of Massachusetts, October 13-16; Amherst, MA.

van Assendelft, A., Fors'en, K. O., Keskinen, H., and Alanko, K. 1979. Humidifier-associated extrinsic allergic alveolitis, *Scand. J. Work Env. Health* **5**(1), 35.

Van Huesden, S., and Mans, L. G. J. 1978. Alternating measurement of ambient and cabin ozone concentrations in commercial jet aircraft, *Aviation Space Env. Med.* **49**, 1056.

Vaughan, V. C., Vaughan, H. F., and Palmer, G. T. 1922. *Epidemiology and Public Health,* Vol. I: *Respiratory Infections.* St. Louis, MO: C. V. Mosby.

Verschueren, K. 1977. *Handbook of Environmental Data on Organic Chemicals.* New York: Van Nostrand Reinhold.

Vesilind, P. A. 1975. *Environmental Pollution and Control.* Ann Arbor, MI: Ann Arbor Science.

Villano, C. E. 1980-1981. *Energy Saving Device Newsletter.* Metropolitan Denver District Attorneys' Office of Consumer Fraud and Economic Crime.

von Pettenkofer, M. 1873. *The Relations of the Air to the Clothes We Wear, the House We Live In and the Soil We Dwell On* (A. Hess, trans.). London: Bruinswick.

Voors, A. W., Stewart, G. T., Gutekunst, R. R., Moldow, C. F., and Jenkins, C. D. 1968. Respiratory infection in marine recruits, *Am. Rev. Respir. Dis.* **98,** 801.

Vos, K., and Thomson, D. B. 1974. Particle size measurement of eight commercial pressurized products, *Powder Technol.* **10,** 103.

Wade, W. A., III, Cote, W. A., and Yocom, J. E. 1975. A study of indoor air quality, *J. Air Pollution Control Assoc.* **25,** 933.

Wahren, H. 1980. Swedish indoor air standards. In *Technical Workshop on Formaldehyde* (B. Meyer and J. Fandey, eds.), Vol. 1, p. 53. Washington, DC: CPSC.

Walberg, K., and Sallvik, K. 1977. Changes in climate and animal reactions during a breakdown in the ventilation system in pig and broiler houses, *Swed. J. Agric. Res.* **7**(2), 121.

Walcott, R. 1982. Formaldehyde Indoor Air Levels in Mobile Homes. Report, HUD Office of Policy Analysis, unpublished.

Walker, J. F. 1964. *Formaldehyde,* 3d ed. Huntington, NY: Robert E. Krieger.

Walker, J. Q., Jackson, M. T., and Maynard, J. B. 1977. *Chromatographic Systems.* New York: Academic Press.

Wallace, L. A. 1979. Use of personal monitor to measure commuter exposure to carbon monoxide in vehicle passenger compartments. Paper 79-59.2 presented at 72nd Ann. Meeting, Air Pollution Control Assoc., Cincinnati, OH.

Wallace, L. A., Zweidinger, R., Erickson, M., Cooper, S., Whittaker, D., and Pellizzari, E. 1981. Monitoring Individual Exposure—Measurements of Volatile Organic Compounds in Breathing-Zone Air, Drinking Water, and Exhaled Breath. EPA.

Wallace, L. A., and Ott, W. R. 1982. Personal monitors: a state of the art survey. *J. Air Pollution Control Assoc.* **32**(6), 333.

Wallach, C. 1981. Control of indoor air pollution by hypernegionization (HNI) and passive electret filtration. Preprint, *Int. Symposium Indoor Air Pollution, Health and Energy Conserv.,* Univ. of Massachusetts, October 13-16; Amherst, MA.

Walsh, M., Black, A., Morgan, A., and Crawshaw, G. H. 1977. Sorption of sulfur dioxide by typical indoor surfaces including wool carpets, wallpaper and paint, *Atmos. Env.* **11,** 1107.

Wang, J. Y. 1975. *Instruments for Physical Environmental Measurements (with Special Emphasis on Atmospheric Instruments),* Vol. 1. Milieu Information Service.

Wang, J. Y. 1976. *Instruments for Physical Environmental Measurements (with Special Emphasis on Atmospheric Instruments),* Vol. 2. Milieu Information Service.

Wanner, H. U. 1974. Moglichkeiten und Grenzen der Reinraumtechnik in der Medizin, *Chem. Rundschau* **27**(37), 9.

Wayne, L. G., Bryan, R. J., and Ziedman, K. 1976. Irritant Effects of Industrial Chemicals: Formaldehyde. HEW (NIOSH) Publ. 77–117.

Weber, A., Fischer, T., and Grandjean, E. 1979. Passive smoking in experimental and field conditions, *Env. Res.* **20**(1), 205.

Weber-Tschopp, A., Fischer, T., and Grandjean, E. 1976. Objektive und subjektive physiologische Wirkungen des Passivrauchens, *Int. Arch. Occupational Env. Health* **37**, 277.

Weber-Tschopp, A., Fischer, T., and Grandjean, E. 1977. Reizwirkungen des Formaldehyds auf den Menschen, *Int. Arch. Occupational Env. Health* **39**, 207.

Weiss, N. S., and Soleymani, Y. 1971. Hypersensitivity lung disease caused by contamination of an air conditioning system, *Ann. Allergy* **29**, 154.

Wellock, C. E. 1960. Epidemiology of Q-fever in the urban East Bay area, *Calif. Health* **18**, 72.

Wells, W. F. 1955. *Airborne Contagion and Air Hygiene: An Ecological Study of Droplet Infections.* Cambridge, MA: Harvard Univ. Press.

Werjefelt, B. R. L. 1980. Aircraft Humidifaction. Report to FAA and CAB. See also: 45FR 41962.

Weschler, C. J. 1978. Characterization techniques applied to indoor dust, *Env. Sci. Technol.* **12**, 923.

West, J. B. (ed.) 1980. *Pulmonary Gas Exchange,* Vol. 2: *Organism and Environment.* New York: Academic Press.

Whipp, B. J., and Mahler, M. 1980. Dynamics of pulmonary gas exchange during exercise. Chapter 2 in *Pulmonary Gas Exchange* (J. B. West, ed.), Vol. 2. New York: Academic Press.

White, J. R., and Froeb, H. F. 1980. Small-airways dysfunction in nonsmokers chronically exposed to tobacco smoke, *New England J. Med.* **302**, 720.

White, J. W., and Kragulski, M. (ed.) 1972. *Proc. 2d Conf. Natural Gas Research and Technol.* (sponsored by American Gas Association and Institute of Gas Technology, Atlanta, GA, June 5–7).

Willard, H. H., Merritt, L. L., Jr., and Dean, J. A. 1974. *Instrumental Methods of Analysis,* 5th ed. New York: Van Nostrand.

Williams, J. (ed.) 1978. *Carbon Dioxide, Climate and Society.* Oxford: Pergamon Press.

Williams, K. (ed.) 1977. *Planning and Building the Minimum Energy Dwelling.* Solana Beach, CA: Craftsman Book.

Williams, R. E. O. 1967. Spread of airborne bacteria pathogenic for man. In *Airborne Microbes* (P. H. Gregory and J. L. Monteith, eds.). New York and London: Cambridge Univ. Press.

Willits, N. H., and Ott, W. 1981. The effect of intersections on the CO exposures of motorists, *Environmetrics-81 Proc.,* p. 204. Philadelphia, PA: Soc. Ind. Appl. Math.

Wilson, J. G. 1977. Fetotoxicity, *Federation Proc.* **36**, 1698.

Wilson, M. J. G. 1968. Indoor air pollution, *Proc. Roy. Soc. (London)* **A307**, 215.

Wilson, R., Colome, S. D., Spengler, J. D., and Wilson, D. G. 1980. *Health Effects of Fossil Fuel Burning: Assessment and Mitigation.* Cambridge, MA: Ballinger.

Winchester, J. W., and Nelson, J. W. 1979. Sources and Transport of Trace Metals in Urban Aerosols. EPA-600/3-79/019; NTIS PB 294 838.

Windholz, M., Budavari, S., Stroumtsos, L. Y., and Fertig, M. N. (eds.) 1976. *The Merck Index: An Encyclopedia of Chemicals and Drugs,* 9th ed. Rahway, NJ: Merck.

Winslow, C. E. A., and Herrington, L. P. 1949. *Temperature and Human Life.* Princeton, NJ: Princeton Univ. Press.

Winslow, C. E. A., and Browne, W. W. 1914. *Mon. Weather Rev.* **42,** 452.

Woebkenberg, N. R., Mostardy, R. A., Ely, D. L., and Worstell, D. 1981. Carboxyhemoglobin and methemoglobin levels in residents living in industrial and non-industrial communities, *Env. Res.* **26,** 347.

Wolfe, R., and Clegg, P. 1979. *Home Energy for the Eighties.* Charlotte, VT: Garden Way Publishing.

Woods, J. E. 1979. Ventilation, health, and energy conservation, *ASHRAE J.* **21**(7), 23.

Woolf, C. R., and Suero, J. T. 1971. The respiratory effects of regular cigarette smoking in women, *Am. Rev. Respir. Dis.* **103,** 26.

World Health Organization. 1975. *Methods Used in the USSR for Establishing Biologically Safe Levels of Toxic Substances.* (Papers presented at a WHO Meeting, Moscow, December 12-19, 1972.) Geneva.

World Health Organization. 1976. Selected Methods of Measuring Air Pollutants. WHO Offset Publ. 24. Geneva.

World Health Organization. 1979. Health Aspects Related to Indoor Air Quality. Euro-Reports and Studies 21. Copenhagen.

World Health Organization. 1981. *IARC Monographs on the Evaluation of the Carcinogenic Risk of Chemicals to Humans,* Vol. 25: *Wood, Leather and Some Associated Industries,* Lyon, France: International Agency for Research on Cancer.

World Meteorological Organization. 1974. *Operations Manual for Sampling and Analysis Techniques for Chemical Constituents in Air and Precipitation.* Part I: The minimum programme at regional air-pollution stations; Part II: Guidelines for baseline stations and regional stations with extended programs. WMO No. 299. Geneva.

World Meteorological Organization. 1977. Special Environmental Report 10: Air Pollution Measurement Techniques. WMO No. 460. Geneva.

Wright, C. G. 1981. Insecticides in the air of rooms following their application for pest control, *Bull. Env. Contamination Toxicol.* **26,** 548.

Wright, D. N., Vaichulis, E. M. K., and Chatigny, M. A. 1968. Biohazard determination of crowded living-working spaces: Airborne bacteria aboard two naval vessels, *Am. Ind. Hyg. Assoc. J.* **29,** 574.

Wright, R. H. 1978. The perception of odor intensity: Physics or psychophysics?, Part II, *Chemical Senses and Flavour* **3**(2), 241.

Wynder, E. L., and Hoffmann, D. 1967. *Tobacco and Tobacco Smoke.* New York: Academic Press.

Wynder, E. L., and Hoffmann, D. 1972. *Environment and Cancer.* Baltimore: Williams & Wilkins.

Wynder, E. L., and Hoffmann, D. 1979. Tobacco and health, *New England J. Med.* **300,** 894.

Yabroff, I., Myers, E., Fend, V., David, N., Robertson, M., Wright, R., and Braun, R. 1974. *The Role of Atmospheric Carbon Monoxide in Vehicle Accidents.* SRI Project 2673; APRAC Project CAPM-12-69; NTIS PB 233 318. Menlo Park, CA: Stanford Research Institute.

Yaglou, C. P., Riley, E. C., and Coggins, D. I. 1936. Ventilation requirements, *ASHRAE Trans.* **42**, 133.

Yaglou, C. P., and Witheridge, W. N. 1937. Ventilation requirements, *ASHRAE Trans.* **43**, 423.

Yamanaka, S., and Hirose, H. 1979. Nitrogen oxides emission for domestic appliances, *Atmos. Env.* **13**, 407.

Yanagisawa, S. 1980. Air quality standards: National and international, *J. Air Pollution Control Assoc.* **23**(11), 945.

Yanagisawa, Y., and Nishimura, H. 1981. Badge-type personal sampler for measurement of personal exposure to NO_2 in ambient air. Preprint, *Int. Symposium Indoor Air Pollution, Health and Energy Conserv.*, Univ. of Massachusetts, October 13-16; Amherst, MA.

Yano, H., and Oikawa, S. 1974. Air cleaning plan for a business building (No. 2 emission of NO_x in rooms). *Japan Soc. Heating, Air Conditioning and Sanitary Eng.* Preprint, p. 67.

Yeh, C., Wang, G. Kuo, P., and Tsai, S. 1977. House dust and pesticide residues pollution, *T'ai wan I Hsueh Hui Tsa Chih* **76**(4), 324.

Yocom, J. E., Cote, W. A., and Clink, W. L. 1969. Summary Report of A Study of Indoor-Outdoor Air Pollution Relationships to the National Air Pollution Control Administration. Contract CPA-22-69-14. Hartford, CT.

Yocom, J. E., Clink, W. L., and Cote, W. A. 1970. Indoor-Outdoor Air Quality Relationships. Paper presented at 63d Ann. Meeting, Air Pollution Control Assoc., St. Louis, MO, June 14-18.

Yocom, J. E., Cote, W. A., and Clink, W. L. 1970. A Study of Indoor-Outdoor Air Pollutant Relationships, Vol. I and II. Research Corporation of New England Publ. 195338.

Yocom, J. E., and Cote, W. A. 1971. Indoor-Outdoor Air Pollutant Relationships for Air-Conditioned Buildings. *ASHRAE Trans.* **77**, 61.

Yocom, J. E., Clink, W. L., and Cote, W. A. 1971. Indoor-outdoor air quality relationships, *J. Air Pollution Control Assoc.* **21**, 251.

Yocom, J. E., Cote, W. A., and Clink, W. L. 1974. Measurement of air pollution penetration into building interiors, *Anal. Instrumentation* **9**.

Yocum, J. E. 1976. The indoor environment: A new frontier?, *Proc. Int. Conf. Env. Sensing Assess.* **1**, 70.

Yocom, J. E., Cote, W. A., and Benson, F. B. 1977. Effects on indoor air quality. Chapter 3 in *Air Pollution,* 3d ed., Vol. 2. New York: Academic Press.

Yoshinaga, A., and Miller, J. 1981. Acid rain on Hawaii, *Science* **212**, 1014.

Young, R. A., Berk, J. V., Hollowell, C. D., Pepper, J. H., and Turiel, I. 1981. Indoor Air Quality and Energy Efficient Ventilation Rates at a New York City Elementary School. LBL Report 11828; UC-95d.

Zepinski, M. 1981. Preliminary Report on Concentration of Chlorocarbons in Ambient Air and in Breath of College Students in Two Cities. Report for EPA.

Zimmerli, B. 1977. Biocide substances in indoor air, *Chem. Rundschau* **30**(49), 8, 11.

Zimmerli, B., and Zimmermann, H. 1979. Simple method for estimating concentrations of noxious substances in indoor air, *Mitt. Geb. Lebensmittelunters Hyg.* **70**(4), 429.

Ziskind, R., 1978. Carbon Monoxide Intrusion into the Passenger Compartments of Sustained-Use Vehicles. Science Applications Inc. interim report to Congress. Washington, DC: EPA.

Ziskind, R. A. 1980. Carbon Monoxide Intrusion into the Passenger Compartments of Sustained-Use Vehicles. Final report, EPA.

Ziskind, R. A., Fite, K. R., and Mage, D. T. 1981. Personal exposure to carbon monoxide: Nine-person study. Preprint, *Int. Symposium Indoor Air Pollution, Health and Energy Conserv.*, Univ. of Massachusetts, October 13–16; Amherst, MA.

Zlatkis, A. (ed.) 1978. *Advances in Chromatography 1978.* (Proc. 13th Int. Chromatography Symposium, St. Louis, MO, October 16–19.) Houston, TX: Univ. of Houston Press.

Zwaardemaker, H. 1895. *Die Physiologie des Geruchs.* Leipzig: Engelmann.

Zweidinger, R. A. 1977. Organic Emissions from Automobile Interiors. EPA-600/7-77-149.

Zweidinger, R. A., Erickson, M., Cooper, S., Whittaker, D., Pellizzari, E., and Wallace, L. A. 1981. Monitoring Methods Development: Direct Measurement of Volatile Organic Compounds in Breathing-Zone Air, Drinking Water, Breath, Blood and Urine. EPA report.

Author Index

Author Index

This Index lists all authors quoted in the text. When a name is followed by another name, rather than a page number, the person is co-author of a paper listed in the Bibliography.

Abbott, J. C., 188, see also Fulwiler, R. D.
Abeles, F. J., 141
Achenback, P. R., 140, see also Coblentz, C. W.
Adams, J. D., see Brunnemann, K. D.; Hoffmann, D.
Alanka, K., see van Assendelft, A.
Albrechtsen, O., 48
Aldrich, F. D., see Harris, J. C.
Allan, G., 203, see also Ozkaynak, H.
Allan, R. J., 172
Alleau, R., 26
Allen, J. R., see Lipschutz, R. D.
Allen, R. J., 172, 176, 177, see also Hawthorne, A. R.; Selway, M. D.
Alter, H. W., 156, see also Hartley, R. J.
Altman, R., see Thun, M. J.
Amdur, A. O., 239
Amoore, J. E., 33, 134, 135
Anaclerior, R. N., see Lin, C.-I.
Anders, J. M., 26
Andersen, I., 103, 167, 203, 217, 239
Anderson, H. A., 156, 203, 265, 278, see also Dally, K. A.; Eckmann, A. D.; Hanrahan, L. P.
Anderson, H. W., see Johnson, C. J.
Anderson, J. L., see DeRoos, R. L.
Andrews, B. A. K., 142
Angus, T., 45
Annis, J. C., 99, 189
Annis, P. J., 99, 189, see also Annis, J. C.
Anthon, D. W., see Lin, C.-I; Miksch, R. R.
Appel, B. R., 153
Applehans, F., see Savage, E. P.
Apte, M. G., see Girman, J. R.; Traynor, G. W.
Archer, P. G., see Brandom, W. F.
Archer, V. E., 239, see also Brandom, W. F.; Lundin, F. E., Jr.
Argus, M. F., see Arcos, J. C.
Armet, D., see Lebowitz, M. D.

Arnest, R. T., 300
Arthur, R. D., 211
Atherton, M. J., 219
Attia, M., 91
Atwater, W. O., 19, 20
Avellini, B. A., see Shapira, Y.
Ayers, S. M., 246

Baars, A., see Harke, H. P.
Baca, T. E., see El-Ahrat, P. H.
Bach, W., 58
Bachand, R. T., see Stewart, R. D.
Bailey, 261
Bale, W. R., 107, see also Darlow, H. M.
Balocco, L., 302
Banaszak, E. F., 106, 254
Bandera, E., see Burge, H. P.
Banks, R. S., see DeRoos, R. L.
Barad, C. B., 284
Bardana, E. J., 223, 278
Baretta, E. D., see Stewart, R. D.
Barniville, G., see Fitzgerald, G. R.
Barr, R., 21
Barras, R. C., 58, see also Morris, A. L.
Barrentine, B. F., 211, see also Arthur, R. D.
Barrett, R. E., 103, 172
Barrie, L., 58
Bartels, T. T., 141
Barthel, W. F., 126, 148
Bartlett, D., Jr., 271
Bartos, J., 149, see also Pesez, M.
Bates, D. V., see Mercer, T. T.
Bates, M., 244
Batten, B., see Mintz, S.
Baughman, G. R., 239, see also Thaxton, P.
Baumann, H., 279
Bay, B., Jr., see Bridbord, K.
Bazydlo, D. P., 149, see also Compton, B.
Beall, J. R., 263
Beard, J. H., see McFadden, J. E.
Beatty, G. H., 145, 152, see also Foster, J. F.

399

Beck, G. J., 246
Beierlein, W., *see* König, H. L.
Belanger, B. C., 140
Bellin, P., 173
Benarie, M., 182
Benedict, F. G., 19, 20, *see also* Atwater, W. O.
Benes, M., 270
Bennett, J. V., *see* Thacker, S. B.
Benoist, L., *see* Miquel, P.
Benson, F. B., 162, *see also* Eaton, W. C.; Henderson, J. J.; Yocom, J. E.
Benzinger, T. H., 34, 46
Berg, D., *see* Ross, H.
Berge, A., 217
Berger, J. M., 279
Bergey, D. H., *see* Billings, J. S.
Bergland, L. G., *see* Cain, W. S.
Bergman, W., Jr., *see* Schaplowsky, A. F.
Berk, J. V., 68, *see also* Hollowell, C. D.; Young, R. A.
Bernard, H. R., *see* Speers, R., Jr.
Bersey, J. T., *see* Erickson, M. D.
Bertsch, W., *see* Liebich, H. M.
Beuthner, H., *see* Druckrey, H.
Beyea, J., 296, *see also* Harrje, D. T.
Bidleman, T. F., 143, *see also* Simon, C. G.
Biersteker, K., 167
Billick, H. I., 330
Billings, J. S., 8, 296
Binder, R. E., *see* Bouhuys, A.
Bird, T. J., *see* Grieble, H. G.
Birenzvige, A., *see* Brookman, E. T.
Birmingham, D., 241
Bishop, Y. M. M., *see* Ferris, B. J.; Samet, J. M.; Speizer, F. E.
Bjornson, S., *see* Sterling, D. A.
Black, A., 304, *see also* Walsh, M.
Black, J., *see* Fitzgerald, G. R.
Blackman, G. S., *see* Hawthorne, A. R.
Blackwell, F. O., 104, 257, *see also* Las, Y. L.
Blackwell, R., *see* Stanton, M. F.
Blanchard, R. L., 193
Blomsterberg, A. K., 296
Bloom, A. D., *see* Brandom, W. F.
Blumstein, G. I., *see* Spiegelmann, J.
Bocanegra, T. S., 284
Bock, F. G., 284
Boegel, J. V., *see* Hollowell, C. D.
Boegel, M. L., *see* Lepman, S. R.
Boethius, A. M. S., 247
Boggess, W. R., 104
Bohn, R., *see* Maxwell, C. M.
Bohr, C., 244
Boise, J. R., *see* Burge, H. P.
Boleij, J. S. M., *see* Lebret, E.
Bond, R. G., 296

Booz, Allen & Hamilton, 69
Boren, H. G., 262
Boushey, H., *see* Sheppard, D.
Borts, I. H., *see* Hendricks, S. L.
Bouhuys, A., 246, *see also* Beck, G., Jr.; Binder, R. E.
Boveri, W., 266
Bowen, R. P., 208
Boyer, R. S., 254
Bragg, G. M., 299, *see also* Rajhans, G. S.
Brain, J. D., 250
Brandom, W. F., 261
Brandt, C. S., *see* Heck, W. W.
Braun, R., *see* Yabroff, I.
Breidenbach, H., *see* König, H. L.
Brenchley, D. L., 140
Brennan, S. T., *see* Long, K. R.
Breslin, A. J., 155, *see also* George, A. C.
Breslow, N. E., 246
Brewis, 262 (*see* U.S. Surgeon General)
Breysse, P. A., 103, 172, 202, 204, 278, *see also* Johnson, C. J.
Brice, P., 173 (*see* Ott, 1981)
Bridbord, K., 100, 104, 198, 211
Brieger, H., *see* LaBelle, C. W.
Briscoe, W. A., 243
Bromberg, S. M., 320, *see also* Shearer, S. D., Jr.
Broome, C. V., 106, 190, 254
Bross, I. D. J., *see* Decoufle, P.
Brown, L. R., *see* Rothman, L. S.
Browne, W. W., 190, *see also* Winslow, C. E. A.
Brubaker, P. E., *see* Bridbord, K.
Brun, M. J., 69
Brunekreef, B., *see* Lebret, E.
Brunnemann, K. D., 101, 184, 301, *see also* Hoffmann, D.
Brunner, A., 106
Brutseart, W. F., *see* Hess, C. T.
Bryan, R. J., *see* Wayne, L. G.
Bryan, W. R., 233
Buchbinder, L., 192
Buchland, F. E., 193
Buckley, R. D., *see* Hackney, J. D.
Budavari, S., *see* Windholz, M.
Budde, W. L., 146, 148
Budnitz, R. J., *see* Hollowell, C. D.
Bulloch, C., 300
Bunch, J. E., 149, 150, 151, *see also* Pellizzari, E. D.
Burch, D. M., 63, *see also* Treado, S. J.
Burge, H. P., 109
Buron, W. E., 190, *see also* Miller, R. L.
Burton, R. M., *see* Eaton, W. C.; Moschandreas, D. J.
Butcher, S. S., 140

Author Index

Cadle, R. D., 153
Caiazza, R., *see* Maxwell, C. M.
Cain, J. E., *see* Arthur, R. D.
Cain, W. S., 12, 81, 83, 134, 140, 211, 300, 301, 316, *see also* Duffee, R. A.
Cairns, T., 148, 276
Calabrese, E. J., 239, 240, 263, 327
Caldwell, D. E., *see* Henderson, J. J.
Campbell, I. R., 257
Candas, V., 47
Caplan, K. T., 153
Carbone, R. D., 202, 204, 279
Carmody, M., *see* Fitzgerald, G. R.
Carrington, T. S., 292
Carson, R., 118
Carter, J., 308
Carter, N., *see* Irvine, R. W.
Cartwright, J., 250
Casparius, R. E., *see* Hess, C. T.
Cetrulo, C. L., *see* Kurzel, R. B.
Chamberlain, A. C., 252
Chamberlin, R. I., *see* Kanarek, D. J.
Chaddock, J. B., 6, 81
Chapin, F. S., 159
Chaplin, S., *see* Horie, Y.
Charlson, R. J., 140, *see also* Butcher, S. S.
Chastein, J. B., *see* Feigley, C. E.
Chatigny, M. A., *see* Wright, D. N.
Cheremisinoff, P. N., 140
Childers, R. E., *see* McKee, H. C.
Chown, G. A., 208, *see also* Bowen, R. P.
Christensen, H. E., 260, 273
Christie, P., 244, *see also* Bates, M.
Chuong, B. T., 182, *see also* Benarie, M.
Clark, C., *see* Sterling, D. A.
Clark, H. O., 68, *see also* Sizemore, M. M.
Clark, R. E., *see* Grot, R. A.
Clark, R. P., *see* Lewis, H. E.
Clark, W. E., *see* Lundgren, D. A.
Clarke, F. B., *see* Birky, M. M.
Clegg, P., 69, *see also* Wolfe, R.
Clink, W. L., *see* Yocom, J. E.
Coblentz, C. W., 140
Coburn, R. F., 271
Coggins, D. I., *see* Yaglou, C. P.
Cohn, M. S., 284, 285, 288
Cole, R. S., *see* Schutte, W. C.
Cole, W. R., *see* Speers, R., Jr.
Colle, J., 330
Colle, R., 129, 130
Colley, J. R. T., 284
Collier, C. R., *see* Hackney, J. D.
Colligan, M. J., 246
Collins, K. R., 46
Collins, R. D., 211
Colome, S. D., *see* Turner, W. A.; Wilson, R.
Coltin, 255

Colton, F. R., *see* Grieble, H. G.
Compton, B., 149
Condit, C., 26
Conright, K. L., *see* Kreiss, K.
Conway, R. A., 280
Cooley, A. M., *see* Kinne, H.
Coombs, E. G., *see* Hess, C. T.
Cooper, H. B. H., Jr., 140
Cooper, S., *see* Wallace, L.; Zweidinger, R. A.
Corkhill, R. T., *see* Colley, J. R. T.
Corn, M., 140, 187, 252, *see also* Hammad, Y. Y.; Leadbetter, M. R.
Cornet, G., 26
Cortese, A. D., 173
Cote, W. A., 114, 164, 171, 174, *see also* Wade, W. A., III; Yocom, J. E.
Couch, R. B., 254
Courtney, R. G., 68, 80, 81
Courtney, W. J., *see* Moschandreas, D. J.
Covelli, H. D., 106
Cowherd, C., Jr., *see* Maxwell, C. M.
Cox, M., *see* Frank, N. H.
Cox, R. N., *see* Lewis, H. E.
Crabtree, J. H., 300, *see also* Mayrsohn, H.
Craker, L. E., *see* Abelles, F. B.
Crawshaw, G. H., 301, 304, *see also* Walsh, M.
Creech, V. C., 24
Creip, L. H., 302
Criscitiello, A., *see* Ayers, S. M.
Cromer, R., *see* Honigman, B.
Croome-Gale, D. J., 296
Crosby, H. J., *see* Honma, M.
Crowley, T. P. 106, 190
Crump, N. L., 141, *see also* Bartels, T. T.
Culot, R., 299
Curs, A., 203

Dadaian, J. H., 252
Dailey, R., 221, 285, 286, 287
Dally, K. A., 203, 204, 206, 278, *see also* Anderson, H. A.; Eckmann, A. D.; Hanrahan, L. P.
Daniels, A., 58, *see also* Bach, W.
Danielson, T. A., 299
Darcy, F. J., 188, *see also* Fulwiler, R. D.
Darlow, H. M., 107
Dasgupta, P. K., 147
Daum, S. M., *see* Anderson, H. A.
David, N., *see* Yabroff, I.
Davies, C. N., 252, 255, *see also* Heyder, J.
Davies, D. J., *see* Mes, J.
Davies, J. E., 97, 99, 189, *see also* Davis, J. H.
Davies, P. O. A. L., 140
Davis, J. H., 279
Davis, T. G., *see* Lao, Y. J.

Davis, W. T., 140, *see also* Noll, K. E.
Davison, R. L., 98, 189
Daws, L. F., 192
Dawson, D. J., *see* Reznikov, M.
Dean, J. A., *see* Willard, H. H.
Debeir, M., 146
DeBruin, A., 240
DeCesare, K., *see* Dasgupta, P. K.
De Chaumont, 315
Decker, C. E., *see* Harrison, J. W.
Decorpo, J. J., 300
Decoufle, P., 241
Decries, D. M., 211, *see also* Collins, R. D.
deGaff, H., *see* Biersteker, K.
Deimel, M., 203
Deitz, V. R., 299
Denniston, J. C., *see* McCarroll, J. R.
Denyszyn, R. B., *see* Harrison, J. W.
DePaso, D., *see* Moschandreas, D. J.; Stone, R.
Derham, R. L., 174
DeRoos, R. L., 297, 300, 302
Desaedeleer, G. G., 188
Desrosiers, A. E., 193
Devitt, T. W., *see* Briggs, T. M.
Diaconescu, M. L., *see* Mincu, P.
Dias, D. D., *see* Trattner, R. B.
Dickey, L. D., 245
Dickgiesser, N., 107, 254
Dickinson, J. B., *see* Lipschutz, R. D.
Didion, D., 296, *see also* Goldschmidt, V. M.
Dietz, A. G. H., 304
Dimich, H., *see* Sterling, T. D.
Dimov, N., 146, *see also* Muchtarova, M.
Dmitriev, M. T., 175, *see also* Gubernskii, Y. D.
Dockery, D. W., 181, *see also* Spengler, J. D.; Turner, W. A.
Dockery, J. D., *see* Ferris, B. G., Jr.
Doemeny, L. J., *see* Caplan, K. T.
Doll, R., 189, 211, 263, 267
Dorman, R. G., 299
Dossat, R. J., 296
Dotto, L., 116
Dougherty, J. M., *see* Hawthorne, A. R.
Douglas, R. L., 193
Doyle, H. N., *see* Tabershaw, I. R.
Dravnieks, A., 141
Drew, R. T., *see* Laskin, S.
Druckrey, H., 261
Dubin, F. S., 69, 296
Duffee, R. A., 81, 83, 134, 301
Duhamel, M., *see* Morveau, M.
Dull, H. B., *see* Thacker, S. B.
Duncan, J. R., 165
Dunn, J. D., *see* Cain, W. S.
Dunning, J. A., *see* Heck, W. W.
Durham, M. D., *see* Lundgren, D. A.

Dutkiewicz, J., *see* Allan, G.
Dutt, G. S., 296, 318, *see also* Harrje, D. T.
Dworetzky, L. H., *see* Giles, W. H.
Dzubay, T. G., 152

Eads, W. G., 296
Eaton, R. D. P., *see* Schaefer, O.
Eaton, R. S., 299
Eaton, W. C., 174
Eckmann, A. D., *see* Dally, K. A.; Hanrahan, L. P.
Eckmann, B. H., 106, 203
Edgar, R. T., *see* Rea, W. J.
Edmonds, R. L., 106, 108, 190, 193
Edmondson, D. L., *see* Spedding, D. J.
Edmondson, W. F., *see* Davies, J. E.
Ehrhart, J., *see* Candas, V.
Eichelberger, J. W., 146, 148, *see also* Budde, W. L.
Eichholz, G. G., 154
Eickhoff, T. C., 106, *see also* Thacker, S. B.
Eisenbud, M., 58, 132, 133, 193
Eisenreich, S. J., 198
Elfers, L., 148
Elkins, R. H., 105, 170, 174, *see also* Macriss, R. A.
Elliott, L. J., 208
Elliott, L. P., 101, 170, 184, 212
Ellis, E. C., 145
Ely, D. N., *see* Woebkenberg, N. R.
Emerick, R. H., 80, 296
Eng, W. G., 247, 273, 279
Enterline, P. E., 265
Erickson, M. D., 149, *see also* Pellizzari, E. D.; Wallace, L.; Zweidinger, R. A.
Espinoza, L. R., 284, *see also* Bocanegra, T. S.
Evans, C. A., *see* Davison, R. L.
Evans, R., *see* Ayers, S. M.
Evelyn, J., 21

FAA, 173, 177, 184, 284
Fahrni, P., 202
Fairchild, E. J., 260, 273, *see also* Christensen, H. E.
Falk, H., 126, 273
Fallon, R. J., 95, *see also* Morris, J. E. W.
Fan, L. T., 35, *see also* Nakanishi, E.
Fandey, J., *see* Meyer, B.
Fanger, P. O., 29, 34, 46, 47, 95, 180, 245, *see also* Berg-Munch, B.; Olesen, B. W.
Fanning, L. Z., 204, *see also* Anthon, D. W.; Lin, C.-I.; Miksch, R. R.
Farber, S. A., *see* Desrosiers, A. D.
Farber, S. M., 193
Feigley, C. E., 143
Feldman, J., 312, 314
Fend, V., *see* Yabroff, I.

Author Index

Fenyves, E. J., *see* Rea, W. J.
Fergusson, D. M., 284
Ferrand, E. F., *see* Ayers, S. M.
Ferris, B. G., Jr., 96, 142, 270, *see also* Speizer, F. E.; Spengler, J. D.
Fertig, M. N., *see* Windholz, M.
Fincher, E. L., 192
Findlay, W. O., 306
Finney, 238
Fink, J. N., 106, 254, *see also* Hodges, G. R.
First, M. W., *see* Hinds, W. C.
Fischbein, A., 257
Fischer, B., 23
Fischer, P., *see* Pohl-Rüling, J.
Fischer, T., 101, 170, 184, 283
Fish, B. R., *see* Mercer, T. T.
Fisher, A. A., 278
Fisher, T., *see* Weber, A.
Fisk, W. J., 296
Fite, K. R., *see* Ziskind, R. A.
Fitzgerald, G. R., 274
Flachsbart, P. G., 169, 219, *see also* Ott, W. R.
Flesh, R. D., *see* Duffee, R. A.
Flindt, M. L. H., 106
Florey, C. V., *see* Goldstein, B. D.
Flowers, E. S., 265
Flügge, K., 8, 26
Fochtman, E. G., *see* Stockman, J. D.
Fofanov, V. I., 35, *see also* Shepelev, E.
Foote, R. S., 109, 257
Fomin, A. G., 94
Forrence, L. E., *see* Abelles, F. B.
Fors'en, K. O., *see* van Assendelft, A.
Foster, A. R., *see* Lewis, H. E.
Foster, J. F., 145, 152, 153
Foster, M., 30, 31
Fox, F. L., *see* Heck, W. W.
Frank, C. W., *see* Long, K. R.; Schutte, W. C.
Frank, N. H., 142, 204
Franklin, B., 16
Fraser, D. W., 106, 190, 254, *see also* Broome, C. V.; Thacker, S. B.
Fraumeni, J. R., Jr., 259, 266, 267, *see also* Lee, A. M.
French, J. G., *see* Bridbord, K.
Friedman, H., *see* Spiegelmann, J.
Fritsch, A. J., 303
Froeb, H. F., 262, *see also* White, J. R.
Fuerst, R. G., 158
Fugas, M., 155, 218, 219
Fukuda, M., *see* Miyazaki, T.
Fulwiler, R. D., 188
Fumarola, D., 254
Furst, 266

Gadsby, K. J., *see* Harrje, D. T.

Gallagher, R. E., *see* Schaplowsky, A. F.
Gammage, R. B., 208, 209, *see also* Hawthorne, A. R.; Matthews, T. G.
Gans, J. C., *see* Hers, J. F. P.
Gardner, R. M., *see* Boyer, R. S.
Gary, V. F., 203, 279
Gaudette, L., *see* Tabershaw, I. R.
Geisling, K. L., 147, *see also* Meyer, B.
Gelperin, A., 247
General Electric Co., 170
Genin, A. M., *see* Fomin, A. G.
Gentilizza, M., 167
George, A. C., 127, 132, 155, 260
Geraci, C. L., *see* Kim, W. S.
Gerber, R. M., 299
Gerzoff, S., 308
Gesell, T. F., 127, 130, 300, *see also* Duncan, D. L.; Prichard, H. M.
Giedion, S., 26
Gilbert, J. R., *see* Hanna, M. G., Jr.
Giles, G., 300
Giles, W. H., 170
Gillis, J. R., *see* Rothman, L. S.
Gilmartin, E. J., *see* Allan, G.
Gip, L., 107
Girman, J. R., *see* Lipschutz, R. D.; Traynor, G. W.
Giulietti, M., 279
Givoni, B., 62
Gjessing, D. T., 140
Godin, G., 103, 172
Godish, T., 203
Goldman, A., *see* Rothman, L. S.
Goldman, R. A., *see* Shapira, Y.
Goldman, R. F., *see* McCarroll, J. R.
Goldschmidt, V. W., 296
Goldstein, B. D., 174, 270
Goldstein, J. R., *see* Eckert, E. R. J.
Goldwater, L. J., 189
Goodhue, L. D., 114
Gonzales, M. G., *see* Kreiss, K.
Graham, J. A., *see* Gardner, D. E.
Grandjean, E., *see* Fischer, T.; Weber, A.
Granier, M., 106, 254
Graunt, J., 21
Gravesen, S., *see* Løwenstein, H.
Gray, D., *see* Gary, V. F.
Gray, J. M. L., 146
Green, A. R., 255, 256
Green, B. G., 30, *see also* Stevens, J. C.
Green, M. A., 302, *see also* Creip, L. H.
Greenberg, M., *see* Rea, W. J.
Greene, T., 308
Gregory, P. H., 8, 97, 106, 107, 190, 191, 192, 193
Grieble, H. G., 106, 190
Griesback, V., *see* Ayers, S. M.
Griffis, L. C., 203, *see also* Pickrell, J. A.

Griffith, L. G., *see* Grieble, H. G.
Griscom, J. H., 23, 25, 39
Groah, W., 325
Grob, R. L., 146
Gronka, 109, 257
Gross, P., 250, *see also* Hatch, T. F.
Grot, R. A., 6
Gubernskii, Y. D., 203, 204, *see also* Dmitriev, M. T.
Gudzinowicz, B. J., 146
Gudzinowicz, M. J., *see* Gudzinowicz, B. J.
Gullino, P., 235
Gunnison, A. F., 145, *see also* Palmes, E. D.
Gunter, B. J., 257
Gutekunst, R. R., *see* Voors, A. W.
Guy, Dr., 23

Hachenberg, H., 146
Hackney, J. D., 239, 272
Hagopian, J. H., *see* Partridge, L. J.
Hahne, R. A., *see* Long, K. R.
Haines, J. E., 296
Hake, C. L., 273, *see also* Stewart, R. D.
Haldane, J. S., 26
Halpern, M., 181, 182
Hamilton, A., 273
Hammad, Y. Y., 188
Hammond, 262, 282 (*see* U.S. Surgeon General)
Hammons, A. S., 110, 273, *see also* Holleman, J. W.
Hanetho, P., *see* Berge, A.
Hanrahan, L. P., 203, *see also* Anderson, H. A.; Dally, K. A.; Eckmann, A. D.
Hardy, H. L., 273, *see also* Hamilton, A.
Harke, H. P., 184
Harley, J. H., 193, 260
Harper, G. J., 254, *see also* Norris, K. P.
Harper, R. J., Jr., *see* Andrews, B. A. K.
Harrington, J. B., 192
Harris, F. S., Jr., *see* Lundgren, D. A.
Harris, J. C., 279
Harrison, J. W., 146
Harrje, D. T., 296, 318, *see also* Blomsterberg, A. K.; Dutt, G.
Hartford, J. H., *see* Solomon, R. L.
Hartford, J. W., 252, 266
Harting, F. H., 260
Hartley, R. P., 156, 287, 308, *see also* Meyer, B.
Hartwell, T. D., *see* Pellizzari, E. D.
Harvey, M. C., 188
Hass, D. R., 107, *see also* Jopke, W. H.
Hasselblad, V., 268
Hatch, T. F., 250, *see also* Mercer, T. T.
Hausler, W. J., *see* Hendricks, S. L.
Hawthorne, A. R., 205, 208, 209, *see also* Matthews, T. G.

Hayhurst, E. R., *see* Kober, G. M.
Haynes, J. L., *see* Olsen, D.
Hazleton Laboratories, 268
Heavner, R., *see* Moschandreas, D. J.
Hecht, S. S., *see* Hoffmann, D.
Heeren, R. H., *see* Hendricks, S. L.
Heitner, K. L., 175, 176, *see also* Shair, F. H.
Held, H. R., *see* Hendricks, S. L.
Helfenbein, D., *see* Horie, Y.
Heller, S. R., 146, *see also* Milne, G. W. A.
Helsing, K. J., *see* Comstock, G. W.
Hembree, J. H., Jr., *see* McGinnis, J. P.
Henane, M., 47
Henderson, J. J., 106, 162
Hendricks, S. L., 254
Henning, H., 26, 32
Herget, W. F., 140, *see also* Stevens, R. K.
Herrman, A. A., *see* Stewart, R. D.
Hers, J. F. P., 232
Hershaft, A., 32
Hesketh, H. E., 97
Hess, A. L., *see* Hess, C. T.
Hess, C. T., 131, 299
Hess, W., *see* Harting, F. H.
HEW, 97, 172, 262, 281
Hewitt, W., 157
Heymann, G., *see* Flugge, K.
Higgins, E. A., 98, 177, 273, 299
Hildes, J. A., *see* Schaefer, O.
Hilgemeier, M. W., 279
Hill, A. C., 244
Hill, T. J., *see* Janssen, J. E.
Hinkle, L. E., Jr., 228, 229
Hirayama, T., 281
Hirose, H., *see* Yamanaka, S.
Hittman Associates, 215
Ho, H. P., *see* Brunnemann, K. D.
Hobbs, C. H., 203, *see also* Pickrell, J. A.
Hodges, G. R., 106, 190
Hoen, P. J. J., *see* Lammers, J. T. H.
Hoetjer, J. J.
Hoffer, E. L., *see* Appel, B. R.
Hoffmann, C. E., *see* Beasley, R. K.
Hoffmann, D., 258, 284, *see also* Brunnemann, K. D.; Schmeltz, I.; Wynder, E. L.
Hofmann, W., 261
Holaday, D. A., 239, *see also* Lundin, F. E., Jr.
Holcombe, J. K., 297, 302
Holland, W. W., *see* Colley, J. R. T.
Holleman, J. W., 110, 119, 231, 273
Hollowell, C. D., 68, 69, 103, 105, 131, 140, 142, 164, 166, 170, 174, 176, 178, 196, 197, 207, 215, 263, 294, 296, *see also* Anthon, D. W.; Fisk, W. J.; Girman, J. R.; Lepman, S. R.; Lin, C.-I.; Miksch, R. R.; Roseme, G. D.; Schmidt, H. E.;

Traynor, G. W.; Turiel, I.; Young, R. A.
Holt, P. F., 250, 252
Honigman, B., 272
Hopkins, L. C., *see* Landrigan, P. J.
Horie, Y., 105
Hornbostel, C., 304
Horvath, C., 146, 157
Horwood, L. J., *see* Fergusson, D. M.
Hosein, H. R., *see* Bouhuys, A.; Binder, R. D.; Mintz, S.
Hosko, M. J., *see* Stewart, R. D.
Houghton, F. C., 43, *see also* Yaglou, C. P.; Fanger, P. O.
Houten, L., *see* Decoufle, P.
Hove, G. M., 247
Howard, J. N., Jr., *see* Eaton, W. C.
Howell, T. C., *see* Hawthorne, A. R.; Matthews, T. G.
Howes, J. E., Jr., 152, *see also* Foster, J. F.
Hrach, F. J., *see* Newman, W. H.
Hu, E. P. C., *see* Gardner, D. E.
Huang, S., 312, *see also* Feldman, J.
Huey, H. J., *see* Cain, W. S.
Huff, J. E., 273, 274
Hughes, R. L., *see* Covelli, H. D.
Hulbrand, S. A., *see* Krost, K. J.
Hultqvist, B., 132
Humble, C., *see* Ferris, B. G., Jr.
Humphreys, M. A., 39
Hund, C. M., 193, *see also* Kusuda, T.
Hunt, C. M., 63, *see also* Burch, D. M.; Treado, S. J.
Hunt, F., Jr., *see* Frank, N. H.
Hunt, G. G., *see* Limaye, D. R.
Huntington, E., 246
Huppert, M., *see* Eckmann, B. H.
Hurewitz, D., 107, *see also* Levetin, E.
Hutchison, J., 24
Hwang, C. L., 35, *see also* Nakanishi, E.
Hyder, A., *see* Limaye, D. R.
Hyland, L., *see* Boyer, R. S.

Ichikawa, 23
Imperato, P. J., 254
Inculet, I. I., 182, 189, *see also* Lefcoe, N. M.
Ingersoll, J. G., 156, 172, 299, 301, 316, *see also* Boegel, M. L.; Hollowell, C. D.
Iondi, E. M., *see* Fisher, A. A.
Ishido, S., 190
Isseroff, B., *see* Cain, W. S.

Jackson, M. T., *see* Walker, J. Q.
Jacobs, M. B., 189
Jacobs, M. G., *see* Goldwater, L. J.
Jacobson, I., 106, 190, *see also* Scott, C. C.
Jaeger, R. J., 173, 272
Jann, P. R., 81, 83, 134, *see also* Duffee, R. A.

Jarke, F. H., 113, 142, 143, 189, 207, 211, 263
Jaung, R., *see* Berlad, A. L.
Jenkins, C. D., *see* Voors, A. W.
Jennings, W., 146
Jennison, M. W., 106, 188
Jensen, P. L., 247, *see also* Andersen, I.
Johansson, I., *see* Berglund, B.
Johns, W. E., 86, *see also* Meyer, B.
Johnson, A. R., *see* Rea, W. J.
Johnson, C. J., 94, 172, 272
Johnson, J. A., *see* Linko, Y.-Y.
Johsnon, R. H., Jr., *see* Duncan, D. L.
Johnston, J. W., Jr., 32, *see also* Turk, A.
Jonassen, N., 48
Jopke, W. H., 107
Joshi, S., 99, 172, 188
JRB Asociates, 274
Judd, M. D., 146, *see also* Debeir, M.
Junge, C. E., 193

Kahn, A., 103, 172
Kalandidi, A., *see* Trichopoulos, D.
Kalika, P., 297, 302, *see also* Holcombe, J. K.
Kalinova, N. V., *see* Gubenskii, Y. D.
Kaloyanova, F., *see* Martin, A. E.
Kaluzny, R. L., 69
Kamada, K., *see* Ishido, S.
Kambulin, N. A., 109, 254, *see also* Nikolov, S. K.
Kanarek, D. J., 266
Kanarek, M. S., *see* Dally, K. A.; Hanrahan, L. P.
Kaneko, F., 169, 174, *see also* Miyazaki, T.
Kanof, N. B., *see* Fisher, A. A.
Karol, M. H., *see* Sykora, J. L.
Karr, R. M., 106, 254, *see also* Salvaggio, J. E.
Katz, M., 141
Kawarabayashi, 107
Kays, W. M., 296
Kazemi, H., *see* Kanarek, D. J.
Keefe, T. J., *see* Savage, E. P.
Keleti, G., *see* Sykora, J. L.
Kelley, J. S., 272
Kelner, L., *see* Erickson, M. D.
Kennedy, B., 278, *see also* O'Quinn, S. E.
Kerr, H. D., 270
Keskinen, H., *see* van Assendelft, A.
Keyes, D. L., 164
Khan, T., *see* Myers, G. E.
Kimaier, N., *see* König, H. L.
Kimmel, H. S., *see* Trattner, R. B.
King, T. K. C., 243, *see also* Briscoe, W. A.
Kiyoko, K., *see* Seisaburo, S.
Kjaer, J., *see* Valbjørn, O.
Kleeman, J., *see* Covelli, H. D.

Klein, H.-J., 295
Klock, L. E., *see* Boyer, R. S.
Klotz, F., 302
Knight, V., 106
Knoll, G. F., 154
Knudson, R. J., *see* Lebowitz, M. D.
Kobayashi, D., 302, *see also* Sterling, T. D.
Koch, E. I., 173
Koepsel, W. C., 103, 173
Koerts, F., *see* Hoetjer, J. J.
Kolhari, A., *see* Briggs, T. M.
Kolupaeva, M. V., 109, 254, *see also* Nikolov, S. K.
Kon, L., *see* Fischbein, A.
Konig, H. L., 48
Kopple, J., *see* Rabinowitz, M.
Koss, 283 (*see* U.S. Surgeon General)
Kothny, E. L., *see* Appel, B. R.
Kragulski, M., 103, 172, *see also* White, J. W.
Kraybill, H. F., 201, 260, 263, 264, 266
Kreith, F., 69
Kreiss, K., 203, 302, 304
Krinkel, D. L., *see* Hollowell, C. D.
Krost, K. J., 149
Kuhn, R. R., *see* Takeushi, K.
Kulle, T. J., *see* Kerr, H. D.
Kunze, M., 395, *see also* Klein, H.-J.
Kuo, P., *see* Yeh, C.
Kupel, R. E., *see* Kim, W. S.
Kuramoto, M., *see* Mayrsohn, H.
Kuschner, M., *see* Laskin, S.
Kurt, T. L., *see* Honigman, B.
Kurzel, R. B., 118
Kusuda, T., 193, 297
Kuznetsova, T. I., *see* Savina, V. P.

LaBelle, C. W., 239, 240
LaFleur, P. D., 140
Lakat, M. F., *see* Thun, M. J.
Lamm, S. H., *see* Berger, J. M.; Tabershaw, I. R.
Landau, W. L., *see* Covelli, H. D.
Landrigan, P. J., 257
Lane, D. J., 263
Langer, H., 173, *see also* Rueden, H.
Langton, N. H., 35
Lao, Y. J., 170, 272
Las, Y. L., 104, 257
Laskin, S., 263
Lategola, M. T., *see* Higgins, E. A.
Laurenzi, G. A., 252, *see also* Dadaian, J. H.
Lave, L. B., 245, 246, 269, 270
Lavine, M., *see* Dutt, G.
Lavoisier, A., 16, 18, 34, 296, 315
Lawry, L. K., *see* Gunter, B. J.

Leadbetter, M. R., 140, 187, 252
Leaderer, P., *see* Cain, W. S.
Lear, C. W., 299
Leather, G. R., *see* Abelles, F. B.
Lee, A. M., 266
Lee, F. S.-C., *see* Schuetzle, D.
Lee, R. E., Jr., 143, *see also* Giles, W. H.
Lefcoe, N. M., 182, 189
Leggo, J. H., *see* Reznikov, M.
Lehmann, G., 42
Lehmann, W. F., 205, 218
Leithe, W., 140
Lemaire, R., 300
Lepman, S. R., 299
Levetin, E., 107
Levin, H. J., 97, *see also* Spirtas, R.
Lewis, H. E., 106
Libert, J. P., *see* Candas, V.
Licht, D., *see* Ayers, S. M.
Lidwell, 191
Lilis, R., *see* Anderson, H. A.
Lilley, F. W., *see* Spengler, J. D.
Limaye, D. R., 68
Lin, C.-I., 152, 204, 300, *see also* Anthon, D. W.; Hollowell, C. D.
Lindvall, T., *see* Berglund, B.
Linko, Y.-Y., 279
Linn, W. S., *see* Hackney, J. E.
Linnaeus, C., 134
Linteris, G., *see* Dutt, G.
Lippmann, M., *see* Lundgren, D. A.
Lipsitt, E. D., *see* Cain, W. S.
Lister, J., 26
Lloyd, J. W., 239, *see also* Lundin, F. E., Jr.
Locklin, D. W., *see* Barrett, R. E.
London, A. L., 296, *see also* Kays, M. W.
Long, C. G., Jr., 69, 296, *see also* Dubin, F. S.
Long, K. R., 88, 142, 203, 204, *see also* Schutte, W. C.
Looney, B. B., *see* Eisenreich, S. J.
Loring, W. C., 228, 229, *see also* Hinkle, L. E., Jr.
Lowrey, A. H., 182, 184, 185, 186, 187, 220, 284, *see also* Repace, J. L.
Luciano, J. R., 140, 302
Luginbyhl, T. T., 273, *see also* Christensen, H. E.
Lumley, J., *see* Pennington, J. H.
Lundgren, D. A., 153
Lundin, F. E., 239
Lundqvist, G. R., 167, 247, *see also* Andersen, I.
Lyman, J., 312, *see also* Feldman, J.

Mack, M., 24, *see also* Creech, C.
MacKnight, N. M., *see* Sayer, W. J.

MacLeod, K. E., 135, 211, 212
MacMahon, B., *see* Trichopoulos, D.
Macriss, R. A., 105, 175, *see also* Elkins, R. H.
Maddalon, R. F., 152
Maeda, 218
Mage, D. T., 143, 144, *see also* Lee, R. E., Jr.; Ziskind, R. A.
Mahler, M., 244, *see also* Whipp, B. J.
Maki, A. G., *see* Rothman, L. S.
Malberg, J. W., *see* Savage, E. P.
Maldinado, E., *see* Janssen, J. E.
Malloy, J. F., 304
Manning, P., 289
Mano, S. H., *see* Mayrsohn, H.
Manoharan, A., 189, *see also* Goldwater, L. J.
Mans, L. G. J., 273, 279, *see also* Van Huesden, S.
Mantel, N., 237
Mantovani, A., 107
March, A. W., *see* Palmes, E. D.
Margeson, J. H., 158, *see also* Ellis, E. C.; Fuerst, R. G.
Margolis, J. S., *see* Rothman, L. S.
Markham, L., 3
Markun, F., *see* Rundo, J.
Marlow, W. H., *see* Lundgren, D. A.
Marret, D., *see* Lebrun, J.
Marsch, A., 189
Marshall, F. J., 279
Martell, E. A., 127, 252, 260, *see also* Radford, E. P.
Martin, A. E., 246
Martin, H. F., *see* Gudzinowicz, B. J.
Martin, J. E., *see* Covelli, H. D.
Marx, P., 138
Maset, M., *see* Morveau, M.
Masurel, N., *see* Hers, J. F. P.
Mathur, J., *see* Everett, J. J.
Matthews, T. G., 208, *see also* Hawthorne, A. R.
Maunsell, K., 253, *see also* Rimington, C.
Maxwell, C. M., 140
Maxwell, K., *see* Boyer, R. S.
May, K. R., 95, 132, 154, 254
Mayhead, G. J., 54, *see also* Oliver, H. R.
Maynard, J. B., *see* Walker, J. Z.
Mayrsohn, H., 207
Maziarka, S., *see* Martin, A. E.
McCarthy, S., *see* Colome, S. D.
McDade, J. E., *see* Thacker, S. B.
McDermott, H. J., 296
McDonald, A. T., *see* Fox, R. W.
McEwen, F. L., 119, 149, 211, 275
McFadden, J. E., 146, 306, *see also* Moschandreas, D. J.
McIlhaney, M. L., *see* Kerr, H. D.

McIntire, M. S., *see* Ange, C. R.
McKee, H. C., 145
McKenzie, J. M., *see* Higgins, E. A.
McKinney, A. S., *see* Landgrigan, P. J.
McLean, R. L., 302
McNall, P. E., Jr., 193, 302, 303, 316, *see also* Colle, R.; Kusuda, T.
McQuiston, F. C., 140
Mehlman, M. A., *see* Kraybill, H. F.
Meier, A., *see* Roseme, G. D.
Meinke, W. W., 140
Meleshko, G., 35, *see also* Shepelev, E.
Melia, F., 284
Melia, R. J. W., 270, *see also* Goldstein, B. D.
Mellegaard, B., *see* Berge, A.
Melton, C. F., *see* Higgins, E. A.
Mercer, T. T., 252
Mergard, E. C., 257, *see also* Campbell, I. R.
Merritt, L. L., Jr., *see* Willard, H. H.
Mes, J., 119, 126, 263
Metcalfe, C. E., *see* Hawthorne, A. R.; Matthews, T. G.
Meyer, B., 27, 58, 85, 86, 87, 88, 103, 119, 125, 147, 148, 167, 202, 203, 204, 207, 221, 223, 224, 225, 279, 327
Meyer, M. B., *see* Comstock, G. W.
Meyers, B. F., 125, 211, 259, 279
Michaelsen, G. S., *see* DeRoos, R. L.
Midgett, M. R., 152, *see also* Mitchell, W. J.
Miksch, R. R., 113, 147, 200, *see also* Geisling, K. L.; Hollowell, C. D.; Meyer, B.; Schmidt, H. E.
Miller, B. S., *see* Linko, Y.-Y.
Miller, E., *see* Stanton, M. F.
Miller, J., 58, *see also* Yoshinaga, A.
Miller, R. L., 190, *see also* Burton, W. E.
Miller, S. E., *see* Barrett, R. E.
Miller, T. L., 140, *see also* Noll, K. E.
Mills, C. C., *see* Riley, R. L.
Milne, G. W. A., 146, *see also* Heller, S. R.
Mincer, H. H., *see* McGinnis, J. P.
Miquel, P., 26, 107, 154, 192
Mirick, W., 299
Mitchell, 270
Mitchell, C. A., 279, *see also* Binder, R. E.; Bouhuys, A.; Schoenberg, J. B.
Mitchell, F. L., *see* Brown, S. S.
Mitchell, K. I, 300
Mitchell, R. S., *see* Phibbs, B. P.
Mitchell, S. W., *see* Billings, J. S.
Mitchell, W. J., 152
Mitre Corp., 140
Mitsubishi Electric Corp., 298
Miura, Y., 167
Miyaji, E., *see* Kobayashi, M.
Miyazaki, T., 184, 211
Mogabgab, W. J., 254

Mohler, J. G., *see* Hackney, J. D.
Moldow, C. F., *see* Voors, A. W.
Møhave, L., 167, *see also* Andersen, I.
Møller, J., *see* Mølhave, L.
Mollier, R., 51
Monteith, J. L., 191, 193, *see also* Gregory, P. H.
Moore, D. J., 140, *see also* Davies, P. O. A. L.
Moore, J. A., *see* Huff, J. E.
Moran, J. C., 272, *see also* Johnson, C. J.
Morgan, A., 304, *see also* Walsh, M.
Morgan, M. S., 265
Morgan, R. W., *see* Shettigara, P. T.
Moriates, S., *see* Ferrand, E. F.
Morken, D. A., 261
Morkin, K. M., *see* Duncan, J. R.
Morresi, A. C., 140, *see also* Cheremisinoff, P. N.
Morris, A. L., 58
Morris, J. E. W., 95
Morris, S. C., *see* Morgan, M. S.
Morrison, J. H., *see* Schaplowsky, A. F.
Morrow, P. E., 252, *see also* Mercer, T. T.
Morschauser, C., 69, 325
Morse, S. S., *see* Moschandreas, D. J.
Morton, J., 32, *see also* Hershaft, A.
Moschandreas, D. J., 99, 142, 169, 171, 172, 174, 175, 182, 204, 207, 216, 220, 294, 299, 322, *see also* McFadden, J. E.
Moss, R. W., *see* Randolph, T. H.
Most, R. S., 279
Mostardy, R. A., *see* Woebkenberg, N. R.
Moulton, D. G., 32, *see also* Turk, A.
Moyer, R. H., *see* Moschandreas, D. J.
Muchtarova, M., 146
Muir, D. C. F., 252
Mullan, B. J., *see* Lewis, H. E.
Müller, E. A., *see* Lehmann, G.
Müller, F., 118
Murphy, G., *see* Riley, E. C.
Muschenheim, C., *see* Holman, C. W.
Muth, F., *see* Druckrey, H.
Muysers, K., 146
Muzet, A., 47, *see also* Candas, V.; Fanger, P. O.
Myers, E., *see* Yabroff, I.
Myers, G. E., 203, 206, 207, 217
Myerson, B., 303
Myrbäck, K.-E., *see* Rylander, R.

Nadel, J., *see* Sheppard, D.
Nagaoka, M., 203, *see also* Myers, G. E.
Nagelschmidt, G., 250, *see also* Cartwright, J.
Nakadoi, T., *see* Kaneko, F.
Nakagawa, T., *see* Ishido, S.
Nakanishi, E., 35

Narasaki, H., 101, 301
Narat, J. K., 235
NAS, 32, 69, 97, 141, 147, 173, 204, 207, 240, 242, 243
Nass, C. A. G., *see* Biersteker, K.
National Council for Radiation Protection, 227
National Council on Radiation, 58
Natusch, D. F. S., *see* Davison, R. L.
Nayak, P. R., *see* Partridge, L. J.
Nazaroff, W. W., *see* Boegel, M. L.; Hollowell, C. D.
Neal, A. D., 182
Nefedov, I. U., 35, 94, 141
Nelson, J. W., 257, *see also* Moschandreas, D. J.; Winchester, J. W.
Nero, A. V., 227
Nettesheim, P., *see* Hanna, M. G., Jr.
Nevins, R. G., *see* Gagge, A. P.; Rohles, F. H.
Nevrala, D. J., *see* Etheridge, D. W.
Newton, A. S., *see* Schmidt, H. E.
Newman, W. H., 300
Newmark, F. M., 108
Newton, A., *see* Miksch, R. R.
Ng, D. Y. C., *see* Elkins, R. H.
Nielsen, B., 46, *see also* Fanger, P. O.
Nielsen, P. A., *see* Valbjørn, O.
Nikel, C. M., 245, 302, 303, *see also* Pfeiffer, G. O.
Nikolov, S. K., 109, 254
Nishi, Y., 39, 43, *see also* Gagge, A. P.
Nishimura, H., *see* Yanagisawa, Y.
Noll, K. E., 140
Nonat, A., 182, *see also* Benarie, M.
Norris, K. P., 254
Norton, S. A., *see* Hess, C. T.
Novak, D., *see* Sykora, J. L.
Noyes Data Corp., 140
NRC, 98, 100, 104, 112, 113, 116, 141, 148, 186, 190, 197, 251, 254, 257, 261, 265, 271, 273, 274, 278, 308, 334
NSF, 141, 334
Nunlist, R., 207, *see also* Meyer, B.

Oatman, L., 20, 279, *see also* Gary, V. F.
O'Callaghan, P. W., 68
O'Dwyer, W. F., *see* Fitzgerald, G. R.
Oelschlager, F., *see* Limaye, D. R.
Oglesbay, F. B., *see* Schaplowsky, A. F.
O'Grady, F. W., *see* Pennington, J. H.; Riley, R. L.; Speers, R., Jr.
O'Hare, D., 253
Öhrström, M., *see* Rylander, R.
Oikawa, K., 146
Oikawa, S., 302, *see also* Yano, H.
Oka, M., *see* Kaneko, F.
Okamoto, S., *see* Takeushi, K.

Author Index

Olgyay, A., *see* Olgyay, V.
Olgyay, V., 62
Oliver, H. R., 54
O'Quinn, S. E., 278
Orlando, J. A., *see* Limaye, D. R.
Orlova, N. S., *see* Gubernskii, Y. D.
Ormstad, E. B., *see* Berge, A.
Osborne, S., 205, 208, 209
Østerballe, O., *see* Albrechtsen, O.
Ostrander, W. S., 68, *see also* Sizemore, M. M.
Ott, W. R., 108, 143, 144, 159, 160, 169, 170, 171, 173, 183, 219, 220, 221, 222, *see also* Flachsbart, P. G.; Repace, J. L.; Wallace, L. A.; Willits, N. H.
Overstreet, M., *see* Briggs, T. M.
Ower, E., 140
Ozkaynak, H., 217

Pack, D. H., 141
Paine, S. C., *see* Johnson, C. J.
Palmer, G. T., 253, *see also* Vaughan, V. C.
Palmes, E. D., 145
Pandolf, K. B., *see* Shapira, Y.
Parker, A., 35
Parker, J. D., 140, *see also* McQuiston, F. C.
Parker, J. F., Jr., 94
Parkhurst, R. C., 140, *see also* Ower, E.
Parr, V. B., *see* McKee, H. C.
Peatman, J. B., 69
Pekich, R., 272, *see also* Johnson, C. J.
Pellizzari, e. D., 143, 148, 149, 150, 151, 198, *see also* Erickson, M.; Krost, K. J.; Wallace, L.; Zweidinger, R. A.
Pelton, D. J., *see* Moschandreas, D. J.
Pennington, J. H., 106, 190
Pepper, J. H., *see* Young, R. A.
Percival, Dr., 310
Periera, N. C., 35, *see also* Nakanishi, E.
Perera, F. P., 97, 152, 153
Perlman, D., *see* Cain, W. S.
Perry, R., 140
Pesez, M., 149
Peters, H., *see* Harke, H. P.
Petersen, G. A., 173, *see also* Derham, R. L.
Peterson, J. E., *see* Stewart, R. D.
Petrocci, M., *see* Fischbein, A.
Pettenkofer, M. v., *see* von Pettenkofer
Pfeiffer, G. O., 245, 302, 303
Phibbs, B. P., 265
Philbin, E. J., *see* Gunter, B. J.
Phillips, S., *see* Flachsbart, P. G.
Pickrell, J. A., 142, 203, 205, 225
Pierson, D. A., *see* Long, K. R.
Pilotte, J. O., *see* Moschandreas, D. J.
Pimental, N. A., *see* Shapira, Y.
Pimm, P., *see* Sebben, J.
Pleus, R., *see* Gary, V. F.

Plondke, N. J., *see* Rundo, J.
Pocchiari, F., 125
Poet, W. E., *see* Martell, E. A.
Pohl, E., *see* Hofmann, W.
Pohl, J., 130
Polk, L. G., *see* Schaplowsky, A. F.
Pomeroy, N. P., 95, 254, *see also* May, K. R.
Port, R., 237
Poston, J. W., 154, *see also* Eichholz, G. G.
Pott, P., 23
Powell, R. J., 306
Prater, T. J., *see* Schuetzle, D.
Preobrazhenskii, L. I., *see* Dmitriev, M. T.
Pressler, C. L., 88, 147, 208, 209
Price, J. P., *see* Limaye, D. R.
Prichard, H. M., 300, *see also* Gesell, T. F.
Prichard, J. H., 127, 130
Priestley, J., 16, 315
Proctor, D. F., 247, 250, *see also* Andersen, I.; Brain, J. E.
Pruett, J. G., 265
Pye, J., 24
Punttila, A., *see* Seppanen, O.

Quackenboss, 218
Quinlivan, S. C., 152, *see also* Maddalon, R. F.

Rabinowitz, M., 104, 257
Radford, E. P., 246, 252, *see also* Spivey, G. H.
Raffonelli, A., *see* Davies, J. E.; Davis, J. H.
Rainer, D., *see* DeRoos, R. L.
Rajhans, G. S., 299
Ramalingam, A., 193, *see also* Rati, E.
Ramazzini, B., 23
Rankin, J., *see* Anderson, H. A.
Rao, A. K., 97, *see also* Schrag, M. P.
Rappaport, S. M., *see* Geisling, K. L.
Rati, E., 193
Razydlo, D. R., *see* Compton, B.
Real Estate Research Corp., 62
Rector, H. E., 299, 322, *see also* Moschandreas, D. J.
Reed, C. E., 256
Reed, J. W., *see* Andrews, B. A. K.
Rehn, L., 23
Reich, G., *see* Davis, J. H.
Reid, L. M., 250, *see also* Brain, J. D.
Reimold, F., *see* Ayers, S. M.
Renzetti, A. D., Jr., *see* Boyer, R. S.
Repace, J. L., 100, 122, 153, 170, 171, 172, 182, 183, 184, 185, 186, 187, 220, 284
Repetto, R., *see* Sexton, K.
Reppert, M. H., 296
Reznikov, M., 106
Rice, C., *see* Fischbein, A.

Rich, T. L., *see* Lao, Y. J.
Richardson, A. C. B., 299
Richter, H., 148, *see also* Elfers, L.
Riley, D. J., 254
Riley, E. C., 106, 254, *see also* Yaglou, C. P.
Riley, R. L., 254, *see also* Riley, E. C.
Rimington, C., 108, 253, 256
Roberts, B. M., 296, *see also* Croome-Gale, D. J.
Robertson, M., *see* Yabroff, I.
Robinson, N., 169 (*see* Meyer, 1977)
Roffael, E., 203, 207
Rogers, D., 157, *see also* Touchstone, J. C.
Rogozen, M. B., 193
Rohles, R. H., 43, 44, 45, 46
Roose, R. W., 68, 296
Roscoe, 315
Rosebury, T., 149
Roseme, G. D., 296, *see also* Fisk, W. J.
Rosenberg, S. H., *see* Neal, A. D.
Rosenfeld, A., *see* Roseme, G. D.
Rosenthal, D., *see* Erickson, M. D.
Rosenzweig, A. L., 106, 190
Ross, E. D., *see* Allen, R. J.
Ross, H., 247
Rossano, A. T., Jr., 140, *see also* Cooper, H. B. H., Jr.
Rossi, R. C., 299, *see also* Gerber, R. M.
Rossiter, W. J., 204, 327
Roth, E. M., 35, 94
Rotheim, E., 114
Rothman, L. S., 141, 283
Rowe, D. R., *see* Elliott, L. P.
Rowland, 116
Rowlands, R. P., 167, *see also* Spedding, D. J.
Ruecker, M. R., 146
Rueden, H., 173
Rueppel, M. L., *see* Beasley, R. K.
Rumack, B. H., *see* Harris, J. C.
Rundo, J., 132
Russenberger, H. J., 106, 192, *see also* Brunner, A.
Rutkowska, I., 203, 204
Ryan, B., *see* Oskaynak, H.
Rye, W. A., 279
Rylander, R., 106, 190, 254, 255
Ryon, M. G., 110, 273, *see also* Holleman, J. W.
Ryzhkova, V. E., *see* Nefedov, I. U.

Saalfeld, F. E., *see* Decorpo, J. J.
Saba, M. P., *see* Crabtree, R. E.
Sabersky, R. H., 173, *see also* Derham, R. L.; Petersen, G. A.
Saccomanno, G., *see* Brandom, W. F.
Saeltzer, A., 24
Saisho, A., *see* Sheppard, D.

Saldana, M., 254, *see also* Riley, D. J.
Sale, 245
Sallvik, K., 296, *see also* Walberg, K.
Salvaggio, J. E., 106, 254
Samet, J. M., 270
Samuels, R., *see* Moschandreas, D. J.
Sansone, E. B., 177
Saracci, R., *see* Huff, J. E.
Sargent, F., 94, *see also* Johnson, R. E.
Sarkozi, L., *see* Fischbein, A.
Satish, J., 69, 99, 172, 188
Savage, E. P., 210, 211, 272
Savina, V. P., 24, *see also* Nefedov, I. U.
Sawicki, C. R., 147, *see also* Sawicki, E.
Sawicki, E., 147, *see also* Mulik, J. E.
Sawyer, R. N., 99, 100, 265
Sayer, W. J., 154
Saylor, D. O., 27
Sazynska, M., *see* Rutkowska, I.
Schakenbach, J. T., *see* Moschandreas, D. J.
Scales, J. W., 140
Scaringelli, F. P., 158, *see also* Fuerst, R. G.
Schaefer, G. L., 249, *see also* Eckmann, B. H.
Schaplowsky, A. F., 272
Schauerte, W., *see* König, H. L.
Scheere, A. R., *see* Kreiss, K.
Schilling, R. S., *see* Bouhuys, A.
Schinz, H. R. V., 24
Schipma, P., *see* Lieberman, A.
Schleuter, D. P., *see* Hodges, G. R.
Schlüter, G., 138, *see also* Marx, P.
Schmähl, D., 234, 235, 237, *see also* Druckrey, H.; Port, R.
Schmeltz, I., 283
Schmidt, C. D., *see* Boyer, R. S.
Schmidt, G., *see* Baumann, H.
Schmidt, H. E., 198, 279, *see also* Miksch, R. R.
Schneider, H. J., *see* Liebich, H. M.
Schneider, M. T., *see* Matthews, T. G.
Schneidermann, M. A., 237, *see also* Mantel, N.
Schodek, D. L., *see* Kim, M. K.
Schoenberg, J. B., 279, *see also* Bouhuys, A.
Schøler, M., *see* Olesen, B. W.
Scholz, E., 192, *see also* Russenberger, H. J.
Schönborn, C., 106
Schosman, E., *see* Collé, J.
Schossler, L., *see* Limaye, D. R.
Schrag, M. P., 97
Schuette, F. J., 143
Schuetzle, D., 148, 149
Schultz, C., *see* Harke, H. P.
Schulze, H.-D., 147
Schumacher, P. M., 152, *see also* Spicer, C. W.
Schutte, W. C., 142, 204

Schwartz, B., *see* Løwenstein, H.
Scott, B. W., 146
Scott, C. C., 106, 190
Scribner, B. F., 140, 146, *see* Meinke, W. W.
Sears, D. R., 63
Sebben, J., 170, 184
Sehmel, G. A., 177, 179, 180
Seisaburo, S., 106, 190
Selber, 101
Selikoff, I. J., 263, *see also* Anderson, H. A.; Fischbein, A.
Selway, M. D., 176
Seskin, E. P., 246, 269, 270, *see also* Lave, L. B.
Sessions, B. W., *see* Mitchell, K. L.
Sevcova, H., 261
Sevcova, J., *see* Sevcova, H.
Sewall, H., 241
Seymour, J. W., 304, *see also* Myers, G. E.
Shair, F. H., 175, 176, *see also* Derham, R. L.
Shannon, F. T., *see* Fergusson, D. M.
Shapiro, R. E., 392
Sharko, J. R., *see* Limaye, D. R.
Shaver, B., 146, *see also* Ruecker, M. R.
Shea, G., 32, *see also* Hershaft, A.
Shearer, S. D., Jr., 320
Sheehy, J. W., 169, 174
Shepard, C. C., *see* Thacker, S. B.
Shepard, R. J., 282
Shepart, E., *see* Stone, R.
Shepelev, E., 35, 94
Shephard, R. J., *see* Sebben, J.
Shepherd, R. J., *see* Godin, G.
Sheppard, D., 268
Shettigara, P. T., 265
Shimkin, M. B., *see* Bryan, W. R.
Shinskey, F. G., 296
Shirtliffe, C. J., 208, *see* Bowen, R. P.
Shivpuri, D. N., *see* Riley, R. L.
Shooter, R. A., *see* Speers, R., Jr.
Sidorenko, G. I., 107, 254
Siegmund, E. G., *see* Cairns, T.
Silano, V., *see* Pocchiari, F.
Silberstein, S., 193, 299, *see also* Kusuda, T.
Silke, B., *see* Fitzgerald, G. R.
Silver, F., 245, 302, 303
Silver, I. H., 192
Silverman, F., *see* Mintz, S.
Simon, C. G., 143
Sittig, M., 140
Sizemore, M. M., 68
Slein, M. W., 177, *see also* Sansone, E. B.
Slike, K. A., *see* Collé, J.
SMACCNA, 296, 335
Smeaton, J., 27
Smidt, U., 146, *see also* Muysers, K.
Smiley, R. E., *see* Rea, W. J.

Smith, E. M., 239, *see also* Lundin, F. E., Jr.
Smith, L., 24
Smith, M. B., 273, 280
Smith, N. J., 172
Smith, P. M., *see* Timm, W.
Smith, P. W., 106
Smith, R. D., *see* Andrews, B. A. K.
Smith, R. W., *see* Lao, Y. J.
Smyth, W. F., 157
Sobeck, E., 312, *see also* Feldman, J.
Socolow, R., *see* Dutt, G.
Sokolov, N. L., 94, *see also* Nefedov, I. U.
Soleymani, Y., 254, *see also* Weiss, N. S.
Solomon, P., 30
Solomon, R. L., 252, 266
Solomon, W. R., *see* Burge, H. P.
Solotorovsky, M., *see* Buchbinder, L.
Solowey, M., *see* Buchbinder, L.
Sophocleus, G. J., 272, *see also* Kelley, J. S.
Sorenson, S. D., *see* Caplan, K. T.
Sothern, R. D., *see* Mayrsohn, H.
Sparaciro, C. M., *see* Pellizzari, E. D.
Sparros, L., *see* Trichopoulos, D.
Spedding, D. J., 98, 167, 189
Speers, R., Jr., 95, 107, 254
Speizer, F. E., 174, 270, *see also* Ferris, B. G., Jr.; Samet, J. M.; Spengler, J. D.
Spendlove, J. C., 106
Spengler, J. D., 103, 142, 153, 169, 172, 173, 174, 218, 270, 294, *see also* Bellin, P.; Colome, S. D.; Cortese, A. D.; Dockery, D. W.; Ferris, B. G., Jr.; Ju, C.; Samet, J. M.; Speizer, F. E.; Sundell, J.; Turner, W. A.; Wilson, R.
Spicer, C. W., 152
Spiegelmann, J., 192
Spirtas, R., 97
Spitzer, H., *see* Lehmann, G.
Spivey, G. H., 246
Spore, R. W., 190, *see also* Miller, R. L.
Sprague, D. E., *see* Rea, W. J.
Stahl, W. H., 32
Stanislawczyk, K., *see* Decoufle, P.
Stanton, M. F., 265
Stark, J. W. C., *see* Moschandreas, D. J.
Stedman, R. L., 100, 102, 261
Steinhäusler, F., *see* Hofmann, W.
Stephenson, G. R., 119, *see* McEwen, F. L.
Sterling, D. A., 190, *see also* Sterling, E.
Sterling, E., 9, 10, 81, 299, *see also* Sterling, T. D.
Sterling, T. D., 184, 186, 284
Stern, A. C., 140, 299, 306, 308
Stevens, J. C., 30
Stevens, R. K., 140, *see also* McClenny, W. A.
Stewart, G. T., *see* Voors, A. W.
Stewart, R. D., 271, 273

Stick, S., 284
Stillwell, D. E., 253, see also Rimington, C.
Stitt, B. D., 156, 316, see also Ingersoll, J. G.
Stolwijk, J. A. J., 43, 203
Stone, K. R., see Spengler, J. D.
Stone, R., 279, see also Moschandreas, D. J.
Stranden, E., 193
Strang, J. E., see Crabtree, R. E.
Straub, C. D., 296, see also Bond, R. G.
Strauss, W., 299
Stricoff, R. S., see Partridge, L. J.
Stroumtsos, L. Y., see Windholz, M.
Stuart, J. H., 146, see also Scott, B. W.
Stuart, W. H., see Thacker, S. B.
Suero, J. T., 262, see also Woolf, C. R.
Sugden, T. M., 116
Sullivan, R. J., 169, 171, 174, 176, 207
Sullivan, W. N., 114
Sultan, L. U., see Riley, R. L.
Summer, W., 136
Sundin, B., 205, 207, 218, 325, 327
Sundin, R. E., see Phibbs, B. P.
Suzuki, Y., see Kobayashi, M.
Swaardemaker, 134
Swan, A. V., see Melia, F.
Swidersky, P., see Kerr, H. D.
Switzer, P., 173
Szalai, A., 159, 160, 161
Szczepanski, C. Z., 297

Takeushi, K., 97, 257
Tashima, M. K., see Geisling, K. L.
Tasai, T. F., see Thacker, S. B.
Tatsuko, N., see Seisaburo, S.
Taylor, B., see Fergusson, D. M.
Taylor, D. K., see Savage, E. P.
Tejada, S. B., see Schuetzle, D.
Tessari, J. E., see Savage, E. P.
Thacker, S. B., 106
Thaxton, P., 239, 257
Thiem, T. H., see Dmitriev, M. T.
Thomas, A. A., 35, 155, 198
Thomas, J., see Sevcova, H.
Thompson, B., 146
Thompson, 182 (see Neal, 1978)
Thomson, D. B., see Vos, K.
Thomson, W. A. R., 47, 190, 245, 248, 249
Thornton, J. D., see Eisenreich, S. J.
Thorshauge, J., see Olesen, B. W.
Thun, M. J., 278
Thurston, B. E., see Turiel, I.
Timby, E. A., 300
Timmermans, F. J. W., see Schaefer, O.
Timmons, M. L., see Harrison, J. W.
Tipping, R. H., see Rothman, L. S.
Tishman Research Corp., 81
To, T., see Stone, R.

Tockman, M. S., see Comstock, G. W.
Toigo, A., see Grieble, H. G.
Tomatis, L., see Huff, J. E.
Tomczyk, C., see Palmes, E. D.
Tomita, B., 207
Touchstone, J. C., 157
Tovey, E. R., 106
Trattner, R. S., 257
Traynor, G. W., 68, 142, 318, see also Girman, J. R.; Hollowell, C. D.; Lipschutz, R. D.
Tredgold, T., 8, 296
Trichopoulos, D., 284
Truhaut, R., 238
Tsai, S., see Yeh, C.
Tsitovich, S. I., 35, see also Shepelev, E.
Tucker, B., see Moschandreas, D. J.; Stone, R.
Turiel, I., 293, 296, see also Fisk, W. J.; Hollowell, C. D.; Roseme, G. D.; Young, R. A.
Turk, A., 32, 216
Turley, C. D., see Brenchley, D. L.
Turner, B. B., 300
Turner, L. D., see Mitchell, K. L.
Turner, W. A., 153, see also Spengler, J. E.
Turton, D., see Mes, J.
Tutu, N., see Berlad, A. L.
Tyrrell, D. A. J., 193, 254, see also Buchland, F. E.

Uehara, C. F., see Sheppard, D.
Ullrey, J. C., see Dasgupta, P. K.
Ulsamer, A. G., 263, see also Beall, J. R.
UNSCAR, 250, 260, 335
U.S. Bureau of Mine and Mine Safety, 143
USDA Forest Service, 89, 90
U.S. Department of Commerce, 55, 58
U.S. National Center for Health Statistics, 281
U.S. Surgeon General, 163, 186, 230, 262, 267, 283, 312

Vaichulis, E. M. K., see Wright, D. N.
Vail, S. L., see Andrews, B. A. K.
Vainio, H., see Niemelä, R.
Valbjørn, O., see Fanger, P. O.
van Assendelft, A., 256
Vandenberg, R. A., 106, see also Tovey, E. R.
Van Huesden, S., 273, 279
Vaughan, H. F., 253, see also Vaughan, V. C.
Vaughan, J. A., see Higgins, E. A.
Vaughan, V. C., 253
Verner-Carlson, B., see Rylander, R.
Verschueren, K., 207
Vesilind, P. A., 299

Author Index

Viadana, E., *see* Decoufle, P.
Viele, M. R., *see* Newman, W. H.
Villano, C. E., 318
Villarreal, E., *see* Stone, R.
Vogt, J. J., *see* Candas, V.
von Pettenkofer, M., 19, 134, 188, 192, 296, 315, *see also* Pettenkofer, M. v.
Voors, A. W., 232
Voronin, G. I., *see* Fomin, A. G.
Vos, K., 190

Waddell, R. D., *see* Pellizzari, E. D.
Wadden, R. A., *see* Allen, R. J.; Neal, A. D.; Selway, M. D.
Wade, W. A., III, 171, *see also* Cote, W. A.
Wagner, G., *see* Muir, C. S.
Wahren, H., 306
Wahrendorf, J., *see* Port, R.
Wainer, R. A., *see* Kanarek, D. J.
Walberg, K., 296
Walburn, S. G., *see* Krost, K. J.
Walcott, R., 142, 204
Walker, J. Q., 146, 147
Wallace, J. R., *see* Davison, R. L.
Wallace, L. A., 143, 144, 148, 157, 170, 171, 182, 198, 209, 275, 276, 277, *see also* Mage, D. T.; Repace, J. L.; Zweidinger, R. A.
Walsh, M., 304
Wander, J., *see* Limaye, D. R.
Wang, G., *see* Yeh, C.
Wang, J. Y., 140
Wanner, H. U., 69, 99, 172, 188, 192, 300, 302, *see also* Brunner, A.; Huber, G.; Joshi, S.; Klotz, F.; Russenberger, H. J.; Satish, J.
Ward, G. H., *see* Eaton, W. C.
Ward, J. R., *see* Moschandreas, D. J.
Warner, 167, 168
Wayne, L. G., 278
Weber, A., 30, 170, 184, 278, 283, *see also* Weber-Tschopp, A.; Fischer, T.; Kanarek, D. J.
Weber-Tschopp, A., 283, *see also* Weber, A.
Weinkam, J., *see* Sterling, T. D.
Weeke, B., *see* Albrechtsen, O.
Weiss, N. S., 254
Weiss, S. T., *see* Schenker, M. B.
Wellock, C. E., 254
Wells, W. F., 106, 247, 254
Weschler, C. J., 152
Wesolowski, J. J., *see* Appel, B. R.
West, R. E., 69, *see also* Kreith, F.
West, T. F., 116, *see also* Sugden, T. M.
West, V. R., *see* Parker, J. R., Jr.
Wetherill, G., *see* Rabinowitz, M.
Wharton, L. M., *see* Powell, R. J.
Wheeler, H. W., *see* Savage, E. P.

Whipp, B. J., 244
White, J. R., 262
White, J. W., 103, 172
Whittaker, D., *see* Wallace, L. A.; Zweidinger, R. A.
Whittemore, A. S., 246, *see also* Brewslow, N. E.
WHO, 56, 190, 245, 259, 267, 270, 335
Willard, H. H., 140
Williams, K., 68
Williams, K. H., Jr., *see* Thacker, S. B.
Williams, R. E. O., 192, 254, 278
Williams, S. R., *see* Pellizzari, E. D.
Willits, N. H., 173, *see also* Ott, W. R.
Wilson, D. G., *see* Wilson, R.
Wilson, H. W., *see* Sayer, W. J.
Wilson, M. J. G., 167, 189
Wilson, R., 212
Wilson, R. H. L., *see* Farber, S. M.
Winchester, J. W., 188, 257, *see also* Desaedeleer, G. G.; Moschandreas, D. J.
Winden, F., 106, *see also* Schönborn, C.
Winslow, C. E. A., 190
Winslow, S. G., 265, *see also* Pruett, J. G.
Witheridge, W. N., *see* Yaglou, C. P.
Wittstadt, F., *see* Riley, R. L.
Wixson, B. G., 104, *see also* Boggess, W. R.
WMO, 141, 335
Woebkenberg, N. R., 271, 282
Wolf, G., *see* Arcos, J. C.
Wolfe, R., 69
Wolfson, M., *see* Ferris, G. B., Jr.
Wong, O., *see* Tabershaw, T. R.
Wong, W. S., *see* Sheppard, D.
Woo, J. K., *see* Meyer, B.
Woodbury, M. A., 203 (*see* Anderson, 1981)
Woods, J. E., *see* Janssen, J. E.; Rohles, F. H.
Woolf, C. R., 262
Worley, J. W., *see* Beasley, R. K.
Worstell, D., *see* Woebkenberg, N. R.
Wright, C. G., 210, 211
Wright, D. N., 192
Wright, G. R., *see* Godin, G.
Wright, R., *see* Yabroff, I.
Wright, R. H., 32
Wright, T., *see* Limaye, D. R.
Wyatt, J. R., *see* Decorpo, J. J.
Wynder, E. L., 258, 262, *see also* Schmeltz, I.

Yabroff, I., 103, 173
Yaglou, C. P., 5, 8, 43, 66, 95, 134, 162, 301, 315
Yamagiwa, 23
Yamaoka, S., *see* Kaneko, F.; Miyazaki, T.
Yanagisawa, S., 306
Yano, H., 302, 304
Yarmac, R. F., *see* Brenchley, D. L.

Yeh, Y.-J., 302, *see also* Berlad, A. L.
Yin, S., 252, *see also* Dadaian, J. H.
Yocom, J. E., 115, 117, 142, 146, 167, 168, 169, 170, 171, 172, 173, 174, 182, 190, *see also* Cote, W. A.; Wade, W. A., III
Yonushonis, W. P., 239, *see also* Thaxton, P.
Yoshinaga, A., 58
Young, D. S., *see* Brown, S. S.
Young, L. D. G., *see* Rothman, L. S.
Young, R. A., 83, 140, 297
Young, R. J., *see* Perry, R.

Zabransky, J., *see* Moschandreas, D. J.
Zahomek, J., *see* Benes, M.
Zakharchenko, M. P., *see* Dmitriev, M. T.
Zaloguev, S. N., *see* Nefedov, I. U.
Zampieri, A., *see* Pocchiari, F.
Zapalac, G. H., 156, 316, *see also* Ingersoll, J. G.
Zawacki, T. S., *see* Elkins, R. H.
Zeidler, 118
Zepinski, M., 275, 276, 277
Ziedman, K., *see* Wayne, L. G.
Zimmer, J., *see* Elkins, R. H.
Zimmerli, B., 148, 201, 211
Zimmermann, H., 148, 211, *see also* Zimmerli, B.
Ziskind, R., 219
Zlatkis, A., 146, *see also* Liebich, H. M.
Zwaardemaker, H., 26
Zweidinger, R. A., 143, 209, 210, *see also* Erickson, M. D.; Wallace, L. A.
Zweig, G., 149, *see also* Compton, B.

Subject Index

Subject Index

Important acronyms and units are explained in the Appendices.

AATCC, 148, 330, 333
abbreviations, 333-334
absentee rate, in offices, 9
acceptable indoor air quality, 316-317
acceptable practices, HUD manual, 320
acephate, 210
    level, 210
acetone, odor, 135, 198-199, 317
ach, ventilation units, 331
acid rain, 58
ACGIH, 141, 152, 241, 265, 327, 333, 338
    analytical methods, 141-142
acrolein, 102, 136
acronyms, 333-334
acrylo compounds, in UFR, 279
actinolite, 99
activity pattern, 2, 159-162
acute bronchitis, 247-248
adenovirus, 106
adhesive, spray, 117
AEC, 329, 333
AECB, 132, 299, 300, 333
aerosols, 96-98, 254
    sprays, 114-115, 302
aerobiology, 192-194
aflatoxin, 267
AGA, 333, 339
agent orange, 125, 148, 211, 274
agent white, 124
agricultural dust, 179
AIA, 68, 289, 291, 292, 323, 333, 339
AIHA, 299, 333
air, chemical composition, 53
    conditioning, 11, 47, 64
        cars, 28
        effect on dust, 181
        health effects, 247
        maintenance, 9
        Roman villas, 24
    diffusion, rate, 157

    exchange rate, 61
    filter efficiency, offices, 81
    freshener sprays, 115, 117
    friction layer, 59
    humidity, 49-50
    Infiltration Center, 338
    ions, 48
    mines, in, 22
    movement, 43, 47
    pollution regulations, European, 21
        Canadian, 308
        U.S., 308
    quality criteria, EPA, 97
    quality, documents, 308
    quality judgment, 30
    recirculation rates, 297
    stale, 19
    transmission, building materials, 89
    traveler, climate effects, 246-247
    velocity, comfortable, 46
    velocity meters, 139-140
    vertical mixing, 56
    water content, 49
airborne half-life of dust, 179
airborne radiation, 127
aircraft, 59
    air control, 300
    air standards, 311
    humidity, 165-166, 295
    military air standards, 311
    ozone level, 177, 273
    particulates, 184
    smoke irritation, 283-284
    smoking regulations, 313-314
airport ventilation rates, 297
airway resistance, 243
Aitken particles, size, 98
alcohol and smoking, risk, 283
aldehydes, as allergens, 256-257
alder pollens, 108

aldrin, 121, 264, 275
Allegheny Airlines, smoking regulations, 313
allergens, 107-108, 252, 256-257
allergies, classification, 256-257
allyl bromide, in air, 150
allyl isocyanate, 280
altitude, effect on health, 247-249
Alkali Works Regulation Act, 21
alkaloids in tobacco smoke, 100
alkanes in tobacco smoke, 102
alkylating compounds, 201
alumina, activated, 303
aluminum foil insulation, 318
alveoli, 241-243
alveolitis, 256
ambient air, particulate level, 250
    pollution, control, 299-300
    quality monitors, 140-141
    standards, 308-310
    vertical mixing, 57
ambient moisture, vertical level, 60
ambient organic vapors, 198
ambient pollution, exposure, integrated, 215-227
ambrosia, settling, 191-192
amino acids in allergens, 108
ammonia, 273
    acceptable indoor level, 317
    ambient air, in, 105
    control, 303
    copy machine, 303
    determination, 146
    eye irritant, 279-280
    field measurement, 142
    health effect, 199
    sweat, in, 94
    workplace limits, 311
ammonial odor, 135
amosite, 99
amputee, climate effect on pain, 249
anatomy, respiratory, 242
Anderson impactor, 152
androsterone, 96
angina pectoris, from smoking, 282
aniline, in cars, 209-210
animal feces, dust, 252
animal hairs, 252
animal testing, toxicity, 240-241, 275
annoyance, by smoking, 283
annual heat requirement, residential, 80
anopheles, mosquito, 118
ANSI, 145, 325, 333, *see also* ASHRAE, ASTM
ANSI-ASTM analytical methods, 145-146
ant (and roach) spray, 117
antagonism of carcinogens, 239
anthophyllite, 99

anthrax, 192, 254
antibodies, allergy, 256-257
anticoagulants, 124
antimoney, 110
antiperspirant, 223
apartments, 72
APCA, 141, 333, 338
APHA, 141, 333
arginine, 94
Arkansas-type wall construction, 77
armored tanks, temperature, 45
arsenate, toxicity, 236, 275
arsenic, 24, 111, 239, 264
Arthus reaction, 256
asbestos, 11, 24, 99-100, 330
    acceptable indoor level, 317
    ban, 326
    body burden, 231
    carcinogen, 264, 285-288
    control, 304
    dose, 227, 231
    dose response, 231
    exposure limits, 327
    levels, 100, 186-189
    NIOSH limit, 311
    risk, assessment, 280-288
    smoking, and, 283
    standards, 308, 327
asbestosis, 265, 288
    among homemakers, 265
ash pollen, 108
ASHRAE, 8, 14, 29, 39, 40, 42, 43, 44, 51, 52, 55, 60, 62, 82, 140, 296, 315, 325, 333
    acceptable indoor levels, 297-298, 317
    analytical procedures, 141-142
    standard 55-1981, 29
    standard 62-1981, 66-67, 297-298, 317-318
asparagus odor, 135
aspen, pollen, 108
aspergillus fumigatus, 109
asphalt, vapor transmission, 85
asthma, effect of ions, 48
ASTM, 141, 145, 325, 333
astronauts, base temperature, 33
    core temperature, 46
    internal temperature, 34
    oxygen application, 36
ASVHAE, 26, 333, *see also* ASHRAE
athlete, oxygen exchange in lung, 244
atmospheric pressure, and radon, 130
ATP, adenosine triphosphate, 244
attic fans, 304
attic temperatures, 292
auditoriums, smoking laws, 314
auramine, as carcinogen, 263

Subject Index 417

automobiles, air conditioning, 64
    CO level, 173
    dust resuspension, 180
    upholstery, vapors, 209-210
autopsy rooms, formaldehyde, 202
average heating degree days, 55
azinophos, 121
azo dyes, 23

baby diapers, 96
bacillus, anthracis, 254
    legionella, 11
    subtilis, 109
    tuberculum, dose-response, 236
bacteria, air, in, 106
    analysis, 154
    concentrations, 107
    settling after sneeze, 191
    size, 98
badges, monitoring, 144
banana, odor, 136
BaP, 211, 263
    analysis, 148
    synergism, 239
barbiturates, dose-response, 236
barracks, ventilation, 314
BART, particulate level, 186
basement, ventilation rates, 298
bathroom, cleanser, 279-280
    infections, see Darlow, 1959
    odors, 96
    temperature standard, 2
    ventilation rates, 298
batting, insulation, 318
BB, body burden, 231
bed, microclimate, 47
bed linen, formaldehyde in, 278
bed-making, microbe release, 192
bedroom temperatures, 2
    England, 2
bedroom ventilation rates, 298
beech, pollen, 108
behavior factors in health, 229
bendiocarb, level, 210
benomyl, 126
benzaldehyde, in UFR, 279
benzene, acceptable indoor levels, 317
    breath, in, 274-277
    health effect, 199
    hexachloride, 123
    indoor levels, 2
    office, in, 198
    regulations, 308, 329
benzidine, 263, 310, 329, 330
benzo-a-pyrene, see BaP
benzylchloride, detection limit, 151

BEPS, building energy performance standards, 322-323
bergamot oil, control, 329
beryllium, toxicity, 266
best available technology, codes, 321
bingo games, dust, 184
biocides, 119-126
biology lab, formaldehyde, 202, 278, 287
biological warfare, model, 215
biotransformation, T, 231
birch pollen, 108
bird droppings, 254
birds, microbes, 254
bituminous shale, radon, 131
black box exposure model, 216
black globe, 43
bleachers, space requirements, 74
blood, pH, 244
Bloom's syndrome, 266
BOCA, 323
body burden, 276-277
    analysis, 156-157
    effect on dose-response, 231
body odor, 2, 33, 95, 134, 296
    acceptance, 301
    decay, 102
body surface, heat loss, 39
Bohr integral, 243
boiled meat, odor, 135
Bourden gauge, 140
bowling alleys, ventilation rates, 297
brake linings, asbestos, 99
breath, ambient contaminants, 275-277
breathing rate, human, 35
brick, 27
    Rn emission, 195
    vapor transmission, 85
    walls, 295
bromidrosis, 135
bromine, 113
bromobenzene, 150
bromoform, 150
bronchial asthma, 256
bronchitis, 247-248, 253
bronchus, 242-243
    dust absorption, 251
Btu, heat units, 331
bubble baths, 223
building, codes, 320
    energy performance standards, 322
    materials, 11, 83-91
        air transmission, 89
        asbestos, 99
        formaldehyde analysis, 147-148
        radon analysis, 156
    maintenance, 69

occupancy periods, 160–162
paper, vapor transmission, 85
stock, U.S., 70
temperature, regulations, 324
unoccupied, 324
bus, space requirements, 83
butane, 114
butanol, ventilation, 301
Butte, MT, Rn level, 194
butterflies, oxygen demand, 36
butyric acid, odor, 135

CAB, Civil Aeronautics Board, 333
  smoking regulations, 313, 333
cabbage, odor, 135
CABO, Council of American Building Officials, 324
cadmium, 110, 264
calcium arsenate, 275
California air resources board, 153
California medfly, insecticide, 121–122
calorimeter, 17
camphor, odor, 136
cancer, affected organs, 23, 258
  cells, doubling time, 235
  clinical manifestation, 234–239
  definition, 258
  risk, asbestos, 285–287
    formaldehyde, 284–285
    radon, 287
    smoking, 282–283
    pesticides, 264
  threshold theories, 236–239
Caplan's syndrome, 266
captafol, 126
captam, 126, 275
carbamates, 122–123
carbaryl, level, 210
carbohydrates, as allergens, 257
carbon dioxide, acceptable indoor level, 317
  analysis, 146
  balance, human, 35
  dose-response, 236
  human, production, 17
  human, tolerance, 37
  indoor level, 166–167
  lung, in, 36, 243
  ventilation requirement, 5
  workplace limit, 311
carbon disulfide, 123
carbon filters, 303
carbon monoxide, 103–104, 308–309
  air standards, 272
  blood, in, 271–272, see also COHb
  body burden analysis, 156
  control, 299
  determination, 145

environmental impact, 318
level, 169–173, 186, 272
  garages, 169
  office buildings, in, 169
smoke, in, 262
total exposure models, 24, 171, 219–220, 222
ventilation, 301
workplace limit, 311
carbon tetrachloride, 264, 274
carcinogens, 22
  classification, 260
  dose response, 234–239
  synergism, 238–239
carpenters ant, 120
carpet shampoos, formaldehyde level, 203
carpets, electricity, 48
  formaldehyde, 205
  odor, 91, 304
cascade impactor, 152
castles, air quality in, 25
cat flea collar, 117
cat trainer, 117
cattle feeding areas, 109
cattle, microbes, 254
caulking, 319
CCl_4, carbon tetrachloride, 201
CD_{95}, curative dose, 236
CDC, 329, see also Center for Disease Control
ceiling, heat loss, 78
ceiling, temperature profile, 76
cellulose, insulation, 87, 318
  dust control, 304
cemeteries, gases, control, 299
cemeteries, methane in, 113–114
Center for Disease Control, 247
  organization, 329
central heating, 13, 63
CEQ, 329
cernospheres, 154
Chambre Syndicale de Paris, 100, 265
CHAMP, National Health Air Monitoring,
charcoal grills, carcinogens, 267
Chattanooga, NO_x, 271
CH_2Cl_2, dichloroethylene, 201, 274
$CHCl_3$, trichloroethylene, 201, 274
CH_3Cl, 201, see chloroform
cheese, odor, 135
chemotherapy, 263
CHESS, 151, 229, 248
chest pain, due to smoke, 262
chicken sheds, 109
chickenpox, 254
children, respiratory health, 284
chimney sweeps, 23
chinook, 249

Subject Index

chlamydia, 254
chloracne, 274
chlordane, 120
    acceptable indoor level, 317
    detection limit, 151
    toxicity, 275
chlorinated hydrocarbon analysis, 148–150
chlorine, 113
    acceptable indoor level, 317
    chloroacetamines, 124
    chlorobutene, sampling, 150
    chloroform, $CHCl_3$, 201
    in breath, 209, 274–277
    intake, 274
chloropierin, 122
chlorotetracycline, 201
chlorox, 117
chlorpyrifos, 210
chocolate odor, 135
chromium, 24, 112, 264
chromotropic acid method, 147
chrysanthemum flowers, 119
church organ, tin pest, 68
churches, air quality, 25
cigarette odor acceptance, 301
cigarette odor, persistence, 301
cigarette smoking, trends, 163, 301, *see also* smoking
CIIT, formaldehyde animal study, 112, 278, 284, 333
ciliated epithelium, 242–243
Cincinnati, 53, 56
cirrhosis, 238
city buses, smoking regulations, 313
cladosporium, 107, 109, 190–191
classroom, heat balance, 82
cleaning fluid, eye irritation, 279–280
Clean Air Act, 96, 299
    standards, 308
clean air, particulate level, 250
climate, conditions, U.S., 291
    control, 289–300
    global, 54
    health, and, 245–249
    improvement, passive, 291
    zones, 54, 290, 323
climatic cycles, 54
climographs, 245–247
clinical ecology, 245–247
clo, clothing insulation units, 332
clothing, 39–40
    insulation value, 29, 40
    moisture absorbance, 230
    static charge, 48
coal, dust, workplace limit, 311
    home fuel, as, 64, 164
    mining, metabolic heat, 42

tar, 23, 263
    NIOSH limit, 310
cobalt, 112
codes, mandated, 320–323
    voluntary, 323–324
coffee odor, 135
COHb, in blood
    health effects, 271
    levels, 282
cold air, health effects, 249
color judgment, 31
Colorado, lung cancer rate, 267
Columbia, bronchitis, 247
combination effect, carcinogens, 238–239
combustion, 16
    gases, 11, 98
        control, 303
    plumes, 189
comfort, 2, 29–30
    ideal, 46
    range, 38, 46, 293
    regulation, 29–30
    standard, ASHRAE 55-1981, 325
commercial aircraft, *see* aircraft
commercial buildings, in U.S., 70
common cold, 254–255
Community Health Environmental Surveillance Study (CHESS), 230
commuter traffic, CO level, 169
commuter trains, smoking, 313
complaints, indoor air quality, 7
concert halls, 32
conchae, 242
Concorde, airplane, 116
concrete, Rn emission, 195
concrete, vapor transmission, 85
condensation, control standards, 321
    indoor, 295
condominiums, 72
conference room, 83, 186
construction workers, 239
consumer fraud, 318
consumer products, formaldehyde level, 20, 203–205
Consumer Product Safety Commission, CPSC, 329, 333
consumption, *see* tuberculosis, 25
contact lenses, 176, 279
convective heat, 39, 41
cool air, heating, of, 53
cooking, 61, 203
    odors, 134–136, 302
cooling degree days, 55
cooling requirements, in schools, 82
copper, 110
    sulfate, 125, 275
copy machines, 175, 304

core temperature, 2, 30, 46
corrosion, in walls, 91
cosmetic sprays, 114–116, 198–201, 303, 329
cosmetics, 224
cosmic radiation, 126–127, 227
cotton, dyes, 223
    formaldehyde, in, 278
    low shrinkage, 223
coughing, aerosol generation, 188–189
Council of Environmental Quality, 329
court rooms, air quality, 25
coxiella burnetti, 254
CPSC, 88, 147, 204, 278, 329, 333, *see also* Consumer Product Safety Commission
    ban of friable asbestos, 326
    ban of UFFI, 308, 325–326
crawl spaces, herbicides, 124
crayfish, oxygen demand, 36
creatinnine, 96
cresol, workplace limit, 311
criteria documents, 309
criteria pollutant, standard levels, 309
crocidolite, 99
cross-draft, 25
croton oil, carcinogen, 238
croup, 230, 258
crysotile, 99
cumulative dose, 231–232
cyanide in cigarette smoke, 262
cyclone separators, 302

2,4-D, 124
dancing, metabolic heat, 42
DDT, 119, 122, 211
    ambient levels, 211
    body burden, 118, 119–120, 263, 271
    dose-response, 231
    history, 116–119
    toxicity, 275
death, leading causes of, 44, 253, 281
deep sleep period, 47
defective chromosome, 266
Defense Department, smoking regulations, 314
defoliant levels, 211
demographic health factors, 229
dental office, mercury, 109–110
deodorant sprays, 114–116, 198–201, 330
dermatophagorides pteronyssinus, 109
detection limit, organic vapors, 150–151
detoxification, metabolic, 238
dew point, 50
diabetes, breath odor, 135
diagnosis of bronchitis, 247–248
diapers, 330
diazinon, 121, 210, 275
dibromoethane, in cars, 209–210
dichlorobenzene, in breath, 274–277

dichloromethylene, in air, 201
dichlorophenoxy acetic acid, 124
dichlorvos, 122, 275, *see also* insecticides
dichotomous samples, 152
dieldrin, 264
diesel gases, analysis, 148
diesel emission, CO, 169
    particulates, 184
dioxane, odor, 136
dioxin, 24, 124, 125, 189, 211, 330, *see also* TCDD
diphacinone, 124
diphtheria, 253
diquat, 124, 275
disinfectants, eye irritation, 279–280
disulfoton, 122, 275
dithiocarbamates, 126
diurnal, climate cycles, 55–57
    exposure cycles, 222
    heating cycles, 70, 79
    temperature cycle, 79, 292
dizziness, 10
DMBA, dimethylbenzamine, carcinogen, 238
DNA, modification, 238–239
DNOC, 124
doctors, formaldehyde exposure, 278, 287
DOD, smoking regulations, 313
DOE, 68, 69, 88, 205, 296, 304, 329, 333
dog repellant, 114, 115, 117
doors, infiltration, 65, 78
dose-response, 23
    curve, 230–234, 236–241
Down's syndrome, 266
draft gauges, 139–140
drain cleaner, 117
drapery, formaldehyde level, 205
drip-splash, 106
dry air, in aircraft, 247
dry-bulb temperature, 43, 55
Duco cement, 280
dust, 11, *see also* particulates
    ambient air, in, 53
    enzyme, 188
    fate of, 250
    highway, 97–98
    hotels, in, 91
    inhalation, 55, 249–263
    level, 177–196
    removal, 182
    resuspension, 179–181, 302
    screening, in lung, 250
    settling, 58, 181, 302
    skin scales, 189
    workplace limits, 311
dyspnea, 244–245

earth field, electrostatic, 249
earth gases, control, 299–300

Subject Index

ECETOC, 333, 352
eddy currents, 84
edema, eye, 247
effective temperature, 43
Egypt, bronchitis, 247
Ehrlich index, 236
electric, base board heaters, 71
  fields, effect, 48
  heating, 64
  wall board heating, 293
electricity, as home fuel, 164
electrostatic air filters, 300
electron microscopy, 152
electrostatic copier, 303
elemental analysis, field, 143
elm pollen, 108
embalmers, cancer risk, 279, 285, 287–288
Emergency Building Temperature Regulations, U.S., 324
endosulfane, 121, 275
endrin, 121, 275
energy, audits, 318–319
  balance, humans, 39
  conservation
    information, 318
    programs, 318–320
    standards, 69, 316–320
  laws, U.S., 316–318
  need for cooling air, 52
  savings by insulation, 318
England, bronchitis, incidence, 247–248
enthalpy of air, 44
  moist air, 50–53
environmental disease, 245–247
environmental impact statements, 318
Environmental Protection Agency, 328, *see also* EPA
enzyme laundry dust, 109
EPA, 14, 56, 97, 101, 140, 141, 167, 169, 171, 187, 200, 202, 205, 207, 224, 230, 241, 248, 268, 269, 271, 272, 285, 288, 299, 308, 333
  equivalent methods, 143
  organization, 328
  standard analytical methods, 143
epichlorohydrin, 150
epidemiological data, 23, 309
epidemiology of formaldehyde, 278–279
equilibrium water pressure, 50
ERDA, organization, 329, 334
ESCA, 152, 334
esophagus, cancer, due to smoke, 262, 282
ET, effective temperature, 43, 44
ethane, 114
etheral odors, 134
ethion, 122
ethylene bromide, 123

ethyl ether vapor, health effect, 199
exercise, effect on tidal volume, 244
  effect on metabolism, 42
exhaled air, composition, 36
exhaled dust, 250
exposure levels, 23, 163
  factors, 213
  models, 215–227
exterior plywood, 85, 148
eye irritation, 10, 247, 283
eyesight, 31

FAA, 173, 177, 184, 284, 334
factory built housing, 27, 320–322
fallout, nuclear, 133
farmer's lung, 109
fatigue, stale air, 246
fat-soluble substances, body burden, 231
fecal odor, 135
feces, dust, 252
Federal Air Pollution Control Act, 22
federal standards, radon, work, 131
Federal Trade Commission, 322
fenestration, 91–92
fention, 122
fertilizer, UFR, 223
FESYP formaldehyde test, 148, 334
fiberglass ceiling panel, formaldehyde, 205
fiberglass, dust, 187–188, 304
fibrosis, lung, 265
  progressive, 266
field audit strategy, 142
field monitoring labs, 142–143
film badges, for radioactivity, 156
fireplaces, 293
  infiltration, 65
fish, odor, 135
  oxygen demand, 36
flame ionization detector, 146
flatus, 96
flea collars, 122
fleas, control, 303
floor, heat loss, 78
  space for occupants, 81, 83
  temperature profile, 76
floral odor, 134
Florida, Rn levels, 194
fluorescent analysis, 148
fluoride, toxicity, 275
fluorine, 112–113
fluoroanthrene, 212
fluorocarbons, 115–117, 273–275, 330
  ban, 310
fluorspar miners, radiation exposure, 261
fog, definitions, 98
Föhn, 248
folpet, 126

Food and Drug Administration, 328
food dyes, 267
food odors, 135
foot and mouth disease, 106, 254
foot powders, formaldehyde in, 223
forced air heating, 70
forced ventilation, 65, 293, 295
forest fires, particulate level, 179
forest products, grading, 325-326
formaldehyde, acceptable indoor levels, 317
    ambient level, 201-203
    analysis, 147-148
    automobiles, from 221
    biotransformation, 232
    cancer in rats, 278
    cancer risk, 239, 285, 288
    chemical filters, 303
    consumer products, from, 205
    environmental impact, 318
    epidemiology, 278-279
    exposure, ambient, 221-227
    health effects, 276-280
    observed levels, 201-207
    release, from insulation, 88
        mobile homes, 322
        models, 217-218
    risk rates, 284-288
    solutions, control, 329
    sources, indoors, 203-205
    standards, 307, 326, 330, *see also* NPA-HPMA, 1982, and Sundin, 1982
    stove and heaters, from, 164
    tobacco smoke, in, 102
    total dose, 226-227
    total exposure, 286
    wall cavities, in, 209
    wood fires, from, 165
FR, Federal Register, citation, 87
Freon-11, 116
Freon-12, 280
fresh air ventilation, 13
friable asbestos, 99-100
FSP, fine suspended particulates, 97, 334
Fuller's earth, health hazards, 265
fumes, particle size, 98
fumigants, 123
fumigation, 11, 57
fungicides, 125
fungus, 11
    spores, 190
    walls, in, 91
furfural, 279
furfuryl alcohol, workplace limit, 211
furniture, 91
    polish, 115, 117, 198
furs, fumigation, 223

Gaeke tubes, 143
ganisters, 241
GAO, General Accounting Office, 14, 308
garages, 80
    CO level, 171, 272
    ventilation rates, 297
garden sprays, 117
garlic odor, 135
gas chromatography, 12, 146
gas as home fuel, 164
gas stove, emissions, 164
gasoline, odor, 136
General Accounting Office, 14, 308
glass, buildings, solar heat, 293
    curtain walls, 27, 84
    fiber insulation, 87
    fibers, 100
global climate, 54
glue-sniffing, 280
goblet cells, 241-242
gold mines, 45
Goodpasture's syndrome, 256
grab bags, 143
Grotta del Cane, 16
ground water, gas control, 300
    radon level, 131
gymnasium, ventilation, 297
gypsum, vapor transmission, 85

hair lotion, 114-115, 117
hairspray, 114-116, 198-201, 302, 329
half life, radon, 128-129
halocarbon levels, 207, 212
hay fever, 108
hazardous waste, disposal sites, 211, 329
headache, 10, 282
health effects, methylene chloride, 199
    perchloroethylene, 199
    factors, environmental, 229
    standards, 306-316, 317, 325-328
    vs. animal data, 309
Health & Morals of Apprentices Act, 310
healthy climates, 247-249
hearing, 32
heart failure, death rates, 281
heat, budget, human, 39, 47
    conservation, during night, 70
    effect on death rate, 246
    exchange, balance, human, 39
    exchanger, 296, 298
    flux, in body, 47
    imbalance, office, 80
    load, offices, in, 81
    loss, residential, 78
    ramps, 293
    requirement, mobile homes, 80

response, old people, 46
stimuli, internal, 46
stroke, 43
transport in clothing, 39
heaters, indoor, regulations, 329
heating, average degree days, 55
   energy, 6
   inadequate, 9
   moist air, 52
   requirements, in schools, 82
   wasted energy, 69
hemlock pollen, 108
hemoglobin complex, carbon monoxide, 156, see COHb
Henderson-Hasselbach equation, 245
heptachlor, 120, 264, 275
   detection limit, 151
herbicides, 24, 124–125, 274
   sprays, 117
hexamethylene tetramine, 221
hexane, 329
HEW, see HHS, 334
HHS, 328–329, 334
hickory pollen, 108
highway dust, 97–98
hippuric acid, 96
histidine, 94
histological examinations, 234
histoplasma capsulatum, 254
histoplasmosis, 254
home fuel use, trends, 164
home weatherization program, 316
homemaker, time budget, 4, 159–160
homemakers, formaldehyde exposure, 288
Homer's syndrome, 263
homicide, death rates, from, 281
hormones, 96, 267
hospital, bacterial infections, 192
   dust levels, 182
   formaldehyde use, 223
   odor, 12
   ventilation, 302
hot air, cooling of, 52
hot design day, 55
hotel, dust control, 302
   odor in rooms, 80, 91
house, calls, by doctors, 247
   doctor, energy program, 318
   dust, 188–189
      composition, 97
      health effects, 252–255
   fly, insecticides, 119–123, 210, 275
   mites, 189
   plants, 26
household, cleansers, components, 199
   irritants, ocular, 279–280
   solvents, components, 199

sprays, 114–117, 198–201, 302, 329
statistics, 4
housing stock, U.S., 5
HPMA, 334
HUD, 63, 64, 75, 78, 79, 88, 162, 204, 289, 290, 334
   codes, 320–323
human, activities, 25, 166
   activity patterns, 159–162
   body, water loss due to work, 2
   respiratory system, 241–245
   skin, surface area, 39
   wastes, 26, 300–301
humidifiers, microbes in, 190
humidity, air, 49–50
   cause of pollutant, 88, 93, 165–166, 208, 294, 322
   control, 13, 294, 295
   cycle, diurnal, 292
      season, 292
   effect on death rate, 246
   exhaled air, of, 36
   human metabolism, from, 166
   indoor, 294
   measurement, 139
   night, 59
   ratio, 50
   see effective temperature scale (ET)
hurricane forces, 54
hydrocarbons, 114
   analysis, 146
   levels, 207–211
   standards, 308
hydrochloric acid, as carcinogen, 235
hydrogen cyanide, 123
hydrogen sulfide, analysis, 146
   filters, 303
   odor, 135
hygroscopic particles, in lung, 251
hyperplasia, 238
hyperpnea, 245
hypersensitivity, 256–257, 279
hyperventilation, 245
hypochlorite, eye irritation, 280
hypothermia, 46, 239

ICBO, 323, 334
ICC, Interstate Commerce Commission, 313
iceskating rink, ventilation, 297
IgA, immunoglobulin, 254
IgE, 256
IgG, 256
IgM, 256
idiosyncrasy, formaldehyde, 279
ill buildings, 7–9, 294
immune response, 256
immunoglobulin, 254

incense, particulate level, 184
indole, odor, 135
indoor, activity patterns, 159-162
    air circulation, 67
    Air Quality Program, U.S., 307
    air, regulation, 22
    air standards, 306-308
    climate, definition, 60-61
    formaldehyde level, average, 206
    humidity, 294
    illness, 7-9
    moisture condensation, 86
    pollutants, variety, 11
    pollutant limits, 67
    pollutant standard, limits, 67
    swimming pools, 68
industrial air measurements, 141-142
industrial effluents, 22
infant, air quality, 25
infections, 25, 253-256
infective agents, 253-256, *see also* bacteria, microbes, pseudomonas, spores, virus
influenza, 253
infiltration, 6, 13
    definition, 65, 296
    measurements, 66
    rates, residential, 6
    sources, 65
    standards, mobile homes, 321
inland climate, 54
insecticide sprays, 114-117, 198-201, 264, 275, 302, 329
instrumental analysis, particulates, 152
insulating value, clothing, 39
insulation, effect on air quality, 294
    effect on ventilation, 294
    energy savings, 318
    mandatory, 318-320
    material standards, 318
integrated exposure, 230-231
integrating radon meters, 155-156
Interagency Regulatory Liaison Group, 284, 329, 334, *see also* IRLG
intercity buses, smoking, 313
interior plywood, adhesive, 148, *see also* UFR
internal heat accumulation, 41
internal sensors, 30
internal temperature, 46
International Energy Agency, 65, 318
interpretation of air analysis, 157
interspecies extrapolation, 240
inversion, 56-57
iodine, 113
    isotopes, 133, 310
IRLG, 308
    formaldehyde study, 284
    organization, 329

iron miners, radiation exposure, 261
iron oxide, health effects, 266
irritation, as cause of cancer, 267
irritation by organic vapors, 199
ISC, intersociety committee, 145
isocyanate insulation, 318
isocyanates, 256, 279
    insulation, 318
isopropanol, 280
itai-itai disease, 110

Japan, formaldehyde standard, 148
    indoor air regulations, 22
    textile tests, 330
jogger, 35
jogging, 245

kepone, 121, 264, 275
Kerala State, radon, 132
keratin, in dust, 252
kerosene heaters, 164, 302
kitchen, CO level, 170
    infiltration in, 65
    nitric oxide level, 174-175
    odors, 134-136
    ranges, 164, 174, 302
    ventilation, 294, 298
known carcinogens, definition, 260
krypton, radiation limit, 310

L, pollutant, elimination rate, 231
Labor, U.S. Department of, 329
laboratory exposures, formaldehyde, 287
laboratory, Lavoisier's, 18
laboratory, space requirements, 83
lachrimators, 22
lacquer, vapor, 199
Lake Athabaska, radon, 132
LaMar University, air and breath levels of PCBs, 276
lapse rate, in air, 56-57
larynx, cancer rates, smokers, 262, 282
laser velocimeter, 153
latent dose response curves, 233
Latex fabric, formaldehyde level, 205
LBL, 140-141, 334
LD_5, lethal dose, 236
lead, 104-105, 329
    ambient standards, 308
    analytical methods, 146
    arsenate, 118, 275
    body burden, 104, 257
    dose-response, 231
    dust, in, 252
    newsprint, in, 257
    NIOSH limit, 310
leather, formaldehyde use, 223

Subject Index

legionnaire's disease, 11, 106, 254-255
leucine, from diapers, 96
leukemia, from smoking, 283
leukocytes, 254
leukoplakia of the esophagus, 283
lice, control, 302
life expectancy, average, 2
lifestyle, health effects, 230
light fixtures, as heat source, 81
light bulbs, ozone level, 175-176
lindane, 264
    detection limit, 151
linear dynamic exposure model, 216-217
lipoid tissue, 24
    organics, in, 156-157
listerol, 280
Liverpool, lung cancer incidence, 267
living conditions, extreme, 45
living room ventilation rates, 298
load controls, furnace, 318
lobectomy, 263
lofting, 57
log homes, PCP levels, 273
Love Canal, 211
lung, 242-243
    air, chemistry, 36
    bacterial, half-life, 255
    cancer rates, 282-283
        passive smoking, for, 284
    cancer, regional effects, 267
    dust filter, as, 250
    fate of dust, 250
    tissue, repair, 255
    ventilation, 39
    vital capacity, 241
lye, 280
lymph nodes, brown, 264
lysol, 117, 280
lysozyme, 254

MAC, maximum allowable concentrations, 136
macrophages, 252
magnesium, 111
maintenance costs, offices, 81
makeup air, 297
malaria, 118
malathion, 121-122, 275
    level, 211
maleic anhydride, in cars, 209-210
mandatory insulation, 318-320
mandatory material standards, 325-326, 328-329
manganese, 111
manufactured housing, 72-80, 320-322
    formaldehyde level, 204-205
maple syrup, formaldehyde in, 279

marathon runner, 35, 40
marijuana herbicides, 124
mass balance, metabolic, 35
mass flow, in heating air, 52
mass spectrometry, 146
material standards, 325-327
measles, 106, 253, 254
mechanical filters, 302
Mediterranean Fruit flies, 121-122
megamouse experiments, 234
melamine resins, 223
melanoma, 266
meningitis, 253
men's clothing, insulation, 40
mercury, 109-110
    analytical methods, 146
    coal combustion, from, 109-110
    dental office, 109-110
    dose-response, 236
    filters for, 303
    fungicides, 125-126
    light, ozone level, 176
    synergism, 239
mesotheliomas, 265, 266
met, metabolic heat units, 39, 332
metabolic, calorimeter, 46
    cancer defenses, 238
    efficiency, 38
    oxygen needs, 2
    pollutant control, 300-301
    rate, 16, 19-20
        for workers, 42
    water, condensation, 38
metabolism, human, 29-31
methane, 113-114
    analysis, 146
    diapers, from, 96
    gas, control, 299
    regulation, 310
methanol, odor, 136
    offices, in, 81
    vapor, health effect, 199
methoxychlor, 120
methyl bromide, 150
    chloride, 150
    formate, odor, 136
    mercaptan, odor, 135
    parathion, 211
methylene chloride, health effect, 199
methylene glycol, analysis, 147
mica, health hazards, 265
mice, oxygen demand, 36
microbe removal rate, 192
microbes, 26, 105-190, 253, 257
    air-conditioners, in, 254-256
    control, 300-301
    fate in lung, 255
    settling speed, 191

microbial dust, 190-193
microbial sampling, 153-154
microclimates, 56-57
micropolyspora faerni, 109
Middle Ages, air quality, 24-26
military air standards, aircraft, 311
military barracks, infections, 192
Mine Safety Act, 310
mineral, batts, 87
    blankets, 318
    fiber insulation, 87, 318
miner's lung, 264
minimum energy design budget, 322
minimum property standards (HUD), 320
minimum ventilation rates, 297
Minnesota, clean air laws, 307
    indoor air laws, 313
mirex, 121, 264
    toxicity, 275
mist in air-conditioners, 106
mistral, 248
mites, in dust, 97
mobile homes, 14, 72-80
    design, 78
    diurnal formaldehyde, 206
    floor plan, 74
    formaldehyde level, 204-207, 286-288
    space requirements, 83
    standards, 320-322
    warranties, 322
    wood surface, 86
moist air, thermodynamics, 48-53
moisture, 85, 93, 294
    cause of pollution, 88, 93, 165-166, 208, 294, 322
    condensation, 69
    content, wood, 89
    control, 294-295
    transmission, materials, 85
molds, 91
    allergenic, 253
mole fraction of humidity, 50
molybdenum, 111
monazite sand, 132
monitoring, comfort, 138
monitors, personal, 144
mopping, 48
mortality, 21
    rates, leading causes, 253, 281
mortgage rates, mobile homes, 321
mortuary, formaldehyde, 202
mosquito repellant, particulates, 184
mosquito spray, 117
moth spray, 117
mouthwash, 223
MPS, minimum property standards, 320
multichamber exposure model, 217

multilith, odor control, 304
municipal air pollution regulations, 22
municipal water, odor, 136
musician, metabolic rates, 42
mussels, oxygen demand, 36

naled, 122
NAPCA, 105, 334, *see also* EPA
naphthalene, in cars, 209-210
NASA, air standards, 141, 311
nasal cavity, 241-243
nasopharyngeal dust absorption, 250
National Air Surveillance Network, 151
National Cancer Institute, 328
national energy act, 317
National Institute of Environmental Health Sciences, NIEHS, 328
National Toxicology Program, 264, 328, 330
natural gas, ventilation, 303, *see also* gas
natural ventilation, 16
    definition, 65, 296
nausea, 10, 282
Navy, air standards, 311
NBS, National Bureau of Standards, 19, 216
    building codes, 324
    smoke spread model, 217
negative ions, 48
neoplasm, definition, 258
nephelometer, 153
neutrophils, 254
New York City, radon levels, 131
    Ventilation Commission, 315
New Zealand, Maori cancer rate, 267
newsprint, lead in, 257
NFPA, 74, 325, 334
nicotine, 184, 275
NIEH, 328
night clubs, CO level, 170
night time, heat conservation, 70
niobium, 112
NIOSH, 140, 141, 143, 207, 241, 265, 273, 297, 329, 334
    analytical procedures, 141-142
    criteria documents, 310
    formaldehyde level, 284, 327
    workplace limits, 311
nitric oxides, 105
    ambient standards, 308
    control, 299
    health effects, 270-272
    smoking, from, 282
nitrogen oxide, 66, 173-175
    determination, 145
    levels, observed, 174-175
    workplace limits, 311
nitrosamines, 149, 303, 330
Nix-Bix, 324

Subject Index

NO, nitrous oxide, see NO_x
NO_2, nitric oxide, 105, *see also* NO_x
NO_x, nitrogen oxides, 105
no-pest strip, 122, 211
nocturnal humidity cycles, 55
nonane level, 207
NPA-HPMA formaldehyde test, 148, 325, 327, 334
NRC, indoor air report, 308
NRCC, smoke spread model, 217
nuclear explosions, 127, 133-134
nuclear power plant, safety, 307, 310
nylon fabric, formaldehyde level, 205

oak pollen, 108
occular irritants, 279-280
occupancy, effect on smoke, 185
occupational, air standards, 310-313
    medicine, 23
    threshold, 241
Occupational Safety Act, 310
octane, 114
    levels, indoor, 207
    workplace limit, 311
odor, 11, 32, 134
    control, 301-302, 304
    indoor/outdoor, 12
    masking, 33
    regulations, 32
    standards for, 32
    standard references, 134
    ventilation effect, 301
office buildings, 80-81
    comfort conditions, 324
    energy use, 81
    occupancy cycles, 160
    organic vapors, 198
office, machines, as heat source, 81
    smoking laws, 314
    trailers, formaldehyde, 288
    ventilation, standards, 297
    work, temperature for, 46
    workers, 9, 14, 245
oil, home fuel, 64, 164
    refineries, odor, 136
        vapor levels, 207-209
olfactory epithelium, 241-243
onions, odor, 135
operative temperature, 46
oral cancer, due to smoke, 262
organic pollutants, variety, 198-200
organic vapors, 11, *see also* individual chemicals
    ambient, 198
    analysis, 146
    chromatogram, 200
    levels, 197-212
organophosphorus insecticides, 121-122
organotin herbicides, 126
orris root, control, 329
OSHA, 329, 334
    workplace limits, 313
        asbestos, 327
        carbon monoxide, 312-313
        formaldehyde, 284, 327
osmogenic materials, *see* odor
osmogenic messages, 30
outdoor/indoor concentrations, allergens, 108, 192
    asbestos, 188, 227
    BaP, 212
    cadmium, 110
    carbon dioxide, 166
    carbon monoxide, 103, 170-172
    dust, 177-181, 183, 187-188
    formaldehyde, 201, 204-206
    humidity, 49-53, 165-166
    lead, 104, 219
    manganese, 219
    mercury, 109
    microbes, 106, 192
    nitric oxides, 105, 175
    odor, 12
    organics, 200, 207
    ozone, 176
    PCP, 211
    particulates, 177-181, 183, 187-188
    pesticides, 118
    radiation, 127
    radon, 131-133, 194, 196
    sulfur dioxide, 168
outdoor pollutant control, 299-300
oven cleaners, 115, 117, 302
oxygen, balance, human, 35
    consumption, 17
    demand, human, 2, 35
    lung, in, 26, 36, 243
    needs, animals, 36
    requirements, 5, 36, 243
oxyhemoglobin, 245
ozone, 11, 48, 67, 103, 175-177, 330
    acceptable indoor levels, 317
    ambient standards, 308
    decomposition rate, 176
    exposure, 203
    filters, 303
    indoor standards, 307
    layer, atmospheric, 116, 176
    level, 176
    odor, 136
    offices, in, 81
    standards, 307

tolerance, 239
total exposure, 220
workplace limit, 311

PAH, analysis, 148
  body burden, 263
  chromatogram, 148
paint, remover, vapor, 199
  solvents, 304
  spray, 117
  vapors, 199
  vapor transmission, 85
Palmes, tubes, 143
PAN, peroxyacetyl nitrate, in smog, 175
paneling, formaldehyde emission, 205
paper mill odor, 136
paper plates, 205
paraquat, 124, 274-275
pararosaniline method, 143-144
parathion, 121, 275
parental smoking, and NO_x, 270
particleboard, 85-86, 279
  adhesives, 148, 202-203
  formaldehyde release from 205, 217
  standards, 325
  vapor control, 304
particle size, comparison, 98
particulates, 11, 96-101, see also dust, microbes
  ambient level, 179
  analysis, 151-154
  chemical composition, 178
  concentrations, 21
  exposure, 183
  industrial sources, 99
  kitchen fumes, in, 178
  mass distribution, 151
  natural sources, 179
  size, 178
  smog, in, 178
  standards, 308-309
  ventilation, 301
passive dosimeter, 144-145
passive smoking, health effects, 283-284
  regulations, 313
  risk, 283-284
passive solar design, 290-293
pathogens, lifetime in lung, 255
pathologists, formaldehyde exposures, 287-288
pathology labs, 224
PCB, 330, 335
  ambient levels, 211
PCP, pentachlorophenol, 126, 211, 275, 330, 335
  body burden, 263
  log homes, in, 273

pedestrian, dust resuspension, 180
people, as heat source, 81
peptides, as allergens, 257
perception, influence on health, 229
perchloroethylene, health effect, 199
perfumes, 11, 21, 33
periodic processes, 38
perlite insulation, 318
perm, moisture units, 85, 331
permanganate, 303
permeance, 85
personal monitoring devices, 143-145
personality, as health factor, 229
perspiration, 29, 41
Peru, bronchitis, 247
PFR, phenolic adhesive resins, 202-203, 279, 304
pesticides, 11, 116-126, 264, 275, 329
  control, 304
  history, 116-119
  levels, 207, 210
  limits, 328
  particle size, 98
  sampling, 149
  sprays, 114-117, 198-201, 302, 329
  vapor pressure, 210
pet collars, 119-122
pets, dust control, 302
phagocytes, 254
phenol, 126
  formaldehyde resins, 85-86, 148, 205, 304
  UFR, in, 279
  wood fires, from, 165
p-phenylene diamine, control, 329
Philadelphia chromosome, 266
phosphate land, radon level, 193-197
photochemical formaldehyde, 221
photocopiers, 11, 81, 176, 198
photometry, 144
picloram, 124
piezobalance, 144, 153
piezoelectric hydrometer, 139
pigeon control, 124
piperazine, as allergens, 256
Pittsburgh, lung cancer, 267
PIXE, 152
plant dust, control, 302
plantain pollen, 108
plaster, vapor transmission, 85
plutonium, atmospheric, 133
plutonium, air limits, 310
plywood, 85-86, 279
  adhesives, 202-203
  ammonia treatment, 304
  odor, 304
  standards, 325
pneumococcal pneumonia, 106

pneumoconiosis, 264–266
pneumonia, incidence, 253
pneumonic plague, 106
poison ivy spray, 117
polarography, 146
police cars, carbon monoxide, 103
poliomyelitis, 253
pollen, settling speed, 190
pollution, data interpretation, 157
pollution profiles, models, 217–227
polonium, toxicity, 111
    decay, 128–129
    toxicity, 111
polyaromatic hydrocarbons, PAH, 148, 303
    analysis, 148
polychlorinated biphenyls, 148–149
polynuclear hydrocarbons, carcinogens, 264, 267
polysaccharides, allergens, 252–253
polystyrene insulation, 87
polyurethane insulation, 87, 304, 318
poor concentration, indoors, 10
Pope and Dunmore hydrometer, 139
population density, effect on SO_2, 167–168
pork, odor, 135
positive ions, 48
potassium hydroxide, as carcinogen, 235
potential carcinogens, 260
pottery work, health hazards, 265
Pressley method, for biological dust, 154
primary air standards, 308
primary pollutants, EPA, 96
printers, health, 23
priority pollutants, EPA, 96
prison, space requirements, 83
probit transformation, 233–235
productivity, and comfort, 245
propane, 114, 311
propoxur, level, 210
pseudomonas aeroginosa, 254–255
pseudomonas legionella, 11, 106, 254–255
psychological factors, stale air, 245
psychrometer, 139
psychrometric chart, 51
psychrometry, 49–53
public buildings, smoking laws, 314
Public Health Service, 328, *see also* HHS
pucinia graminis, 107
pulmonary tuberculosis, 106
pumpkin odor, 135
putrefaction, 105
PVC, polyvinyl chloride, level, 209
pyrethrins, toxicity, 275

Q-fever, 106, 193, 254
quad, heat unit, 331

R, thermal resistance, 84
    definition, 84
    dose-response, 230–235
radiation, body burden, 126–127
    cancer risk rate, 261
    detectors, 144, 154–156
    dose, natural, 227
    effect, 47–49
    indoor standards, 30, 307
    regulations, 307, 330
radiative heat, glass wall buildings, in, 70, 293
    indoor, 41
    loss, 39, 41, 70
radical surgery, 263
radioactive dust, 193–197
    level, 179–180
radioactive lead, 104
radioactive rain, 58
radioactivity meters, 144, 154–156
Radium Hot Springs, radon, 132
radon, 11, 327
    acceptable indoor levels, 317
    average indoor levels, 196
    control, 299–300
    daughters, 128–129, 130
    decay series, 128–129, 130
    emission from building materials, 195
    environmental impact, 318
    exposure levels, 193–197
    infiltration, 193–197
    level, ambient, 130
    level, normal, 194–197
    risk, assessment, 288
    total dose, 227
    total exposures, 287
ragweed pollen, 1, 107–108
    settling, 191–192
railroad, smoking regulations, 314
rain, radioactive, 58, 133
rainstorm and climate, 54
Raman spectroscopy, 152
ramblers, 71–72
rat, cancer testing, 264, 284–287
    disease carrier, 116–119
    pesticide toxicity, 275
    rodenticides, 123–124
Raynard's syndrome, 284
RCS, residential conservation service, 317–319
real-time radon meters, 155
recirculated air, 70
recognized carcinogens, 260, *see also* NTP, 1982
rectal temperature, astronauts, 34
red blood cell, size, 98
refrigeration, units, 52

regulations, 306–329
    municipal, 22
regulatory sweating, 40, 43
reheat systems, 69
relative humidity of air, 49–50
rem, definition, 261
rep, definition, 85
research, indoor air, 308
residential building stock, U.S., 70
residential comfort, 64
residential conservation service, 317–319
residential construction, 27
residential formaldehyde level, 203–207, 286
residential housing, 71–80
resmethrin, *see* insecticides
respirable fibers, 326
respirable particles, field measurement, 149–156
respiration, 16–17
    chemistry, 243–245
respiratory, calorimeter, 19–21
    cancer, target organs, 257
    distress syndrome, 245
    fluid drainage, 254
    infections, submarines, 95–97
    needs, history, 15–17
    quotient, RQ, 19, 36
restaurant, particulate levels, 186
resuspended dust, 302
rhinitis, allergy, 256
rhinovirus, 106
rickettsia, 254
risk assessment, 280–288
Rn, *see* radon
road traffic, dust, 180
rodenticides, 123–124
rodents, as microbe carriers, 254
Roman villa, air conditioning, 24–25
Ronel, toxicity, 275
roof overhang, 92
roof shingles, 99
rug shampoos, 302
rum, cancer rate, 283
rural dust, 98–99
rural formaldehyde exposures, 288

saliva, vapors, and aerosols, 96
saltation, dust, 179
Salzburg, radon level, 131
samples, deterioration during shipment, 158
sandblasting, silicosis, 265
sauna bath, 41
SBCCI, 323
scales, skin, 95
scarlet fever, 253
Schiff reaction, 147
Schneeberg, radon levels, 260

school buses, carbon monoxide, 103
school children, air quality, 25
school trailers, formaldehyde levels, 288
schoolroom, 82
schools, asbestos level, 187
    heat flux, 82
    odors, 12
    ventilation rates, 297
scintillation counters, 155
Scotch Guard, 280
    see climate, health, 248
sea spray, in lung, 252
    particle size, 98
Seattle, humidity, 56
seaweed, 201
sebum, 95, 97, 253
secondary air standards, 308
sedentary metabolism, 29, 35
selenium, 111
senses, human, 29–34
sensory recognition, 30
sensory variety, 33
serpula merulius lacrymans, 109
Seveso, 24, 189, 211, *see also* Dioxin and TCDD
shampoos, formaldehyde, 279
shaving cream, vapor, 198
shaving foam, 115, 117
sheep, microbes, 254
shellac, vapor, 199
shirt, men's formaldehyde, 205
shoe polish, 280
shoes, fumigation, 223
siding, construction, 77
silicate, sodium, 280
silicosis, 265
silver polish, 280
sinus infections, 247
six-city study, 96
skin, allergies, 223
    cancer, 249, 267
    heat loss, 39–41
    scales, bacteria, 107
        health effects, 253
        removal, 192
    surface, 39
    water transmission, 30
    wetted, 46
skunks, control, 123
sleepiness, 10
sleeping, metabolic rate, 42
sleeping sickness, 118
sleepwear, formaldehyde, 330
    TRIS, 330
smallpox, 106, 254
smell, sense of, 26
smelters, SO_2-levels, 169

smog, 56
    definitions, 96-98
    London, 21
    NO_x, health effect, 270
smoke, cigarette, 11
    definition, 96-98
    retention rate, 189
    ventilation models, 217
    ventilation rate, 301
smoking, aircraft, in, 184
    buses, in, 313
    carbon monoxide levels, 172
    dose-response, 231
    dust level, 181-184
    eye irritation, 247
    formaldehyde, 203, 288
    inhaled volume, 284
    laws, 312-314
    particulates, 181-184
        settling, 185
    passive, see passive smoking
    regulations, 312-314
        history, 312-314
    risk assessment, 281-284
    tobacco, 61
    ventilation rate, 185
    ventilation requirement, 5
sneeze droplets, transport, 106-107
    microbe level, 193
    settling of bacteria, 191
sneezing, 188
socio-economic health factors, 229
soil gases, control, 300-301
soil, sink for CO, 172
solar, design, passive, 290-293
    heat gain cycles, 292
    homes, 63
    load, 75
    radiation, 54-55
soot, analysis, 152
space heaters, carbon monoxide, 272
    emission, 164
    unvented, 5, 327
spacecraft, 94, 245
    air analysis, 146
    organic vapor levels, 198
space flight, air standards, 311
specific humidity, 50
splash-drops, 106
spores, settling speed, 190
spores, transport, 106-107
sports arena, PaB levels, 212
sports arena, smoking, 184
spot remover, 117
spray propellants, 114-116, 198-201, 302, 329, see also fluorocarbons
spruce pollen, 108

squash, metabolic rate, 42
SST, radiation exposure, 127
stale air, 19, 25-26, 296
    psychology, 245
standardized samples of contaminant, 158
standards, AATCC-112, 1978, 330
    ACGIH, 315, 327, see individual chemical
    ambient air, 308-310
    ASHRAE 55-1981, 325
        62-1973, 297-298
        62-1981, 297-298, 315-317
    ASTM, 327
    best available technology, 322
    building energy performance (BEPS), 322-323
    building temperature regulation, 324-325
    DOE, 322-326, 329
    enforcement, 309
    EPA, 308-310, 328
    HUD, 320-322
    indoor air, 306-308
    monitoring methods, 137-138
    NIOSH, 310-312, 327, 329
    OSHA, 310-312, 327, 329
    pollutant levels, see individual chemical
    procedures for data, 309
    safety margin, 309
    ventilation, 8, 297-298, 315-316
    warmth, 2
staphylococci, 254
    in spacecraft, 95-97
state attorneys, association, 318
static electricity, 48
Stefan's law, 41
stem-cell, evolution, 266
sterols, in tobacco smoke, 102
stimuli, variety, 33
stirred settling chamber, model, 215
stomata, 252
stone mason, dust, response to, 241
stonecutters, silicosis, 265
storm windows, 318
straw, microbes, 254
straw, thatched, 26
strip mining, radon, 130
strontium, 112
strontium-90, air, 133
students, time budget, 4
styrofoam, insulation, 304
submarines, 94
    air analysis, 146
    air standards, 311
    oxygen use, 35
subway station, particulates, 184
suffocation, 21
sulfate, determination, 144
    emissions, natural, 179

level, 181–182
sulfur compounds, odor, 135
sulfur dioxide, 22, 66, 103
    ambient level, 167
    analytical methods, 143
    control, 299
    episodes, 101
    health effects, 230, 268–270
    indoor levels, 167–169
    odor, 136
    primary standard, 269
    reduction, by insulation, 318
    sources, 101
    standards, 308–309
    toxicity, 239, 330
    workplace limit, 311
sulfur, elemental, 119
sulfur hexafluoride, 66
sulfur, toxicity, 275
summer, diurnal, wall temperature, 79
sun angle, 92
sun exposure, heat, 41
sun, heat source, as, 82
surface creep of dust, 179
surface response,
surgical dressing, 114
surgical powder, UF, 279
suspension of dust, 179
sweat, evaporation, 46
sweat, regulatory, 94
sweating, 31
    onset, 46
Sweden, indoor air regulation, 22, see Sundin, 1982
Sweden, radon levels, 131
swimming pool disinfectant, 280
swimming pools, ventilation rates, 297
sycamore pollen, 108
symptoms, non-specific, 246
synthetic fibers, insulating value, 40

T, metabolic removal rate, 230–232
2,4,5-T, 124–125, 275
    toxicity, 275
tachypnea, 245
talcosis, 265, 266
talcum powder, particle size, 98
tallow, odor, 136
tar, analysis, 152
tall buildings, 27, 70
    climatic zones, 80
tank, armored, air temperature, 45
taste, 32
tax status of mobile homes, 73
taxa, fungus, 109
TCDD, 24, 124–125, 330, see also dioxin
    Seveso, 189–190

TD_{50}, 234–239
tears, aerosol, 96
tellurium, 111
temperature, acceptable range, 293
    control, 293–294
    cycle, diurnal, 292
        seasonal, 292
    gradient, vertical, 29
    measurement, 138–139
    profile, of walls, 76
    receptors, location, 31
    rectal, 34
    regulation, standards, 308, 324
    response, 34
    sense, 30–31
    units, 332
    zones, 56
temples, Japanese, 26
TEPP, 122
termiticides, 119–123, 210–211, 264, 275
terrestrial radiation, 127
tetanus toxoid, formaldehyde in, 223
2,3,7,8-tetrachloro-dibenzo-p-dioxin, see TCDD and dioxin
tetrachloroethylene, 150
    breath, in, 209, 274–277
tetrachlorophenoxy acetic acid, 124
textile, formaldehyde in, 278, 288
testile, standard formaldehyde method, 148
thallium, 112
theaters, air quality, 25
theaters, ventilation rates, 297
thermal comfort, 27, 29, 40–42, 43–44
    measurement, 138
thermal efficiency, 68–70
thermal insulation, mobile homes, 73–76
thermal performance, mobile homes, 75–79
    walls, 74
thermodynamic wet-bulb temperature, 50
thermodynamics, 17
    of air, 48–53
thermostat, economizers, 299
threshold limit values (TLV), 310
thunderstorms, health effects, 249
thyroid, carcinoma, 266
tile, Rn level, 195
time budget, 2, 159–162
tin pest, 68
titanium, 112
TLV, 310
tobacco smoke, 100–101, 102
    control, 310–312
    dust level, 181
    odor decay, 162
    particle size, 98
    synergism, 239
tobacco smoking, 26, see also smoking

Subject Index

toiletries, 303-304
toluene, 280
   in indoor air, 198
ton, refrigeration unit, 52
total exposure, 171, 183, 230-235
   models, 215-227
   formaldehyde, 224-227
total suspended particulates (TSP), definition, 97
touch, 30-32
   receptors, 30
tourists, comfort regulation, 29-30
toxaphene, 121, 275
   level, 211
Toxic Substances Control Act, 328
toxic waste sites, 328
toxicity, 235-241
   natural radon, of, 236
   testing, 240-241
trace elements, dose-response curves, 236
track-edge film, 155
traffic noise, 32
traffic, particulate levels, 184
transit, air quality, 27
transit, particulate levels, 184
Transportation, U.S. Department of, 329
tremolite, 99
trichloroethane, in breath, 274-277
trichloroethane, workplace limit, 311
trichloroethylene, 280
   odor, 136
trichlorofluoromethane, 116
trichlorophenoxy acetic acids, 124
TSCA, Toxic Substance Control Act, 328
tsetse fly, 118
TSP, total suspended particulates, 97
Tsiroglou method, radon, 155
tubercle bacillus, dose-response, 237
tuberculosis, 1, 25, 315
   incidence, 253
   due to silicosis, 265
   spore, concentration, 192
Tucson, air cooling, 53
tumor, definition, 258
   manifestation, 234-236
tumor-free latent period, 234-235
tunnel workers, silicosis, 265
turkey sheds, 109
turpentine, health effects, 199
TVA, house doctors, 318
typhoid epidemics, 118
typing fluid, 280

UFFI, urea-formaldehyde foam insulation, 86-88, 202
   ban, 308, 325-326
   formaldehyde emission, 205, 208-209
   standards, 318, 325-327
   surgical powder, 279
UFR, see urea formaldehyde resins
underwear, formaldehyde in, 224, 330
units, definition, 333-334
unnecessary risk, 240
UNSCAR, 250, 260, 335
upholstery, automobile, vapors, 209-210
upholstery, formaldehyde level, 205
uracils, 124
Uranium City, radon, 132
uranium miners, exposure, 193, 239
uranium soil, radon in, 131
   standards, 307
urban asbestos levels, 100
urban dust, 98
urea, in sweat, 94
urea-formaldehyde foam insulation, see UFFI
urea-formaldehyde resins, 85-88, 202
urine, odor, 95-96, 135
urinoid odor, 135
USAF, 56, 138, 335
USBoM, 143
USN, 35, 138, 335
USSR, health standards, see WHO, 1975
USSR, NO_x, 271
uv-light, effect on air, 47

vacuum cleaning, 44, 99, 302
valeric acid, odor decay, 162
validation, analytical methods, 141
vanadium, 112, 266
   toxicity, 266
vapona, 122
vapor barriers, 13, 75, 87, 90-91
   design, 77
vapor resistance, 85
varnish, aerosol, 114, 115, 117
VCM, vinyl chloride monomer, level, 209
vent dampers, 318
ventilation, 38, 65-67, 295-299
   active, 289, 295-299
   definition, 65, 295-296
   measurement, 66
   passive, 289-291
   periods, 160-162
   rate, effect on CO, 172
      effect on smoke, 185
   procedures, 296
   recommended, 297, 298
   requirements, 4, 5, 8
   residences, in, 71
   standards, 315-316
      history, 315
   system, 21
      Roman, 24

unbalanced, 9
vermiculite, insulation, 87, 319
vertical air movement, 47
veterinary sprays, 119-124
vinyl chloride, 24, 150, 264, 329
    cars, in, 209-210
    NIOSH limit, 310
    regulation, 308
vinylidene chloride, in breath, 274-277
virus size, 98
virus vaccines, formaldehyde in, 223
vitamin A, lack, 255
volcano eruptions, dust level, 179
voluntary standards, 326-328

W, humidity ratio of air, 50
wall cabinet, formaldehyde, 208-209
wall construction, 77
wall, heat loss, 78
walls, temperature profile, 76
wall-to-wall carpets, bacteria, 107
walnut pollen, 108
warfarin, 124
warmth, standards, 2
wasp and hornet spray, 117
waste heat, 2
waste utilization, 330
water, balance, human, 35
    condensation, 38
    content, of air, 49
    equilibrium pressure, 50
    exhaled air, in, 36-37
    filters, 302
    leak, 9
    supply, radon control, 299
    toxicity of excessive drinking of, 23, 235
    transmission, skin, 30
    vapor density, 50
    wells, radon in, 131
waxes, 114, 199
WBGT, wet-bulb globe temperature, 43
WCI, windchill index, 43
weak acids, in body burden, 231
weather stripping, 318
weight lifter, 38
welding and soldering vapors, 110
well-mixed tank, model, 215-216
wells, carbon dioxide content, 36
west wall, temperature profile, 76
wet bulb globe temperature, 50

wet bulb temperature, 43
WHO, World Health Organization, 56, 190 245, 259, 267, 270, 335
    air monitoring methods, 141
    indoor activities, 308
whooping cough, 253
Williams-Steiger Act, 310
willow pollen, 108
wind resistance, 88
windchill factor, 43
window cleaner, 117, 199
window, design, 92
window ventilation, 293
windward coats, 54
wigs and beards, artificial, 330
wine cellars, carbon dioxide, 36
winter, design temperature, 75
    diurnal wall temperature, 79
Wisconsin, clean air laws, 307
WL, working level of radiation, 128-130
WMO, 141, 335
women, comfort temperature, 46
women's clothing, formaldehyde level, 205
women's clothing, insulation value, 40
wood, adhesives, *see* UFR and PFR
    heating, 64
    home fuel, as, 164
    moisture content, 89
    products, 84-87
    rot, 26
    stoves, 165, 302
wool, insulating value, 40
    microbes, 254
work conditions, extreme, 45
work required for speech, 17
working level, WL, definition, 128-130
workmen's compensation, smoking, 313
workplace, NIOSH/OSHA limits, 311

x-ray spectroscopy, 152
xylene, 198-199

yellow fever, 118
Yusho incident, 276

zinc, 110
    deficiency, 110
zirconium, 112
zoning laws, mobile homes, 73